SCIENCE BOUGHT AND SOLD

SCIENCE BOUGHT AND SOLD

Essays in the Economics of Science

Edited by

Philip Mirowski

and

Esther-Mirjam Sent

The University of Chicago Press ▪ Chicago and London

The University of Chicago Press, Chicago 60637
The University of Chicago Press, Ltd., London

Printed in the United States of America
11 10 09 08 07 06 05 04 03 02 1 2 3 4 5

ISBN: 0-226-53856-7 (cloth)
ISBN: 0-226-53857-5 (paper)

Library of Congress Cataloging-in-Publication Data

Science bought and sold : essays in the economics of science / edited by Philip Mirowski and Esther-Mirjam Sent.
p. cm.
Includes bibliographical references and index.
ISBN 0-226-53856-7 (cloth : alk. paper)—ISBN 0-226-53857-5 (pbk. : alk. paper)
1. Research—Economic aspects. 2. Science—Economic aspects. I. Mirowski, Philip, 1951– II. Sent, Esther-Mirjam, 1967–

Q180.55.E25 S33 2002
338.4′70014—dc21

2001042486

♾ The paper used in this publication meets the minimum requirements of the American National Standard for Information Sciences—Permanence of Paper for Printed Library Materials, ANSI Z39.48-1992.

Contents

PART VI The Future of Scientific "Credit"

Acknowledgments

We are grateful for the support of NSF grant SBER 9601056, although this constitutes no endorsement of any opinions herein expressed. We also acknowledge the support of Notre Dame's Institute for Scholarship in the Liberal Arts. We also thank the following people for helpful comments: Steve Fuller, Jane Calvert, Kyu Sang Lee, Tommy Scheiding, and the participants at the Skagen Conference on Spontaneous Order and the Economics of Science and the Notre Dame Conference on the Need for a New Economics of Science.

Introduction

Philip Mirowski and Esther-Mirjam Sent

It is a commonplace observation that economists love the Individual; it is just real people that they cannot be bothered about. A wag once added that economists also profess to love Science; it is just real scientists that make them nervous. Their penchant for stark abstractions, gilded icons, and representative agents in a world of messy particulars and intractable diversity is the hallmark of fin de siècle economists, one they like to think they share with their neighbors, the natural scientists.

We think that many natural scientists, on the contrary, more often than not tend to harbor an aversion for economists, although some of their best friends might just happen to be card-carrying practitioners of the dismal science. If they have ever bothered to look into what economists do, they have often come away with reactions ranging from boredom to distaste for what impresses them as an insufficiently scientific inquiry. For precisely that reason, the very idea of an "economics of science" has frequently evoked vertigo and worse among scientists. Deep down, all scientists understand that at some fundamental level, there is some sort of economic process or processes channeling and fortifying their science; it is indisputable that someone, for some reason, has been picking up the tab. Yet the tendency until recently has been to deny that process has any substantive bearing on the real activity of scientific research. We suspect that this rather pervasive disdain may be eroding under the pressure of recent events, and therefore the random scientist may just pick up this book in an effort to clarify her own thinking on these issues. We aim to please; and nothing would please us more than to have natural scientists venture beyond stereotypic philippics about the utter folly of not funding their own favorite research agenda to the hilt.

Of course, we also have other potential audiences in mind. As will be inferred from the roster of authors in our collection, we are inclined to pay as much attention to the communities of science studies scholars and specialists in the history and philosophy of science as we are to economists and to the scientists themselves. There too, we feel that there is room for improvement. In our experience, there has persisted a strangely antihistorical approach to the study of the economics of science among those interested in science policy. In the immediate postwar period, both science and the economy were treated by commentators as timeless generic entities, with overarching "norms of science" supposed uniformly to exercise their regulative sway over scientists from the seventeenth to the twentieth centuries. Both historians and sociologists of science have protested against such invariant norms from the 1960s onward; nevertheless, whereas the processes of scientific research have been openly acknowledged to undergo temporal change and cultural variation in the interim, somehow the economics behind the science was imagined as mired in stasis. Even in sophisticated work on changing structures of scientific organization, such as that by Joseph Ben-David (1971), a species of timeless market was held to underlie the changing parade of scientific institutions. We believe that one novel contribution of this volume will be to insist upon the fact that the *economics* has been changing in noteworthy ways in tandem with the sciences.[1] Indeed, the reason we have produced an anthology of approaches to the economics of science is to demonstrate that it would be a serious error to appeal to a monolithic dogma called "*the* economics of science," either in history or in modern science policy.

And then there is our conventionally obvious audience. A contemporary economist who might chance upon this volume will undoubtedly be looking for meditations upon one of his favorite themes, namely the significance of Science for the abstract Representative Individual and, conversely, the centrality of this Individual in the progress of Science. Be assured once more, dear reader, that we aim to please: in the follow-

1. We intend this to refer both to the school or theoretical tradition that is accessed in order to explain the provisioning and organization of science, as well as the social structures and institutions within which scientific careers are embedded. In this regard, the ambiguity present in the phrase "economics of science" is appropriate, for when it comes to science, the conditions of its support cannot be readily separated from what we think the set of institutions is *for*. To those aware of our individual work in the history of modern economics (Mirowski 1989, forthcoming; Sent 1998) and the history of science our insistence upon the interplay and joint historicity of economics and the natural sciences should come as no surprise.

ing pages one will find many nameless rational actors plying their trade, and a fair amount in the way of statistics of aggregate scientific agents producing and distributing their wares. Nevertheless, one of the primary reasons we undertook to put together this volume is that we believe (and we are not alone) that something rather drastic and profound has been happening to the social organization of science in America and Europe at the end of this century, if not throughout the world; and that an inordinate fascination with a timeless representative scientific agent serves mainly to distract attention from these changes; and therefore a serious reconsideration of the "economics of science" is long overdue. (In saying this, we do *not* think the standard trope of triumphantly announcing the immanent arrival of a "new economics of science" is anywhere near an adequate response.) We will approach this claim from many angles in this volume, but the best way we can think of to embark upon our journey is to render this thesis as palpable and immediate as we know how—that is, to tell a story.

1. The Unfortunate Case of Petr Taborsky

The quintessential American success story starts off, just as this one, with an immigrant coming to America, the land of opportunity and freedom, partaking of the benefits of our educational system, and then turning her natural abilities to innovation and hard work in pursuit of a better life. The tale of Petr Taborsky, born in Prague in 1962, begins in just that way.[2] Taborsky's family fled the events of 1968 in Czechoslovakia for the United States. Taborsky was a precocious child and an inveterate inventor, filing for his first patent at age sixteen. In the fall of 1986, Taborsky enrolled as an undergraduate at the University of South Florida (USF), pursuing a double major in biology and chemistry; needing financial aid, he began working as a part-time student lab assistant in a civil engineering laboratory at the USF run by Dr. Robert Carnahan.[3] USF, although a public university, like many others strongly en-

2. The sources for this story are Sanchez 1996; a series of reports in the USF student newspaper *The Oracle* found at http://www.oracle.usf.edu/archive; Jaroff 1997; the AP wire service story dated June 18, 1996, entitled "Ex-Student Sent to Chain Gang"; and the web site of the Student Coalition for Handling Intellectual Property, http://www.ij.net/S-Chip/petr.

3. The shifting importance of foreign-born students in American science at the end of the century is just one of the many aspects of this story that renders it representative. In 1995, 23 percent of American Ph.D.s in the sciences and engineering were foreign born, although the proportion was much higher in certain fields like civil engineering, where it was 50 percent. See National Research Board, *Science and Engineering Indicators 1998,* 3–19. All subsequent citations to this source will assume the format *S&EI [YEAR].*

courages its faculty to attract and conduct research sponsored by corporations and other external private entities. Not only does this augment basic state funding for university activities, but attracting such grants also enhances the reputation and status of the university in the wider world. USF had found itself impelled over the years to develop various institutional structures in order to negotiate the public/private interface, not to mention the reconciliation of the obligations of the faculty to their university with their obligations to their sponsors and grantors. For instance, USF had established a Division of Sponsored Research that supervised roughly fifty million dollars in grants in the academic year 1988–89 and was charged with overseeing the assignment of property rights arising from such research. However, as at so many other universities in this period, USF had to improvise many procedures and policies toward this innovative sponsored research as it went along, in part due to lack of prior experience with this kind of research funding, partly because of the prior legal and commercial expertise of the corporate funders, and in part simply to keep up with the more irrepressible entrepreneurial innovations of some of its more active faculty members.

At first, Carnahan was impressed with Taborsky's abilities and encouraged him to pursue a master's degree in civil engineering, which Taborsky embarked upon in August 1987. Meanwhile, Carnahan and USF managed to land a small portion of large study of wastewater treatment procedures offered by a 1986 consortium of water and power utilities, the immediately relevant participant being Florida Progress Corporation. Florida Progress awarded a three-month $20,000 contract to Carnahan at USF to determine the capacity of some bacteria to clean calcium deposits from a granular clay called clinoptilolite for reuse in wastewater treatment. Carnahan assigned Taborsky to this research account to perform the laboratory tests. At the end of the three months the testing was completed, and therefore Carnahan removed Taborsky from the account and wrote up a project report. At the termination of the grant, Taborsky embarked upon his master's project on the physical properties of the aluminosilicates (resembling commonplace kitty litter), separate from any question of bacterial action on the clay. In this interval, Florida Progress expressed no interest in funding further research, although Taborsky did perform tasks for Carnahan, primarily laboratory assistance at $8.50 per hour, on other accounts. Taborsky summarized the findings of his own project for Carnahan in a May 5, 1988, report, in preparation for completion of his degree.

In July 1988, while working on his own ideas in the lab, Taborsky discovered that heating aluminosilicates like clinoptilolite at temperatures of to 850°C would vastly improve the abilities of the clay to be reused in treating wastewater. Since conventional wisdom was that heating this clay above 600 degrees would destroy it, no one had looked into the effect of extreme heat on its ability to reject calcium deposits and absorb ammonium ions. Previous work on the bacteria project had prompted Taborsky to realize that this process would be potentially valuable to Florida Progress and other such water utilities. He discussed this with Carnahan, who informed Taborsky that he could not expect to recoup any benefits from this discovery, which Carnahan suggested would belong to USF. Carnahan also advised that Taborsky could be prosecuted and incarcerated if he attempted to file a patent on the discovery or attempted to publish the result. The blow was softened, however, by the further suggestion that Florida Progress might hire Taborsky as an engineer.

Taborsky met with representatives of Florida Progress in September 1988 and described his thesis work. In December, Florida Progress offered Taborsky a job, which Taborsky declined, believing (rightly or wrongly) that all Florida Progress wanted was legal control over the discovery. However, as part of the job application, Taborsky did sign what he was told was a routine confidentiality agreement. On January 6, 1989, Taborsky filed a patent application on his discovery, after leaving USF and resigning from his lab job, without taking his finals or in any other way wrapping up his obligations there; but he did abscond with the lab notebooks in which he had first described his discovery. Carnahan, finding the notebooks missing, tried to get hold of Taborsky by phone, leaving a message on his answering machine threatening him with jail if he did not return the notebooks. Carnahan then reported the theft to university police, charging that Taborsky had filched at least thirty-two trade secrets from himself, USF, and Florida Progress. On September 27, Carnahan and Florida Progress filed a competing patent on the aluminosilicate process.

Up to this point, the reader might suspect that, however unfortunate the details and ferocious the recriminations, such patent disputes are nothing especially novel or noteworthy in the rough-and-tumble world of corporate research and development and that inventions are especially prone to legal contests over ownership. Combine this with poor interpersonal dynamics and one principal who was dead-set in his fervently held conviction of moral superiority, and regrettable conflicts will inevitably ensue. However, such offhand impressions would not begin to appreciate the special characteristics of this case, nor focus upon attributes that bear important lessons for contemporary economics of science. This incident constitutes a veritable witches' brew of all the ingredients that make up the "new regime" of science funding and organization in America at the beginning of the twenty-first century. Into the pot were tossed vulnerable foreign students, warped career paths, visions of virtual riches (Wilson 2000), ill-prepared university administrators, vague case law, intangible intellectual property rights, reengineered corporations bent on capturing competitive advantage, and self-seeking faculty entrepreneurs in a fragmenting university, not to mention the ersatz kitty litter. The older Mertonian ethos of noble, disinterested "communistic" science had no purchase here; but even simplistic notions of "exploitation" would equally lead us astray.

The most striking aspect of subsequent events was the intransigence with which USF sought to assert what it (or, at minimum, its research administrators) conceived as its prerogatives in this case. Note well Taborsky's particular vulnerabilities: a student with a paucity of standard intellectual credentials, occupying an underlaboring role in the laboratory, and an immigrant whose pending application for citizenship could itself be put at risk. Recall further that Taborsky had never signed any intellectual property agreements with the university, but only with Florida Progress. USF, far from seeing itself as sharing interests with one of its own students, followed the inclination of its vengeful faculty member in prosecuting Taborsky to the utmost. Not only did USF file criminal charges, but it also sought permission from its regents to file a civil suit as well. In a downward spiral of recrimination, Taborsky lodged a countersuit, alleging conspiracy, violation of civil rights, and racketeering. There is some suggestion that USF pursued the criminal case because it might be used to block award of the patent to Taborsky. In January 1990, Taborsky was convicted of grand theft and theft of trade secrets, sentenced to fifteen years' probation, and ordered to turn over all research materials to USF. Former USF president Francis Borkowski wrote a letter to the judge in the case urging a prison sentence for Taborsky, asserting that he was "beyond rehabilitation," and more ominously, that his actions had threatened the good relationship that had been fostered between USF and its corporate sponsors.

From Taborsky's vantage point, USF had gone well beyond the pale in seeking to prosecute him in this matter. For instance, Taborsky alleged that some administrators falsely maintained that he had signed a confidentiality agreement with USF prior to the Florida Progress project, that Carnahan had backdated his own confidentiality agreement with Florida Progress, and that the jury in the criminal trial was denied access to USF's own "Property and Procedures Manual." Whatever the truth of these allegations, on January 24, 1991, the United States Patent Office denied the Carnahan application and awarded patent 5082813 to Taborsky for the aluminosilicate process; he was subsequently awarded two further patents. Regarding this ruling as personal vindication, Taborsky refused to turn over his notebooks to Carnahan and USF, stating that they would deny him access and use of his own notes; he regarded the order to sign his patent over to USF by a district court judge a travesty. The USF responded by taking him to court again for violation of his parole agreement; in 1992 he was sentenced to three and a half years in prison for the infringement. Exhausting all appeal procedures, he was incarcerated in 1993, including an eight-week interlude on a chain gang; it was at this juncture that the press got hold of the story.

Reading about the case in a local newspaper, the Florida corrections secretary got him removed from the chain gang; then the governor got into the act and offered a full pardon. By this time, Taborsky, in full righteous dudgeon, flat out refused the pardon, on the grounds that accepting a pardon would be an admission of guilt. As Taborsky said in his radio interview: "I'm seeking justice and seeking the truth. What actually happened? What were the contracts for? Where did the money go? And those sorts of things. Those things will vindicate me. They will show that I did nothing wrong."[4] Taborsky's own story is not over yet. He was released from prison after serving eighteen months of his sentence and managed to get a judge to terminate his probation. He has filed an appeal to overturn his original conviction, and the civil suit has yet to be settled. Taborsky, on the warpath, says he will settle for nothing less than complete and total vindication.

Was this merely an iconic "failure to communicate" after the manner of Cool Hand Luke, or was it something larger? One fears that the cinematic aspects of "working on the chain gang" account for much of the publicity that brought this case to public consciousness (however fleeting); but far from being an aberration, troublesome cases such as this one have become a fact of life in modern science. While few have gone to the lengths of squandering vast resources on legal fees for a criminal case as did USF, most modern research universities now find they must take a certain level of civil prosecution over intellectual property rights

4. Sanchez 1996.

in stride. For instance, the University of Michigan paid $1.67 million in damages to a scientist who maintained that her work had been stolen by a superior; and that was in the field of psychology, and not a really high-stakes research area like biotechnology or computer technologies.[5] Some faculty have taken to suing their own universities for shortchanging them on royalty agreements, as the case of Jerome Singer and Lawrence Crooks in their dispute with the University of California over magnetic resonance imaging. Most large research universities now find a persistent nagging parade of problems of science regulation, construed broadly to encompass not just property disputes but also "fraud," negotiation of contractual agreements with industry and the federal government, and the monitoring of human subjects and hazardous materials research, not to mention the portmanteau category of "research ethics" sufficient to justify the maintenance of a full-time "Office of Scientific Integrity"—although terminology has not yet been stabilized in this area, so the bureau might also travel under the more colorless but upbeat designation of "Office of Technology Transfer." And as good economists, our thoughts immediately turn to ways to quantify the costs and benefits of this endeavor. Interestingly enough, although individual universities do keep fairly close tabs on the costs of various breakdowns of their own operation of the scientific enterprise, they are rarely (if ever) shared between institutions, much less rendered the subject of academic study. One observer has made a very impressionistic back-of-the-envelope calculation that quotidian regulatory problems of science at the university level absorb something like 2 percent of the total research budget, although worst-case scenarios such as the aforementioned Michigan dispute might cause that figure to go higher.[6] Another source estimates that "dozens of major universities—Brandeis, West Virginia, Tufts, and Miami among them—actually spent more on legal fees in fiscal year 1997 than they earned from all licensing and patent activity that year" (Press and Washburn 2000, 48).

It is our impression that this novel administrative arm of the university, combined with incidents resembling the Taborsky debacle that have

5. See Philip Hilts, "University Forced to Pay $1.6 Million to Researcher," *New York Times,* August 10, 1997, A-13. For other such cases, see Hilts, "Jury Award Voided in Scientific Research Case," *New York Times,* February 2, 1997, A-17. Indeed, the corporate penetration of university research in molecular biology has proceeded to such an extent that it is getting to the point that such disputes are treated as newsworthy only to the extent that the acrimony spills outside the cloistered halls of the university and into a more public arena: see, for instance, Horton 1999.

6. Wright 1997.

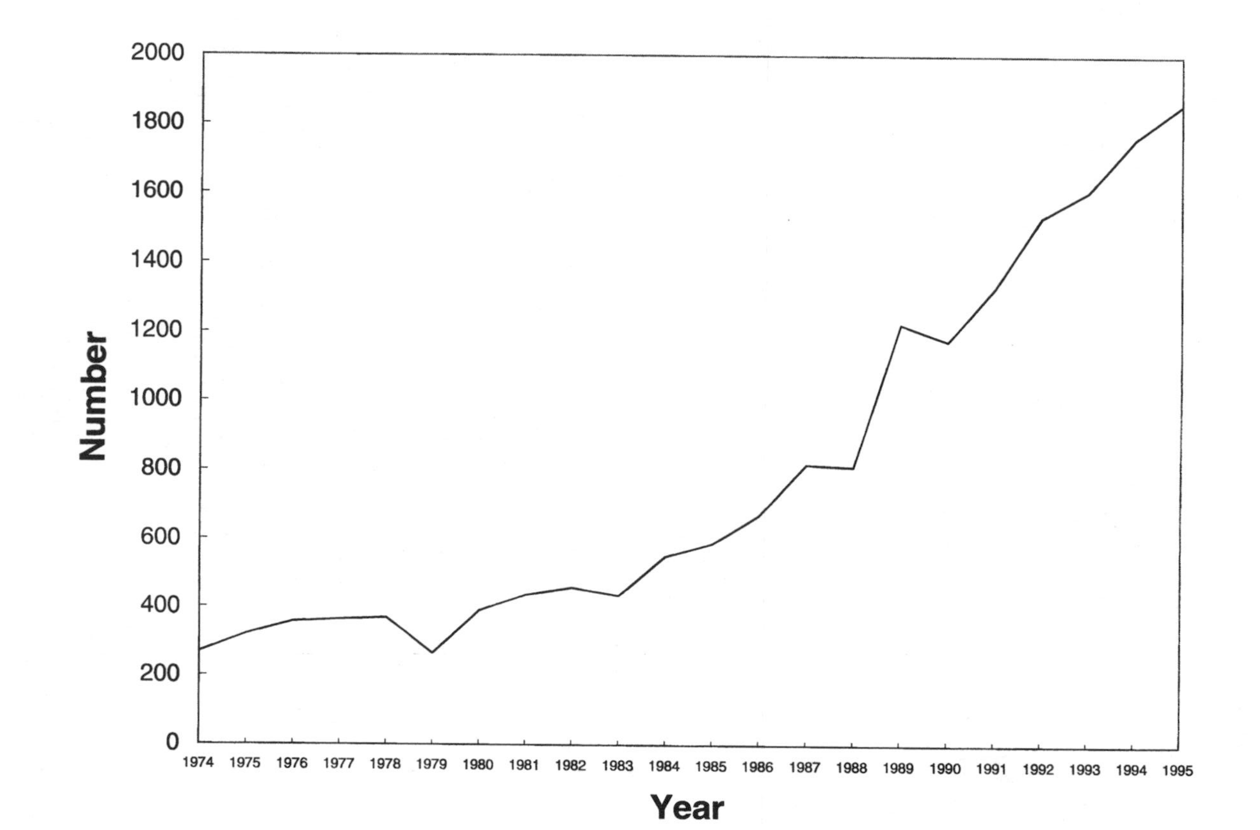

FIGURE I.1. Number of Patents Granted to U.S. Academic Institutions
SOURCE: *S&EI 1996*, appendix pp. 249–51 (table 5-41) for 1974–1991 and *S&EI 1998*, pp. A-337–A-343 (table 5-57) for 1992–1995.

summoned it forth, are themselves symptomatic of much larger changes happening within the structure of the scientific enterprise, and as such, should be of more direct concern to a suitably contemporary economics of science. But that would necessitate viewing the Taborsky incident as something more than an unfortunate fluke or a local oddity, as a bellwether instead of an aberration. It is part and parcel of the phenomenon illustrated in figure I.1, where universities have been reengineered in the recent past to become active producers of patented knowledge.

What is sorely needed for a better understanding of this incident is some historical background in order to gauge the extent to which the relevant economic structures of scientific research have indeed changed dramatically over the last century, which includes the extent to which universities have ceased to serve as repositories for wisdom and have become profit centers for the generation of intellectual property.

2. Scientists Who Came out of the Cold (War): An American Tale

The lesson that things change over time is hardly novel or earthshaking; rather, the proposition that there are some discernible regularities to perceived change is the starting point of all analysis. Science has indeed undergone profound transformations in organization and content over the course of the twentieth century; and yet, the only sort of change that many commentators on the health and well-being of Science deign to recognize is that there has been but more of the same. In other words, in this simple but widespread view, progressively larger phalanxes of scientists produce progressively more "knowledge" that is poured into a vast communal storage tank of information, available to be tapped into by anyone willing to invest in the training required to successfully drink from the wellsprings. Of course, from this perspective, bigness itself can create social problems—problems of coordination, problems of organization, and if science grows "too fast," problems of funding and support. These problems, it has been suggested, exist merely as the unfortunate by-product of the demonstrated success of science and as such have nothing whatsoever to do with the nature of the knowledge produced, the culture that produced it, or the nature of the understanding of science in its multiple manifestations. Partisans of this account favor locutions like "If it ain't broke, don't fix it." One observes these inclinations in what has been called the "public understanding of science" movement, where "the public" is repeatedly quizzed and badgered on its ability to recognize correct scientific propositions and willingness to report favorable attitudes toward science in general.[7] These inclinations were also popularized by one of the pioneers of "scientometrics," Derek de Solla Price, and have been elaborated more recently by John Ziman (1994). In this approach, many of the seemingly dysfunctional attributes of the prosecution of modern scientific research are attributed to the sheer magnitude of the modern scientific enterprise. Big Science, it is claimed, must eventually run into diminishing returns, if only because it has grown so much more rapidly than the underlying population base, economic base, or even the cognitive capacity of any individual to comprehend its achievements. "Nowadays planning a new research program is much more like . . . setting up a factory to manufacture a new product line" (Ziman 1994, 47). But mass production and bureaucratic rationality are said to be the unfortunate price that must be paid for ratcheting up the scale of scientific research; paraphrasing Joseph Schumpeter: it is ham-fisted and soulless, but it delivers the goods.

However commonsensical such quasi-economic language of scale economies might initially appear to those with social science backgrounds, we must beg to differ that such an approach is ahistorical, mechanistic in the extreme, entangled in problems concerning the appropriate specification of production functions for science, and therefore fundamentally misleading. Although it would seem the height of consistency to apply a species of autonomic technological determinism to science itself, we argue that it contributes little to an understanding of the present predicament of science, and much less to specific cases of severe malfunction, such as the Taborsky incident. In its place, we seek to suggest an analytical classification of structures of science organization in the twentieth century that is more intentionally connected to the relevant geographic, historical, and economic contexts; one that does not trade in a passive fatalism and instead serves to direct our attention to more finely detailed issues concerning specifics of social organization in the economics of science. What this implies is that whereas it is undeniable that the content of science aims to transcend all geographical and temporal bounds, the actual prosecution of scientific research cannot seriously aspire to any such transcendence. Accepting that premise for the moment, principles of selection must be invoked in order to restrict analysis to a limited set of structures of social organization of scientific

7. See, for instance, S&EI [1998], chap. 7. There we learn, among other interesting tidbits, that Japanese citizens tend to report very high anxiety over the negative effects of science and technology, at much higher levels than American citizens.

inquiry. In this volume, we are concerned with what have been predominantly "American" theories of the economics of science—that is, postwar neoclassical theories—and that consequently dictates some brief familiarity with the history of American regimes of science funding as a prerequisite. We want to stress that this will not ultimately preclude consideration of science funding and organization in other countries—rather, exploration of the demonstrably parochial character of unfounded assertions concerning the existence of a universal "economics of science" (a characteristic American vice) must take precedence, before one explores alternative national idioms of the organization of research.

In brief, based upon a cross-section of recent work in the history of science and science policy, we would like to propose that there have existed three very distinct regimes of science organization and funding in the United States in the twentieth century, and that each regime has borne a special relationship to the contemporary organization of science in other countries.[8] While they may overlap in certain particulars, we will place the temporal divisions of these regimes at the following boundaries: early twentieth century to 1940; from World War II through the Cold War; and (roughly) 1980 to the present. For convenience, we will call them the (1) protoindustrial regime, (2) the Cold War regime, and (3) the globalized privatization regime. Our object in describing these regimes is not to provide a full-fledged history of science organization in the United States—those can better be found in our references—but rather to set the stage for our larger argument, namely, that each regime comprised a distinct set of structures that have in practice summoned quite differing versions of an "economics of science" to justify and account for their regularities. As promised, we regard funding structures and theoretical accounts of their efficacy as inextricably interlinked. The stubborn quarantine of science policy discussions from the history and sociology of the organization of science must cease if there is to be a sensible economics of science. One of the aims of this volume, and of

8. The description of these regimes is distilled out of a large literature, some highlights of which are Kohler 1991, Reich 1985, Stine 1986, Sarewitz 1996, Hart 1998, Brooks 1996, Kleinman 1995, Gruber 1995, Mowery and Rosenberg 1998, Noll 1998, Guston and Kenniston 1994, Reingold 1995, Gibbons et al. 1994, Ziman 1994, Kline 1995, and Branscomb, Kodama, and Florida 1999. For reasons of space, we must curtail any elaborate historical description of the first two regimes in this volume. The evidence concerning the specific shape of the third of our three regimes is summarized in this volume in the article by Slaughter and Rhoades (chapter 1) and the pieces contained in our section on the "contours of the globalized privatization."

this introduction then is to situate some of what are generally conceded to be the classic texts representing these alternative approaches within the various American regimes of science organization and funding.

2.1 The Protoindustrial Regime

The first, protoindustrial regime of science funding and organization dates from an era when American science was widely regarded as inferior in many respects to European science and when a few American universities were entering their initial phase as incubators of scientific research. In this era, most colleges and universities existed almost exclusively to perform the service of education and the propagation of the liberal arts, based upon Scots and British models of liberal education.[9] As our label intimates, most scientific research and development in this period was to be found in a few large American corporations. The reasons that some large corporations such as General Electric, DuPont, American Telephone and Telegraph, and Eastman Kodak fostered in-house research capacity had much more to do with the need for routine testing capacities and with the fin de siècle merger wave and American antitrust and patent policies than with any belief in the necessity of innovation or the commercial value of science, as is now widely acknowledged in the historical literature.[10] America lagged behind European practice only by a few decades, since the modern literature dates the inception of corporate research labs for the purpose of product innovation in Europe to roughly the 1880s (Fox and Guagnini 1998–99, 215, 251). In the public sphere, the U.S. federal government role in supporting research was comparatively small and consisted primarily of promotion of agricultural research through a network of agricultural extension stations, or of subsidizing specialized research in government-run labs

9. Our characterization of this situation as "protoindustrial" is supported by one of the earliest texts in the economics of science, Thorstein Veblen's *Higher Learning in America* (1918). There Veblen complains that universities are increasingly being run according to business principles; and although he casts this in invidious comparison to the Germanic model, in fact it paralleled the organization of science in American corporate and government laboratories at that time. For an instructive instance of the prehistory of protoindustrial science in the United States before the great wave of consolidation, see Lucier 1995; for an overview of the protoindustrial landscape, see Carlson in Krige and Pestre 1997 and Hounshell in Rosenbloom and Spencer 1996. The European history of protoindustrial laboratories is comprehensively surveyed in Fox and Guagnini 1998–99.

10. "Industrial research laboratories were first established in the U.S. primarily to protect large corporations from competition" (Reich 1985, 239). On the uses of research labs to ward off antitrust prosecution and foreign competition, see Reich 1985, Mowery and Rosenberg 1998, and Dennis 1987.

tied to motives of nation building. Examples of the latter would include the Coast Survey (which employed the earliest author in this anthology, Charles Sanders Peirce), the U.S. Geological Survey, the Bureau of Chemistry of the Agriculture Department, and the National Bureau of Standards. In many respects, American policy reformers sought (with indifferent success) to mimic science policies innovated in Germany, especially in the promotion of state-funded higher education combined with state-run research institutes, and with good reason, since German science was thought to be the best in the world (Lenoir 1998). What the German system had started under the aegis of the Humboldtian reforms was a closer integration of (graduate) teaching and research, which extended to the institution of a laboratory-based pedagogy (Fox and Guagnini, 1998–99). American students seeking advanced academic training were therefore urged to spend time in German universities, in the absence of suitable American infrastructure. The professionalization of academic disciplines was then in its earliest stages; and in most fields, a career consisting primarily of research was simply not a viable option. Unless one worked for a federal or a corporate lab, the life of an American scientist was a hard one. Only toward the very end of the period did a handful of private foundations (such as Rockefeller and Carnegie; see Kohler 1991) begin to innovate new forms of scientific patronage, aimed at building up a few selected universities as research institutions and revising the previous construction of the research grant as a temporary dole parceled out as charity to poverty-stricken academics. Thus both corporate science and the nascent structure of academic careers in America were almost entirely defined by the captains of industry and their managers.

As noted by Reingold (1991), Americans had great trouble coming to terms with the nascent idea of public funding for a scientific elite. By the 1920s, there arose a substantial cultural trend that regarded industrial concentration and technological advancement as two sides of the same coin, a dynamic resulting in technological displacement, and in the 1930s, even widespread unemployment. Moreover, Continental Europeans (with the British emphatically excluded) tended to treat their scientists as having a status patterned upon their previous aristocracies and thus took for granted their role within the state; but this option was foreclosed in the American context. Suspicion of elites tended to shade over into skepticism over the very premise that there should exist a cadre of researchers who would do their thinking for the benefit of the larger populace. This was captured by some comments in the *New York Times* of 1885:

> Like other men [scientists] are self-seeking, ambitious, and have their personal ends to gain. Can we assume that they are morally any better than their neighbors; or that, if they get possession of place and power, they will not use and pervert them to the promotion of their selfish objects? (quoted in Kevles 1995, 54)

Hence, we should note that there is nothing particularly novel or radical about the American penchant for regarding the scientist as a rational self-interested agent; nor is the notion of a market-driven science especially innovative or pathbreaking.[11] Indeed, it constituted the intellectual underpinnings for the bulk of science support in America at the turn of the last century. The point we wish to stress is that the science supported under this regime was relatively modest and rarely attained world-class status. Innovative American scientists tended to be autodidacts and loners (Peirce himself serving as an extreme example) even if they managed to ascend to a university position, and many did not. It took a dramatic change of regime to propel American science to the front ranks of world science.

Because this spotty situation barely qualified as a "system," it should perhaps come as no surprise that there was very little literature dating from this era that could qualify as propounding a self-conscious "economics of science"; and with the exception of the short article by Peirce, we have found nothing suitably relevant to include in this volume.[12]

2.2 The Cold War Regime

It is now almost universally acknowledged that World War II stands as the watershed of American science and that the system of science funding and management in the United States propelled it to world

11. Here we take issue with a strain within the science policy community that attempts to portray a single Mode 1/Mode 2 watershed in the organization of modern science between communalist and corporate science. The locus classicus of this theme is Gibbons et al. 1994.

12. The only candidates would have been the aforementioned *Higher Learning in America* (1918) by Veblen, and perhaps some Marxist writing on the relationship between base and superstructure within the context of some version of historical materialism. Because what little funding there was emanated from the trust-dominated industrial sphere, most of this literature sported a certain muckraking attitude.

dominance for the second half of the century.[13] This is not to say that the feat was accomplished consciously and intentionally; nor is it to ignore the fact that the Depression and Nazi-era disruption conveniently destroyed the hitherto-dominant German university system and drove a generation of stellar scientific talent to seek asylum in the United States. For obvious reasons, this is not the place to rehearse the familiar narrative of how World War II became known as the "physicist's war" and how the postwar politics of the bomb and the Cold War locked America into a military-dominated system of science funding.[14] Rather, we merely wish to indicate the ways in which the various components of the Cold War regime that were forged in the fires of World War II tended to fit together, and the ways in which this system was virtually unprecedented among the other economically developed nations, at least until the 1980s. Our purpose, to reiterate, is to set the stage for the conditions giving rise to a nascent analytical discussion of the "economics of science" in the United States in this period.

The centerpiece of this regime was the massive federal presence in science planning and funding. The dominance can be demonstrated by means of many quantitative measures, showing federal expansion from the immediate postwar period to the mid-1960s, and then subsequent contraction, as illustrated in figure I.2 below.

But it was also a structural dominance, with the wartime practice (innovated at the Office of Scientific Research and Development [OSRD]) of research support in the format of government "contracts" being granted through universities and industrial firms on a "nonprofit" but fully reimbursed basis, strengthening both of these institutions by using them to channel support to individual scientists, rather than simply hiring the scientists as civil service employees of a federal laboratory system. This infusion of cash jump-started the cultivation of Big Science on a scale previously unimaginable, with massive instrumentation and hierarchical teams of interdisciplinary scientists and engineers roughly patterned upon the successes of the MIT Radiation Laboratory and the

13. "The postwar R&D system, with its large well-funded research universities and Federal contracts with industry, had little or no precedent in the pre-1940 era and contrasted with the structure of research systems of other postwar industrial economies. In a very real sense, the US developed a postwar R&D system that was internationally unique" (Mowery and Rosenberg 1998, 12).

14. On this history, see Kevles 1995, Kevles in Galison and Hevly 1992, Rhodes 1986, Reingold 1995, Leslie 1993, Morin 1993, Hacker 1993, Brooks 1996, Lowen 1997, Mowery and Rosenberg 1998, Kleinman 1995, Hart 1998a.

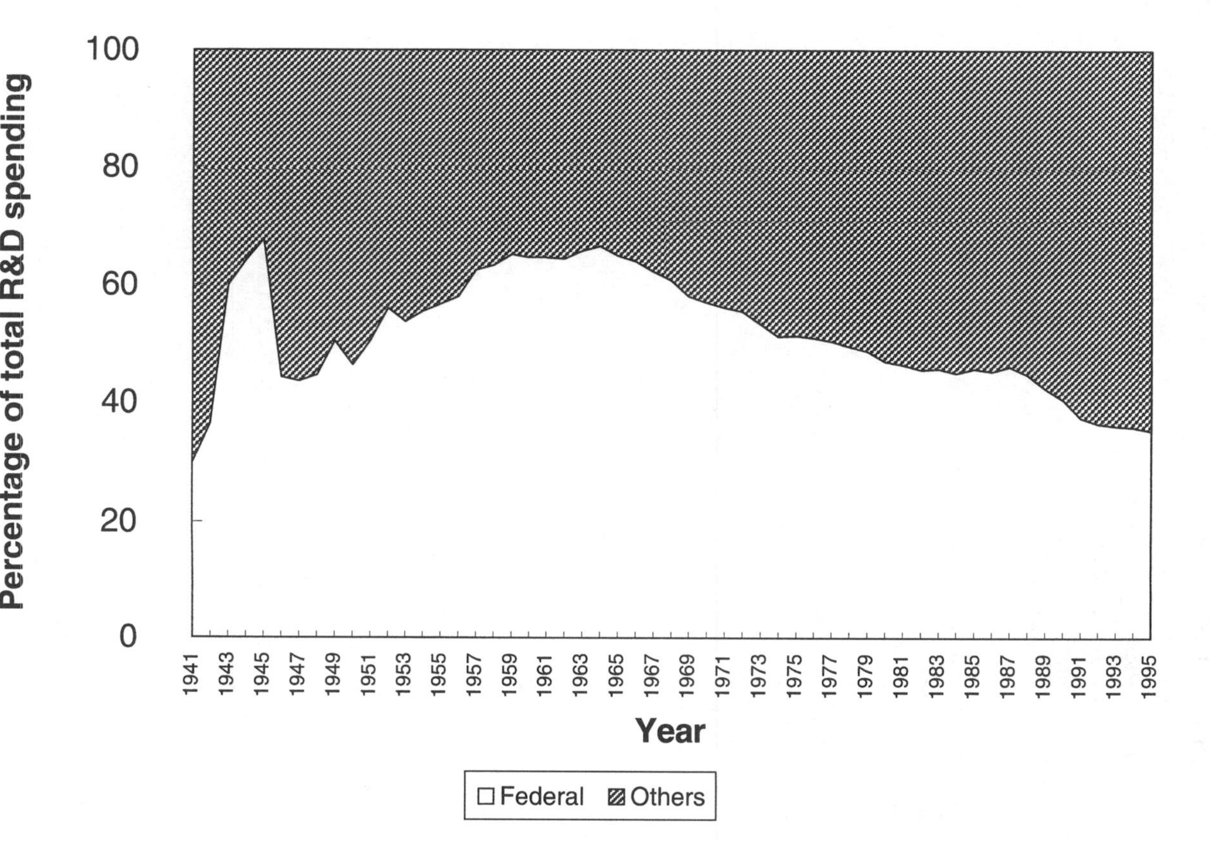

FIGURE I.2. National R&D Expenditures, by Sources of Funds (Federal vs. Others)
SOURCE: *The First Annual Report of the National Science Foundation: 1950–51,* p. 30, appendix 6 [estimated data] for 1941–1952, and *National Patterns of R&D Resources, 1995 Data Update,* table B-3, available at http://www.nsf.gov/sbe/srs/s2195/start.htm for 1953–1995.

Manhattan Project.[15] Curiously enough, this was combined with a strong reinforcement of the centrality of academic discipline as the arbiter of legitimacy of training and career success of the scientist. The vehicle for reconciling these seemingly contrary trends was the reconfigured postwar American university, where federal science policy created a situation in which teaching was now openly avowed to be complementary to research (but not in those massive introductory courses staffed by graduate students); where overhead costs on contracts helped fund the operating expenses and the graduate population; where the GI bill and the creation of the temporary occupation known as the "research assistantship" forged a whole new career trajectory for novice scientists and underwrote a massive expansion in graduate education. If some especially favored scientists still chafed under this shotgun marriage of teaching and research and disciplinary identity, a hybrid largely absent in most other developed economies, then a novel quasi-governmental entity known as the "think tank"—or more colloquially, a university campus without students—was created to cushion the irritation for the favored few.[16]

It would be a mistake, however, to see American science policy as focused solely or even primarily upon universities. Federal and military science policies also fostered what has been called a "stealth industrial policy" in the United States.[17] It was an ideological imperative in America that the government not be seen as favoring certain industries in contradiction to the marketplace, a ruse that Cold War security considerations substantially facilitated. Nevertheless, most federal R&D funding was channeled through private corporations, even at the peak of university support, skewing the direction of technological exploration in selected industries. Private industry (broadly defined) has always been far and away the largest performing sector of R&D in the United States, as revealed in table I.1.

Indeed, these figures understate the magnitude of federal subsidy of

15. See, for definition of "Big Science," Galison and Hevly 1992, Galison 1997, Heilbron and Seidel 1989.

16. The paradigm instance of the postwar think tank was the RAND Corporation, split off from the Douglas Aircraft Corporation and supported initially by Air Force funding. On the phenomenon of think tanks, see Stone, Denham, and Garnett 1998. The early history of RAND is covered in Smith 1966 and Collins 1998.

17. The notion that the United States did have a sub rosa version of industrial policy has gained increasing attention among political scientists and other policy analysts. See, for instance, Etzkowitz 1994; Markuson and Yudken 1992; Teske and Johnson 1994; Hart 1998a, 227–29.

TABLE I.1. National Expenditures for Total R&D, by Performer

Year	Total	Federal Govt.	Industry	U&C	U&C FFRDCs	Nonprofits
1953	5,124	1,010	3,630	255	121	108
1954	5,644	1,020	4,070	290	141	123
1955	6,172	905	4,640	312	180	135
1956	8,364	1,041	6,605	372	194	152
1957	9,775	1,220	7,731	410	240	174
1958	10,711	1,374	8,389	456	293	199
1959	12,357	1,639	9,618	526	338	236
1960	13,520	1,723	10,509	646	360	282
1961	14,320	1,878	10,908	763	410	361
1962	15,392	2,096	11,464	904	470	458
1963	17,059	2,279	12,630	1,081	530	539
1964	18,854	2,838	13,512	1,275	629	600
1965	20,044	3,093	14,185	1,474	629	663
1966	21,846	3,220	15,548	1,715	630	733
1967	23,146	3,396	16,385	1,921	673	771
1968	24,605	3,494	17,429	2,149	719	814
1969	25,629	3,501	18,308	2,225	725	870
1970	26,134	4,079	18,067	2,335	737	916
1971	26,676	4,228	18,320	2,500	716	912
1972	28,476	4,589	19,552	2,630	753	952
1973	30,718	4,762	21,249	2,884	817	1,006
1974	32,863	4,911	22,887	3,022	865	1,178
1975	35,213	5,354	24,187	3,409	987	1,276
1976	39,018	5,769	26,997	3,729	1,147	1,376
1977	42,783	6,012	29,825	4,067	1,384	1,495
1978	48,128	6,810	33,304	4,625	1,717	1,672
1979	54,939	7,418	38,226	5,366	1,935	1,994
1980	62,596	7,632	44,505	6,063	2,246	2,150
1981	71,869	8,426	51,810	6,847	2,486	2,300
1982	80,018	9,141	58,650	7,323	2,479	2,425
1983	89,143	10,582	65,268	7,881	2,737	2,675
1984	101,167	11,572	74,800	8,620	3,150	3,025
1985	113,818	12,945	84,239	9,686	3,523	3,425
1986	119,555	13,535	87,823	10,927	3,895	3,375
1987	125,376	13,413	92,155	12,152	4,206	3,450
1988	132,889	14,281	97,015	13,462	4,531	3,600
1989	140,981	15,121	102,055	14,975	4,730	4,100
1990	151,544	16,002	109,727	16,283	4,832	4,700
1991	160,096	15,238	116,952	17,577	5,079	5,250
1992	164,493	15,690	119,110	19,794	5,249	5,650
1993	165,849	16,556	118,334	19,911	5,298	5,750
1994	169,100	17,200	119,700	20,950	5,250	6,000
1995	171,000	16,700	121,400	21,600	5,300	6,000

NOTE: U&C = universities and colleges; FFRDCs = federally funded research and development centers. All the data are measured in current dollars.

SOURCE: *National Patterns of R&D Resources: 1994,* An SRS Special Report, NSF 95-304, Division of Science Resources Studies, National Science Foundation at http://www.nsf.gov/sbe/srs/s2194/dst1.htm for 1953–1991 and *S&EI 1996,* appendix p. 107, table 4-4 for 1992–1995.

corporate science, since a fair proportion of it occurred through tax rebates and third-party payments. (It is an artifact of the Cold War regime that there exist no dependable consolidated R&D accounts for the entire federal government, even down to the present. Things have, if anything, gotten worse with the decline of the Cold War regime, with the Department of Defense no longer providing detailed breakdowns of its own R&D spending by academic field after 1993.)

This industrial policy extended well beyond monetary grants and subsidies, however. In sharp contrast to the protoindustrial regime, the Cold War regime was characterized by a very weak legal structure of intellectual property protection combined with a very active antitrust posture. The net consequence of this mode of science organization was that many of the scientific and technological breakthroughs achieved by corporate labs such as Bell Labs, Xerox Parc, RCA Sarnoff, Merck Rahway, and IBM Yorktown were not adequately capitalized upon by their huge corporate sponsors, but were instead turned into downstream marketable commodities by small start-up firms, themselves often formed by fugitives from those very same corporate labs.[18] This "communal" approach to appropriation of the fruits of subsidized research was also encapsulated in the Department of Defense policy of a "second-source rule" for suppliers of high-tech weaponry and devices, duly sweetened by cost-plus contracts. This rather cavalier attitude toward technology transfer from the laboratory to the marketplace was one of the prime hallmarks of the Cold War regime, one that could trace its provenance to the looming presence of the military in science funding.

The mutual reinforcement of this stealth industrial policy and the postwar ideology of the "freedom" of the scientist is another phenomenon, like the existence of a stealth industrial policy and the structure of intellectual property, that we believe has not yet been adequately explored in the science studies literature. The scientists most heavily embroiled in military funding had to submit to the classification and clearance procedures of the state; in exchange, the state would promise not to micromanage their research agenda. Control, while not completely internalized, was certainly rendered unobtrusive; and the ability to ap-

18. One such famous instance, that of the transistor, is related by Riordan and Hoddeson 1997. The capture of Xerox Parc technology by Apple Computer is legendary in the business history literature. This aspect of industrial policy was pointed out by Mowery and Rosenberg (1998, 44). The interplay of antitrust and science policy is also stressed in Hart 1998b.

peal to freedom of expression was a critical component of the ideological rivalry of the period.[19] Since a major feature of the Cold War regime was the maintenance of ongoing university ties of scientists doing government-sponsored work, they were exhorted also to publish in the "open" literature to meet their disciplinary obligations and bolster their credentials if they exercised prudent discretion. (The existence of completely classified "scientific journals" stands as one of the more extreme anomalies of that era.) The "uses" of various discoveries therefore became more radically separated from their original elaborations (as well they might, given the rather imprecise concepts of intellectual property), especially in the formats in which they were disseminated.[20] The economics of these disciplinary outlets were themselves often obscured through such indirect devices as page charges, submission fees, and wildly inflated library subscription rates. Crudely, it became possible for nominal academic scientific stature to be denominated in terms of public intangibles like disciplinary "credit" or "eminence," all the while the money was being allocated according to somewhat different criteria. Even though science was being closely managed by research officers, at first in the wartime OSRD and after the war in the Office of Naval Research, DARPA (Defense Advanced Research Projects Agency, and other permutations), the Atomic Energy Commission, and elsewhere, the scientists eventually learned to come to terms with any residual sense that there might fester some conflict between their own freedom of inquiry and larger decisions to channel research in certain directions. If the exigencies of national security did not appear sufficiently compelling, the researcher could always take succor from the ethos of "pure science" within the academy. Indeed, this became the background to the public face of science policy enshrined in Vannevar Bush's famous 1945 mani-

19. Many in policy positions were quite adamant that it was the scientists who had to adapt to political reality. See Bureau of the Budget controller Harold Smith to Admiral Julius Furer, quoted in Owens 1994, 536: "The real difficulty, I think, has been that the physical scientists are worried about governmental controls largely because most of them—as they make clear to me by what they do and do not say—do not know even the first thing about the basic philosophy of democracy. . . . However, most of them have learned to accept government funds with ease, and I think they can adapt themselves to governmental organization with equal ease."

20. One of the major trends in recent science studies has been to seek to recover the military context of postwar scientific discoveries that have been treated as disembodied "pure" science. See, for instance, Forman 1987, Edwards 1996, Mendelsohn in Krige and Pestre 1997, Forman and Sanchez-Ron 1996, Kay 2000, Sent forthcoming, Mirowski forthcoming.

festo, *Science: The Endless Frontier,* namely, the "linear model" of "basic" science → "applied" science → "development" → production. The alert reader will have detected that, until now, we have not acknowledged the existence of any hard and fast distinction between basic vs. applied science and technology in this introduction, for reasons that should now become clear. We regard the endless fascination in science studies (and, as we shall shortly observe, economics) with boundary maintenance between "basic" and "applied" science to be itself an artifact of the Cold War regime (Kline 1995). This doctrine, so taken for granted within modern orthodox economics and much of orthodox science policy, was hardly even present in economic writings prior to World War II.

The Bush report has been analyzed repeatedly, and perhaps to excess, in the literature on the history of science policy. Our only concern here is to suggest that it played an important conceptual and ideological role in the Cold War regime, even if it did not end up serving as a blueprint for the actual structures of science funding and management that were eventually instituted in the United States in this period.[21] The idea that there was some necessary but unproductive form of scientific research that required state funding for its very existence, and that the economic growth of the nation would suffer in its absence, whereas applied R&D could be safely left to the corporate sector to organize, in conjunction with the previously described stealth industrial policy that had precipitated out of the immediate postwar political process, provided the ideal cover for the absence of accountability of military science planning. Although often pitched at a rarefied level of abstraction seemingly free of any parochial considerations, it has only become clearer in retrospect that it was a product of local conditions prevalent in the spe-

21. This distinction will shortly play a role in our account of the evolution of successive "theories" of the economics of science. For instance, see the comment of Mowery and Rosenberg (1998, 31): "Anticipating subsequent economic analysis, Bush argued that basic research was the ultimate source of economic growth." It is now generally understood that Bush did *not* seek to have his OSRD experience extended into the postwar period because he believed in something much more closely resembling the "first" regime was superior; rather he was outmaneuvered on this issue. See, for instance, Zachary 1997, Reingold 1995, Hart 1998a. The eventual compromise on the shape of the National Science Foundation was itself heavily informed by military imperatives, focused upon the "open" science practiced in universities, and has always stood as a relatively small component in the overall federal approach to science policy. Because economists have neglected these points, their commentaries on the Bush report tend to misconstrue its significance. See, for instance, Holton 1998 and the papers in Barfield 1997.

cific postwar regime in the United States, and in fact bore little relevance for science policy in other countries in the same time frame.[22]

2.3 The Globalized Privatization Regime

Whether it be the cataclysmic downsizing of physics in the last two decades, or sweeping changes in the rules of the game for academic entrepreneurship, or the radical restructuring of research universities, everyone now realizes to a greater or lesser degree that the Cold War structure of science management is rapidly going the way of the whalebone corset and the phonograph record. Tectonic shifts of science funding between various sciences, as indicated in table I.2, have been accompanied by drastic reorganization in the very structures of scientific funding and conduct. The papers collected in this volume attest to that fact, from varying angles and perspectives.

From many different vantage points, it should now become apparent that the Cold War regime of science policy could not have persisted over the longer term. The convenient fiction of a clear separation between "pure" and "applied" science could not be long maintained. There were simply too many internal contradictions and repressed economic considerations in what had initially seemed a politically viable set of compromises. First, it was inevitable that the heavy subsidies provided by the federal government to the universities and corporations would sooner or later have run into political and economic obstacles in the U.S. political culture. Most Americans had never really relinquished their suspicions concerning coddled intellectual elites (Hollinger 1995), and the trope of the danger of the self-interested and therefore untrustworthy scientific expert could easily be revived as a democratic plea for greater accountability in value for money. Furthermore, because the R&D budget of the federal and state governments was so widely dispersed among numerous agencies and programs, and so devoid of coordination and effective interest-group mobilization (with the possible exception here of biomedical research), they made an inviting target if and when budgetary stringencies would prompt belt-tightening measures. A high-profile example of this growing vulnerability came with the cancellation of the Supercon-

22. One wonders, for instance, how different the economic trajectory of postwar Japan might have been if the Japanese had heeded Bush's assertion that "[a] nation which depends upon others for its new basic scientific knowledge will be slow in its industrial progress and weak in its competitive position in world trade, regardless of its mechanical skill" (Bush 1945, 19).

TABLE I.2. Federal R&D Obligations by Field, Basic Research

Year	Federal Total	National Total	% Physics	% Life Sciences	% Social Sciences	Federal/ National Total
1963	1,152	1,965	19.7	32.2	2.1	0.586
1970	1,926	3,567	17.6	36.19	3.32	0.54
1971	1,980	3,698	17.73	37.73	3.54	0.535
1972	2,187	3,829	16.55	39.69	3.66	0.571
1973	2,232	4,051	15.73	39.78	3.58	0.551
1974	2,388	4,439	15.08	43.22	3.14	0.538
1975	2,588	4,827	14.65	43.12	2.86	0.536
1976	2,767	5,291	14.02	44.16	3.14	0.523
1977	3,259	5,925	14.33	42.44	2.95	0.55
1978	3,699	6,841	14.03	42.93	3.35	0.541
1979	4,193	7,736	12.78	45.12	3.1	0.542
1980	4,674	8,651	14.29	43.95	3.15	0.54
1981	5,041	9,741	14.58	44.12	2.72	0.518
1982	5,482	10,658	14.43	46.08	2.19	0.514
1983	6,260	11,859	13.66	46.18	2.2	0.528
1984	7,067	13,176	13.03	46.52	1.88	0.536
1985	7,819	14,510	12.28	48.43	1.8	0.539
1986	8,153	16,885	12.3	47.33	1.39	0.483
1987	8,942	18,213	11.99	48.78	1.45	0.491
1988	9,474	19,381	12.73	47.52	1.55	0.489
1989	10,602	21,477	13.16	46.37	1.46	0.494
1990	11,286	22,556	13.06	45.88	1.28	0.5
1991	12,171	26,629	13.52	44.65	1.32	0.457
1992	12,490	27,044	12.87	46.77	1.12	0.462
1993	13,400	28,125	11.95	46.93	1.45	0.476
1994	13,523	28,934	11.11	47.86	1.36	0.467
1995	13,877	28,642	10.86	47.57	1.49	0.484
1996	14,464	29.574	10.69	47.56	1.47	0.489
1997	14,942	31,212	10.45	48.21	1.48	0.479
1998	15,862	NA	NA	48.47	1.51	NA
1999	16,914	NA	NA	49.2	1.57	NA

SOURCE: Table 35 in *Federal Funds Survey, Detailed Historical Tables, Fiscal Years 1951–99* at http://www.nsf.gov/sbe/srs/nsf99347/htmstart.htm and *S&EI 1998,* appendix p. A-125, table 4-7.

ducting Super Collider in October 1993, an event that marks the dethronement of physics as the unchallenged champion of the Cold War regime (Sarewitz 1996), even though, as shown in table I.2, physics as a funding priority had eroded even earlier.

The supposed immunity of the quotidian prosecution of science from economic considerations was further compromised when universities themselves stopped being perceived as otherworldly ivory towers removed from politics and were caught in the unseemly act of trampling all over each other in competition to get federal research funds specifically earmarked to their campuses, lobbying representatives not on grounds of some abstract scientific peer review but rather for geographic or other political justifications (Brainard and Cordes 1999). Another doomed holdover from the Cold War regime was the extreme concentration of research funds in a small number of elite institutions—Bush's own MIT being the most favored recipient—a skewed distribution that could not persist once excluded universities realized that they could actively enter the political sweepstakes for federal funds and effectively challenge previously hidden old-boy networks. Indeed, peer review was one of the early casualties of the breakdown of the Cold War regime, since the insistence upon functioning internal standards of quality control could not begin to assuage external demands for responsibility, relevance, and accountability.[23]

Third, the Cold War premise of "science policy in one nation" could not continue to be maintained in a world that was growing less bipolar and more economically developed. Contacts between scientists across the Iron Curtain could be monitored in the name of military security; but as European and Asian firms regrouped and became economically significant, they conceived a desire to tap into the scientific and technological developments that had become such a prominent feature of postwar American prosperity. Since their own indigenous research infrastructures were so divergent from that found in the United States—frequently with one set of institutions dedicated purely to instruction, another different set for state-sponsored research, and a third set charged with state planning of industrial research—they initially had to send their most promising students to the United States to partake of the novel developments within the framework of the unfamiliar pedagogy. American universities were initially inclined to welcome the newcomers, but this had perverse unintended effects on American science policy. As the American research infrastructure grew so prodigiously up through the 1960s, it became apparent that the pool of indigenous candidates for scientific careers would not keep pace; and so foreign students were

23. Paul Forman's paper in this volume (chapter 2) takes up the discussion of the shift from self-generated and self-referential norms of excellence to appeals to social responsibility in modern science and relates it to larger cultural issues of postmodernism and the demise of the national security mindset. Shaun Hargreaves Heap discusses how the English version of accountability for state-funded universities assumed the format of "Research Assessment Exercises."

increasingly recruited into American academe and industrial research. Although wonderfully beneficial for American culture, this did tend to create problems for the prevalent political rationale for the government funding of higher education and its integration within the research system, which tended to be phrased in terms of investment in national human capital. It became increasingly difficult to justify the subsidized training of foreign students, many of whom would return to their home countries in order to staff the major economic and political competitors to the American system; and this does not even take into account the undercurrent of xenophobia undermining political support for government-subsidized education. As figure I.3 reveals, some disciplines began to be dominated by foreign nationals by the 1990s.

The net result was that nationalist pride and xenophobia were effectively undercut as rallying cries for American science policy from many different directions, and the emotional center of the national discourse about science congealed into a fear that convergence of other economies to U.S. standards and practices (inevitable to some extent in any case) was undermining American economic "competitiveness." Hence, Slaughter and Rhoades in this volume (chapter 1) refer to the complex of events characterizing the third regime as the rise of the "competitiveness R&D coalition."

The fourth countervailing tendency, materially the least gradual and therefore the most obvious in retrospect, was the utter collapse of the Soviet Union as the premier rival in the Cold War system. Because of its fragmented and partially classified nature, it had been previously impossible to gauge the extent of the importance of the national security imperative for the framework of science organization in America, at least until the fall of the Berlin Wall. Subsequently, many analysts have come to an increasingly nuanced appreciation of just how much the unquestioned assumption of an implacable technologically advanced enemy ramified throughout the practices and presuppositions of American science. The immediate fallout was the contraction of direct defense-related R&D after 1988 while nondefense expenditures failed to take up the slack, as shown in figure I.4.

However, shrinking federal research budgets were just the tip of the iceberg that had punctured the supposedly unsinkable vessel of Cold War scientific research.

The collapse of the national security imperative heralded an array of incursions upon universities; it has by now become clear that it had stood as their line of defense against many forms of encroachment originating

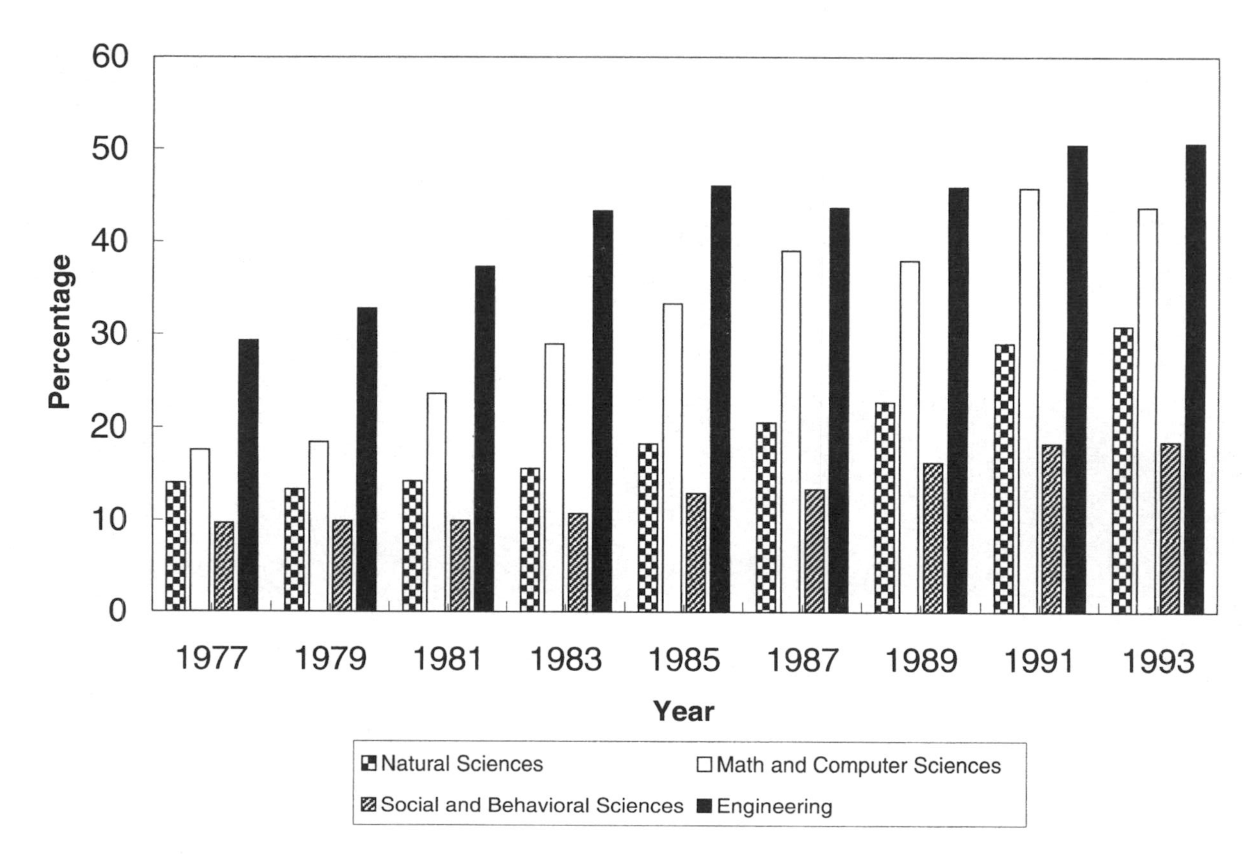

FIGURE I.3. Percentage of Science and Engineering Doctoral Degrees Awarded to Foreign Students
SOURCE: *S&EI 1996,* appendix pp. 58–59, table 2-29.

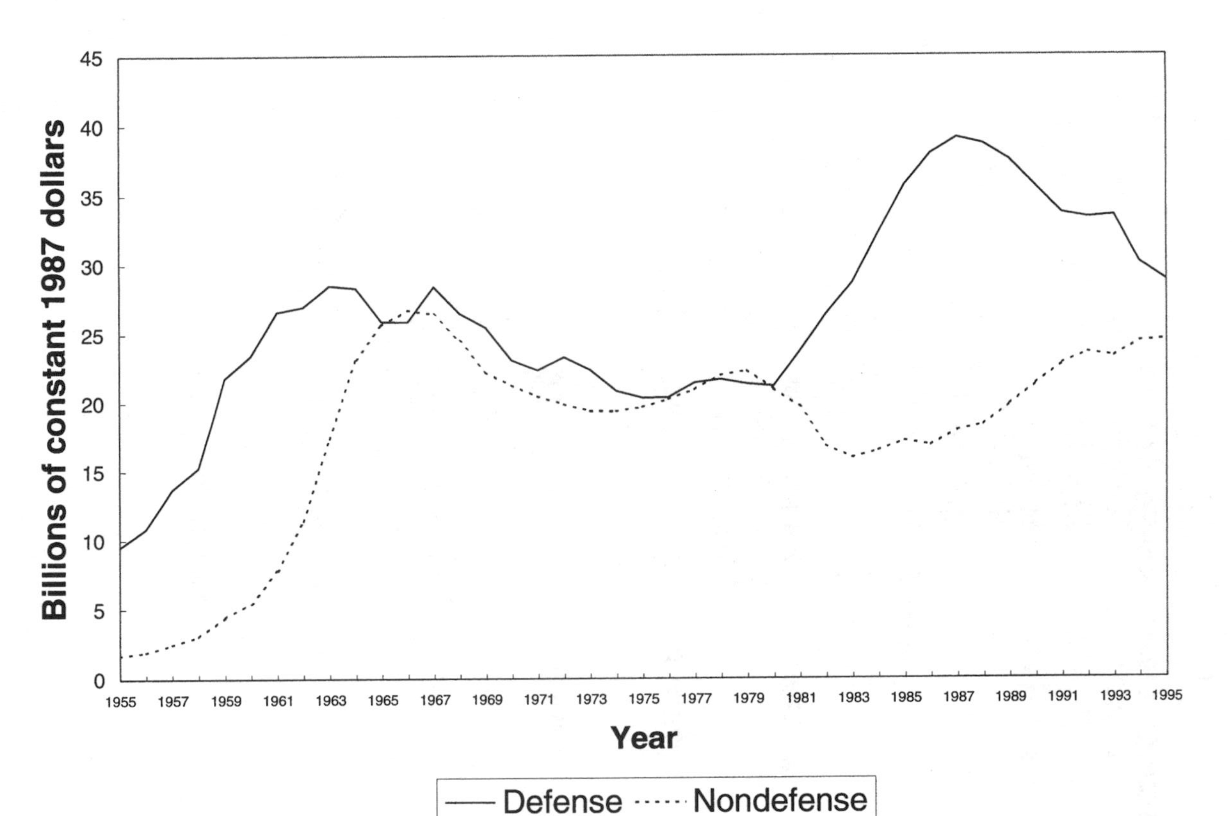

FIGURE I.4. Federal R&D Funding, Defense vs. Nondefense

SOURCE: Tables 25a–25g in *Statistical Tables on Federal Research and Development (R&D) Funding by Budget Function: Fiscal Years 1995–97 [Early Release Tables]* available at http://www.nsf.gov/sbe/srs/fbf95-97/budget.htm and GDP implicit price deflators: 1953–95 available at http://www.nsf.gov/sbe/srs/s2194/dst1.htm.

from disgruntled elements of their clientele. Soon after the fall of the Berlin Wall, universities came under attack for their lax accounting standards and supposed abuse of indirect cost provisions that had been forged back during World War II.[24] Various high-profile fraud accusations were lent some credence by hastily constituted government inquisitions; and periodic eruptions of litigation, such as that described in section 1 of this chapter, made it seem as though universities were not quite altogether capable of keeping their own houses in order. After years of quiescence, the quality and amount of teaching at major research universities came in for scathing criticism, as did the professorate in general.[25] Admittedly, there was some basis in fact for the upsurge of dissatisfaction, since the universities had reacted to the decline in federal subsidies as a signal that the combined teaching-and-research model was no longer being supported by the government and responded to their own fiscal crisis by resorting to "gypsies," part-time adjuncts and other improvised nontenured job categories of "unfaculty" as a means of maintaining their position in the research sweepstakes while simultaneously offering the accustomed broad array of coursework. But the irony of the attack on the professoriat was that it came just as the job description for that profession was being reengineered within the new regime.

As Paula Stephan and Sharon Levin argue in this volume (chapter 14), reconfigured career paths of laboratory scientists in particular have induced a more collaborative and hierarchical social structure of research. Growth areas of the university are no longer found in the traditional departmental structures, but rather in interdisciplinary research units, often constituted around a particular applied research topic and outside corporate funding. The phenomenon of "research parks" is merely the geographical manifestation of this disestablishmentarian movement. But as research increasingly assumes a privatized character, so too do the teaching functions of the university. A very important de-

24. This refers to the Dingell committee hearings in the House, beginning in 1991. The high-profile Stanford "scandal" is discussed in Guston and Keniston (1994, 177–93). The early history of indirect and overhead cost provisions can be found in Gruber 1995.

25. One can regularly find the following sorts of outbursts in mainstream outlets, such as the *New York Times Magazine:* "Nevertheless, says Charles Clotfelter, an economist at Duke University, 'higher education is the biggest service industry that hasn't gone through a substantial, gut-wrenching restructuring.' What might that mean? Start with research. Once you get past the top 20% of research being done, the rest is—how shall I put it?—idle, tenure-earning junk with little or no social value. Is the *Journal of Plankton Research* really indispensable? Cracking down on dubious research would go a long way toward making college affordable for the masses" (Miller 1999, 49).

velopment is the increasing use of computers and the Internet to pursue the thoroughgoing automation of university teaching, under the colorless rubric of "distance learning" (Schiller 1999, chap. 4). Controversial issues of intellectual property, once only the concern of the research scientist, have thus now invaded the classroom, as argued here in the article by David Noble. The commoditization of the act of teaching has proceeded apace since his article, with firms such as Blackboard, IntraLearn, and WebCT supplying software to facilitate distribution of the courses outside of their universities of origin, and the Department of Defense chiming in with its own plan for standardization of Internet instruction, the "Advanced Distributed Learning Initiative" (Carr 2000). The alliance of new technologies and a new entrepreneurial culture has given rise to all sorts of automated teaching curricula subject to the privatization frenzy of IPOs and commercial takeover bids, which themselves serve to further fragment the university into profit centers and loss leaders.[26]

Academic career paths have themselves been downgraded, if not deskilled (as argued herein by Fuller), with the standard career in the laboratory sciences encompassing two or even three "postdoctoral" positions prior to attaining the assistant professorship that aspirants used to count on right out of graduate school under the previous regime. Since the very construct of "academic freedom" had itself been a key structural aspect of the Cold War regime, it was only a matter of time before the practice of granting academic tenure itself came under attack, and the academic freedom defense of self-directed inquiry grew to sound increasingly like an exorbitant luxury, if not a hollow catechism. Some universities sought to close down entire academic departments, and the sciences were not always exempted. Centrifugal forces began to separate out what federal largesse had held together for four decades.

Moreover, it was not only the universities that were experiencing an unanticipated gale of creative destruction. Many sectors of corporate America also initially felt blindsided by the collapse of the Cold War system. The industries that had most benefited from the stealth industrial policy were the first to feel the chill winds of restructuring. Close on the heels of the fall of the Berlin Wall, many of the corporations with the most illustrious in-house R&D units decided upon the withdrawal of military subsidy they could no longer afford such forward-looking subsidiaries and sought to either downsize the units or spin them off as free-

26. Updates on these issues can be found in Young 2000 as well as on the web in regular updates posted by *The Chronicle of Higher Education.*

standing firms.[27] The alternative model for R&D has increasingly been to seek collaborations where directed research can be outsourced from the corporation. It has been estimated that in 1997 roughly 10 percent of all U.S. companies outsource some portion of their R&D, and in research-intensive industries like pharmaceuticals, the proportion is more like 37 percent (Borchardt 2000). Some of these contract research organizations (CROs) benefiting from the R&D spin-off can be found in those research parks abutting American universities; but increasingly, CROs are to be found in areas like the former Soviet Union or Ireland, where labor costs are significantly lower and the temporary character of employment for research staff is much more accepted.

Not unexpectedly, the other salient aspects of the stealth industrial policy were also summarily reversed (Branscomb, Kodama, and Florida 1999). Antitrust policy veered in a much more lenient direction, while legal strictures on intellectual property rights grew much more stringent. Corporate legal staffs thrived while research units were downsized. In the new regime, the watchword was "privatization" of functions and entities that had previously enjoyed governmental subsidy; what this meant for corporate R&D is that it could be separated out as a modular profit center subject to thoroughgoing restructuring as it was subordinated to the competitive strategic position of the new multinational firm. The mutual interaction of the induced economic vulnerability of the universities with the novel corporate drive to reinvent contract research, itself the outcome of a change in federal government policy, has resulted in a new science policy for a global privatized economy of information.

As in all economic formations, there are golden opportunities and there are raging disasters. One opportunity has been for academic entrepreneurs to pursue these franchised research projects and to use them as leverage to change the governance structures of the structurally weakened university. Academic disciplinary boundaries, and even the budgeting and planning functions of deans and administrators, are eroded by what are essentially commercial deals being negotiated by laboratory heads and directors of research institutes for long-term funding. University-industry research parks are merely the latest topological manifestation, and part-time faculty CEOs a social innovation, of the structural makeover of the university as a site of privatized research ca-

27. Bell Labs was spun off from AT&T as Lucent Technologies in 1996. Other firms simply chose to severely curtail their in-house research units. See Uchitelle 1996 and Sweet 1993. Some corporate research directors discuss recent events in (Rosenbloom and Spencer 1996).

pacity.[28] But these novel sources of support of science are not an unmixed blessing. Not only do tangled commercial/academic contretemps like the Taborsky affair surface with increasing frequency; once the research function has been uniformly privatized throughout academia, it will become increasingly impossible to insist upon any residual academic control over the conduct of the research, as illustrated by the 1998 agreement between the University of California-Berkeley and Novartis, giving the latter an active veto in the research committee of the university's department of plant and microbial biology (Press and Washburn 2000). This loss of independence is already endemic in the lucrative area of clinical drug trials for pharmaceutical concerns. Drug companies, finding that clinical trials of new drugs both take too long and are too expensive when conducted at university hospitals, have turned to entrepreneurs who recruit general-practice physicians to themselves recruit their patients as clinical subjects and to conduct basic experimental protocols (Eichenwald and Kolata 1999). Beyond the thorny ethical issues raised by the perverted incentive structures in this system, it should stand as testimony to the fact that, if universities privatize research, then there is no guarantee they will long remain the low-cost provider of corporate research services. In an increasingly globalized setting, short-term research contracts can be as volatile as short-term capital flows.

3. Tracking the Economics of Science with Gun and Camera

Economists have frequently been caught in a bind when confronting the social phenomenon of scientific research. Their initial temptation has been to treat science as just another commodity, on a social and epistemic par with poetry and pushpin. This is the first reaction of anyone who asserts that science is just a special case of the greater "marketplace of ideas"; since the market is thought to allocate resources in an optimal manner, there is no need for anything as pretentious as an "economics of science." The free operation of "Open Science" and individual competition for the applause of peers is all that is required, and it follows that the idea of science policy is utterly otiose. This conviction, practically second nature for a neoclassical economist, has been a snare

28. For a survey of the development of university-corporate alliances, see Cohen et al. in Noll 1998 and Branscomb, Kodama, and Florida 1999. The prospects for further industry alliances are regarded from different perspectives in Rosenbloom and Spencer 1996. A critical approach restricted to the field of biotechnology is provided by Krimsky 1991, Blumenthal et al. 1996, and Thackray 1998.

and a pitfall for those who have turned their attentions to science.[29] Quite baldly, economists are not free to treat science like putty clay or pancakes. The reasons are fourfold: to begin, neoclassical economists have looked upon the natural sciences, and physics in particular, with something akin to envy and admiration, relative to their own debased status (Mirowski 1989), giving the lie to any such epistemic leveling at the level of practical discourse. Second, natural scientists rarely see their own world the same way, and bear such scorn for economists' aspirations to scientific status that they could never wholeheartedly acquiesce in an unvarnished "invisible hand" account of intellectual endeavor or indeed of the present predicament of the academy. Third, as the contribution by Wade Hands to this volume (chapter 19) explains so pithily, there lurks the problem of paradoxical self-reference whenever an economist evokes the marketplace of ideas, leading to interminable discussions of whether the "market for economics" is flawed or not; and finally, while neoclassical theory aspires to have much to say about desire, it really has no special resources to pronounce upon truth and cognition.

One consequence of this intolerable bind was that, up until World War II, economists had next to nothing to say about the social structure of science per se. True enough, they may have touched upon something called disembodied "knowledge" or market innovation here and there; but the vicissitudes of research organization and funding were deemed beyond their purview.[30] Interestingly enough, prior to World War II it was frequently outsiders to the field, such as Charles Babbage and Charles Sanders Peirce, who proffered essentially economic models of the social organization of science. The physical chemist and philosopher

29. Some examples: Kealey 1996; Wible 1998; Radnitzky 1987. A telling example of how this simplistic approach is rapidly brought up short in the real world can be found in Ehrenberg 1999, 104:

> One of the first things that economists teach undergraduate students is that relative prices matter and that as the relative price of something increases, one should substitute away from that commodity. Hence, it appeared quite obvious to me that, to the extent that the increased relative cost of the physical sciences and engineering is permanent . . . we should seriously consider reducing our investments in engineering and the physical sciences and redirect these saved resources to other areas. To even suggest this in the presence of faculty from those fields would have marked me as a very dangerous person in the administrative hierarchy.

30. The history of the profound shift from the definition of neoclassical economics as the allocation of scarce means to given ends to the organizing principle that economics should be concerned with the agent as information processor is covered in Mirowski forthcoming.

Michael Polanyi was one of the first to make a classical liberal case for the "Republic of Science," although his quest to link this to a market metaphor was a failure on his own account (see Mirowski 1997). These attempts to broach the possibility of an economics of science were scattered and ill focused and had little or no impact upon the actual practices and prognostications of economists, much less scientists themselves. We might point out that, at least in the United States, there was little if any demand for such an analysis, since during the protoindustrial era science funding was primarily subordinate to other corporate goals and the prospect of large-scale public support of science simply was not even on anyone's agenda. Scientists themselves were only embarking upon their process of academic professionalization, and their place in the academy was not yet secure. Economic growth was not linked to the provision of epistemic novelty by the organized concerted efforts of researchers in most schools of economics, with the exceptions of the work of Joseph Schumpeter and a few others in the 1930s; instead, growth tended to be attributed to surpluses arising from various "natural" endowments. Any way one looked at it, an economics of science was a commodity without a consumer, a manifesto without a constituency—a superfluous entity.

All that changed drastically with the advent of World War II and what we have called the Cold War regime. Not only were the funding structures and organization of the physical sciences utterly revamped; the very composition of the economics profession also was irreversibly transformed. One of us has argued in a number of articles that the defining moment of the American economic orthodoxy was World War II, from which point one can date the definitive supercession of the Institutionalist school of economics by the neoclassical school.[31] For the purposes of the present volume, the salient aspects of this history are these: that physicists and mathematicians themselves had innovated a novel set of techniques and tools for modeling issues of command, control, communications, and information transmission during the war under the general rubric of operations research (OR); that OR was in part an attempt to address issues of the funding and integration of scientific research and consultation into the military command structure, which constituted one of the defining features of World War II in America; that

31. This account starts with the narrative of the 1930s found in Hands and Mirowski 1998 and continues with the relationship of American neoclassical economics to war work and operations research in Mirowski 1999; the postwar period is covered in detail in Mirowski forthcoming. The watershed of the war was the topic of a number of papers collected in Morgan and Rutherford 1998.

various neoclassical economists were inducted into these techniques and tools in the course of their war work; that through their sustained OR ties in the postwar period, they were brought face to face with problems of technological change, military funding of R&D, and the construction of plans and strategies in the face of inadequate knowledge and uncertainty; and that the military alliance became a resource in the defeat of the Institutionalist school in the American university. This encounter of economists with military research regime would have far-reaching implications for the postwar profession, and also for the existence of an economics of science.

It may be one of the ironies of the story that we relate here that neoclassical economists, science policy scholars, and the sociologists of science may indeed have had more in common (at least in this regard) than they have been previously willing to admit. In stark outline, the distinctive formats of OR found in various cultural settings in World War II had quite a differential impact upon the way in which the study of science organization would be approached in the postwar period. In Britain, "operational research" was initially inseparable from the "Social Relations of Science" movement of J. D. Bernal, Patrick Blackett, and Solly Zuckerman (McGuckin 1984). There, due to their political location, science policy and planning were loosely wedded to a Marxian approach; thus it provoked the formation of an opposition movement called the Society for Freedom in Science spearheaded by Michael Polanyi, John Baker, and Friedrich von Hayek. This reaction growing out of the "socialist calculation controversy" provided the context for the early development of notions of markets as conveyors of information (rather than simpler allocation devices) associated with those latter figures. Some have asserted that the overt left-wing bias of operational research in Britain spelled its doom for science policy (Mendelsohn in Krige and Pestre 1997), but perhaps it is more correct to say that the contretemps over planning served merely to remove the subject from the explicit realm of orthodox academic economics, with it ending up instead split between a narrow technical specialization in OR as management consulting and a different branch of OR assuming the professional identity of a "sociology of science."

The situation in America was altogether different. There, largely under the tutelage of some key figures such as John von Neumann, Vannevar Bush, James Conant, and Warren Weaver, the military planning of science managed to evade any socialist connotations and therefore avoid political donnybrook; it proceeded to recast OR itself to encompass a number of analytical technologies never favored in Britain, such as game theory, computer simulation, linear and dynamic programming, and systems analysis. The last consisted primarily of projecting the future shape of various nonexistent weapons systems (hydrogen bombs, ICBMs, nuclear airplanes, military satellites), including their technological capabilities and strategic implications. Neoclassical economists, far from being revulsed by such procedures, were induced to become pivotal players in the nascent professionalization of OR in postwar America. It was out of the insistent controversies dogging postwar Pentagon procurement practices, and therefore OR and systems analysis, particularly at the RAND Corporation, that the first self-conscious theoretical papers in America on the "economics of science" were conceived, as has been deftly documented by the historian David Hounshell (2000). Thus, the Cold War regime of science funding not only bequeathed Americans a distinct and novel set of economic structures for the support of science, but it also identifiably nurtured a certain specific variant of economic doctrine to buttress and support the kinds of science policy it had produced. The Cold War left Britons with an altogether different sort of system of military funding of research, essentially decoupled from the universities (Edgerton 1996), and as a consequence of the original Bernalist social relations of science movement, it was blessed (or cursed?) with an academic sociology of scientific knowledge that kept relatively aloof from the military system of patronage. In Britain, the sociology of science therefore became the locus of science scholarship in the academic community, especially beginning with the Edinburgh school's refusal to abide a strict content/organization dichotomy; whereas in America, a chastened Mertonian sociology provided a species of non- or antieconomics of science, which refused to become embroiled in issues of scientific content and management, and thus discreetly declined to actively engage with issues of military planning in science policy. To this day, vibrant sociology of science speaks with a British (and even Continental) accent, whereas America became the preeminent bastion of an economics of science.

Still, once the gravamen of a serious economics of science was raised within American neoclassical economics, it had trouble becoming fully integrated into the core of the subject, and thus it suffered further trials and tribulations. Although the economics of technical change grew to be a quasi-respectable subfield of either growth theory or the theory of production functions, the "economics of science" tended to be relegated to an almost subterranean existence within the profession, at least until the 1990s. One reason for this delayed development is that the sharp distinction between "basic" and "applied" science, which had become an unquestioned dogma in science policy and an indispensable component of the neoclassical approach in the immediate postwar period, essentially sanctioned the quarantine (and neglect) of the economic considerations impinging on basic science in deference to those economic forces supposedly governing the diffusion and rate of growth of a distinct technological change. Although we shall further trace the genesis of the basic/applied divide to the Cold War regime below, it should not pass unnoticed that the suppression of economic consideration of "science" as functionally subordinate to that of "technology" conveniently short-circuited all manner of thorny conceptual paradoxes inhibiting the development of a professionalized economics of science, as discussed by Hands and noted at the beginning of this section. In many ways, this rather convenient quarantine paralleled the similar artificial distinction forged within the philosophy of science at roughly the same juncture between the "context of discovery" and the "context of justification."[32] If the genesis of new ideas was treated as fundamentally ineffable, the stuff of genius and serendipity, then it would follow that the only phenomenon that could be subject to rational analysis was verification, quality control, and the diffusion of those ideas. "Knowledge" had to be both decontextualized and rationalized to be confined to the realm of the economic (and, we would insist, the philosophic) in the immediate postwar period, so as not to be mired in what was then deemed the "psychologistic fallacy" in the 1950s (Kusch 1995); and the way this was done was to insist upon its "thinglike" qualities, as distinct from issues of personal and social cognitive processes.

Yet even this evanescent debut of a neoclassical economics of science in America should not be regarded as a fait accompli in any sense. The very insubstantiality of its marginal existence within economics rendered it vulnerable to any number of sweeping revisions in microeconomic (or, to a lesser degree, macroeconomic) orthodoxy in the American economics profession. Since it had never really attained the status of an academic specialty, it was constantly thrust into the uneasy position of having to prove its bona fides by allying itself with various subsets of neoclassical theory, as well as maintaining some coherence with neoclassical conceptions of technological change. The profound rupture between the Cold

32. This comment is based upon some recent unpublished work on the history of the philosophy of science by Don Howard.

War and global privatization regimes of actual science organization since the 1980s has only further undermined any pretensions to strong continuity of analytical tradition. We contend that the shape of "the economics of science" was both at the mercy of specific science management structures prevalent in the American context *and* subordinate to the changing conceptions of the nature of theoretical orthodoxy within the American economics profession. Hence, our second task in this introduction is to try to make some sense of the postwar sequence of landmark papers in the economics of science anthologized herein by the analytical technique of relating them to (a) the evolving content of microeconomic orthodoxy and (b) the problems thrown up by the sequential appearance of two rather distinctive regimes of science organization prevalent in the postwar American context. Again, we need to stress that it is necessarily the American context that is most relevant to attaining an understanding these papers, since until very recently these papers were produced solely within the American profession.

Because there is no canonical history of postwar American economic orthodoxy to which we can direct the curious reader, we opt to divide the papers up according to what we view as three phases of "high theory" microeconomic concerns within the economics profession: (1) welfare economics and production theory as organized within Walrasian general equilibrium; (2) game theory and the cognitive turn in microeconomics; and (3) theories of computational complexity, networks, and bounded rationality. These have tended to recast the dominant image of science prevalent in the neoclassical community in each era as, respectively: (1*) the provision of a public good; (2*) presenting a problem of agency and incomplete contracts in a world of strategic uncertainty; and (3*) giving rise to an unintentional emergent order out of stochastic processes plus local interactions of cognitively flawed agents.

3.1 Science as a Problem of Production and Welfare Economics

The two papers in part II on science conceived as a production process are representative of the initial configuration of the economics of science during the Cold War regime in the postwar United States. Although they represent different idioms coexisting within neoclassical theory—Richard Nelson reflecting Samuelsonian welfare economics and Kenneth Arrow representing Cowles Commission Walrasian theory—both share the characteristic insistence upon science as producing a "thing" called knowledge, and all promote the central question of a formal economics of science as whether or not the "thing" produced by scientists qualifies as a public good and therefore as economically deserving of public subsidy. Because the entry point from which these theorists approached science was through the question of commodification, the analysis that grew out of these concerns tended to treat science itself as just another generic production process: Kenneth Arrow's favorite metaphor for research was, notably, that of mineral extraction or mining. Moreover, this tradition sought to erect a sharp distinction between science and technology, or pure and applied science, or basic and applied research. "Basic science" produced knowledge that was considered to be conveyed unidirectionally to "applied" or technological contexts, where it was portrayed as subject to further processing into conventional economic goods: this became known (perhaps with derogatory inflection) as the "linear model" of the relationship of science to the economy, as discussed above. By insisting upon the generic market as the general paradigm for all modern social organization, it sought to subsume science as just one special case of this generic social structure. Neoclassicism codified rational action; science was just thought to be the highest form and best instantiation of human rationality. As Nelson put it, with the usual 1950s gender oblivion, "men have always been, at least in a limited way, scientists."

It behooves us to resist this siren song of atemporal generality and instead be reminded of the ways in which this construct resonated with the initially precarious postwar situation of American Walrasian economics, as well as the Cold War funding regime. The Cowles Commission and its patron, the RAND Corporation, were at the time exploring the ways in which Walrasian theory and "decision theory" could be brought into line with OR and systems analysis as part of the military regime of science management. Both Arrow and Nelson were consultants at RAND when they wrote the papers included in this volume. The Cowles objective in the 1950s was to promote general equilibrium as an "institution-free" theory of social organization and furthermore to portray the rational agent as an "intuitive statistician," combining Neyman-Pearson statistical inference procedures with the maximization of utility (Mirowski forthcoming, chap. 5). Arrow, in particular, sought to combine decision making under uncertainty, information as a commodity, and the underfunding of science as a special case of market failure in the presence of uncertainty into one tidy package. Briefly, in an uncertain world, Arrow maintained that knowledge is effectively transformed into a commodity.[33] According to Arrow, various special characteristics of this commodity, such as uncertainty, indivisibilities, nonexcludability, and nonappropriability, dictated that it be recast as a particularly troublesome thing, a "public good."[34] The linear model implicitly came into play at this juncture, since the possibility of defining a metric of distance away from the more conventional production process (basic knowledge being used to "produce" applied knowledge being used to produce physical commodities) implied that basic knowledge would be "especially unlikely to be rewarded."[35] Next, conventional neoclassical welfare economics of the 1950s suggested that such basic knowledge would be underproduced relative to some virtual welfare maximum in the absence of public subsidy. This was a reprise of Arrow's favored argument that divergences from full Walrasian general equilibrium, dubbed (somewhat misleadingly) "market failures," were the true and only legitimate justifications for government interventions in the marketplace. The quest, then, was to achieve certain welfare goals by judicious government intervention in the research production process.

By this rather circuitous route, various neoclassical economists convinced themselves of what must have seemed a bald oxymoron to outsiders: "science" was really little different from any other human activity and thus best conceptualized as yet another market process, virtual or otherwise; and yet, simultaneously, science was also a problematic special case, a locus of endemic "market failures" that required lavish and persistent shoring up if it were to function on a credible scale. In a curious way, this paradoxical doctrine dovetailed with the other main development of the era, namely the rise of Keynesian macroeconomics. There, attempts to account for growth of national economies by quantification of conventional productive inputs had come a-cropper; one solution, proposed by Robert Solow, was to associate the "residual" with a species

33. Because this was a world purely abstract and institution free, with markets suspended in a void, no serious attention was given to property rights or cognitive structures, much less actual scientific practices.

34. Ontological freedom in defining the commodity to be anything one pleased, entirely disengaged from any empirical or logical considerations, was a Cowles practice codified in another key text, Gerard Debreu's *Theory of Value.*

35. The idea that there could be any sort of sequential ranking of the inputs of production in what was confessedly a thoroughgoing static model was just one of the ways in which the shotgun marriage of the process of research and Walras was unconsummated. One might indeed detect the faint echoes here of the Austrian tradition, which had been concerned with uncertainty and production in a manner antithetical to the Cowles approach.

of "technological progress" that shifted the production function upward relative to the same set of inputs. While we need not rehearse all the various controversies from the 1960s onward over growth accounting, the effect of this literature was to equate "technological change" with "spending on research and development" as a prerequisite for prognostications about measurement of the ways in which aggregate dollar amounts of spending on science and R&D generally would translate into macroeconomic improvement. This was the genesis of that all-time favorite statistic of the science policy wonk, the ratio of R&D spending to GNP or some other derivative macroeconomic aggregate. While not strictly conformable in all technical aspects, the public good/market failure microeconomic story resonated quite nicely with the Keynesian growth theory story, since both treated knowledge as a commodity whose economic impact could be conflated with the dollar amounts paid to vaguely demarcated activities dubbed "R&D" that supposedly eventually resulted in technological change on the shop floor. American Keynesians thought macroeconomic stabilization required increased government expenditure; American Walrasians thought scientific research required government subsidy; and both were united under the banner of correcting "market failures." As for those who thought this all a contemptible slide down the road to serfdom, such as various partisans of the Chicago school, they could still participate in the discussion by asserting the contrary proposition that the conventional market did adequately allocate knowledge as a commodity, or else that scientific knowledge did not actually exhibit the relevant attributes of a public good.

To understand the attractions of this position, we shall insist that one must venture beyond these relatively arid discussions between a few select economists and indicate how the broad policy prescriptions of "science as a production process" served to buttress the Cold War science regime. First and foremost, brief comments by both Nelson and Arrow reveal that the looming subtext of these debates was indeed the military reorganization of scientific research in the Cold War era.[36] But we need

36. "Much, though not all, of the government contribution to basic research is national defense–oriented." (Footnote in original: "This paper will not consider the vital question of whether the Department of Defense is spending enough on defense-oriented basic research.") Nelson 1959, 298, citation omitted; p. 153 herein. "The rapid growth of military research and development has led to a large-scale development of contractual relations between producers and a buyer of invention and research. The problems encountered in assuring efficiency here are the same as those that would be met if the government were to enter upon the financing of invention and research in civilian fields." Arrow 1962, 624; p. 179 herein.

not accept their protestations that there was no difference between civilian and military support of research; indeed, one way to comprehend their approach is to see it as a *justification* for the special arrangements that had grown up around the Cold War regime of science planning. The thrust of this analysis was to portray "basic" research as irreducibly opaque and uncertain and therefore requiring lavish government subsidy on welfare grounds, but simultaneously not being subjected to more conventional democratic structures of accountability that one might find in the more market-oriented "downstream" applications. In a sense, it ended up being the preferred economic counterpart to Vannevar Bush's manifesto *Science: The Endless Frontier* and simultaneously serving as an apologetics for the reigning division of labor between universities and industrial labs in the midst of the stealth industrial policy. There is no better brief summary of the postwar military's own view of its conduct of science management for defense purposes than this first incarnation of an economics of science in the papers of Arrow and Nelson.

The treatment of science as a commodity bore other indirect Cold War benefits. The analytical stress of this school upon the problematic aspects of knowledge as a commodity had a tenuous but nonetheless real relationship to the fact that intellectual property rights were *not* well defined or closely policed during the Cold War regime; this was not simply an inadvertent oversight, as we have discussed in the previous section. The characteristic trope of these postwar economists was to envision science as "nothing but" a market, and then sequester those problematic aspects of markets that did not seem to fit the characterization—here, the definition of the commodity—securely couched in some putative timeless predicament of human knowledge, rather than in some adventitious and transient social institutions linked to actual structures of scientific management and funding. This problem of the conceptualization of "ownership" in science as a function of historically identifiable institutions is the subject of the paper by Biagioli herein. Furthermore, the shift to macroeconomic aggregates of R&D spending relative to GNP as the central quantitative tools of the science policy analyst were skewed to make the U.S. economy look especially progressive given the gargantuan levels of American defense budgets, rather conveniently diverting attention from very real concern over whether all that weapons research really would eventually bear beneficial spillover effects for the civilian standard of living. The question of the magnitude of spillovers from military research projects festered as a low-grade controversy in science policy throughout the entire Cold War period. And finally, by diverting all attention away from the actual social structures of contemporary scientific research (and especially the security clearances that many of our protagonists themselves had to undergo), the economists could readily chime in with the general populace in praising the freedom of Open Science in America, as a prelude to invidious comparison with regimented science behind the Iron Curtain. Treating science as just another production process tended to reinforce a simplistic "marketplace of ideas" gloss on what was effectively a disdain for scrutiny of the actual process of scientific research. Hence, the papers on science as a production process are Cold War artifacts through and through.

3.2 Science as a Matter of Implicit Contracts and Cognitive Information Processing

The intellectual image of science as the production of public goods for the common weal never has entirely faded away from economic discourse—indeed, one might venture that no concept in economics ever becomes so obsolete that it cannot with a little effort be recycled for another generation. However, it seems to us indispensable for science studies scholars to realize that, when they set out to attack the public good scenario in the economics of science, they are thrashing a horse on its last legs, if not already in the knacker's yard.[37] It might redound to the credit of American economists were they to admit to outsiders that their later models of an economics of science are not at all conformable to their earlier welfare economics–inspired images of knowledge as a commodity. It would take a book in itself to do simple justice to the transition from the "production" scenario of science prevalent through the 1970s to the "cognitive/contracts" scenario embodied in the papers on science conceived as a problem of information processing herein. The situation might seem further complicated by the fact that one of our exemplars—that of Charles Sanders Peirce, the great American outlier—dates from more than a century earlier. In the interests of brevity, we merely sketch some causal factors, which await the critical scrutiny of future historians.

Those familiar with postwar neoclassicism are well aware that the center of gravity of orthodoxy has shifted from the early touchstone of Walrasian general equilibrium to game theory, and in particular, Nash equilibria for noncooperative games (Rizvi 1994). Starting in the 1980s, strategic reasoning began to be deployed in applied areas of economics such as industrial organization and law and economics. We shall simply

37. We have in mind recent articles, such as Callon 1994 and Fuller 1991 and 2000a.

take this fact as an unexamined primitive in our subsequent arguments. A second premise is the utter demise of the Keynesian/neoclassical synthesis in macroeconomics. A third premise, perhaps less familiar than the first two, is a progressive shift of theory away from the static allocation of objects and toward increasingly complicated specifications of the cognitive capacities of the rational economic agent (Mirowski forthcoming; Sent 1998 and forthcoming). Under the aegis of the computer, a cognitive revolution has been sweeping many of the sciences; this has shown up in economics in the 1980s as a fascination with rationality as a strategic and epistemic problem for the utility model. Again, we shall merely take this observation as a given. What is interesting for our present purposes is that, under the banner of these three trends, it became less and less acceptable to treat knowledge as a simple thing[38] and thus as an unquestioned commodity; indeed, pursuant to the influence of the computer, one begins to notice a subtle distinction materializing in economics, where information could still be treated as a thing that is conveyed between parties, but knowledge was regarded as more akin to a state of epistemic virtue, which could be characterized only as the outcome of a process. This subtle transformation bears immediate relevance for the scaffolding of any neoclassical "economics of science," as the reader of the previous section will appreciate. If "science" no longer is regarded as straightforwardly producing a thing, but rather fostering the existence of a complex of cognitive states, then the onus of the economics of science is displaced from a fascination with technology and the "public good" welfare characteristics of the output and redirected toward questions of the optimal organization of the actual process of inquiry in the face of strategic uncertainty, with preference given to approaches constructing the analysis as a search for optimal contractual relationships. A fourth causal factor, to which we will return at the end of this section, is the structural transition out of the Cold War regime and into a global privatization regime, something that altered the original analytical imperative for justifications of the public subsidy of science. These four factors are the building blocks of our interpretation of the papers in our information processing section, and in particular the work of Paul David and Partha Dasgupta as representative of this "new" economics of science, and the work of Philip Kitcher as its rational choice analogue within the philosophy of science literature.

38. "Knowledge is not a homogeneous commodity. There are different kinds of knowledge and no obvious natural units in which they can be measured" (Dasgupta 1988, 2).

The economic approach that grew out of these developments, then, tended to treat science itself as a matter of individual information processing. As in the previous phase, markets continued to be seen as the general paradigm for all modern social organization, with science remaining as just one special case of this generic social structure; but now, in a twist upon the old chestnut of a marketplace of ideas, that subsumption was founded upon the idea that the neoclassical market model provided a general paradigm for information dissemination and computation in science. However, with this shift the implicit distinction between information and knowledge cited above then began to come into play. If science resulted in knowledge, then perhaps what really mattered was how research conjured up something less thinglike than "information"; various computer metaphors for cognitive processes were then brought into the modeling mix. Researchers developing this version of an economics of science were more than willing to acknowledge the functional importance of tacit knowledge, traditional practices, and less-than-explicit or -codified norms in the realm of scientific research, something that had also been stressed earlier in a different context by Michael Polanyi and Friedrich von Hayek in their denial of the very possibility of science planning, as well as in some precincts of the more recent sociology of science (Collins 1985). Indeed, we find in the excerpts of the paper by Paul David and Partha Dasgupta included in this volume (chapter 7) an identification of the distinction between tacit and codified knowledge with that between "pure" and "applied" science, and furthermore, a suggestion that the mix of the two in any particular circumstance is a function of such economic variables as transmission costs, differential reward structures, and the costs of codification.[39]

A signal characteristic of this version of an economics of science became the repudiation of the linear model of basic → applied science,[40] which often stands in as a surrogate for the repudiation of the entire previous public goods framework. The reason often given for this about-face is that the sequence of links between the original basic discovery and the final market commodity is too serpentine and indirect to under-

39. It may be important to note that in David and Foray 1995, Paul David appeared to retreat from the rather simplistic conflation of tacit/codified with pure/applied science. We tend to read this as a further retreat from the notion that it is the characteristics of the commodity/output that somehow determine the optimal funding structures of science.

40. "Everyone knows the linear model of innovation is dead" (Rosenberg 1994, 139). "The linear model has been repudiated by scholars of innovation" (Sarewitz 1996, 97). See also Brown 1998, 39–40.

write any backward imputation of valuation. What this admission does tend to accomplish is to open the door to the proposition that there exist other, more immediate goals or objectives of the individual scientist, objectives that should instead be inserted into the utilitarian cost-benefit calculation. Charles Sanders Peirce, that visionary eccentric, essentially pioneered this style of analysis in 1879, when he suggested modeling a major objective of scientists as the minimization of quantitative probable error. Although this sounds eerily similar to many versions of modern decision theory, those familiar with Peirce's elaborate philosophical architectonic will readily appreciate that he himself could never be confused with any species of neoclassical utilitarian. Indeed, the paper reprinted here was one of his numerous attempts to argue that his pragmatism suggested that science was a process of communal reduction of error, resulting in a convergence to truth in the long run. A modern reader will not be impressed with his rather arbitrary manipulation of utility and cost functions to arrive at his reassuring results; nevertheless, Peirce displayed canny insight at the end of this paper, when he admitted that the aggregate of scientists may have as their shared communal goal the ascertainment of truth, but that individual scientists may be motivated to seek personal distinction, and that "the economics of the problem are entirely different."

The modern school of the cognitive/contracts approach to the economics of science—the self-proclaimed "new economics of science" that adopts the language of the agent as information processor—takes this possible divergence between individual and social goals as its major point of departure. Scientific goals, once treated as unproblematic, now become the crux of the debate. One consequence of the spread of game theory in economics has been to elevate the hermeneutics of suspicion to an automatic analytical principle; its influence upon the economics of science has been to embolden economists (and, indeed, some philosophers) to entertain the notion that individual scientists may not all be the epistemic angels and selfless lovers of truth so glorified in earlier hagiographies of science; nevertheless, the tendency of this school has been to assert that *something* about the behavior of scientists or the social organization of science still manages to promote the goal of truth and predominantly preserve the time-honored virtues. It may be vaguely unsettling to watch economists preach that private vices produce public virtues in science, as Hands himself indicates, but we would be remiss if we did not mention that this particular trope has proved immensely attractive to all manner of other intellectuals, especially in the context of the fin de siècle Science Wars.[41] Most immediately relevant for this volume, philosophers such as Philip Kitcher sought to reprimand their rivals in the science studies community who upheld the view that the nonepistemic goals of individual scientists, such as professional success or political aspirations, could divert the epistemic aims of science away from time-honored virtues such as the acquisition of truth. In response, philosophers of science started launching ripostes against perceived postmodernist extremists in the Science Wars by appropriating models from neoclassical economics, in order to assert their own versions of a "naturalized epistemology" and to account for the social structures of science.[42] More than once it was asserted that these game-theoretic models revealed how social structures did *not* matter ultimately for the successful attainment of the true or ultimate goals of science, in obvious parallel to some strains of economics that argued that alternative attributions of property rights (in the absence of pesky transactions costs) would not change efficient market outcomes. Individual rationality was deemed to triumph inevitably over myriad social obstacles. Thus it came to pass that one version of the economics of science was recruited as the first line of defense against the depredations of postmodern skepticism in the rather dreary reprise of the older "two cultures" controversy. What seemed to have been overlooked by both sides is the thesis suggested here by Paul Forman, that some aspects of postmodern skepticism could themselves have been regarded a function of changing structures of science organization and funding.

While neoclassical models may always be dragooned into performing ideological boundary work in many cultural controversies, our present concern is rather to understand the central tendencies of a particular set of theoretical developments, and then to relate them to contemporary changes in the structures of science policy and provisioning. To that end, we opt here to summarize the extensive writings of Paul David, undoubtedly the premier proponent of what has come to be called the Stanford

41. The so-called Science Wars, precipitated by a "hoax" by Alan Sokal that gained notoriety out of all proportion to its importance, has been dominated by those seeking to banish all manner of activities that they deem detrimental to the health of science. Representative works are Gross and Levitt 1994 and Koertge 1998. A somewhat more temperate account can be found in Hacking 1999.

42. On various versions of neoclassically socialized epistemology, see Bartley 1990; Goldman and Shaked 1991; Goldman 1999; Sent 1996; Kitcher 1990 and chapter 8, this volume; Mirowski 1996; Radnitzky 1987; Rescher 1989.

school of the economics of science.[43] David, following the footsteps of earlier economists, aspires to be regarded as a defender of "pure science" versus market-oriented research, but his initial quandary is that "the linear model is dead," or, in the early words of David and Dasgupta (1987, 542), "it seems less and less promising to separate research in science from that in technology on the basis of the characteristics of the knowledge generated by these activities." The response of David and Dasgupta is not to totally abandon the basic/applied distinction, but rather assert that it is socially created in pursuit of an optimal solution to the strategic problem of a divergence between private incentives and social welfare in the consequences of research. By contrast to the previous "production" approach, David and Dasgupta assert that knowledge can be differentially appropriable, but there persist principal-agent problems in controlling researchers.[44] Bluntly, incentives have to be restructured to force scientists both to disclose their discoveries with alacrity—that is, to turn private tacit "knowledge" into interpersonal fungible "information"—and to induce their peers to perform the vetting and validation functions that "scientific outsiders" would themselves find too time and resource intensive to conduct on a need-to-know basis. The purported solution to this problem is to foster something called "open science" according to a priority/credit system, and to sequester other scientific research subordinated to "technology" to be conducted along more secretive and proprietary pecuniary valuations. After the manner of the "new industrial organization" literature, David and Dasgupta proffer a prisoner's dilemma game setup and suggest the Pareto optimal solution constitutes the basis for a "new institutional economics": that is, somehow cultures in "the West" stumbled upon the one best way to structure science *for all time* so that it mutually reinforces the expansion of the market. Note well there is no facile reference to an efficient marketplace for ideas ab initio in David and Dasgupta: the optimal organization by their lights implies a putative *nonmarket* incentive system for

43. The Stanford school, loosely construed, encompasses the work of economic historians Paul David, Gavin Wright, Nathan Rosenberg, and their students such as David Mowery. Our summary of David's thematic is based upon David and Dasgupta (1987; chapter 7, this volume) as well as his papers (1993b, 1994, 1998a, 1998b).

44. There is a tendency of some scholars in science studies to attempt to appropriate this principal-agent literature and bend it to the purposes of a sociology of science. See, for instance, Guston 1999 and Turner (chapter 12, this volume). It might clarify matters if the sociologists became more acquainted with the sorts of presumptions built into these game-theoretic models so that they might better assess the extent to which they clash with conventional sociological preconceptions.

science, abutting a market system for more conventional commodities. The incongruous aspect of this characterization is that they cast the mix of market and nonmarket systems as a choice over a continuum: that is, the proportion of market to nonmarket institutions is somehow itself subject to the selfsame optimization calculation conventionally used in neoclassical economics to describe market operation. Hence, what is initially cast as a nonmarket process is actually surreptitiously given a market interpretation.

Although their recourse to noncooperative games is clearly a function of the shift of orthodox microtheory away from Walrasian general equilibrium in the profession at large, there remain many logical elisions that hobble the use of game theory in this particular context, both in the description and motivation of the existence of certain cognitive practices and social institutions, not the least of which are the standard presumptions of hyperrationality and common knowledge as the basis for the Nash equilibrium.[45] If scientists really ever did fit that characterization, it would be hard to understand most of their daytime behavior spent uncovering the secrets of the universe and attempting to persuade their colleagues that they had indeed found some of them out. There is also the nagging matter of what manner of superplayer is "choosing" the optimal mix of tacit and privatized science for the scientists themselves. (We have already mentioned the concession in David and Foray 1995 that the tacit/codified continuum cannot be readily mapped onto the public/privatized distinction, nor indeed that of disclosure/secrecy.)

While these critiques are undeniably salient for someone concerned with modeling strategies, what we would instead prefer to consider in this introduction is the manner in which David's other papers reveal the extent of his ambitions for the proposed "new economics of science." David is well aware that there are many other types of social analysis afoot in science studies and the history of science, and that these scholars would not placidly entertain simple stories of universal optimal science organization with equanimity.[46] Indeed, because he is nearly unique among economists in his appreciation for trends in the analysis of science

45. Further critiques of the deployment of noncooperative game theory to discuss cognitive and social formations can be found in Hargreaves Heap and Varoufakis 1995; Rizvi 1997; and Mirowski forthcoming.

46. "Subscribers to the theory of rational utility-maximizing behavior will want to know what induces researchers to seek admission to these 'clubs'; and conform by and large to club rules, by sharing knowledge they have not yet divulged" (David 1998b, 128). This argument was formalized in Mirowski and Sklivas 1991. One of the better critiques of David's approach located within science studies can be found in MacKenzie 1996, chap. 3.

outside of economics proper, David's later papers often provide the best critique of his earlier work. Furthermore, in his role as an economic historian, David also appreciates that the model inscribed in the David and Dasgupta paper is excessively static and abstracted from any historical specificity, tending to belie any claims that it is capable of explaining real institutions and social structures of science. To ameliorate those perceived drawbacks, David has proposed a "stage theory" interpretation of the rise of modern science (1991; 1998a). There he argues that there was a premodern stage in which secrecy and suspicion rendered almost all knowledge "tacit" and precluded anything like a cumulative process of scientific inquiry. He then posits the existence of a transitional stage, which he associates with the historiography of the Scientific Revolution, where aristocrats bestowed patronage on savants as another form of conspicuous consumption. The problem confronting patrons with the rise of the new learning, and especially the importance of mathematical expression, was that they themselves were incapable of judging the excellence of their house savants. David regards this as an instance of the problems of information asymmetries and principal-agent problems found in the modern industrial organization literature. The solution, presaged in the David and Dasgupta paper, was to allow the savants to judge each other according to reputation and socially attributed priority claims thrashed out in an open forum, with the patrons essentially excluded from evaluation but persisting in their willingness to pay the bills for the pleasure of their company.[47]

In our view, one way to understand the perspective from Stanford on the economics of science is to realize that they envision two essentially incommensurate systems of organization and valuation coexisting side-by-side in the modern world of scientific research. The first posited social structure they envision is a highly idealized invisible college of scholars who operate purely according to their own whims and inclinations, whose stature rests entirely upon disciplinary reputation and intellectual credibility, and whose evaluations of the quality of research are so tacit and maintained in such multilateral conformity by the relevant reference

47. David admits (1991) the inspiration for this story derives from Michael Spence's model of signaling and screening. In that paper, he also rejects the thesis found in Biagioli 1993 that patronage *produced* credibility in a savant as much as it was conditional upon prior credibility. David's screening model would indeed fail if the act of funding polluted the supposedly separate and independent value nexus of professionally awarded credit. This illustrates just one way in which David's work is frequently pitted in unacknowledged opposition against the themes found in modern science studies.

group that the actual process of producing warranted knowledge can largely be left out of the picture. As David has written (1998b, 120), it was

> not surprising that the "new economics of science" found it most natural to start by reworking the area of organizational analysis originally ploughed by Mertonian sociology of science, looking at the implications of certain institutional arrangements for allocative efficiency in the production of generic information . . . but not troubling itself over issues of socio-cognitive interaction that have occupied the sociology of scientific knowledge.

The second posited formation envisioned by the Stanford school is the everyday corporate reality of proprietary information and market-driven research, where the coin of the realm is not scientific fame but cold hard cash and success is denominated in tangible products and patents. Paul David and the Stanford school insist that we all need both, although their justification for this insistence is largely external to their models themselves. A world consisting solely of corporate science would be one where novel results were diffused much more slowly, they insist; because David denies that it would affect the *content* of science, speed of conveyance of discovery is the only operant variable. Since this seems a rather trivial hook upon which to hang their entire argument for the necessity of the persistence of academic science, David and Dasgupta also raise the possibility that a world of uniform corporate science would forgo the training and screening externalities that are generated from within the academic sector: in other words, corporate patrons face the same difficulties of asymmetric information and selection as their Enlightenment aristocratic predecessors, and elite universities (like Stanford) are better at recruiting and choosing and socializing candidates into the norms of science than would be corporate research managers. Hence, the bottom line of their prognostications is that (at least some) university research should be subsidized with no strings attached.[48] It is

48. David and Dasgupta themselves signal in a few places that this argument might be countered by the suggestion that tyro scientists not be subsidized in the academic sector, but rather borrow against their anticipated future earnings in their corporate careers to fund their education, and then let the corporate sector "cherry-pick" the more successful academic scientists in midcareer. This would recapitulate the usual Chicago deconstruction of liberal public goods arguments by simply redistributing the property rights and appealing to perfect capital markets. While they evince some discomfort with this option, they realize it clearly has some relevance as a description of what we earlier called the globalized privatization regime.

not science *as such* that requires public subsidy in their model, but rather certain elite universities in their role of providing scientific screening and education services to the larger society.

There can be no doubt that the doctrines of the Stanford school, as well as many other game-theoretic analyses of science, have been provoked by experience with the end of the Cold War regime and the onset of global privatization. These authors tend to position themselves as defenders of the university from various commercial and political onslaughts, such as those we have described above, while trying to use modern economics to locate the correct or efficient boundaries between public and privatized science. Yet, because of their desire to maintain conceptual continuities with the previous neoclassical tradition, the thrust of their program is backward looking, seeking to preserve a relationship between the postwar university and the corporate sector that is rapidly disappearing. But this strategy is fundamentally unavailing: by ignoring almost everything that was characteristic of science funding and management in the Cold War regime, in the final analysis it bears little relevance to the ultimate dissolution of that regime. Attenuated intellectual property, the stealth industrial policy, the military imperative (incidentally unmistakably present at Stanford: see Lowen 1997), the academic separation from commercial funding, the fragile construct of the teaching-and-research career path: all this and more is entirely absent from their model. Indeed, one might aver that their game-theoretic models could equally stand as a brief for the utter liquidation of the American mass-education university system, one where most scientific research is spun off into corporate or quasi-corporate settings, and most teaching outside of a few high-priced ivies became externally privatized as computerized distance education. Not only is there no attempt to analyze the profound changes taking place in intellectual property, industrial policy, university governance, the relative standing of the individual sciences, and the utter devaluation of the teaching-and-research career path; rather awkwardly, there is no serious consideration of any conditions that might demarcate science from any sort of generic "learning" by a generic rational actor. It is not an exaggeration to suggest that the only thing that renders someone a scientist in their model is possession of a certain configuration of risk and time preferences.

Furthermore, whereas both Paul David and Philip Kitcher acknowledge the existence of rival analyses of the operation of modern science, especially those emanating out of the Social Studies of Knowledge movement (SSK), science policy centers and science studies units, it seems apparent to us that they assert their equilibrium characterizations of their abstract university ideal of circa 1960 as prophylactics against those fine-grained studies of the impact of funding and organization upon scientific process and content, such as those that we summarized in the previous section of this introduction. "Credit" is the gaping black hole at the heart of their models, a cognitive imponderable that supposedly sports many of the interpersonal attributes of money without actually capitulating to Mammon. It seems here that, for all the mathematical sophistication, we have not progressed all that much from the original quasi-market account found in Polanyi's paper "The Republic of Science." David and Kitcher, it seems, are resistant to acknowledging, or perhaps oblivious to, the raft of research in the sociology of science that demonstrates that the categories of "priority" and "discovery" are frequently agonistic with outcomes that bear a tenuous relationship to true responsibility and are rarely independent of the economic and social status of the claimant (Brannigan 1981); they have missed the commonplace notion in SSK that complete disclosure has *never* been the rule in "open science" throughout its history;[49] and it goes without saying that no agent's full panoply of rationality in their models is ever transformed by their experience as a scientist. Because they start with a neoclassical agent purportedly capable of solving any problem, their appeals to "tacit" or local knowledge represent little more than artificially induced frictions superimposed upon a global optimization carried out by hyperrational individuals: it has no recognizable connection with any modern developments in cognitive science or social epistemology. Michael Polanyi, who worried much more about the significance of tacit knowledge in science, concluded that serious consideration of the tacit dimension would conflict with such an imperious utilitarianism.

From our perspective, this "implicit contracts" version of an economics of science has all the drawbacks of inertia often attributed to old generals: far too absorbed with fighting the last war, to the neglect and detriment of the present war. While it is undoubtedly driven by the dual motivations of defending the Cold War university structure and appearing to keep up to date with modern changes in microeconomics, so far it has had very little of substance to contribute to the pressing problems of understanding the current regime of globalized privatization of science.

49. The classic text in this literature is Collins 1985; for a game-theoretic model that states this in terms economists might appreciate, see Mirowski and Sklivas 1991.

3.3 Science as the Outcome of an Interactive Network of Cognitively Challenged Agents

We would be less than candid if we concluded our survey of the economics of science on such an unremittingly downbeat note. There are daunting challenges facing scientists in the twenty-first century; but fortunately there are more than a few scholars who realize this, and the papers in the part of this volume on science conceived as a network represent the efforts of those who seek to rethink the entire "market-place of ideas" approach. Of course, these forays are not so closely tied to the fortunes of the microeconomic orthodoxy in America as were the previous two versions of an economics of science, and therefore it is not so easy to lump them all together under some shared rubric. However, it does seem that science studies scholars located outside of the economics profession proper are especially concerned nowadays to interrogate the market metaphor and seek to uncover whatever may indeed be apposite and illuminating about the ideas contained in the contemporary economists' rucksack. Each representative of this modern trend, in his own fashion, subjects different aspects of the orthodoxy to an external audit: Michel Callon the industrial organization aspects of university-firm alliances; John Ziman the metaphor of scientific credit as market exchange; and Steve Turner the principal-agent relationship between the patron and the scientist. It will become apparent to the reader that each of these contributors approaches the favored tropes of the neoclassical economist with some measure of hesitation and distrust, but nevertheless ends up extracting some positive lessons for the globalized privatization regime.

However, it appears to us that there are also indications of a larger conceptual drift in the ways in which an economics of science will be conceived of in the near future, a shift that began to surface at the New Economics of Science conference at Notre Dame in March 1997, the conference at which the papers in this section were first presented. What these writers all seemed to share was a conviction that science should not be approached as some species of completed timeless entity, like some triumphant realization of an apotheosis of self-sufficient rationality and Pareto optimality. Rather, many writers are presently engaged in a quest for a description of something a little more nearly in the process of becoming; something comprised of individual scientists equipped with cognitive capacities falling well short of the superhuman neoclassical agents engaged in solipsistic activities; something whose telos is not given in advance, but itself coevolving with circumambient economic and political structures. These writers are often attracted to notions such as path dependence, spontaneous emergence of order, and evolutionary epistemology.

The mode of expression of this incipient novel trend among both the economists and the science studies community has been to cast their models in terms of networks of fallible, even cognitively limited agents engaged in some communal endeavor or quest. One observes this tendency in Bruno Latour and Michel Callon's advocacy of what they call "actor-network" theory (Latour 1987), as well as in the paper by Steven Durlauf and William Brock reprinted herein. (See also David 1998b.) What is heartening about this seeming convergence of approaches advocated within science studies and those in economics is that the turn toward the treatment of substantive rival scientific theories in all their diversity as a fitting topic for the social characterization of science, something notably absent in the prior two traditions of an economics of science. Plainly, if there were no rivalry and dissension in science, and there persisted no multiple paths to exploring nature and society, then there would be no problem of knowing what sorts of research to support and prioritize. For this reason alone, network theories are a welcome departure for an economics of science. Furthermore, the explicit recognition that scientists are frequently no better at prodigious baroque calculations of self-interest than are the superhuman caricatures of rational expectations theory could only encourage recourse to serious study of how scientists actually make decisions concerning both their careers and their research methodologies (Shadish and Fuller 1994). Perhaps this development might even betoken an appreciation for the fact that solutions to the problems of knowledge and discovery are often first posed as innovations in structures of social order. However salutary these consequences, we believe that this nascent modeling trend still has some distance to go before it has demonstrated the capacity to illuminate the problems and challenges of the third regime of globalized privatized science.

There is a large literature of critique of the actor-network framework; we cannot deal here with its possible ties to science policy.[50] Here we opt to concentrate instead on the Brock and Durlauf model as the network approach cast in a recognizably economic idiom. In this paper the primary concern seems to be the issue of whether "social" considerations, here a simple metric of conformity to the opinion of peers, can somehow

50. See McClellan 1996; Fuller 2000b, 365–72; Lee and Brown 1994; Law and Hassard 1999.

come to dominate the search for "truth," also modeled as a unique scalar metric. Clearly the "individuals" in this model do not have the capacity to mimic the full-blown rationality of the neoclassical model, and therefore simplistic welfare notions are much attenuated. Yet the idea enshrined in the mathematics that there exists some unique ranking of the validity of scientific theories and, more implausibly, that this ranking is freely available to all participants displays an unwillingness to seriously entertain the most significant work in the history and philosophy of science of the last four decades. Nevertheless, the overriding intention of Brock and Durlauf is to develop a simple model by which scientists at specific locations sample the opinions of their nearby neighbors and then examine the resulting statistical dynamics of the population in its acceptance or rejection of theories.

It should be noted that there is one curious aspect of the model that itself raises larger philosophical issues about the aims and procedures of future economics of science. Although it is not admitted in this paper, the model proposed therein is a slightly modified version of a model found in statistical mechanics, that of Ising spins and phase transitions using mean field techniques (see Weisbuch 1991, chap. 8). Without going into details, brief familiarity with the physics will reveal that most of the specific assumptions of the model, ranging from the binary character of the "theory choice" to the probability measures attributed to the utility functions are artifacts of the fine points of a minimum energy condition of a standard ferromagnet. As in so many other cases in the history of neoclassical economics (Mirowski 1989), the issue will not be so much to challenge the exact isomorphism between the physical model and the social phenomenon (do you sometimes feel you are trapped in a magnet?), but rather to ask, Why is this metaphor deemed compelling or meaningful? For instance: Is it more likely that your median modern solid-state physicist will embrace a formal model in an economics of science if he is being compared to a molecule in a magnet? Do natural scientists prefer to apply their own parochial disciplinary models to their own personal experience to assist in their understanding of how science works? Will physicists tend to gravitate toward physical models of the scientific enterprise, geologists to metaphors of continental drift, biologists to more concertedly evolutionary models (Hull 1988)? And if that is the case, then isn't the role of the economist rather an attenuated one of merely cheering on the project of metaphorical transfer but ultimately desisting from proposing anything that is really more intrinsically "economic" in scope? Are economists to be confined to providing a thin "economic" veneer for what is more correctly regarded as small communities projecting their parochial understandings of Nature onto Society?

It would seem that in the case of Brock and Durlauf, that is essentially what has happened. For where, in the final analysis, is the "economics" content in their economics of science? They begin by suggesting that they will evaluate the role of the "social" in actual theory choice in science; but by the end of the paper they admit, "[W]e are skeptical that there is any generic empirical regularity to be found in the role of social factors in the evolution of scientific theories, rather careful case-by-case historical studies need to be conducted." But then, what functions does this model perform for us? Indeed, there is nothing even remotely "economic" about their portrayal of their limited scientific agents: nothing about funding, nor intellectual property, nor the structure of pedagogy, nor the relationship of ideas to fungible technologies or devices. In the article by Turner, we at least confront directly the problems faced by those outside the narrow specialist research community in trying to form an arm's-length opinion about the validity and worth of a research program, which includes practices akin to "bonding" that facilitate such decisions; whereas in Brock and Durlauf, scientific opinions are formed by a simple weighted sampling procedure, already familiar from an earlier vintage of mechanistic decision theory.

From a more historically sensitive vantage point, it would seem that the problem of acceptance or dissent facing the scientists cannot be so readily expressed in a decision-theoretic framework, even one where "rationality" appears much less hubristic. Don Lavoie has sagely observed that

> market participants are not and should not be price takers any more than scientists should be theory takers. In both cases a background of unquestioned prices or theories is relied upon by the entrepreneur or scientist, but the focus of the activity is on disagreeing with certain market prices or scientific theories. (Lavoie 1985, 83)

This warning should serve as a prophylactic against confusing theory acceptance with either market purchases or statistical sampling procedures. Unlike the Brock and Durlauf model, this mode of inquiry would not aim to explain how agents all individually came to coordinate and agree on the content of "good science"; rather, it would explore how social organizations (such as the NSF, or the university department, or the professional academic society) channel irreducible miscommunication, noise, and dissent into manageable error and (sometimes) spontaneous order. Network metaphors often conjure up visions of computer hookups; but perhaps a better heuristic would be provided by systems configured to reconcile discordant signals in a system of accounts.[51]

Whatever one thinks of the specific network models of science, we would be hard-pressed to point to ways in which they have yet to be deployed to address the shattering implications of the current regime of globalized privatized science. Indeed, one of the most repeated complaints about Latourian actor-network analysis is that it leaves everything pretty much just the way it found it, *pace* the protests of the actual author Bruno Latour (1999). While some might welcome an economics of low ambitions and even lesser achievements, we believe the future lies instead in an economics of science that takes as its mandate dogged confrontation of the big questions facing science in the new century. These would include: What will happen to the university once research and teaching are spun off as separate privatized self-contained endeavors? When most research is being carried out under the auspices of some corporate agency or funding agreement, how will people come to factor that into their own assessment of the validity or dubiousness of the published account? Once scientific journals all get transformed into electronic archives, what will that do to the structure of the professional organizations whose raisons d'être were the care and maintenance of said erstwhile journals? When certain scientific specialties such as subatomic particle physics or algebraic topology (or the history of economic thought!) fail to find their corporate patrons, what will happen to them? Can universities continue to support these sad homeless creatures just off the revenues from their past patents, or will the successful sciences demand these funds as their rightful tribute? Indeed, can reengineered universities survive as the repositories for everything that corporations have considered and found useless? What is the role of various governmental funding agencies such as the National Science Foundation or the National Institutes of Health in the chastened circumstances of the new globalized privatized regime?

While the history we have proffered in this introduction does not give grounds for boundless confidence, we do feel justified in closing with two observations. First, economists will eventually come round to the idea that they will have to discuss the social structures of science as they

51. For an example of a formal network model of reconciliation of measurement of interdependent physical constants with error that illustrates this approach, see Mirowski 1996.

really exist and not how they would like to think of them in some misty monastic academic ideal. And second, the models that will shape the discourse will necessarily have to be cast into the idioms prevalent within the avant garde of the theoretical wing of the economics profession if they are to attract widespread comment. The days of science portrayed as a black-boxed "marketplace of ideas" are over.

REFERENCES

Arrow, Kenneth. 1962. *The Rate and Direction of Inventive Activity.* Princeton: Princeton University Press.

Barfield, Claude. Ed. 1997. *Science for the 21st Century: The Bush report revisited.* Washington: AEI Press.

Bartley, William W. 1990. *Unfathomed Knowledge, Unmeasured Wealth.* LaSalle, IL: Open Court.

Ben-David, Joseph. 1971. *The Scientist's Role in Society.* Englewood Cliffs, NJ: Prentice Hall.

Bender, Thomas, and Schorske, Carl, eds. 1997. *American Academic Culture in Transformation.* Princeton: Princeton University Press.

Bernal, J. D. 1939. *The Social Function of Science.* London: Routledge.

Biagioli, Mario. 1993. *Galileo, Courtier.* Chicago: University of Chicago Press.

Bloor, David. 1976. *Knowledge and Social Imagery.* London: Routledge and Kegan Paul.

Blumenthal, David, Nancyanne Causino, Eric Campbell, and Karen Louis. 1996. Relationships between Academic Institutions and Industry in the Life Sciences. *New England Journal of Medicine* Feb. 8:368–73.

Borchardt, John. 2000. Playing the Economics Game with Outsourcing. *Modern Drug Discovery* March 3(2):28–29; 31–32; 34.

Brainard, Jeffrey, and Colleen Cordes. 1999. Pork-Barrel Spending on Academe Reaches a Record. *Chronicle of Higher Education* July 23: A44–A48.

Brannigan, Augustine. 1981. *The Social Basis of Scientific Discoveries.* New York: Cambridge University Press.

Branscomb, Lewis, Fumio Kodama, and Richard Florida, eds. 1999. *Industrializing Knowledge: University-Industry Linkages in Japan and the United States.* Cambridge: MIT Press.

Brooks, Harvey. 1996. The Evolution of US Science Policy. In B. Smith and C. Barfield, eds., *Technology, R&D, and the Economy.* Washington: Brookings.

Brown, Kenneth. 1998. *Downsizing Science.* Washington: American Enterprise Institute.

Bucci, Massimo. 1998. *Science and the Media.* London: Routledge.

Bush, Vannevar. 1945. *Science: The Endless Frontier.* Washington: U.S. Government Printing Office.

Callon, Michel. 1994. 'Is Science a Public Good?' *Science, Technology, and Human Values* 19:393–424.

Carr, Sarah. 2000. Wisconsin Project Seeks to Create a Common Standard for Online Courses. *Chronicle of Higher Education* Feb. 17. www.chronicle.com/free/2000/02/2000021701u.htm.

Cole, Jonathan R., and Stephen Cole. 1973. *Social Stratification in Science.* Chicago: University of Chicago Press.

Cole, Stephen. 1978. Scientific Reward Systems: A Comparative Analysis. *Research in Sociology of Knowledge, Sciences, and Art* 1:167–90.

Cole, Stephen, Jonathan R. Cole, and Gary A. Simon. 1981. Chance and Consensus in Peer Review. *Science* 214:881–86.

Collins, Harry. 1985. *Changing Order.* London: Sage.

Collins, Martin. 1998. *Planning for Modern War: RAND and the Air Force.* Ph.D. thesis, University of Maryland.

Dasgupta, Partha. 1988. The Welfare Economics of Knowledge Production. *Oxford Review of Economic Policy* 4:1–12.

Dasgupta, Partha, and Paul David. 1987. Information Disclosure and the Economics of Science and Technology. In George Feiwel, ed., *Arrow and the Ascent of Modern Economic Theory.* New York: New York University Press.

———. 1988. Priority, Secrecy, Patents, and the Socio-economics of Science and Technology. Stanford CEPT, publication no. 127.

David, Paul A. 1991. Reputation and Agency in the Historical Emergence of the Institutions of Open Science. Stanford CEPR, publication no. 261.

———. 1993a. Intellectual Property Institutions and the Panda's Thumb: Patents, Copyrights, and Trade Secrets in Economic Theory and History. In M. B. Wallerstein, M. E. Mogee, and R. A. Schoen, eds., *Global Dimensions of Intellectual Property Rights in Science and Technology.* Washington: National Academy Press.

———. 1993b. Knowledge, Property, and the System Dynamics of Technological Change. In L. Summers and S. Shah, eds., *Proceedings of the World Bank Annual Conference on Development Economics 1992.* Washington: World Bank Press.

———. 1994. Positive Feedbacks and Research Productivity in Science. In O. Grandstrand, ed., *Economics of Technology.* Amsterdam: Elsevier.

———. 1998a. Common Agency Contracting and the Emergence of Open Science Institutions. *American Economic Review: Papers and Proceedings* 88: 2:15–21.

———. 1998b. Communication Norms and Collective Cognitive Performance of Invisible Colleges. In Navaretti et al. 1998.

David, Paul, and Dominique Foray. 1995. Accessing and Expanding the Knowledge Base in Science and Technology. *STI Review* 16:13–68.

Debreu, Gerard. 1959. *The Theory of Value.* New Haven: Yale University Press.

Dennis, Michael. 1987. Accounting for Research. *Social Studies of Science* 17: 479–518.

Desruisseaux, Paul. 1999. Canadian Professors Decry the Power of Companies in Campus Research. *Chronicle of Higher Education* Nov. 12.

Durlauf, Steve. 1997. Rational Choice and the Study of Science. Santa Fe Institute Working Paper. Santa Fe: Santa Fe Institute.

Edgerton, David. 1996. *Science, Technology, and British Industrial Decline.* Cambridge: Cambridge University Press.

Edwards, Paul. 1996. *The Closed World.* Cambridge: MIT Press.

Ehrenberg, Ronald. 1999. Adam Smith Goes to College. *Journal of Economic Perspectives* 13:99–116.

Eichenwald, Kurt, and Gina Kolata. 1999. Drug Trials Hide Conflicts for Doctors. *New York Times* May 16:A1.

Etzkowitz, Henry. 1994. Technology Centers and Industrial Policy. *Science and Public Policy* 21:78–87.

Feyerabend, Paul. 1978. *Science in a Free Society.* London: New Left Books.

———. 1987. *Farewell to Reason.* London: Verso.

Forman, Paul. 1987. Behind Quantum Electronics. *Historical Studies in the Physical and Biological Sciences* 18:149–229.

Forman, Paul, and Jose Sanchez-Ron, eds. 1996. *National Military Establishments and the Advancement of Science.* Boston Studies, vol. 180. Boston: Kluwer.

Fox, Robert, and Anna Guagnini. 1998–99. Laboratories, Workshops, and Sites: Research in Industrial Europe, 1800–1914. *Historical Studies in the Physical and Biological Sciences.* 29:1, 55–140; 29:2, 193–294.

Fuller, Steve. 1991. Studying the Proprietary Grounds of Knowledge. *Journal of Social Behavior and Personality* 6:105–28.

———. 2000a. *The Governance of Science.* Philadelphia: Open University Press.

———. 2000b. *Thomas Kuhn: A Philosophical History for our Times.* Chicago: University of Chicago Press.

Galison, Peter. 1997. *Image and Logic.* Chicago: University of Chicago Press.

Galison, Peter, and Bruce Hevly, eds. 1992. *Big Science.* Stanford: Stanford University Press.

Gibbons, Michael, Camille Nowotny, Simon Schwartzman, Peter Scott, and Martin Trow. 1994. *The New Production of Knowledge.* London: Sage.

Gleick, James. 2000. Patently Absurd. *New York Times Magazine* March 12: 44–49.

Goldman, Alvin. 1999. *Knowledge in a Social World.* Oxford: Clarendon Press.

Goldman, Alvin, and Moshe Shaked. 1991. An Economic Model of Scientific Activity and Truth Acquisition. *Philosophical Studies* 63: 31–55.

Gross, Paul, and Norman Levitt. 1994. *Higher Superstition: The Academic Left and Its Quarrel with Science.* Baltimore: Johns Hopkins University Press.

Gruber, Carol. 1995. The Overhead System in Government-sponsored Academic Science: Origins and Early Development. *Historical Studies in the Physical and Biological Sciences* 25:2:241–66.

Gruner, Sol, James Langer, Phil Nelson, and Viola Vogel. 1995. What Future Will We Choose for Physics? *Physics Today* December:25–30.

Guston, David. 1999. Stabilizing the Boundary between US Politics and Science. *Social Studies of Science* 29:87–111.

Guston, David, and Kenneth Keniston, eds. 1994. *The Fragile Contract.* Cambridge: MIT Press.

Hacker, B. 1993. Engineering a New Order: Military Institutions, Technical Order, and the Rise of the Industrial State. *Technology and Culture* 34:1–27.

Hacking, Ian. 1999. *The Social Construction of What?* Cambridge: Harvard University Press.

Hands, D. Wade, and Philip Mirowski. 1998. Harold Hotelling and the Neoclassical Dream. In Roger Backhouse, Dan Hausman, Uskali Mäki, and Andrea Salanti, eds., *Economics and Methodology: Crossing Boundaries.* London: Macmillan.

Hargreaves Heap, Shaun, and Yanis Varoufakis. 1995. *Game Theory: A Critical Introduction.* London: Routledge.

Hart, David. 1998a. *Forged Consensus: Science, Technology, and Economic Policy in the US, 1921–53.* Princeton: Princeton University Press.

———. 1998b. Antitrust and Technological Innovation. *Issues in Science and Technology* 15:75–82.

Heilbron, J. L., and R. Seidel. 1989. *Lawrence and His Laboratory.* Vol. 1. Berkeley and Los Angeles: University of California Press.

Hollinger, David. 1995. Science as Weapon in the *Kulturkampfen. Isis* 86:440–55.

Horton, Richard. 1999. Secret Society: Scientific Peer Review and Pusztai's Potatoes. *Times Literary Supplement* Dec. 17: 8–9.

Hounshell, David. 1997. The Cold War, RAND, and the Generation of Knowledge, 1946–62. *Historical Studies in the Physical and Biological Sciences* 27:237–67.

———. 2000. The Medium Is the Message. In Agatha Hughes and Thomas Hughes, eds., *Systems, Experts, and Computers.* Cambridge: MIT Press.

Hull, David. 1988. *Science as a Process.* Chicago: University of Chicago Press.

Jaroff, Leon. 1997. "Intellectual Chain Gang." *Time,* Feb. 10.

Kay, Lily. 2000. *Who Wrote the Book of Life?* Stanford: Stanford University Press.

Kealey, Terrence. 1996. *The Economic Laws of Scientific Research.* New York: St. Martin's.

Kevles, Daniel. 1995. *The Physicists.* Rev. ed. Cambridge: Harvard University Press.

Kitcher, Philip. 1990. The Division of Cognitive Labor. *Journal of Philosophy* 87:5–22.

Kleinman, Daniel. 1995. *Politics on the Endless Frontier.* Durham: Duke University Press.

———. 1998. Untangling Context: Understanding a University Laboratory in a Commercial World. *Science, Technology, and Human Values* 23:285–314.

Kline, Ronald. 1995. Constructing 'Technology' as Applied Science. *Isis* 86: 194–221.

Koertge, Noretta. 1998. *A House Built on Sand: Exploding Postmodern Myths about Science.* New York: Oxford University Press.

Kohler, Robert. 1991. *Partners in Science.* Chicago: University of Chicago Press.

Krige, John, and Dominique Pestre, eds. 1997. *Science in the 20th Century.* Amsterdam: Harwood.

Krimsky, Sheldon. 1991. *Biotechnics and Society.* New York: Praeger.

Kuhn, Thomas. 1962. *The Structure of Scientific Revolutions.* Chicago: University of Chicago Press.

Kusch, Martin. 1995. *Psychologism.* London: Routledge.

Latour, Bruno. 1987. *Science in Action.* Cambridge: Harvard University Press.

———. 1999. *Pandora's Hope.* Cambridge: Harvard University Press.

Lavoie, Don. 1985. *National Economic Planning: What's Left?* Cambridge: Ballinger.

Law, John, and John Hassard, eds. 1999. *Actor Network Theory and After.* Oxford: Blackwell.

Lee, Nick, and Steve Brown. 1994. Otherness and the Actor-Network. *American Behavioral Scientist* 37:772–90.

Lenoir, Timothy. 1998. Revolution from Above: The Role of the State in Creating the German Research System. *American Economic Review: Papers and Proceedings* 88:2:22–27.

Leslie, Stuart. 1993. *The Cold War and American Science.* New York: Columbia University Press.

Levin, Sharon G., and Paula E. Stephan. 1991. Research Productivity over the Life Cycle: Evidence for Academic Scientists. *American Economic Review* 81:114–32.

Lowen, Rebecca. 1997. *Creating the Cold War University.* Berkeley and Los Angeles: University of California Press.

Lucier, Paul. 1995. Commercial Interests and Scientific Disinterestedness: Consulting Geologists in Antebellum America. *Isis* 86:245–67.

MacKenzie, Donald. 1996. *Knowing Machines.* Cambridge: MIT Press.

Markuson, Ann, and Joel Yudken. 1992. *Dismantling the Cold War Economy.* New York: Basic.

McClellan, Christopher. 1996. Economic Consequences of Bruno Latour. *Social Epistemology* 10:193–208.

McGuckin, William. 1984. *Scientists, Society, and the State.* Columbus: Ohio State University Press.

Merton, Robert K. 1973. *The Sociology of Science.* Chicago: University of Chicago Press.

Merton, Robert K., and Jerry Gaston, eds. 1977. *The Sociology of Science in Europe.* Carbondale: Southern Illinois University Press.

Miller, Matthew. 1999. $140,000—and a Bargain. *New York Times Magazine* June 13: 48–49.

Mirowski, Philip E. 1989. *More Heat than Light.* New York: Cambridge University Press.

———. 1996. A Visible Hand in the Marketplace of Ideas. In Michael Power, ed., *Accounting and Science.* Cambridge: Cambridge University Press.

———. 1997. On Playing the Economics Trump Card in the Philosophy of Science: Why It Didn't Work for Michael Polanyi. *PSA 96 Supplement to Philosophy of Science,* 64:4:S127–S138.

———. 1999. Cyborg Agonistes. *Social Studies of Science* October 29:5:685–718.

———. Forthcoming. *Machine Dreams: Economics Becomes a Cyborg Science.* New York: Cambridge University Press.

Mirowski, Philip, and Steve Sklivas. 1991. Why Econometricians Don't Replicate (Although They Do Reproduce). *Review of Political Economy* 3:146–63.

Morgan, Mary, and Malcolm Rutherford, eds. 1999. *From Interwar Pluralism to Postwar Neoclassicism.* Durham: Duke University Press.

Morin, Alexander. 1993. *Science Policy and Politics.* Englewood Cliffs: Prentice Hall.

Mowery, David, and Nathan Rosenberg. 1998. *Paths of Innovation.* New York: Cambridge University Press.

National Science Board. 1998. *Science and Engineering Indicators, 1998.* Arlington, VA: National Science Foundation.

Navaretti, G., P. Dasgupta, K. Maler, and D. Siniscalco, eds. 1998. *Creation and Transfer of Knowledge.* Berlin: Springer.

Noll, Roger, ed. 1998. *Challenges to Research Universities.* Washington: Brookings Institution.

Odlyzko, Andrew. 1999. The Economics of Electronic Journals. *First Monday,* issue 2, 8.

Owens, Larry. 1994. The Counterproductive Management of Science in the Second World War. *Business History Review* 68:515–76.

Polanyi, Michael. 1962. The Republic of Science: Its Political and Economic Theory. *Minerva* 1:54–73.

Press, Eyal, and Jennifer Washburn. 2000. The Kept University. *Atlantic Monthly* March 285:3:39–54.

Price, Derek de Solla. 1963. *Big Science, Little Science.* New York: Columbia University Press.

———. 1975. *Science since Babylon.* New Haven: Yale University Press.

———. 1986. *Little Science, Big Science . . . and Beyond.* New York: Columbia University Press.

Radnitzky, Gerard. 1981. Progress and Rationality in Research. In J. Grmek, R. Cohen, and G. Cimino, eds., *On Scientific Discovery: The Erice Lectures 1977.* Dordrecht: Reidel.

———. 1987. The "Economic" Approach to the Philosophy of Science. *British Journal for the Philosophy of Science* 38:159–79.

———. 1989. Falsification Looked At from an Economic Point of View. In K. Gavroglu, Yorgos Goudaroulis, and Pantelis Nicolaoupoulos, eds., *Imre Lakatos and Theories of Scientific Change.* Boston: Kluwer.

Radnitzky, Gerard, and Peter Bernholz, eds. 1987. *Economic Imperialism.* New York: Paragon.

Reich, Leonard. 1985. *The Making of American Industrial Research.* New York: Cambridge University Press.

Reingold, Nathan. 1991. *Science American Style.* New Brunswick: Rutgers University Press.

———. 1995. Choosing the Future. *Historical Studies in the Physical and Biological Sciences* 25:2:301–27.

Rescher, Nicholas. 1989. *Cognitive Economy: The Economic Dimension of the Theory of Knowledge.* Pittsburgh: University of Pittsburgh Press.

Rhodes, Richard. 1986. *The Making of the Atomic Bomb.* New York: Simon and Schuster.
Riordan, Michael, and Lillian Hoddeson. 1997. *Crystal Fire.* New York: Norton.
Rizvi, S. Abu Turab. 1994. Game Theory to the Rescue? *Contributions to Political Economy* 13:1–28.
———. 1997. The Evolution of Game Theory. Paper presented to Erasmus University Philosophy and Economics seminar.
Rosenberg, Nathan. 1994. *Exploring the Black Box: Technology, Economics, and History.* New York: Cambridge University Press.
Rosenbloom, R., and W. Spencer, eds. 1996. *Engines of Innovation: U.S. Industrial Research at the End of an Era.* Boston: Harvard Business School Press.
Salter, Ammon, and Ben Martin. 1999. The Economic Benefits of Publicly Funded Research: A Critical Review. SPRU Working Paper, no. 34.
Sanchez, Claudio. 1996. "Disputes Rise over Intellectual Property Rights." Report aired on National Public Radio, September 30.
Sarewitz, Daniel. 1996. *Frontiers of Illusion.* Philadelphia: Temple University Press.
Schiller, Dan. 1999. *Digital Capitalism.* Cambridge: MIT Press.
Sedaitis, Judith, ed. 1997. *Commercializing High Technology.* Lanham: Rowman and Littlefield.
Sent, Esther-Mirjam. 1996. What an Economist Can Teach Nancy Cartwright. *Social Epistemology* 10:171–92.
———. An Economist's Glance at Goldman's Economics. *Philosophy of Science,* Proceedings 64:S139–S148.
———. 1998. *The Evolving Rationality of Rational Expectations.* New York: Cambridge University Press.
———. Forthcoming. Military/Artificial Intelligence.
Shadish, W., and S. Fuller, eds. 1994. *The Social Psychology of Science.* New York: Guilford Press.
Smith, Bruce. 1966. *The RAND Corporation.* Cambridge: Harvard University Press.
Smith, Merrit Roe, ed. 1985. *Military Enterprise and Technological Change.* Cambridge: MIT Press.
Stephan, Paula E. 1996. The Economics of Science. *Journal of Economic Literature* 34:1199–235.
Stephan, Paula E., and Sharon G. Levin. 1988. Measures of Scientific Output and the Age-Productivity Relationship. In A. F. J. van Raan, ed., *Quantitative Studies of Science and Technology.* North-Holland: Elsevier Science.
———. 1992. *Striking the Mother Lode in Science.* New York: Oxford University Press.
Steinberg, Jacques, and Edward Wyatt. 2000. Boola Boola: E-Commerce Comes to the Quad. *New York Times* Feb. 13: sec. 4, pp. 1, 4.
Stigler, George J. 1961. The Economics of Information. *Journal of Political Economy* 69:213–25.
Stine, Jeffrey. 1986. *A History of Science Policy in the U.S., 1940–85.* Washington: U.S. Government Printing Office.

Stone, D., A. Denham, and M. Garnett, eds. 1998. *Think Tanks across Nations.* Manchester: University of Manchester Press.

Sweet, William. 1993. IBM Cuts Research in the Physical Sciences. *Physics Today* 96:6:75–79

Teske, Paul, and Renee Johnson. 1994. Moving towards an American Industrial Technology Policy. *Policy Studies Journal* 22:296–311.

Thackray, Arnold, ed. 1998. *Private Science: Biotechnology and the Rise of the Molecular Sciences.* Philadelphia: University of Pennsylvania Press.

Uchitelle, Louis. 1996. Corporate Outlays for Basic Research Cut Back Significantly. *New York Times* Oct. 8: A1, D6.

Veblen, Thorstein. [1918] 1954. *The Higher Learning in America.* Stanford: Academic Reprints.

Werskey, Gary. 1988. *The Visible College.* London: Free Association Press.

Weisbuch, Gerard. 1991. *Complex System Dynamics.* Redwood City, Calif.: Addison-Wesley.

Wible, James. 1998. *The Economics of Science.* London: Routledge.

Wilson, Robin. 2000. They May Not Wear Armani to Class, but Some Professors Are Filthy Rich. *Chronicle of Higher Education* March 3: A16–A18.

Wright, David. 1997. "Testing the Market Metaphor for Science: The Matter of Fraud and Regulation." Paper presented to Notre Dame Conference on the Need for a New Economics of Science, March.

Young, Jeffrey. 2000. David Noble's Battle to Defend the Sacred Space of the Classroom. *Chronicle of Higher Education* March 31: A47–A49.

Zachary, G. 1997. *Endless Frontier.* New York: Free Press.

Ziman, John M. 1968. *Public Knowledge.* London: Cambridge University Press.

———. 1994. *Prometheus Bound: Science in a Dynamic "Steady State."* Cambridge: Cambridge University Press.

———. 2000. *Real Science.* Cambridge: Cambridge University Press.

PART I

Science at the Turn of the Millennium

1

The Emergence of a Competitiveness Research and Development Policy Coalition and the Commercialization of Academic Science and Technology

Sheila Slaughter and Gary Rhoades

ABSTRACT: This article describes the emerging bipartisan political coalition supporting commercial competitiveness as a rationale for research and development (R&D), points to selected changes in legal and funding structures in the 1980s that stem from the success of the new political coalition and suggests some of the connections between these changes and academic science and technology, and examines the consequences of these changes for universities. The study uses longitudinal secondary data on changes in business strategies and corporate structures that made business elites in the defense and health industries consider supporting competitiveness R&D policies. The article identifies and assesses an array of national R&D legislation concerned with competitiveness that was passed in the 1980s and 1990s and that has implications for academic R&D. The effects of competitiveness R&D policies on universities and academic science and technology are appraised by analyzing changes in time-series data (1983–93) on science and technology indicators compiled by the National Science Foundation.

Over the past fifteen years, the policy issues—defense and health—that preoccupied the Washington, D.C.–focused science and technology policy community since World War II have changed substantially. The end of the cold war undermines the time-honored rationale for mission agencies directing substantial funds to corporate and academic research and development (R&D) that sustain defense, threatening the well-being of what Eisenhower (1961, 22) called the "military-industrial complex." The profit-making potential of intellectual property, most dramatically exemplified by biotechnology, has greatly intensified privatization and commercialization in the academy, particularly in university-based medical education and research. Privatization and commercialization of health care seem likely to direct funds

Reprinted with permission of Sage Publications, Inc. From *Science, Technology, and Human Values* 21 (1996): 303–39.

away from practitioners, clinical professors, and some established research fields and toward corporations concerned with the application of medical technologies, diagnostics, and interventions as well as toward fully insured, large patient populations, giving pause to the medical-industrial complex.

Even before the standard justifications used to defend government spending on science and technology came into question, a new rationale for R&D policy began to emerge. The "competitiveness" agenda was proposed as a basis for science and technology policy in the 1980s, during the Reagan and Bush administrations, and found an articulate and ardent champion in President Clinton (Slaughter 1990, 1993a, 1993b; Greenberg 1993a, 1993b; Clinton/Gore 1992). Science and technology policy directed toward competitiveness uses government funds to commercialize science and technology via corporations and R&D agencies. The aim is to increase U.S. shares of global markets and to increase the numbers of high-technology, high-salaried jobs in the domestic economy.[1] With the breakdown of the traditional epics—"winning the cold war," "the fight against disease"—that justify spending on science and technology, the rhetoric of "global competitiveness" is an effort to create a new narrative of heroic proportion that serves similar purposes.

Many participants in the science and technology policy communities write as if competitiveness policies can easily be substituted for military-medical industrial policies so that, after a period in which military and health care industries make various adjustments, a new science and technology regime will be installed.[2] Like many in the science and technology community, we think the shift from a cold war to a competitiveness R&D policy is under way but believe the transition will be lengthy, uneven, and incomplete. At least for a time, the cold war and the competitiveness technoscience regimes will be sustained simultaneously. Given that the consequences of a cold war R&D policy for the academy are

1. Although the individuals and organizations promoting a competitiveness agenda usually see no disjuncture between increasing shares of global markets and increasing numbers of high-technology jobs in the domestic economy, these may be alternative policies—one precluding the other—or at least inconsistent policies (Aronowitz and DeFazio 1994).

2. For examples of policy documents promoting the competitiveness R&D agenda, see Business–Higher Education Forum (1983, 1986b), National Academy of Science, Committee on Science, Engineering and Public Policy (1992, 1993), President's Council of Advisors on Science and Technology (1992), Kaufman and Waterman (1993), Marston and Jones (1992), National Academy of Science, Government-University-Industry Research Roundtable (1992), and Smith (1990).

well understood, in this article we explore the likely consequences of a competitiveness policy on academic science and technology.

The academic science and technology policy literature that takes a substitution approach suggests that universities will fare as well under a competitiveness regime as they did under the cold war regime. This literature, written largely by university managers, university participants in policy-making forums, and spokespersons for professional and scientific associations, assumes that if science and technology are seen as contributing to economic competitiveness, then funding for academic R&D, including basic research, will increase. Second, it assumes that science and technology funding with a commercial rationale will not greatly change scientists and their work. Participants in the policy process recognize that commercial science and technology pose problems in a university setting—for example, secrecy with regard to intellectual property, conflict of interest, conflict of commitment—but they make the case that such problems are not insurmountable and can be handled through a variety of procedural solutions (Etzkowitz 1983, 1989; Louis et al. 1989; Blumenthal et al. 1986). Third, the literature often assumes that academic science and technology are somehow separate from the universities in which they occur. The institutional problems and constraints that might stem from commercial competitiveness as a rationale for science and technology policy are only rarely discussed.

We argue that these assumptions are questionable. We believe that the new emerging pattern of support for academic science and technology policy will be substantively different from the constellation of support that undergirded academic science and technology from World War II until the 1980s. This article (1) describes the new bipartisan political coalition that supports commercial competitiveness as a rationale for R&D, (2) points to selected changes in legal and funding structures in the 1980s that stem from the success of the new political coalition and suggests some connections between these changes and academic science and technology, and (3) examines the consequences of these changes for the universities. This analysis allows us to reassess the assumptions made by many researchers and policy activists about competitiveness as a rationale for science and technology policy.

The theory that guides our analysis is post-Marxist and poststructuralist. Our approach is post-Marxist in its focus on the power of capital, social class, and knowledge as commodity, analytical foci important for examining commercialization of academic knowledge and its implications for universities. Like Marxist and neo-Marxist theory, post-Marxist theories continue to offer a critique of oligopolistic capitalism but—in contrast to them—regards class structures as fairly fluid and pay greater attention to the state as distinct from capital and to human agency, ideology, symbolism, and social constructionism. It no longer looks to centralized socialism as the inevitable and the only viable alternative to capitalism (Aronowitz 1988; Chomsky 1969, 1994; Krimsky 1982; Noble 1976, 1984; Rifkin 1983). Poststructural theory checks the Marxist (whether neo- or post-) tendency to mechanistic structural explanations and grand narratives (Greenberg 1967; Dickson 1984; Latour 1987; Fligstein 1990). Rather than looking only at the foreground where powerful social, political, and economic actors and organizations shape policy, our poststructural approach allows us to pay attention to the background where subversive, recalcitrant, or unengaged actors and organizations pursue their own R&D and academic agendas.

Our design and data are relatively straightforward, as are the problems they present. In considering the composition of R&D political coalitions, we examine longitudinal secondary data on changes in business strategies and corporate structures that made business elites in the defense and health industries interested in promoting and supporting competitiveness R&D policies, but we do not analyze primary data regarding the involvement of particular companies in shaping and lobbying for R&D policies and legislation. We identify and assess an array of national competitiveness R&D legislation passed in the 1980s and 1990s, concentrating on its implications for academic R&D, but we do not perform a systematic content or discourse analysis of legislative texts. In appraising the effects of competitiveness R&D policies on universities and academic science and technology, we rely on analysis of changes in time-series data (approximately 1983–93) on science and technology indicators compiled by the National Science Foundation (NSF). We look for shifts in patents and publications activity, in the balance between support for basic and applied research, and in the targets of support (e.g., individuals, teams, centers). We also draw on other national data sets such as faculty salaries in various fields. The strength of these national data lies in their uniformity over time and in their breadth. The limitations are equally evident, including a lack of clarity about exactly what goes into particular categories and an absence of linkage among the various data sets. In short, the data enable us to provide a "big picture" that suggests substantial change in coalitions, in legislation, and in impacts on academic science and technology, but they may overlook important details and leave unexamined the many cases that are not captured by national trends.

R&D Political Coalitions

Because the business class plays a powerful part in shaping social and economic policy, including academic R&D, our narrative concentrates on major changes in the strategies of defense and health industries that contributed to the emergence of a competitiveness R&D policy. Although business strategies were crucial to change in federal R&D policy, business leaders were far from omnipotent and did not act alone. For the most part, corporate strategists were reacting to destabilizing events—the end of the cold war, the rise of Pacific Rim countries as economic competitors, the possibilities presented by intellectual property. They found allies and partners in state bureaucrats, university administrators, and a number of research professors, to name a few of the groups that joined in the competitiveness coalition. These allies and partners were not dupes or pawns, nor were they, in Mills's (1956) terms, part of an institutionally based, tightly knit power elite or, in Domhoff's (1978, 13) terms, a "power elite" acting as the "operating arm of the ruling class." Instead, they were players in diverse, loosely coupled institutions that could see the organizational utility of a competitiveness policy in a destabilized global environment (Slaughter 1990; Slaughter and Leslie 1997).[3] University administrators and a number of academic scientists and engineers were willing to rewrite the narrative that privileged basic science and champion a more commercial science and technology as long as university R&D continued to be supported.

In narrations about science and technology policy, it was always difficult to include resource providers in the plot of the post–World War II basic science narrative because science was portrayed as autonomous and scientists as beholden only to truth (Slaughter 1993a). Nonetheless, academic researchers depended on the mission agencies, organizations with applied goals, for the vast majority of their federal funds. The NSF, the only federal agency arguably dedicated to basic science, never accounted for more than 20 percent of federal funds for academic R&D in any given year between 1971 and the present. (Indeed, NSF funds declined from a high of 19.5 percent in 1973 to 14.1 percent of all federal funds for academic R&D in 1991 [NSF 1993a, table 5.9].) The mission

3. We find the notion of institutional class more powerful than that of power elites. The concept of institutional class is explained in Slaughter (1990, chap. 8).

agencies supplied universities with 80–85 percent of their federal R&D funds, of which approximately 65–75 percent were designated as basic (NSF 1993b). The academic science and technology community always interpreted "basic" or "fundamental" research to mean that university-based scientists, following the imperatives of their fields, set direction for their research programs independently of the mission agencies (Smith and Karlesky 1977; Wolfle 1972). However, even when funds were tagged as support for basic science, it was not clear how distinct these were from funds for applied science. Accounts of scientists' and engineers' negotiations with the mission agencies suggest that the academic interpretation of basic science was only partially shared by bureaucrats at the Department of Defense (DOD), Department of Energy (DOE), National Aeronautics and Space Agency (NASA), and National Institutes of Health (NIH). These organizations had a much more instrumental definition of basic science. Some historians of science have argued that basic science merely meant unclassified science and that even basic science was powerfully and directly shaped by mission agency goals (Forman 1987; Leslie 1993). If basic or fundamental research was not as independent as the academic science and technology community might wish, then marked changes in the political coalitions forming mission agencies' constituencies suggest that academic science and technology are likely to change when the composition and goals of R&D coalitions change.

Two broad bipartisan R&D political coalitions emerged after World War II; one, around DOD/DOE/NASA/aerospace, came to the fore by the early 1950s and another, around the NIH, developed somewhat later (see table 1.1). The composition of these coalitions was complex but fairly stable until the late 1970s and early 1980s. At that point, a new coalition began to emerge, closely related to the older ones but distinct enough to propose new directions and new mechanisms for R&D, whether inside or outside of universities.

The Defense Coalition

The story of the bipartisan R&D coalition built around DOD/DOE/NASA/aerospace has often been told (Greenberg 1967; Chomsky 1969; Melman 1992; Forman 1987; Herken 1992; Leslie 1993). Defense industries, the armed forces, university administrators, and many scientists, along with supporters of the Keynesian welfare-warfare state, pushed for the expansion of defense and defense-related spending. This funding was generally justified as necessary for winning the cold war,

TABLE 1.1. Political Coalitions Supporting Academic Science and Technology

	Cold War/Health War (1945–1970)	Competitiveness (1975–present)
Political parties	Republicans Democratic cold warriors	Republicans Most Democrats
The State	Armed services	Qualified support from the armed services State and local governments
	Universities (1) Many scientists (2) Fewer students	Universities (1) Fewer scientists (2) Many students
Corporations	DOD industries (1) Contractors (2) Suppliers	Some DOD industries (1) Contractors (2) Suppliers Multinationals (1) Diversified (2) Conglomerates a. Agriculture b. Pharmaceuticals c. Chemicals Corporate medicine Small(er) business
	Pharmaceuticals Independent professionals (1) Physicians (2) Nonprofit hospitals (3) Nonprofit insurance	(1) High technology including health care and biotechnology Services (1) Producer services (2) Health insurance
Goals	Win the cold war and the war against disease	Win global control of markets through privatization and commodification of science-based intellectual property
	Administer high profits to defense and defense-related industries and to pharmaceutical companies and specialist physicians	Establish government subsidies for high technology including the health industries and the insurance and producer service industries
	Subsidize corporate and academic R&D	Move R&D, including university R&D, toward commercial science and technology

NOTE: DOD = Department of Defense; R&D = research and development.

which meant defeating international communism in its many insidious forms (May 1988; Rogin 1987). The goals of this broad bipartisan coalition were to increase R&D funding to (1) win the cold war; (2) administer high profits to defense corporations and, indirectly, their contractors and suppliers; and (3) subsidize corporate and academic R&D. Between the end of World War II and the early 1970s, DOD/DOE/NASA/aerospace provided an increasing share of all federal funds for academic science. In the early 1970s, approximately 26.7 percent of all federal R&D obligations came from these agencies (Greenberg 1967; Forman 1987; NSF 1993b). After the Vietnam war, these funds dipped dramatically, falling to 17.7 percent in the mid-1970s, when corporations, including defense, moved away from concentration on single products, such as military goods, and began to pursue more diversified growth strategies (NSF 1993b).[4]

Student protest against the Vietnam war and against university involvement in defense R&D severely disrupted the flow of funding from DOD/DOE/NASA/aerospace to college campuses (Dickson 1984). But the ties between the military and the academy were ultimately cut more cleanly by the falling rates of profits and productivity that plagued American manufacturing corporations in the 1970s than they were by the antiwar movement. In the global economic arena, Japan destabilized the bilateral international trade arrangements that had dominated the world in the postwar period, precipitating fierce, multipolar global competition (Carnoy et al. 1993; Slaughter and Leslie 1997). U.S. corporations' after-tax profits fell from a high of almost 10 percent in 1965 to a trough of 4.5 percent in the mid-1970s, and recovery was slow and uneven (Harrison and Bluestone 1990, 110, fig. 5.1). Productivity grew at a rate of just under 1 percent from the late 1960s through the 1970s, a far slower rate of growth than that in the 1950s and most of the 1960s (Harrison and Bluestone 1990).[5]

4. By agencies, the shares of all federal funds for academic R&D from 1970 to 1989 were as follows: DOD, 12.8 percent in 1970, 8.4 percent in 1975, 11.3 percent in 1979, 11.6 percent in 1980, 16.7 percent in 1986, 13.7 percent in 1989; DOE, 5.7 percent in 1970, 4.2 percent in 1974, 7.1 percent in 1978, 6.7 percent in 1980, 5.2 percent in 1989; NASA, 8.2 percent in 1970, 3.6 percent in 1979, 3.7 percent in 1980, 5.0 percent in 1989 (NSF 1993a). From the mid-1970s, the DOD/DOE/NASA/aerospace funds began rising again, reaching 22 percent in 1980, 25.9 percent in 1986, and then falling to 23.9 percent in 1989.

5. Productivity greatly increased in the 1990s, but median family income declined, pointing to low-wage productivity, a story that runs contrary to the high-technology prosperity narrative told by the competitiveness coalition (Phillips 1993).

The profits and productivity crises were accompanied by several major changes in the structure of U.S. corporations. In their wake, the American business elite began to rethink its commitment to cold war R&D policies. These changes included diversification, increased commitment to high-technology products, globalization, and a greater involvement with producer services.[6] These changes occurred in corporations that had defense and defense-related product lines as well as in corporations in the medical and related industries. The business class, the most powerful player in R&D politics, confronted these changes as it moved away from the cold war R&D policy coalition and united in a competitiveness R&D coalition (see Useem 1984 for a definition of business class).

By the 1970s, many large U.S. manufacturing corporations had diversified. In 1939, 77.8 percent of the largest firms manufactured a single or dominant product, only 22.2 percent manufactured related products, and none manufactured unrelated products. As late as 1959, 37.7 percent manufactured a single product, 57.4 percent manufactured related products, and only 5.0 percent manufactured unrelated products. By 1979, only 23.2 percent of the 100 largest firms still manufactured a single product, 49.5 percent had diversified to related products, and 27.3 percent had become conglomerates, manufacturing unrelated products, expanding the only way possible given the constraints of U.S. antitrust law on horizontal and vertical integration (Fligstein 1990, 261, table 8.1). This diversification created an opportunity for defense industries to rethink their corporate strategies.

Although conglomeration fostered diversification into unrelated products, increases in profits and productivity did not follow, causing considerable consternation in the corporate community. In the late 1980s, due to the relaxation of antitrust law under Reagan and Bush, corporations moved away from this "firm as portfolio" or conglomerate model to a related products strategy (Bhagat, Shleifer, and Vishny 1990; Davis, Diekmann, and Tinsley 1994). Antitrust laws were reinterpreted administratively and judicially to allow horizontal mergers, allowing

6. Producer services are tied to the rise of commercial capital, which has played an important role in globalization of the political economy. Producer services consist of nonlife insurance and reinsurance, accountancy, advertising, legal services, tax consultancies, information services, international commodity exchanges, international monetary exchanges, and international securities dealing. Producer services are financial instruments and legal tools—as much product as they are service—that are necessary to manage an international economy. Business schools and law schools participate in the development of producer services and train graduates to use them (Thrift 1987; Sassen 1991).

firms to buy up competitors and firms producing related products rather than pushing firms to acquire unrelated products in a conglomeration strategy. Despite the wave of "deconglomeration" in the late 1980s, the clear majority of the largest firms (Fortune 500) continued to operate in more than one industry, a decidedly higher level of diversification than that in the 1950s and 1960s, even though diversification of large firms tended to be into industries that were closely related (Davis, Diekmann, and Tinsley 1994). In the 1990s, then, large corporations were still diversified but focused on related products, often in high-technology industries or industries pursuing intellectual property strategies.

These changes in manufacturing companies' corporate structure—conglomeration in the 1960s and 1970s and deconglomeration accompanied by related products diversification in the 1980s—were paralleled in the defense industry, which was heavily concentrated in aerospace, communications, and electronics (Markusen and Yudken 1992). In the 1970s, defense high-technology firms that manufactured single products or even related products were more likely to exit from the ranks of the 100 largest corporations than were diversified defense or high-technology firms. Undiversified firms such as Control Data, General Dynamics, and Lockheed left the list in the 1969–79 period, while conglomerates with defense divisions such as Honeywell, Litton Industries, LTV, and Minnesota Mining and Manufacturing entered in the 1959–69 period and Rockwell International in the 1969–79 period (Fligstein 1990, 278–79, table 8.7). By the 1980s, defense firms assimilated into corporations pursuing conglomeration strategies were no longer as deeply invested in the cold war R&D coalition as they were when these firms pursued single product or closely related product strategies. Instead, they were more willing to support competitiveness R&D policies that allowed them to draw on federally funded R&D across a wide range of research-intensive high-technology fields (Business–Higher Education Forum 1986a).

Even when the Reagan administration greatly increased defense funding in the 1980s, only 5 (Grumman, General Dynamics, Northrop, Martin Marietta, and McDonnell Douglas) of the top 15 defense contractors were dependent on DOD and NASA primes for more than 60 percent of their sales (Markusen and Yudken 1992, 78, table 4.3; Nimroody 1988, 56, table 3.1). Some prime contractors moved to conversion "by increasing non-defense business with the federal government, especially the National Aeronautical and Space Administration," while more diversified companies, such as General Electric and Raytheon, pursed dual-use strategies or reduced their defense commitments (Nimroody

1988, 57–58). Quite diversified companies, such as Ford and Honeywell, abandoned the defense industry altogether. By 1990, there were fewer defense-dependent corporations, although those that remained were more highly concentrated, as suggested by the Lockheed Martin-Marietta merger (Wulf 1993). Diversification of the majority of large high-technology defense firms made it possible for their leaders to consider growth strategies other than DOD/DOE/NASA/aerospace contracting and to rethink their commitment to the cold war R&D policy coalition.

The NIH Coalition

The story of the bipartisan coalition built around the NIH is more fragmented than that about the cold war, perhaps because it attracted less attention from university-based dissidents protesting against the war in Vietnam or because it initially was built around separate diseases. Prior to the 1980s, the medical-industrial complex was formed by physicians in private practice, nonprofit hospitals and insurance companies, pharmaceuticals, university administrators, and many research professors (Ehrenreich and Ehrenreich 1970). Like the military-industrial complex, the medical-industrial complex formed an R&D coalition that effectively lobbied Congress for more funds for fighting the war against an increasing number of specific diseases, foremost among them heart disease and cancer. Unlike the military-industrial complex, which targeted the executive branch (particularly the president in his capacity as commander-in-chief), Congress was usually the arena in which members of the medical-industrial complex, together with groups of relatives of disease-stricken constituents, lobbied to fight specific diseases or disease clusters. The National Institutes of Health became the mainstay of academic R&D, increasing its share of federal funds for academic R&D from 36.7 percent in 1971 to about 45.0 percent in the mid-1980s (NSF 1993a, table 5.9).

The general goals of this political coalition were similar to those of the military-industrial coalition. The medical-industrial coalition sought to (1) win the war against disease, (2) administer high profits to specialist physicians and pharmaceutical companies, and (3) subsidize corporate and academic R&D.

In the 1980s, the corporate structure of the medical-industrial complex experienced shifts as great as those in the military-industrial complex, although the concrete changes in corporations were quite different. Whereas in the military-industrial complex the relevant corporations were concentrated in manufacturing, the medical-industrial complex included service as well as manufacturing industries: hospitals and insurance companies along with the pharmaceutical, agriculture, and chemical industries. As in the military-industrial complex, the changes in corporate structure of the medical-industrial complex made the corporations involved appreciate the benefits of a broad, bipartisan competitiveness R&D policy.

In the 1970s and 1980s, the structure of the hospital and insurance industries changed from nonprofit to for-profit. Prior to the 1970s,

> the medical-industrial complex referred to the linkages between the doctors, hospitals, and medical schools and the health insurance companies, drug manufacturers, medical equipment suppliers, and other profit-making firms . . . [that were connected to] a medical system that was still made up almost entirely of independent practitioners and local, non-profit institutions. (Starr 1982, 428–29)

After 1970, growing privatization of hospitals, health care services such as health maintenance organizations and preferred providers, and the insurance industry began to replace independent practitioners and local nonprofit institutions (Relman 1980; Starr 1982). In the 1970s and 1980s, nonprofit community hospitals were consolidated and bought up by multi-institutional profit-making corporations. These profit-making hospitals grew rapidly. Independent health maintenance organizations also were increasingly purchased by large corporations. In the insurance industry, Blue Cross and Blue Shield, "voluntary" health insurers, controlled most health premiums until the 1970s. They were replaced by large corporate insurance companies (e.g., life, auto) for which health insurance was a related product. The market shares of Blue Cross and Blue Shield for private insurance fell from a high of 45 percent in 1965 to a low of 33 percent in 1986, when they privatized, losing their tax-exempt status (Navarro 1994). Currently, major insurance companies such as

> Metropolitan, Aetna, Prudential, Connecticut General, and the Blues, among others, are developing managed care plans and acquiring the already existing plans. . . . Eighty-two percent of the delivery system is now under some form of managed care, contracted by, controlled by and/or influenced by insurance companies. (Navarro 1994, 207, 209)

As the health care industry moved from nonprofit status to for-profit status, it began to pursue high-technology health care solutions that focused on diagnostics, protocols, pharmaceuticals, and biotechnology to reduce labor costs and increase profits. Under the relaxed antitrust laws of the Reagan and Bush administrations, the health care industry developed a strategy of horizontal and vertical integration and became very interested in non-cold-war-research-intensive products.

At the same time, the pharmaceutical, agriculture, and chemical industries began aggressively pursuing biotechnology. Pharmaceutical companies began working with dedicated biotechnology firms in the 1980s, using biotechnology "both as a production technology and a research tool . . . [because] biotechnology is likely to be the principal scientific driving force for the discovery of new drugs as we enter the 21st century" (U.S. Congress, Office of Technology Assessment 1991, 94–95; see also Kevles and Hood 1992). Sometimes pharmaceutical firms acquired small dedicated biotechnology firms, but they often used "nexus-of-contracts" strategies to develop temporary research-intensive networks to develop specific products (Powell 1990; Davis, Diekmann, and Tinsley 1994). Similarly, agricultural companies acquired dedicated biotechnology firms and invested in biotechnology (Kenney 1986; U.S. Congress, Office of Technology Assessment 1991), whether to increase yields, lower costs, or create new products. The chemical industry, increasingly challenged by global competition in the 1980s, turned away from bulk chemical production, the markets for which had been captured by European companies. Instead, pharmaceutical, agricultural, and chemical companies began investing more heavily in in-house research and purchasing or developing links with smaller, research-intensive biotechnology firms (U.S. Congress, Office of Technology Assessment 1991). The industrial contribution to medical and related R&D increased from 39 percent in fiscal year (FY) 1993 to 50 percent in FY 1993, indicating the commitment of these industries to research-intensive intellectual property strategies to compete more successfully in a global economy (NIH 1994).

In the 1980s, structural changes in insurance companies and hospitals and in the pharmaceutical, agricultural, and chemical industries prompted a greater interest on their part in a competitive R&D policy coalition. Privatized insurance and hospital companies concerned with cost containment turned to high-technology strategies such as screening, diagnostics, and "magic bullets" (Brandt 1985). Simultaneous diversification and convergence focused the pharmaceutical, agricultural, and chemical industries on biotechnology as the key area for future growth, with profit-intensive human therapies as a main goal.[7]

7. The NIH and NSF were not the only mission agencies to support biotechnology. While the NIH provided the most support in 1991 at $2,275.95 million, the DOD was second at $118.80 million, the NSF was third at $93.80 million, the U.S. Department of

The pharmaceutical, agricultural, and chemical companies, together with the hospital and insurance corporations, created a "new" profit-taking medical-industrial complex that drew heavily on discoveries and technologies based on NIH research. In the 1970s and 1980s, these firms joined together with diversified manufacturing companies that moved away from the cold war R&D coalition. They began to push for broad-gauge competitiveness R&D policies that depended as much on intellectual property as they did on more traditional products. Their goals were to (1) win control of global markets through privatization and commodification of intellectual property; (2) establish government subsidies for high-technology and producer services industries; and (3) move R&D, including university R&D, toward commercial science and technology (see table 1.1).

The Competitiveness Coalition

The transition from cold war to competitiveness R&D policies was made easier by the liberal R&D spending policies of the Reagan and Bush administrations. There were increases in military expenditures and more legislative and financial support for competitiveness (Nimroody 1988; Markusen and Yudken 1992; NSF 1993b). Under Bush, the President's Council of Advisors on Science and Technology (PCAST) began to articulate a competitiveness R&D policy.[8] Policy analysts and policy advisers regularly moved back and forth between the PCAST staff and the Council on Competitiveness (Greenberg 1993a, 1993b). By the late 1980s, even the NSF (1989) began to promote a competitiveness R&D program. Cold war and competitiveness strategies were pursued simultaneously, receiving broad bipartisan support. Clinton borrowed much of the language and many of the concepts developed by Bush's staff in his presidential campaign (Clinton/Gore 1992; Greenberg 1993a, 1993b).

In response to the crises in profits and productivity, the armed services as well as defense contractors began to develop dual-use policies in the 1980s (Slaughter 1990). Dual-use policies served as a bridge between the cold war and competitiveness policy communities. Some agencies within the armed services, such as the Advanced Research Projects

---

Agriculture was fourth at $84.00 million, the DOE was fifth at $62.40 million, followed by the Association for International Development at $43.76 million. Other agencies participated in biotechnology research, but at less than $10.00 million (NSF 1991, 9, table 2).

8. Although his science advisors and staff supported a competitiveness policy, Bush himself did not. He was opposed to anything that could be construed as industrial policy.

Agency (ARPA) and the Office of Naval Research, began to promote research goals that included collaborative partnerships between the military, corporations, and universities geared to products for civilian markets (U.S. Department of Defense, Office of Naval Research 1994).[9]

University managers were enthusiastic supporters of the competitiveness agenda. Their early and eager support for competitiveness was most clearly demonstrated through their participation in the Business–Higher Education Forum, an organization comprised mainly of the chief executive officers (CEOs) of large firms and the presidents of prestigious universities. The organization built consensus among its leadership and their various constituencies for a variety of legislative changes that made possible the easy transfer of technology from universities to corporations and created the opportunity for universities to engage in a wider range of for-profit activity (Slaughter 1990).

The degree of support for a competitiveness R&D policy on the part of scientists and engineers was not clear. Scientists were not enamored of industry-university-government collaborations when these projects diverted funds from individual investigators. However, some scientists and engineers were very supportive of competitiveness policies, and professors generally, whether in science and engineering or other fields, thought that universities and colleges should pursue commercialization aggressively (Campbell 1995). Moreover, the majority of scientists and engineers has not organized for or against a competitiveness agenda in any noticeable way.

Because our narrative about the rise of the competitiveness coalition is one of several we want to present in this article, we have glossed over a number of tensions, problems, and contradictions unresolved by the supporters of this coalition. The relation of the competitiveness R&D coalition to the cold war R&D coalition remains ambivalent. CEOs of large multinational conglomerates are active in both and are unlikely to easily forgo the cost-plus/military-specifications profit system characteristic of defense contracting. Military spending encompassed much more than corporations; federal laboratories, universities, communities centered around military bases, contractors and suppliers, as well as the

9. Several states became deeply involved in competitiveness R&D policies in the late 1970s and 1980s, often anticipating programs later developed by the federal government. However, space constraints compel us to forgo discussion of state R&D policies. For research on the rise of state-university-industry initiatives, see Lambright and Rahm (1991); for research on changes in state legal structures to accommodate state-university-industry initiatives, see Slaughter and Rhoades (1993).

beneficiaries of nonweapons defense procurement are apt to resist the cuts in defense spending that would make possible a full-fledged competitiveness policy. As the industries and constituencies that surround the DOD/DOE/NASA/aerospace mission agencies balance between technoscience regimes that are sometimes competing and sometimes complementary, so too the industries and constituencies focused on the NIH sometimes share regimes and sometimes compete. Perhaps the strongest tension in the medical/biotechnology area is between insurance companies that strive to keep costs down and companies that develop and manufacture costly high-technology products, whether mechanical, electronic, or biological. The earlier basic science narrative glossed over conflicts and tensions among the various branches of the armed services over which weapons systems to support, among various groups in the NIH about which diseases to fight, between universities and scientists about how deeply they should be involved, and about the degree of threat on the military and health fronts. Similarly, the competitiveness narrative provides a coherent story line to explain untidy, complex, and contradictory events.

Accomplishments of Competitiveness R&D

In the late 1970s and early 1980s, the emerging competitiveness R&D coalition began to iterate and reiterate a narrative of science and technology that differed significantly from the cold war saga of winning the fight against communism and from the physician/nonprofit hospital stories about defeating disease. Although the old heroic narratives—the struggle against the ultimate other, the Evil Empire, and the battle to vanquish disease—continued to be invoked, the new narratives about science and technology focused more on economic competitiveness. The public discourse about science and technology involved less frequently the tale of basic science, ensconced in university ivory towers where scientists and engineers developed "seed corn" from which national security, health, and prosperity would grow, and emphasized instead a story about business and industry working closely with science and technology to create commercial products and processes that would make the United States more competitive in global markets (Slaughter 1993a).

In contrast to the cold war narrative, the competitiveness narrative presented the needs of business and industry as paramount. University-based scientists and engineers played only a secondary role. Knowledge was valued not for its own sake, or for what it might someday contribute to economic development, but for its contribution to the creation of products and processes for the market of the moment. As the boundaries between commerce and the university dissolved into partnership agreements with the private sector, universities were no longer portrayed as separate and inviolable ivory towers beyond the sordid concerns of commerce and the petty squabbles of politics. Universities, like business and industry, began to try to make profit on their intellectual property. The government agencies that once had specific science and technology missions now worked together to support science and technology geared to help industry. Universities were valued as much for their training function as they were for their capability for novel discovery.

In the 1980s, competitiveness narratives were told over and over again in countless policy documents, position papers, and congressional hearings.[10] Policy makers found the new narrative compelling, and it was embodied in law and various government rules in the 1980s and 1990s. The legislative accomplishments of the competitiveness R&D coalition (see table 1.2), which incorporate and embellish the competitiveness narrative, have had important implications for academic science and technology.

The Bayh-Dole Act (1980) signaled the inclusion of universities in profit taking. It allowed universities and small businesses to retain title to inventions made with federal R&D funds. In the words of the Congress, "It is the policy and objective of the Congress . . . to promote collaboration between commercial concerns and nonprofit organizations, *including universities*" (emphasis added). Before the Bayh-Dole Act, universities could secure patents on federally funded research only when the federal government, through a long and cumbersome application process, granted special approval.

The Bayh-Dole Act changed the relationship between university managers and faculty as they negotiated ownership of intellectual property through university committees and in state courts and as states developed legislation that addressed ownership issues, further contractualizing faculty-management relations (Slaughter and Rhoades 1990, 1993; Chew 1992; Olivas 1992). As potential patent holders, university

10. For examples of policy documents promoting the competitiveness R&D agenda, see Business–Higher Education Forum (1983, 1986a, 1986b), National Academy of Science, Committee on Science, Engineering and Public Policy (1992, 1993), President's Council of Advisors on Science and Technology (1992), Kaufman and Waterman (1993), Marston and Jones (1992), National Academy of Science, Government-University-Industry Research Roundtable (1992), and Smith (1990). For examples of congressional hearings, see U.S. Congress, House of Representatives, Committee on Science, Space and Technology (1987, 1992, 1993).

TABLE 1.2. Selected Legislation Enabling a Competitiveness Research and Development Policy

Year	Legislation
1980	Public law 96-480, Stevenson-Wydler Technology Innovation Act, as amended in 1986 and 1990
1980	Public Law 65-517, Bayh-Dole Act, and Reagan's 1983 memo on government patent policy
1982	Public Law 97-219, Small Business Innovation Development Act
1983	Public Law, 97-414, Orphan Drug Act, as amended in 1984, 1985, and 1990
1984	Public Law 98-462, National Cooperative Research Act
1986	Public Law 99-660, Drug Export Amendments Act of 1986
1987	Presidential Executive Order 12591
1988	Public Law 100-418, Omnibus Trade and Competitiveness Act
1993	Public Law 103-182, North American Free Trade Agreement
1993	Public Law 230-24, Defense Appropriations Act, Technology Reinvestment Program
1994	Public Law 103-465, General Agreement on Tariffs and Trade

trustees and administrators could suddenly see all research generated by faculty as potential intellectual property. Faculty also could conceptualize their discoveries as private, valuable, and licensable products or processes rather than as knowledge to share publicly with a community of scholars (Rhoades and Slaughter 1991a, 1991b). The Bayh-Dole Act gave new and concrete meaning to the phrase "commodification of knowledge." The act enabled universities to enter the marketplace and to profit directly when universities held equity positions in companies built around the intellectual property of their faculty as well as to profit indirectly when universities licensed intellectual property to private sector firms. In a very real sense, the Bayh-Dole Act encouraged academic capitalism (Slaughter and Leslie 1997).[11]

Competitiveness legislation blurred the boundaries between public and private sectors. Several technology transfer acts, beginning with the Stevenson-Wydler Act of 1980, pioneered the legal and administrative mechanisms for transfers between public and private entities. These acts were aimed primarily at the federal laboratories but also affected universities. For example, in the Federal Technology Transfer Act (1986), the

11. In 1983, by executive order, Reagan included all corporations in the Bayh-Dole Act, changing a law meant to benefit small business and universities to one that included large corporations as well.

federal laboratories could enter into cooperative R&D agreements with "other federal agencies, state or local governments, industrial organizations, public and private foundations, and nonprofit organizations, *including universities*" (emphasis added). Although universities were not a main target of technology transfer legislation, it was important in incorporating universities into the competitiveness agenda because it pioneered the legal structures that shaped collaborative research agreements between public nonprofit organizations and private sector corporations. Collaborative research and development agreements (CRADAs) permitted private corporations to select marketable products and processes from inventories of research performed in federal laboratories and to collaborate in bringing the product or process to market. In return, federal laboratories received a share of the profits through license or royalty agreements. Universities emulated CRADAs and also developed directories of the problems on which their scientists and engineers were working to be shared with the business community. Like the Bayh-Dole Act, the technology transfer acts changed common understandings of what public and nonprofit meant, authorized segments of public and nonprofit organizations to participate in the market, and helped to dissolve the boundaries between university and society.

The place of universities in the competitiveness agenda was underscored by the Small Business Innovation Development Act of 1982. This act mandated that federal agencies with annual expenditures of more than $100 million devote 1.25 percent of their budgets to research performed by small businesses that were deemed the engines of economic recovery. It passed despite the opposition of major research universities (Slaughter 1990). In FY 1989, $18.6 million of the NSF budget went to small business programs (NSF 1989, 3). Although the amounts of money captured for commerce by the Small Business Innovation Development Act were insignificant items in the mission agencies' budgets, they symbolized the inability of the research university lobby to hold its share of federal dollars when it acted outside the purview of the competitiveness R&D coalition.

A couple of pieces of legislation—the Orphan Drug Act of 1983 and the National Cooperative Research Act of 1984—revealed the increasing importance of research in business strategies. Research served multiple functions for corporations. It enabled product development in the case of orphan drugs, promoted government subsidies, and served legal and ideological functions (as in the case of the National Cooperative

Research Act, which weakened national antitrust legislation). Universities could take advantage of the opportunities created by this legislation only if they were willing to embrace the competitiveness R&D agenda.

The Orphan Drug Act provided incentives for developing drugs for rare diseases affecting human populations of under 200,000. By providing tax incentives and market monopolies, this act encouraged biotechnology firms, which drew heavily from academically based, federally funded R&D, to pursue niche markets for vaccines and diagnostics for diseases, such as Huntington's chorea, that struck relatively small groups. Such companies received a 50 percent tax credit for the cost of conducting clinical trials, often performed by universities, as well as the exclusive right to market the product for seven years (U.S. Congress, Office of Technology Assessment 1991). A number of universities profited from the sales of such drugs. The University of California, Los Angeles, for example, profited from the human growth hormone handled through an arms-length corporation, Genetech (Goggin and Blanpied 1986).

The National Cooperative Research Act afforded special antitrust status to R&D joint ventures and consortia. This act was crucial to university-industry collaborations. Previously, the courts had ruled that collaborations at the enterprise level were inappropriate, barring joint R&D efforts by firms in the same industries on the grounds that these constituted restraint of trade. The National Cooperative Research Act made an exception for R&D, enabling broad government-industry-university funding of R&D such as that which occurred with Sematech. In short order, there were more than 100 such ventures (NSF 1989). The National Cooperative Research Act was also a counter in business leaders' strategy to overhaul national antitrust policy, promoting cooperation at home and competition abroad (Dickson 1984; Fligstein 1990).

A series of acts—the Drug Export Amendments Act of 1986, the Omnibus Trade and Competitiveness Act of 1988, the North American Free Trade Agreement (NAFTA) of 1993, and the General Agreement on Tariffs and Trade (GATT) of 1994—embodied the competitiveness coalition's strategy toward global control of intellectual property. The Drug Export Amendments Act allowed drugs that had not yet been approved by the Food and Drug Administration for use in the United States to be exported to twenty-one foreign countries that had regulatory mechanisms. Prior to 1986, U.S. companies that wanted to export new drugs to foreign countries with more rapid regulatory processes than those in the United States had to forfeit their proprietary rights to these drugs to their multinational partners in those countries (U.S. Congress, Office of Technology Assessment 1991). Like the Orphan Drug Act, the Drug Export Amendments Act created a supportive climate for the development of biotechnology, perhaps the most dynamic sector among burgeoning university-industry relationships. The Omnibus Trade and Competitiveness Act (PL 100.418) stressed the growing importance of intellectual property in world markets. This act stipulated that anyone who sells, uses in the United States, or imports without authority a product made by a process under patent protection by the United States was liable to prosecution. Like much of the legislation supported by the competitiveness R&D coalition, the Omnibus Trade and Competitiveness Act increased the protection of intellectual property and raised penalties for violation. The NAFTA extended the regional protection of intellectual property, treating it like any other commodity. It barred discrimination against cultural exports, even when that meant that fragile, partially industrialized cultures could be inundated with U.S. cultural products. The GATT compelled signatories to honor intellectual property laws (patent, trademark, and copyright), which means that Third World countries that previously used Western products ranging from pharmaceuticals to learned books and journals without paying licensing fees or royalties now must pay for them. (In a postcolonial twist, established industrial countries are able to patent biological resources indigenous to newly industrialized countries, ultimately requiring that these relatively poor countries pay to use products based on their own raw materials.)

This legislation, which enabled corporations to enact global competitiveness strategies on a playing field engineered by the United States, did not focus particularly on universities, but it had grave consequences for them. Knowledge produced by university-trained scholars previously had remained the communal possession of an international community of scholars. Now it was to be transformed into products and processes owned jointly by faculty and universities in industrialized countries, thereby committing professors and institutional managers to further commercialization of science and technology. The fact that universities were not central to this well-articulated, legislatively embodied intellectual property strategy underlines the secondary role of the academy in the competitiveness narrative.

The secondary role of universities was well demonstrated by the Technology Reinvestment Project (Defense Appropriations Act of 1993), which exemplifies trends toward centralization of decisions about research across government agencies and collaboration across industrial sectors. The Technology Reinvestment Project is headed by the ARPA but involves the military services, the departments of Commerce, Energy, and Transportation, NASA, and the NSF. Its mission is to merge defense and commercial industrial bases so that the DOD can have access to low-cost critical technologies. The program combines eight statutory programs that fall into three areas: technology development for innovative dual-use products and processes, technology deployment that uses third-party providers to deploy manufacturing and management technologies in small and medium-size firms seeking to develop dual-use capacity, and "Manufacturing Education and Training projects [that] combine university and industry expertise to formulate new programs of education in the science of manufacturing" (U.S. DOD 1994, 71). The ARPA legislation typifies the position of competitiveness legislation in the Clinton administration with regard to academic science: initial mention usually speaks only to a training function for universities. Proposals for these funds must be submitted by teams, partnerships, and consortia of industry, local governments, and institutions of higher learning, with proposers sharing at least 50 percent of the costs. The ARPA initiative builds on legislation that allows free movement of technologies, costs, and profits between previously discrete sectors of the economy. It shares budgetary and mission responsibility across federal agencies that previously had distinct goals and objectives, and it relies heavily on consortia and collaborative efforts. These strategies reduce costs but also commit sectors previously outside the market to commercialization through dual use.[12]

Table 1.3 shows bipartisan support for competitiveness legislation. Of the eleven laws, four were passed by voice vote in both houses of Congress and seven by roll-call vote. Voice votes are taken for various reasons, including members' wish to have no record of their votes or because consensus is so strong that a roll-call vote is inefficient. (See notes to table 1.3 for clarification of the circumstances surrounding some of the votes.)

When roll-call votes were taken, only the NAFTA lacked bipartisan support. Bipartisan support was strongest for legislation that allowed U.S.-based corporations to use public resources to enhance economic

12. While dual use bridges the competitiveness and military-industrial technoscience regimes, dual use at the same time remains a separate option for the military-industrial regime and is pursued as such, depending on the strength of the competitiveness R&D coalition. In other words, policies such as dual use relieve the tensions between the military-industrial and the competitiveness R&D coalitions only on some occasions (see MacCorquodale et al. 1993).

TABLE 1.3. Votes on Competitiveness Coalition Legislation

	House of Representatives			Senate		
Legislation	Democrats	Republicans	Total	Democrats	Republicans	Total
Public Law 96-480[a]	voice	voice	voice	voice	voice	voice
Public Law 96-517[b]	voice	voice	voice	voice	voice	voice
Public Law 97-219	195–30	158–27	353–57	40–0	50–0	90–0
Public Law 97-414[c]	voice	voice	voice	voice	voice	voice
Public Law 98-462[d]	255–0	162–0	417–0	43–0	54–0	97–0
Public Law 99-502[e]	227–0	159–0	386–0	voice	voice	voice
Public Law 99-660[f]	voice	voice	voice	voice	voice	voice
Public Law 100-418	243–4	133–41	376–45	50–1	35–10	85–11
Public Law 103-82 (NAFTA)	102–156	132–43	234–199	27–28	34–10	61–38
Defense	230–24	95–77	325–101	52–2	36–7	88–9
Public Law 103-465 (GATT)	167–89	121–55	288–144	41–14	35–9	76–23

NOTE: NAFTA = North American Free Trade Agreement; GATT = General Agreement on Tariffs and Trade.

[a]House agreed to motion to suspend rules that required two-thirds support from members who are present and passes by voice vote. House later agreed to request unanimous consent to accept bill as amended by Senate.

[b]House agreed to motion to suspend rules and passed by voice vote.

[c]House agreed to motion to suspend rules and passed by voice vote. Backers in both chambers fought an expected vote.

[d]House vote acceptance in House and Senate of conference report, after unanimous passing of each chamber's legislation.

[e]Senate agreed to motion to suspend rules and passed by voice vote. Voice votes in each chamber of conference report.

[f]Senate passed earlier version of bill 91–7 (Democrats 40–7, Republicans 51–0). Senate 1844, which became Public Law 99-660, passed unanimously in committee before being accepted by voice vote in Senate and House.

competition or that permitted nonprofit institutions to enter the marketplace by holding title to intellectual property or by forming alliances with corporations. When regional or local agreements were likely to take jobs or decision-making power away from home, as was the case with the NAFTA, the Democrats were more reluctant to support them. Nonetheless, apart from the NAFTA, all legislation on which formal votes were taken had broad bipartisan support if such support is construed as 70 percent or more of each house voting for passage.

In addition to legislation, administrative interpretations of new laws, rulings by administrative law judges, and litigation in civil courts helped to create a climate that promoted competitiveness R&D policies. For example, the Internal Revenue Service does not tax universities' royalty income, creating a strong incentive for universities to encourage patenting and copyrighting (Martino 1992). In 1980, in *Chakrabarty v. Diamond,* the Supreme Court ruled that living organisms were patentable. In the same year, the Patent and Trademarks Office issued the Cohen-Boyer patent on rDNA to Stanford University. In 1988, the Patent and Trademarks Office issued Harvard University a patent on the transgenic mouse, later globally marketed by DuPont as oncomouse, a laboratory animal for researchers. In 1990, the California Supreme Court ruled that a patient did not have a property right to his body tissues after they were used by researchers to develop a commercially important cell line (U.S. Congress, Office of Technology Assessment 1991). Rule-making modalities other than legislation interact with new statutes to create a dense administrative-legal infrastructure for the new competitiveness policy.

The competitiveness R&D coalition created a new narrative for science and technology policy and instituted changes in legislation that reshaped the rules governing R&D. The rule changes allowed public and nonprofit entities—whether universities, government agencies, or nonprofit research institutes—to enter the market, changing our commonsense understanding of what is public and what is private. Institutions still labeled public and nonprofit were able to patent and profit from discoveries made by their professional employees. Simultaneously, private profit-making organizations were able to make alienable areas of public life previously held by the community as a whole: scientific knowledge, databases, technology, strains and properties of plants, and even living animals and fragments of human beings.[13] Historically, this shift

13. The European Community parliament, led by the Green Party, recently prohibited the patenting of animal and human life.

in ownership rights is on a scale with the enclosures of communal property by large landholders in Great Britain and Latin America with the onset of market economies.

Legislation promoting competitiveness made possible the fluid movement of commodities and capital among private, nonprofit, and public institutions. In academe, this fluidity gave rise to new organizational forms: arms-length agencies run by universities to handle profit-making activities, for-profit corporations created with nonprofit and state funds, collaborative research agreements that were funded by university-government-corporate contributions and relied on a variety of arrangements concerning ownership of intellectual property and disposition of profits. These changes in academic organization complemented changes in corporate structure, facilitating academic interaction with corporations pursuing nexus-of-contract strategies. These changes integrated the state into the production process more directly than before and, to some extent, rendered problematic the older distinctions between the state and the economy.

Effects of the Competitiveness R&D Policy

Although we cannot show direct causal links between the emergence of the competitiveness R&D coalition and changes in academic science and technology, the observed changes are consistent with the coalition's success. We review briefly changes in federal obligations for business and university R&D; in federal funding for the NIH and NSF and their support of individuals, teams, centers, and major facilities; and in all obligations for academic science and technology. We also examine changes in the patent behavior of university faculty, in numbers of scientific and technical articles coauthored by academic and industrial scientists and in faculty salaries. We use 1980 or 1983 as a base point because these were the years when the competitiveness agenda first began to take legislative form, and we compare the data from 1980 and 1983 to those from the 1990s.

The business share in the total federal obligations for R&D remained about the same during the ten-year period, while higher education's share grew slightly, from 13 to 17 percent (see figure 1.1).

Industry increased slightly its share of basic research funds provided by the federal government, while universities' share remained the same. Industry's share of funds designated for applied research remained roughly steady over the decade, while universities' share increased by 6 percent. Both industry and universities increased their share in federal

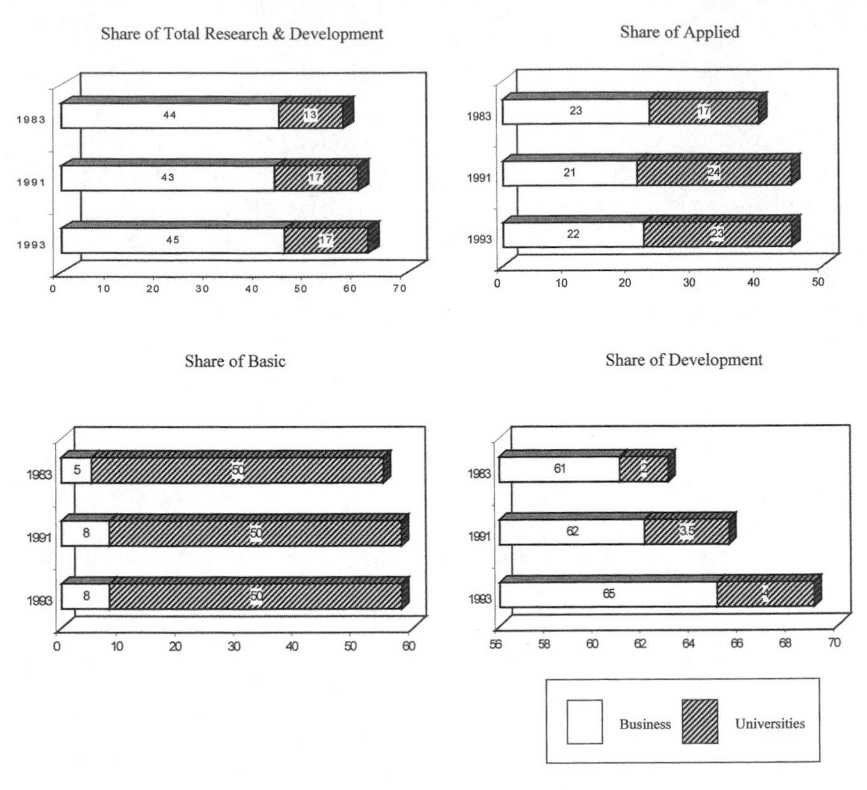

FIGURE 1.1. Federal Obligations for Research and Development, by Character of Work Performed: 1983, 1991, 1993 (percentages)
SOURCE: Adapted from National Science Foundation 1993b, appendix table 4-12.

funding of development work. Industry's share increased by 4 percent, and universities' share increased by 2 percent. The greatest change in all categories was the increase in universities' share of applied research funds, which probably reflects the Reagan administration's emphasis on the Strategic Defense Initiative, a heavily applied program, and an increasing commitment by the mission agencies and the NSF to the competitiveness agenda, particularly to the harnessing of science for development of commercial technology.[14]

14. Although the greatest growth in federal R&D obligations was in applied academic science and technology, the NSF's estimates of that growth are probably conservative. A number of scholars and policy analysts, as well as government officials, are uncertain about the accuracy of the NSF categories. The NSF categories—basic, applied, and development—used for reporting types of research may obscure the degree of growth. Perhaps a more precise set of measures will be forthcoming, given the need of the competitiveness coalition to monitor more closely the development of commercial science and technology

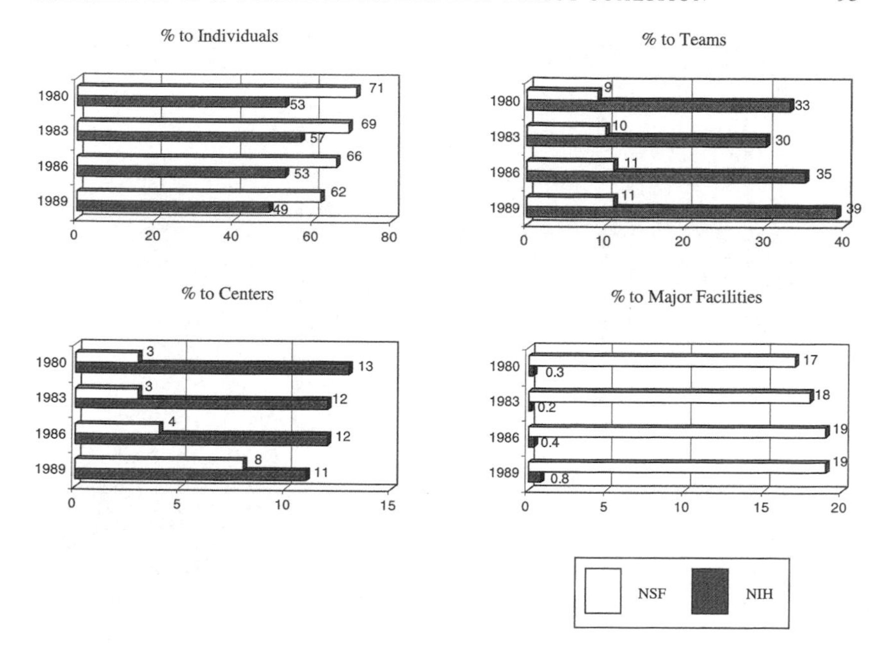

FIGURE 1.2. Federal Funding of Academic Research, by Mode of Support in National Institutes of Health and National Science Foundation: 1980, 1983, 1986, 1989 (percentages)
SOURCE: Adapted from National Science Foundation 1993b, appendix table 4-20.

Federal support for academic research (see figure 1.2) has changed not only in terms of the proportion of funding for basic and applied research but also in terms of whether research funds are awarded to individuals, teams, or centers. At the NSF, the percentage of funds going to individuals dropped from 71 percent in 1980 to 62 percent in 1989, while the percentage going to teams rose by 2 percent and the percentage going to centers rose by 5 percent. The NIH funding moved away from individuals to teams. The share of funds awarded to individuals dropped by 4 percent in the nine-year period, while the share going to teams increased by 6 percent. However, the share going to centers dropped by 2 percent. There were no changes in the percentage of funds that the NIH and NSF provide to major facilities. In other words, a shift away

initiatives. As Gibbons and Panetta (1994, 1) indicated in a memorandum on R&D to the heads of executive departments and agencies, the current categories are regarded as inadequate, and a new "reporting structure will be designed to help assess our programs in supporting the [a]dministration's highest priority programs and will be much more extensive than the three R&D categories now collected."

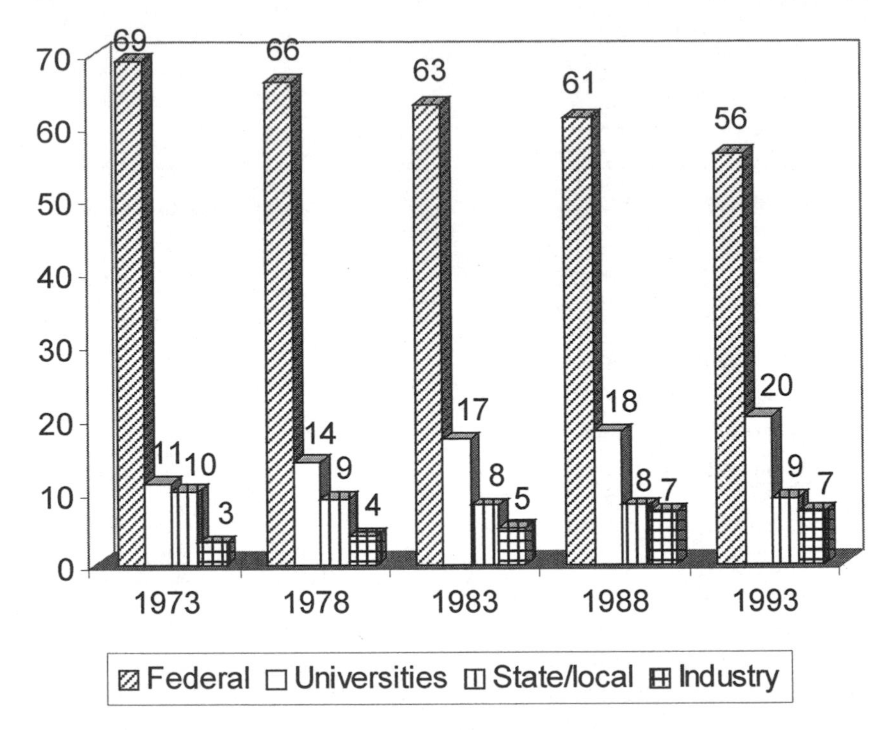

FIGURE 1.3. Support for Academic Research and Development, by Sector: 1973, 1978, 1983, 1988, 1993 (percentage shares)
SOURCE: Adapted from National Science Foundation 1993b, appendix table 5-2.

from individual researchers was the most pronounced change during the period under consideration. This trend may reflect the competitiveness coalition's emphasis on collaborative efforts across sectors geared toward facilitating the transfer of knowledge from public and nonprofit entities to corporations and commercial endeavors.

The individual researcher is not alone in his or her loss of federal share. Overall, federal share of academic R&D has declined markedly in the past twenty years. Whereas in 1973 the federal government funded 69 percent of academic R&D, in 1993 it funded only 56 percent (see figure 1.3).

From 1983 to 1993 alone, the federal share has declined by 7 percent. State and local governments' share remained relatively stable, at about 10 percent, while industry's share of academic R&D increased by 4 percent. As if to compensate for loss in federal share, universities increased their share of support most dramatically, from 11 percent to 20 percent. Although universities may be trying to keep their research profiles high by making up for their loss of federal funds, they may also be using their assets to invest in new commercial research (see Feller and Geiger 1993). They may also be committing more money to R&D to match funds from mission agencies and industry in pursuit of collaborative ventures, such as those funded by the Advanced Technology Program.

Examination of trends in R&D spending suggests that competitiveness R&D legislation may be changing the flow of research dollars to academic science, but it generally says little about how changes in funding shape laboratory life. Over the last decade, some indicators were developed to directly capture changes in scientists' and engineers' behavior. These indicators focus on patenting activity and patent income, on academic-industry coauthorship, and on changes in peer review processes.

After the Bayh-Dole Act changed the patent law, there was an explosion of growth in patenting by universities, indicating their efforts to take advantage of commercial opportunities. In 1980, the number of all academic patents was 390; in 1991, the total number of academic patents was 1,324, an increase of 239 percent. If, instead of examining the total number of patents for all institutions, we ask which institutions gained and which lost patent shares, two interesting patterns emerge. First, commercialization apparently enabled the strongest R&D performers to acquire more patents. In 1980, 290 patents were awarded to the top 100 R&D performers among universities; in 1991, there were 1,112 such patents, an increase of 283 percent. The top 100 R&D performers increased their share of total academic patents from 74 percent in 1980 to 84 percent in 1991. Second, performance with regard to patents did not necessarily parallel R&D performance. Within the top 100 R&D performers, the share of patents awarded to the top 30 performers declined by 3.5 percent while the share of the bottom 70 increased. This pattern is consistent with Feller and Geiger's (1993) finding that there was a pattern of dispersion and shifting of R&D expenditures among the top 200 R&D universities in the 1980s. For example, the share of research expenditures of the top 10 performers fell from 20.2 percent in 1979–80 to 17.9 percent in 1989–90. "[A]mong the largest 60 performers in 1979–80, 20 gained share and 40 experienced declines. Among the next 110 institutions, gainers of share outnumbered losers 64 to 46" (Feller and Geiger 1993, 10). In other words, the shift to a competitiveness R&D policy may have implications for universities' positions in rankings, whether based on standard R&D performance criteria or in terms of new indicators such as numbers of patents.

In 1993, the Association of University Technology Managers began to rank institutions by patent income, allowing us to look not only at number of patents acquired by institutions but also at revenue generated (Blumenstyk 1994). Institutions that ranked high on federal R&D funds tended to rank high on patent income; for example, the University of California system, Stanford University, Columbia University, the University of Wisconsin, and the University of Washington were ranked one through five on both counts. But Iowa State University, Clemson University, and Tulane University were among the top twenty institutions in terms of income generated by patents. This suggests that the ability to generate such income may reshape the hierarchy of academic R&D, perhaps foreshadowing new and multiple modes of competition among universities. These forms of competition are not closely governed by rules and regulations established by the mission agencies or peer review process.

Time-series data suggest that as academics became more involved with the market, they were more likely to collaborate with their industrial counterparts. Across all fields, the number of articles coauthored by academic and industrial scientists rose from 22 percent of all articles written by scientists in industry in 1981 to 35 percent in 1991 (NSF 1993a, table 5.26). This increase may reflect the success of legislation designed to stimulate collaboration between universities and industry in an effort to make the United States more competitive in global markets.

Changes in faculty salaries were yet another effect of the new R&D policies and legislation. The salaries of the faculty in fields able to engage in commercialization and able to move close to the market rose significantly, whereas the faculty unable or unwilling to do so did not prosper. Analyses of changes of faculty salaries by field in the decade between 1983 and 1993 show that faculty and fields with the highest salaries and the greatest percentage increase (70 percent or above) were in fields concerned with technology (engineering and computers), producer services (business and management, law), and health sciences. In all these fields, knowledge can be treated as a commodity (see table 1.4).

The greatest gains were made by engineering, an applied science closely geared to R&D competitiveness policies. The physical sciences and mathematics, the doyens of "pure science," did not make nearly such dramatic gains. Even if salary level rather than percentage of increase is considered, the physical sciences and mathematics are substantially below the top tier, with a difference that ranges between $10,000 and $23,000. The lowest salaries and the lowest percentages of increase are

TABLE 1.4. Percentage Salary Increase, by Field: 1983–1993

Field	Average Full Professor Salary (in dollars)	Percentage Increase
$75,000–$90,000 range		
Law	89,777	71.1
Engineering	77,985	84.6
Health science	77,913	78.7
Business and Management	77,535	79.0
Computers and information science	75,964	74.8
$60,000–$74,000 range		
Physical science	65,914	63.2
Mathematics	63,776	59.7
Psychology	62,567	59.5
Public affairs	62,435	57.6
Social science	62,352	59.9
Library science	61,827	59.5
Interdisciplinary	61,808	59.0
$50,000–$60,000 range		
Architecture	59,322	57.0
Agribusiness	59,178	63.8
Communications	58,933	61.3
Philosophy and religion	58,424	53.7
Foreign language	57,344	52.4
Home economics	57,157	54.3
Letters	56,744	52.3
Education	56,605	55.9
Performing arts	52,495	61.4

SOURCE: American Association of University Professors (1994, table 5). Based on data from National Association of State Universities and Land Grant Colleges.

in the third tier in fields furthest from the market. The difference in the percentage of salary increases between the lowest five fields in the third tier (philosophy and religion, foreign language, home economics, letters, and education) and the five fields in the top tier ranges between 22 and 30 percent.[15]

Between 1983 and 1993, there were changes in the balance of research that the federal government supports, in the sites where research is performed, in the mechanisms through which academic science and technology funds are awarded, in faculty research behavior, and in fac-

15. With the exception of philosophy and religion, these fields have majority female student bodies. For an analysis of the gender implications of the restructuring precipitated in part by competitiveness R&D, see Slaughter (1993b).

ulty rewards. The U.S. policy shift from the cold war and the war against disease narratives to the competitiveness narrative had concrete consequences for academic scientists and engineers. Waging war, whether against communism or cancer, made the federal government free with resources; no expense was spared, and universities as well as corporations benefited from cost-plus contracts, product and process specifications that went far beyond commercial needs, long-term funding, and little oversight of research. The R&D policy supported by the competitiveness coalition calls for much greater emphasis on cost, on products and processes geared for immediate market use, on relatively short-term funding, and on much closer oversight of research. Moreover, R&D is now evaluated in terms of the commercial bottom line: profitability. At the same time, the new policy contributes to cleavage among scientists and within the university. Disciplines associated with the cold war (e.g., physics) lost funding, while those associated with commercial science (e.g., biotechnology) gained funding. Interdisciplinary science and engineering centers committed to commercial endeavors had abundant resources and could operate like small, high-technology research firms while science and engineering departments carried on with the routine teaching of undergraduates (Etzkowitz 1992; Slaughter and Leslie 1997). The split between the "two cultures"—the sciences and engineering, and the arts and letters—is deepening, making it more difficult for academics to support the concept of the university as a community.

Conclusion

The academic science and technology policy literature that takes a substitution approach suggests universities will fare as well under a competitiveness regime as they did under the cold war regime. This suggestion is based on at least three assumptions. First, it assumes that if university-based science and technology are seen as contributing to economic competitiveness, then funding for research, including basic research, will increase. Second, it assumes that funds directed to the enhancement of competitiveness will not greatly change the ways in which scientists work in laboratories. The third assumption is that what happens to academic science and technology has no consequences for the rest of the university. Our analysis of the effects of national competitiveness suggests that these assumptions are unfounded. More than the rhetoric about the ends of science has changed. As policy narratives shift away from winning battles against communism and specific diseases and toward competitiveness, funding patterns, laboratory life, and faculty compensation, practices begin to change, altering the daily lives of many members of research university communities.

Although our research strongly suggests that change is occurring, it leaves many questions unanswered. If changes in the political economic coalition that supports research funding seem to make a difference in the kinds of science and technology that universities support, then identifying these coalitions and understanding patterns of coalition formation are critical to predicting the future direction of academic science and technology policy. We need to know more about the boundaries of the R&D competitiveness coalition and the nature of its links to political economic processes, generally, and to the cold war coalition, specifically. As we suggested earlier, we see the cold war technoscience regime as deeply rooted, difficult to destabilize, and fairly flexible—able to accommodate many competitiveness initiatives through dual-use policies. Perhaps instead of a split between the post–World War II military-medical industrial complex and a single competitiveness R&D coalition, a number of competing competitiveness coalitions are emerging; for example, in the medical arena, a coalition aligned with insurance companies and health maintenance organizations committed to prevention and low-cost treatment might compete with a coalition aligned with biotechnology companies, academic researchers, and specific disease lobbies. In the manufacturing arena, a coalition concerned with traditional manufacturing might compete with one supporting high technology.

Similarly, we need to know more about the goals and strength of the bipartisan coalition supporting competitiveness legislation in Congress. The goals can be determined by studies of party platforms and of campaign and candidate position papers; the strength can be determined by continued tracking of voting records on competitiveness R&D legislation. If the bipartisan coalition is strong, then we would expect support for competitiveness R&D to persist regardless of which party is in power. The recent GATT legislation is a case in point. Although some Republicans spoke out forcefully against the GATT, garnering a great deal of media attention, in fact most Republicans, like most Democrats, voted for this legislation, which is central to R&D competitiveness policies. So too, Republicans bent on trimming the federal government initially spoke out about cutting departments such as Commerce, but we would expect that the Department of Commerce, the site of many recent competitiveness initiatives such as the Advanced Technology Program in the National Institute of Standards and Technology, will be preserved and enhanced. R&D functions could then be moved away from the NSF and the traditional mission agencies and into organizational structures housed in the Department of Commerce, which is focused on economic development. However, mustering voter numbers, the Republican right and small businesspersons might outweigh the strength of large corporations, whether traditional or high technology, forming a bloc that advocates the end of R&D in all areas other than military.

Finally, we need more studies of the effects of changes in national R&D policies on academic science and technology. Changes in national data sets need to be examined and interpreted in light of competitiveness R&D policies (e.g., amounts of funding for particular fields and subfields, numbers of patents by fields and subfields, numbers of students and gender of students receiving degrees by field and subfield); changes in rankings of institutions need to be examined using a variety of scales such as changes in funding, patenting, and student populations. At the institutional level, we need to examine control and management of commercial science and the problems and possibilities it creates for universities. Changes in organizational structures can be tracked fairly easily. As the commercialization of knowledge increases, we would expect to see the creation of interdisciplinary academic organizations that more nearly match the research directions of the manufacturing and service corporations that rely on commodified knowledge. Although these new interdisciplinary units might be small and fluid, and although faculty employed in these units might gain some autonomy by developing intellectual property, we think that their very success may bring about tighter regulation and greater oversight in the name of responsible management and accountability to the public. At the same time, faculty not involved in commercialization of knowledge are likely to be more closely managed and regulated to ensure the fulfillment of the traditional instructional functions of the university. As federal and state funds for research and education are increasingly targeted for specific purposes such as commercial science, institutions will become more and more concerned with devising means to generate discretionary revenues. For example, public service may be redefined as service for a fee, not service for free. Consulting and scholarship may become subject to various institutional taxes. All professors may have to contribute a share of their income-generating or consulting services to the institution, as university-based physicians currently do; similarly, the institution may take a percentage of royalties on scholarly books and articles. All in all, we think that commodification of knowledge will probably result in greater contractualization, regulation, and bureaucratization in research universities.

Although we have focused on changes in policy coalitions in which business elites are the central actors, we think that it is possible for popular social movements to destabilize, reshape, or modify technoscience regimes, as was the case with the student movement's attack on the Vietnam war and the military-industrial complex in the late 1960s and early 1970s. Currently, no social movements are as strong or as focused on R&D policy as the student movement was at that time, but a number of popular groups have a strong interest in R&D policy. Among these are various military conversion groups that would have us turn swords into plowshares, environmental groups that are not enamored of biotechnology, and women's health groups and AIDS activists who pressure the government for greater R&D spending for particular diseases or disease clusters. Any of these social movements could alter the direction of the competitiveness technoscience regime in unforeseeable ways. However, none of these groups is particularly concerned with preserving or reinstating values dear to the hearts of professors, such as the autonomy of the research community or veneration of basic science. Like leaders of the competitiveness R&D coalition, persons in popular social movements concerned with R&D want more applied science and technology or at least fairly goal-directed research, accountability, and pragmatism. All this signals a decline in the importance of basic research and the likelihood of closer links between universities and society.

ACKNOWLEDGMENTS

The research for this article was supported in part by the Ethics and Values Studies program of the National Science Foundation. We were ably assisted by our graduate students, Paulette Cattel, Richard King, and Kimberley Lund.

REFERENCES

Aronowitz, S. 1988. *Science as power: Discourse and ideology in modern society.* Minneapolis: University of Minnesota Press.

Aronowitz, S., and W. DeFazio. 1994. *The jobless future: Sci-tech and the dogma of work.* Minneapolis: University of Minnesota Press.

Bayh-Dole Act. 1980. Public Law 96-517, 96th Congress, 2nd session (12 December).

Bhagat, S., A. Shleifer, and R. W. Vishny. 1990. Hostile takeovers in the 1980s: The return to corporate specialization. In *Brookings papers on economic activity: Microeconomics 1990,* edited by M. N. Baily and C. Winston, 1–84. Washington, DC: Brookings Institution.

Blumenstyk, G. 1994. A 40% increase in income from inventions. *Chronicle of Higher Education* 9 November, A37–A38.

Blumenthal, D., M. Gluck, K. S. Louis, M. A. Stoto, and D. Wise. 1986. University-industry relationships in biotechnology: Implications for the university. *Science* 232:1361–66.

Brandt, A. M. 1985. *No magic bullet: A social history of venereal disease in the U.S. since 1880.* New York: Oxford University Press.

Business–Higher Education Forum. 1983. *America's competitive challenge: The need for a national response.* Washington, DC: Business–Higher Education Forum.

———. 1986a. *An action agenda for American competitiveness.* Washington, DC: Business–Higher Education Forum.

———. 1986b. *Export controls: The need to balance national objectives.* Washington, DC: Business–Higher Education Forum.

Campbell, T. D. 1995. *Protecting the public's trust: A search for balance among benefits and conflicts in university-industry relations.* Ph.D. diss., University of Arizona.

Carnoy, M., M. Castells, S. S. Cohen, and F. H. Cardoso. 1993. *The new global economy in the information age: Reflections on our changing world.* University Park: Pennsylvania State University Press.

Chew, P. 1992. Faculty-generated inventions: Who owns the golden egg? *Wisconsin Law Review* 1992:259–314.

Chomsky, N. 1969. *American power and the new mandarins.* New York: Pantheon.

———. 1994. *World orders old and new.* New York: Columbia University Press.

Clinton/Gore. 1992. Campaign pamphlet. *Technology—The engine of economic development: A national technology policy for America.* National Campaign Headquarters, Little Rock, AR (21 September).

Davis, G. F., K. A. Diekmann, and C. H. Tinsley. 1994. The decline and fall of the conglomerate firm in the 1980s: The deinstitutionalization of an organizational form. *American Sociological Review* 59:547–70.

Dickson, D. 1984. *The new politics of science.* New York: Pantheon.

Domhoff, G. W. 1978. *The powers that be: Processes of ruling class domination.* New York: Vintage.

Ehrenreich, J., and B. Ehrenreich. 1970. *The American health empire.* New York: Random House.

Eisenhower, D. D. 1961. Text of Eisenhower's farewell address. *New York Times* 18 January, 22.

Etzkowitz, H. 1983. Entrepreneurial scientists and entrepreneurial universities in American academic science. *Minerva* 21:198–233.

———. 1989. Entrepreneurial science in the academy: A case of the transformation of norms. *Social Problems* 36:14–19.

———. 1992. Individual investigators and their research groups. *Minerva* 30: 29–50.

Federal Technology Transfer Act. 1986. Public Law 99-501 (H.R. 3773), 99th Congress, 2nd session (20 October).

Feller, I., and R. Geiger. 1993. *The dispersion of academic research during the 1980s: A report to the Andrew Mellon Foundation.* University Park: Pennsylvania State University, Institute for Policy Research and Evaluation, Graduate School of Public Policy and Administration.

Fligstein, N. 1990. *The transformation of corporate control.* Cambridge, MA: Harvard University Press.

Forman, P. 1987. Behind quantum electronics: National security as basis for physical research in the United States, 1940–1960 (part 1). *Historical Studies in the Physical and Biological Sciences* 18:149–229.

Gibbons, J. H., and L. E. Panetta. 1994. Memorandum to the heads of executive departments and agencies (6 May), White House, Washington, DC.

Goggin, M. L., and W. A. Blanpied. 1986. *Governing science and technology in a democracy.* Knoxville: University of Tennessee Press.

Greenberg, D. S. 1967. *The politics of pure science.* New York: New American Library.

———. 1993a. Clinton and science support: There's still a big blank. *Science and Government Report* 23:5.

———. 1993b. Clinton unveils program for industrial technology. *Science and Government Report* 23:1–5.

Harrison, B., and B. Bluestone. 1990. *The great U-turn: Corporate restructuring and the polarizing of America.* New York: Basic Books.

Herken, G. 1992. *Cardinal choices: Presidential science advising from the atomic bomb to SDI.* New York: Oxford University Press.

Kaufman, A., and R. E. Waterman. 1993. *Health of the public—A challenge to academic health centers: Strategies for reorienting academic health centers toward community health needs.* Philadelphia: Pew Charitable Trusts and the Rockefeller Foundation, Health of the Public Program.

Kenney, M. 1986. *Biotechnology: The university-industrial complex.* New Haven, CT: Yale University Press.

Kevles, D. J., and L. Hood, eds. 1992. *The code of codes: Scientific and social issues in the human genome project.* Cambridge, MA: Harvard University Press.

Krimsky, S. 1982. *Genetic alchemy.* Cambridge, MA: MIT Press.

Lambright, W. H., and D. Rahm. 1991. Science, technology and the states. *Forum for Applied Research and Public Policy* 6 (3): 49–60.

Latour, B. 1987. *Science in action: How to follow scientists and engineers through society,* Philadelphia: Open University Press.

Leslie, S. W. 1993. *The cold war and American science: The military-industrial-academic complex at MIT and Stanford.* New York: Columbia University Press.

Louis, K. S., D. Blumenthal, M. Gluck, and M. Stoto. 1989. Entrepreneurs in academe: An explanation of behaviors among life scientists. *Administration Science Quarterly* 34:110–31.

MacCorquodale, P. L., M. W. Gilliland, J. P. Kash, and Andrew Jameton, eds. 1993. *Engineers and economic conversion: From the military to the marketplace.* New York: Springer-Verlag.

Markusen, A., and J. Yudken. 1992. *Dismantling the cold war economy.* New York: Basic Books.

Marston, R. Q., and R. M. Jones, eds. 1992. *Medical education in transition:*

Commission on medical education—The sciences of medical practice. Princeton, NJ: Robert Wood Johnson Foundation.

Martino, J. P. 1992. *Science funding: Politics and porkbarrel.* New Brunswick, NJ: Transaction.

May, E. 1988. *Homeward bound.* New York: Basic Books.

Mills, C. W. 1956. *The power elite.* Oxford, England: Oxford University Press.

Melman, S. 1982. *Profits without production.* New York: Knopf.

National Academy of Science, Committee on Science, Engineering and Public Policy. 1993. *Science, technology and the federal government. National goals for a new era.* Washington, DC: National Academy Press.

National Academy of Science, Committee on Science, Engineering, and Public Policy, Panel on the Government Role in Civilian Technology. 1992. *The government role in civilian technology: Building a new alliance.* Washington, DC: National Academy Press.

National Academy of Science, Government-University-Industry Research Roundtable. 1992. *Fateful choices: The future of the U.S. academic research enterprise.* Washington, DC: National Academy Press.

National Institutes of Health. 1994. *Key facts and history of funding FY 1984–FY 1994.* Washington, DC: National Institutes of Health.

National Science Foundation. 1989. *Industrial participation in NSF programs and activities.* Washington, DC: NSF.

———. 1991. *Biotechnology opportunities: The NSF role.* Washington, DC: U.S. Government Printing Office.

———. 1993a. *Federal funds for research and development: Federal obligations for research by agency and detailed field of science and engineering—Fiscal years 1971–1993.* Washington, DC: NSF (prepared by Quantum Research Corporation).

———. 1993b. *Science and engineering indicators 1993.* Washington, DC: National Academy Press.

Navarro, V. 1994. *The politics of health policy: The U.S. reforms, 1980–1994.* Cambridge, MA: Blackwell.

Nimroody, R. 1988. *Star wars: The economic fallout.* Cambridge, MA: Ballinger.

Noble, D. 1976. *America by design: Science, technology and the rise of corporate capitalism.* New York: Knopf.

———. 1984. *Forces of production: A social history of industrial automation.* New York: Knopf.

Olivas, M. 1992. The political economy of immigration, intellectual property, and racial harassment: Case studies of the implementation of legal change on campus. *Journal of Higher Education* 63:570–98.

Phillips, K. 1993. *Boiling point: Republicans, Democrats and the decline of middle-class prosperity.* New York: Random House.

Powell, W. W. 1990. Neither market nor hierarchy: Network forms of organization. *Research in Organizational Behavior* 12:295–336.

President's Council of Advisors on Science and Technology. 1992. *Renewing the promise: Research intensive universities and the nation.* Washington, DC: U.S. Government Printing Office.

Relman, A. S. 1980. The new medical-industrial complex. *New England Journal of Medicine* 303:963–70.

Rhoades, G., and S. Slaughter. 1991a. Professors, administrators and patents: The negotiation of technology transfer. *Sociology of Education* 64 (2): 65–77.

———. 1991b. The public interest and professional labor: Research universities. In *Culture and ideology in higher education: Advancing a critical agenda,* edited by William G. Tierney, 187–212. New York: Praeger.

Rifkin, J. 1983. *Algeny.* New York: Viking.

Rogin, M. P. 1987. *Ronald Reagan, the movie: And other episodes in political demonology.* Berkeley: University of California Press.

Sassen, S. 1991. *The global city: New York, London, Tokyo.* Princeton, NJ: Princeton University Press.

Slaughter, S. 1990. *The higher learning and high technology: Dynamics of higher education policy formation.* Albany: State University of New York Press.

———. 1993a. Beyond basic science: Research university presidents' narratives of science policy. *Science, Technology, & Human Values* 18:278–302.

———. 1993b. Retrenchment in the 1980s: The politics of prestige and gender. *Journal of Higher Education* 64:250–81.

Slaughter, S., and L. L. Leslie. 1997. *Academic capitalism: Politics, policies and the entrepreneurial university.* Baltimore, MD: Johns Hopkins University Press.

Slaughter, S., and G. Rhoades. 1990. Renorming the social relations of academic science: Technology transfer. *Educational Policy* 4:341–61.

———. 1993. Changes in intellectual property statutes and policies at a public university: Revising the terms of professional labor. *Higher Education* 26: 287–312.

Smith, B. L. R. 1990. *American science policy since World War II.* Washington, DC: Brookings Institution.

Smith, B. L. R., and J. Karlesky. 1977. *The state of academic science: The universities and the nation's research effort.* New Rochelle, NY: Change Magazine Press.

Starr, P. 1982. *The social transformation of American medicine: The rise of a sovereign profession and the making of a vast industry.* New York: Basic Books.

Thrift, N. 1987. The fixers: The urban geography of international commercial capital. In *Global restructuring and territorial development,* edited by J. Henderson and M. Castells, 203–33. Newbury Park, CA: Sage.

U.S. Congress, House of Representatives, Committee on Science, Space and Technology. 1987. *The role of science and technology in competitiveness: Hearings before the Subcommittee on Science, Research and Technology.* 100th Congress, lst session, 28–30 April, no. 22. Washington, DC: U.S. Government Printing Office.

———. 1993. *The National Competitiveness Act of 1993: Hearing before the Subcommittee on Technology, Environment and Aviation.* 103rd Congress,

1st session, 2 March, no. 13, vol. 3. Washington, DC: U.S. Government Printing Office.

U.S. Congress, House of Representatives, Committee on Science, Space and Technology, Subcommittee on Technology and Competitiveness. 1992. *U.S. industrial competitiveness.* 102nd Congress, 2nd session, serial T, September. Washington, DC: U.S. Government Printing Office.

U.S. Congress, Office of Technology Assessment. 1991. *Biotechnology in a global economy.* OTA-BA-494. Washington, DC: U.S. Government Printing Office.

U.S. Department of Defense. 1994. *Advanced projects research agency.* Washington, DC: U.S. Government Printing Office.

U.S. Department of Defense, Office of Naval Research. 1994. *Department of Defense multidisciplinary research program of the university research initiative, fiscal year 1994,* ONR 3594.8. Arlington, VA: Office of Naval Research.

Useem, M. 1984. *The inner circle: Large corporations and the rise of business political activity in the U.S. and U.K.* New York: Oxford University Press.

Wolfle, D. L. 1972. *The home of science: The role of the university.* New York: McGraw-Hill.

Wulf, H. 1993. Arms industry limited: The turning-point in the 1990s. In *Arms industry limited,* edited by H. Wulf, 3–26. Oxford, England: Stockholm International Peace Research Institute and Oxford University Press.

2

Recent Science

Late-Modern and Postmodern

Paul Forman

Introduction

We are in the midst of a radical departure in cultural discourse, an ever more widespread disbelief in, indeed rejection of, that ever-receding "horizon" of true knowledge, true art, true utopia which legitimated most of our cultural and social practice over the past two or three centuries. Instances of this abandonment of the modernist ambition of uniqueness, uniformity, unification, universality confront us everywhere. In every sphere of our social and cultural existences we are obliged to find our way without a vanguard, without any consensus as to which is the "forward" direction. The demise of socialism as political and moral ideal became patent in the past ten years, though in retrospect we recognize the turn to have occurred nearly three decades ago. So also can we now recognize the *weltanschaulich* affinity of this turn with the "anything goes" aesthetic that has overwhelmed the supercilious moralism that was modernism in painting and architecture, and likewise its affinity with the epistemic relativism and latitudinarianism that have in these same three decades overturned the modern philosophic enterprise.

Why then, with such evidence at every hand, do so many knowledgeable observers—and among them many of the most critical—remain skeptical of the reality and durability of postmodernity?[1] Perhaps be-

Reprinted with permission of Harwood Academic Publishers. From Thomas Söderqvist, ed., *The Historiography of Contemporary Science and Technology* (Amsterdam: Harwood Academic Publishers, 1997), 179–213.

1. Such a one was Ernest Gellner (1995), who saw only deplorable cultural manifestations, and clung to the notion of cultural and ethical truth—Western European civilization being the closest approximation thereto—grounded in the indubitable fact of scientific truth. Similarly those banded together in the U.S. in the National Association of Scholars (Heller 1994, Browne 1995).

cause they see what Zygmunt Bauman points out so forcefully: that this is all intellectual "superstructure." (I cling momentarily to this terminological spar from the Marxist wrack.) Beneath it, at the base, in the substructure of social and economic institutions, there is no radical reorientation—indeed, little change at all in direction. Excepting but few and still marginal practices such as recycling of waste, we operate in our daily lives under conditions of aggravated modernity. "Modern practice continues," Bauman writes, "now, however, devoid of the objective that once triggered it off."[2]

> What the inherently polysemic and controversial idea of postmodernity most often refers to (even if only tacitly) is first and foremost an acceptance of ineradicable plurality of the world. . . . By the same token, postmodernity means a resolute emancipation from the characteristically modern urge to overcome ambivalence and promote the monosemic clarity of the sameness. . . . Postmodernity is modernity that has admitted the non-feasibility of its original project. Postmodernity is modernity reconciled to its own impossibility—and determined, for better or worse, to live with it.[3]

Bauman speaks of "postmodernity," but avoids the term "postmodernism." "Postmodernity" is a relatively neutral term, for it is "merely" a label for a constellation of putative actualities defining a historical reality. "Postmodernism" is that too, but qua "-ism" postmodernism is also an ideology—an ideology that, Bauman points out, has arisen under conditions of postmodernity.[4] As ideology, postmodernism claims to transcend its own historical conditions, creating new standards by which to judge all preceding periods and orientations. Here the historian, like the sociologist, is obliged to maintain a degree of reserve, to take hold of postmodernism while yet holding it at arm's length, to construe postmodernism as a functional ideology responsive to the conditions of postmodernity.

Bauman sees postmodernism as an ideology of—heretofore mandarin—intellectuals, an ideology that reflects our increasing impotence as political-cultural "legislators," increasingly unnecessary and irrelevant both to those who hold power and to those who produce our mass cul-

2. Bauman 1991, 98, 272.
3. Bauman 1991, 98.
4. Bauman 1987 and 1992a.

ture. In Anthony Giddens' powerful image, we intellectuals can no longer imagine ourselves guiding—but are now simply *riding*—a juggernaut.[5] Thus, Bauman points out,

> the contemporary reorientation of cultural discourse can be best understood as a reflection on the changing experience of intellectuals, as they seek to reestablish their social function on a new ground in a world ill-fit for their traditional role.[6]

This interpretation is surely right, yet it remains too largely in the realm of the intellectuals' vaunting ambitions, refers too little to the world of practice—practice of the intellectuals qua *Kulturträger* and *Kulturerzeuger.* I hope therefore to make a small contribution to the sociology of postmodernism, by relating this ideology of contemporary intellectuals to three leading characteristics of late-modern knowledge production, characteristics that are, in aggravated form, carried over into postmodernity: first, the ever accelerating overproduction of all cultural goods, including knowledge as perhaps the fastest-growing; second, the proliferation of an instrumentally defined and oriented science, inseparable and increasingly indistinguishable from its technological correlatives (literally, technoscience); and third, the fact, and more particularly the increasingly general acceptance of the fact, that all knowledge production, and most especially technoscience, is bound to particular interests.[7]

Of course, the fact of interested and instrumental knowledge is no new thing. New is the full and complete legitimacy of such knowledge, its parity with those "disinterested" and "purely conceptual" forms of knowledge that previously were complimented by the epithet "true." This multiplication of knowledges—oxymoronic for the modern,[8] and premodern—is the difference that makes all the difference, that sets

5. Giddins 1990.

6. Bauman 1992a, 24.

7. Only belatedly, through John Ziman's (1996) highly pertinent paper, have I become aware of the inquiry of Michael Gibbons et al. (1994) into the nature and consequences of "postacademic science." Their characterization of this mode of knowledge production centers on the concept "transdisciplinarity," and includes, inter alia, the three key characteristics stressed in the present chapter. They fail, however, to see the interdependence of these characteristics, and they fail—perhaps through politesse—to acknowledge that "postacademic science" is not merely supplementing, but is displacing and transforming the modernist mode of knowledge production they call "academic science."

8. [Note added 2001: See my review of P. R. Gross, N. Levitt, and M. W. Lewis, eds., *The Flight from Science and Reason,* in *Science* 276:750–53 (2 May 1997).]

postmodernity off as a distinctive era. Such a transvaluation of knowledge becomes intelligible only when the production of knowledge is viewed against a horizon devoided of all absolutes and destituted of all destinies. Only when we have dispensed with such authentications can interested and instrumental knowledge come to be regarded as fully legitimate—and only now, because regarded as fully legitimate, can its extent be fully acknowledged.[9]

The bulk of this essay is, then, an exploration, against this "disfigured" horizon, of interconnections between our contemporary knowledge production as interested and instrumental and the emergence, on the one hand, of an ends-justify-the-means instrumental rationality in all areas of social and cultural practice, and, on the other hand, of a demand, remarkably even also among scientists themselves, that moral considerations, and more particularly "responsibility," play a central role in science. And, by the way, the prominence in recent decades of the allegation of "incommensurability" of different descriptive or explanatory frameworks—languages, discourses, theories, "paradigms"—finds a natural explanation.

Overproduction of Knowledge as *Kulturträgerkrise*

Modernity as intellectual orientation is conceptually coherent with, and historically inseparable from, modernity as an economic orientation. As Didier Nordon has observed,

> society assumes that all material progress is good; the pure scientists assume that all progress in knowledge is good. Material progress is an accumulation of objects; progress in knowledge is an accumulation of objects of knowledge. In both cases, the actors produce in order to produce.[10]

9. [Note added 2001: This and the following paragraph are added in this republication.]

10. Nordon, *Les mathématiques pure n'existent pas!* (1981), quoted by Restivo 1990, 128–129. Cf. Kline 1980, 334, and Mehrtens 1994, 330, 339, who observes that "the leading norm [of science] is not simply knowledge, but discipline-specific productivity," and that "[t]he self-definition and memory of mathematics force each mathematician to take his *proper* business more seriously than any other, thereby being optimally productive." These issues are a continual theme of Mehrtens' (1990) differentiation between the modern and counter-modern in turn of the twentieth-century mathematics: "Mir scheint, daß der Primat der Produktion so recht erst mit der Moderne zum Ausdruck kommt" (1990, 423). Likewise technology as an end in itself is cognate and correlate in modernity with production as an end in itself. Bauman, following Jacques Ellul, points out that the capacity to do is not created in order to achieve predefined ends so much as the ends are determined by the capacity to do: "have car, will travel" (Bauman 1993, 187–90).

What is described here is the characteristically modernist performatist compulsion, in which the activity—as measured by its output—itself becomes the purpose of the activity.

From the point of view of a consumer society—i.e., a society composed of *passive* consumers of goods and services—there is nothing to complain of in this state of affairs. Overproduction under the conditions of our "postindustrial" economy leads no longer to a glut of a fixed menu of "goods," but rather to an endless augmentation of choices available, and an unceasing stimulation of the consumer (through advertising, mass media, etc.) to extend the range of his/her desires to cover those augmented choices.

If, however, we regard "society" as comprising not passive consumers but discriminating actors, the situation appears essentially different. Then, as Herbert Simon recognized on the threshold of our "information age":

> In a world where information is relatively scarce . . . information is almost always a positive good. In a world where attention is a major scarce resource, information may . . . turn our attention from what is important to what is unimportant. We cannot afford to attend to information simply because it is there. I am not aware that there has been any systematic development of a theory of information and communication that treats attention rather than information as the scarce resource.[11]

Though presumably unacceptable as such to Simon, postmodernism is that theory. For if we give *our* attention to those special subsocieties of active cultivators of "higher" cultural goods—i.e., those engaged, as *Kulturträger* and *Kulturerzeuger,* in the perpetuation and augmentation of a "canon" of whatever sort, be it artistic, literary, scientific, or technical—then, as Randall Collins narrates:

> What we see around ourselves in recent decades has been an enormous expansion of cultural production. There are over one million publications annually in the natural sciences, over 100,000 in the social sciences and comparable amounts in the humanities. To perceive the world as a text is not too inaccurate a description, perhaps not of the world itself, but of the life position of intellectuals: we are almost literally buried in papers. As the raw size of intellectual production goes up, the reward to the average individual goes down, at least in the realm of pure intellectual rewards of being recognized for one's ideas and of seeing their

11. Simon 1978, 13.

> impact upon others. The pessimism and selfdoubt of the intellectual community under these circumstances is not surprising.[12]

For Collins, "the content of the 'postmodern' message is to deny objective truth," and this he sees "as an ideology of cultural producers in a highly pyramided market structure, where nothing in sight seems to touch solid earth." Or, as Frank Ankersmit said, varying the metaphor slightly in motivating his essay on "Historiography and postmodernism": "My point of departure in this article is the present-day overproduction in our discipline . . . the situation in which historiography itself impedes our view of the past."[13] But metaphors apart, consider the obvious *practical* consequences of this overwhelming overproduction of higher cultural "goods": in a situation in which the producers of culture are also the bearers of that cultural tradition, an obligation falls upon those producers to critically consider, and discriminatingly assimilate, the cultural "goods on offer." Under these expectations, overproduction produces a crisis of impotence.

It is, then, not that the *Kulturträger,* overwhelmed by this flood of paper, are metaphorically unable "to touch solid earth," or to get a "view of the past," but rather that they, in the face of this insuperable "literature," must still somehow fashion a practicable mode of cultural production and reproduction. If then we understand postmodernism not primarily as a denial of "objective truth," but as a denial of a unique and universal canon, as primarily an affirmation of pluralism and of a plurality of standards of value, then, yes, we can very well understand postmodernism as an ideological response to the condition of postmodernity, a response that makes a virtue of that multiplication and dispersion of the producers of knowledge and products of their activity—of that superabundance which would otherwise, under modernist axioms, have to be experienced just as desperately as Collins depicts it. And if that denial of the canon is sufficiently radical, then "in a postmodern culture the writer"—I here quote a writer of a book on *Composition as a Human Science*—"needs to read what she needs in order to think, to make sense, not in order to know what is fashionable. There can be no canon of theory, any more than of literature."[14]

I quote, but I do not fail to note that there are serious difficulties

12. Collins 1992, 92–94. Collins refers his numbers to Price 1986. Cf. Gibbons et al. 1994, 94, 103.

13. Ankersmit 1989, 137.

14. As quoted by Schilb 1991, 182.

with this stance, difficulties already in relation to literature, but still more serious difficulties in relation to "the literature" of any scholarly or scientific discipline. Yet if we turn our attention to actual practice of scholarship and research in the postmodern present, we find that the modernist axioms which we historians of science have implicitly sought to fulfill in accounting for science as the "doing" of well-disciplined disciplines have been put aside to a remarkable extent. Lawrence Dowler, associate librarian at Harvard College, starting likewise from the insuperable tide of monographic publications—he has 850,000 volumes published annually, and increasing at 2.5 percent per annum—concludes from his study of its use that "[T]he particular inquiry, rather than the academic discipline, is now the primary engine of research."[15]

Late-Modern/Postmodern Knowledge as Instrumental

More and more "the particular inquiry" is an "instrumental" inquiry. It is instrumental in two senses, linking the practical and the ideological, that seem especially important to postmodernity: a scientific practice dominated by instruments, and a scientific attitude of instrumental rationality. The two together, are, in effect, the theme of Bruno Latour's *Laboratory Life* and *Science in Action: How to Follow Scientists and Engineers through Society.* Latour's account of instruments and inscriptions finds impressive confirmation in the impressions gained in America by a visiting Russian biologist, Leonid Margolis:

> Young scientists start to think that science consists of putting the results produced by one machine into another, and then into the next one, and of arranging thus obtained beautiful pictures and graphs into a publication.[16]

If the young scientist comes to think "technique is exciting,"[17] it is because her seniors as well as her experience tell her that "science and engineering are becoming more alike, in large measure because of instrumentation," quoting Lewis Branscomb, sometime head of the National

15. Dowler 1993. It may be objected that thirty years ago Derek J. de S. Price (1986, 56–57, 63–66, 74–76) had already recognized this problem of overproduction of scientific literature as an aspect of the exponential growth of science through three centuries of modernity, and had, moreover, "solved" this problem by discovering science's outgoing, continual process of growth through fragmentation into "invisible colleges." But even as we acknowledge, once again, Price's remarkable insight and prescience, we should recognize that Price as modernist refused to feature the inevitable disciplinary disunity.

16. Margolis 1992.

17. Gornick 1983, 27–28.

Bureau of Standards, sometime head of IBM's research, here a candidate for the presidency of the American Association for the Advancement of Science.[18]

Twenty years earlier, in the early 1960s, Derek Price, deploying his ingenious bibliometric analysis, had sought to differentiate science, as oriented toward "the literature," from technology, as papyrophobic. But by the early 1980s this demarcation criterion was failing. Extending Price's analysis to patents, Francis Narin suggested that:

> [t]here is more going on today than just a diminution of the time lag between scientific discovery and its use in technology: we suggest that high-tech patents and the related scientific papers are so tightly linked as to be almost indistinguishable, and that this is concrete evidence that high-technology and science itself are almost indistinguishable.[19]

And in reviewing recently the further studies on this question, Narin found that the average number of references in a U.S. patent to the scientific literature "has been increasing steadily from 1975."[20]

There is, of course, a technologizing of science through the deployment of instrumentation that has been essential to its operations and to its successes since the early 17th century. Indeed, Martin Harwit has argued that nearly all astronomical discoveries since the Renaissance are direct results of the introduction of new technical means of observation.[21] It is not necessary, however, to contest this secular feature of modern science in order to contend that late modernity saw a great intensification of this technologizing of the sciences, and that the enormous expansion of science since World War II has come about overwhelmingly through the creation of technoscientific tools.[22] Scientific specialties of astounding practical competence, and even entire scientific fields, have been brought into existence by an instrument or a technology. Indeed, as David DeVorkin has convincingly shown, the rocket was less the tool than the creator of "space science,"[23] which science has never been defined

18. Lewis Branscomb, presidential candidate's statement enclosed with AAAS election ballot, 1985.

19. Narin and Noma 1985, 371.

20. Narin and Olivastro 1992, 243; Narin 1994.

21. Harwit 1981.

22. General arguments to this effect have been made by inter alia Forman 1987, Galison and Hevly 1992, Pestre 1992, Pickering 1989 and 1995a, and Schweber 1989. Studies entering into the details of this technologizing of late-modern science have been made by inter alia Baird 1993, Forman 1995a and 1996, Kay 1988, and Stine 1992.

23. DeVorkin 1992 and 1996.

conceptually, but only instrumentally, namely—in the words of one of its ablest promoters, Homer Newell—as "those scientific investigations made possible or significantly aided by rockets, satellites, and space probes."[24]

In the course of these postwar decades quantitative change became qualitative transformation: the technical means ceased to be merely means and became ends as well.[25] Already in the early 1950s Otto Loewi, a physiologist of the older generation, could deplore the "general tendency of our time to worship methods and gadgets." And, Loewi continued, "This has gone so far that sometimes one has the impression that in contrast with former times, when one searched for methods in order to solve a problem, frequently nowadays workers look for problems with which they can exploit some special technique."[26]

More recently sociologist David Edge has made this "drive toward excellence by purely *technical* criteria" the burden of his study of instrumental innovation in infrared astronomy. One of his subjects began by avowing that "my own particular interests are in star formation," but came eventually to allow that:

> I've found . . . that we always get good reviews when we do something interesting technologically. If we decide to just do science for a couple of years without developing new instrumentation our reviews start getting worse. And I think they just like to see new technology development.[27]

The vastness and potency of technologically oriented science—and scientifically oriented technology—are of course well known. What is not so well appreciated, however, is how differently this fact is now coming to be appraised in—and in consequence of—postmodernity. It used to be that pure science, and abstract theory, had an unquestioned place of honor within the scientific estate. An applied physicist of such exceptional capacity and originality as W. W. Hansen, highly esteemed as the

24. Quoted by DeVorkin 1996, 253.

25. This has been feasible only because the sciences are not bound as are the humanities to an ideology that requires each and every *Kulturträger* to carry—not the whole, but—a significant part of the whole cultural goods of the humanistic discipline. Unable to renounce this featured self-conception—because having nothing like the instrumental role of the sciences to substitute for it—the humanists have been forced to redefine "a significant part" in such a way as to cut the whole down to a manageable size. Hence postmodernism.

26. Quoted by Blume 1992, 90.

27. Edge 1992, 130, 154.

creator of concepts essential to the production and technical exploitation of microwaves—who died in his prime shortly after World War II—was by no means exceptional in that "pure and actually rather abstract research was held by him in almost exaggerated admiration; with typical modesty and humor he sometimes referred to his 'platonic love for pure research.' "[28]

If the American public had not shared this relative ranking of theoretical and practical science in the decades before the Second World War, they certainly did in those following. As President Kennedy said in 1963 when addressing the U.S. National Academy of Sciences at its centennial celebration:

> If I were to name a single thing which points up the difference this century had made in the American attitude toward science, it would certainly be the wholehearted understanding today of the importance of pure science. We realize now that progress in technology depends on progress in theory; that the most abstract investigations can lead to the most concrete results.[29]

Such, of course is what the Academy members wanted to believe and to hear. But the point is that the first president who was prepared to tell them that was also the last president who would be willing to tell them that: just at this moment the tide was beginning to turn.

The engineers rebelled and forced the establishment of a National Academy of Engineering. The proliferation of technoscientific microcultures, each owing its existence more to common tools than to common concepts (to say nothing of a common world view), gradually undermined the veneration enjoyed by the pure, general, and abstract theorists within the scientific estate. Just as Bauman's "cultural legislators," the rough equivalent in the humanistic fields of science's abstract theorists, have become more and more irrelevant to the process of knowledge production, so now an analogous displacement is occurring also in the natural sciences. The postmodern mode of knowledge production—through its instrumentalism, and the priority thus assigned to specific procedures for coping with the complexity of the real world, rather than transcending that complexity through abstraction—has pushed the abstract theorists more and more out of the *scientific* limelight. However much the general public, still addicted to transcendence, continues to

28. Bloch 1952, 125.
29. Greenberg 1967, 254.

wonder at those who mentally "touch God,"[30] the postmodern transvaluation of values (including the devaluation of disciplines) has now largely deprived those soaring theorists of the role of disciplinary culture hero.

That said, it is important to emphasize that cultural logic is never cultural inevitability, that the abstract theorists could very well have retained their standing had that been the wish of their scientific fellows—as it has remained, largely, that of the public at large. Thus here again we can identify the late-modern/postmodern transition with a cultural "decision," taken inside as well as outside the technoscientific world, to cease venerating the pure and abstract.[31] An aggressive expression of the new—postmodern—stance comes, appropriately, from Nicolas Metropolis, a mainstay for four decades of the Los Alamos weapons-science culture:

> [T]he fact is that quite some time ago the tables were turned between theory and applications in the physical sciences. Since World War II the discoveries that have changed the world were not made so much in lofty halls of theoretical physics as in the less-noticed labs of engineering and experimental physics. The roles of pure and applied science have been reversed; they are no longer what they were in the golden age of physics, in the age of Einstein, Schroedinger, Fermi. . . .[32]

Metropolis goes on to place in wartime Los Alamos the mythic act of recognition of the superiority of instruments over thought. By Metropolis' account, this seminal insight originated with Richard Feynman, a brilliant younger member of Hans Bethe's Theory Division:

> Feynman soon came to realize that reliable [mechanical desk-top calculating] machines in perfect working order were far more useful than much of what passed for theoretical work in physics. . . . We spent hours fixing the small wheels until they were in perfect order. Bethe, visibly concerned when he learned that we had taken time off from our physics research to do these repairs, finally saw that having the desk calculators in good working order was as essential to the Manhattan Project as the fundamental physics.[33]

30. Forman 1997.

31. Revealing expositions of the factors undermining the traditional hierarchy of subfields of physics: Schweber 1993b, and Cao and Schweber 1993.

32. Metropolis 1992, 120.

33. Metropolis 1992, 126–127.

Needless to say, the advent of postmodernity was required to liberate from Los Alamos this bit of parochial culture-constituting mythology and give it global significance.

With Metropolis we are, in tone at least, already encountering the instrumental as a comprehensive ideology—an ideology that includes instrumental rationality. In postmodern science, as in war (we now think differently about love), all is fair: "the ends justify the means." Anathematized by liberals in modernity, this maxim is the implicit credo of postmodernity. Nowhere was the modernist rejection of this maxim clearer than in the conventional view of scientific knowledge, continually reaffirmed from the middle of the seventeenth to the middle of the twentieth century, namely that all the distinctive and desirable qualities of scientific knowledge were already immanent in the method by which it was attained, i.e., that the means sanctified the end, rather than the end justifying the means.

The switch in attitude, to an ends-justify-the-means instrumental rationality, is not without precedents, particularly among physicists, over the past hundred years.[34] But it is only recently that the conventional view has been widely and generally abandoned across the spectrum of scientific fields. These altered attitudes have an important bearing upon the much mooted question of "scientific misconduct," the discussions of which have so often missed the mark through assumption of the conventional view that the validity of the knowledge produced follows from and depends upon the validity of the methods employed. But from a rigorously instrumentalist point of view the phrase "scientific misconduct" can have no definite meaning—or rather the only definite meaning of which it is susceptible is "lack of success." Thus in the Imanishi-Kari/Baltimore case there was a clear tendency to exonerate the researchers if the published results—even if invented—prove reproducible.[35] Investigating this "murky borderland," Ullica Segerstraale found that physicists, especially, espouse such an ends-justify-the-means attitude toward scientific practice, and noted its chilling effect on whistle-blowing.[36] All this is indeed postmodern, but not specifically postmodern. Rather, it is the late-modern ethos of production as an end-in-itself, now aggra-

34. Heilbron 1982. Schweber (1986) cites American theoretical physicist E. U. Condon as stating, in 1937, "All is fair in love and war and, I might add, in Theoretical Physics," and as referring to John Dewey in justification of his instrumentalism.

35. Travis 1993, Stone 1995, Kevles 1996.

36. Segerstraale 1990, 15.

vated by lifting the restraint of the modernist internal morality of means.[37]

"Foundations" of "Incommensurability"

The ends-justify-the-means instrumentalism that has come to dominate the cultural practice of our late-modern/postmodern knowledge society has also had substantial impact upon our theorizing about the nature of scientific knowledge. The most commonly encountered periodization of the philosophy of science places in the early 1960s the beginnings of a turn away from the logical positivist view of science as a unitary, hierarchic, conceptual structure that grows by comprehending within itself an ever wider range of natural phenomena and/or adding additional layers to its hierarchy.

This turn away from a disinterested, unitary, positivism in favor of a view of scientific knowledge as discovered/constructed in the pursuit of a *weltanschaulich* preconception, or in the elaboration of a preconceived program of knowledge-production, has been ascribed to Thomas Kuhn's *Structure of Scientific Revolutions* more often and more widely than to any other work. And in the philosophic literature concerned with Kuhn's theory of science, one concept, one issue, dominates discussion, viz., the putative "incommensurability" of alternative, competitive, successive scientific theories. Kuhn himself said of "incommensurability" in his presidential address to the Philosophy of Science Association,[38] "No other aspect of *Structure* has concerned me so deeply in the thirty years since the book was written."[39] While Kuhn, unlike most philosophers

37. Forman 1991. [Note added 2001: Bauman (1992b, 5) offers a pregnant characterization of the "world of instrumental rationality: the world where deeds are lived as a means to ends and justify themselves by the ends which they serve as means." Bauman identifies this as "the modern world"; I argue that—notwithstanding such stark modern aberrations as the gulag, the concentration camp, and "strategic" bombing—Bauman's characterization applies not to the modern but to the postmodern world.]

38. Kuhn 1991, 3.

39. When, in *the* book on Kuhn's book, Hoyningen-Huene (1993, 207) comes to "Incommensurability," he provides some 140 references to the literature, far, far more than he gives at any other point or for any other Kuhnian concept. And Kuhn himself, in the foreword he contributed to H-H's book, added that "[s]ince the publication of *Structure* my most persistent philosophical preoccupation has been the underpinnings of incommensurability." Biagioli (1990; 1993, chap. 4) has proposed an "anthropological"—i.e., transhistorical, social/logical—account of the indispensability of incommensurability (hence, by the same rationalist reasoning that Kuhn employs, the reality of same) "derived from an homology I perceive between Kuhn's concept of paradigm and Darwin's notion of species. . . . According to this Darwinian metaphor, incommensurability would be *necessar-*

engaged in this discussion, did appreciate that the *Zeitgeist* of those three decades had a significant role in this reorientation of the philosophy of science, Kuhn too failed to see how very largely the logical-metaphysical problem of "incommensurability" is really just a vain reflection in philosophy of the ends-justify-the-means instrumentalism of contemporary cultural practice.[40]

The "Binding" of Knowledge[41]

"Possession of property is exclusive; possession of knowledge is not exclusive: for the knowledge which one man has may also be the possession of another." This claim of nonexclusivity of knowledge, which, like the loaves and fishes, is undiminished through being shared, was already long familiar in 1886 when John Wesley Powell presented it to the U.S. Congress.[42] In the following century, however, circumstances of knowledge production and distribution inconsistent with this claim, incipient at that time, continued to multiply—until in the past decade this claim has come to seem quaint and naive in the extreme.

In affirming a conception of knowledge as bound and interested, postmodernity stands in striking opposition to both the classical and the modern conception of science as a liberal pursuit, i.e., the free activity of unfettered minds, the results of which, freely published, will conduce to the freedom of all mankind.[43] Indeed, concerned as we are with notions of knowledge, this is the most distinctive criterion demarcating postmodern from modern science: postmodernity begins where the production of bound and interested knowledge is unequivocally accepted.

---

ily related to the conceptual speciation of a new paradigm." Kuhn (1991, 7) found Biagioli's article "splendid," and continued to say that "[w]ith much reluctance I have increasingly come to feel that this process of specialization, with its consequent limitation on communication and community, is inescapable, a consequence of first principles." The "reluctance" is inseparable from the need for "first principles."

40. "But others were present too," wrote Kuhn (1991, 3), "Paul Feyerabend, and Russ Hansen, in particular, as well as Mary Hesse, Michael Polanyi, Stephen Toulmin, and a few more besides. Whatever a *Zeitgeist* is, we provided a striking illustration of its role in intellectual affairs." What that *Zeitgeist* might be is suggested by Steve Fuller (1992): "Being there with Thomas Kuhn: a parable for postmodern times."

41. I have borrowed the term "binding" from John A. Remington (1988), whose perspicacity regarding the transformation under way in the production of scientific knowledge I have only recently come to recognize.

42. Dupree 1994; 1986, 227.

43. Steven Shapin (1991 and 1994) has emphasized the extremely close connection between the early-modern English conception of the gentleman as personally free—unbeholden—and the creation there in the seventeenth century of the notion of a trustworthy scientific report.

Proceeding in parallel since the late nineteenth century were two, only seemingly distinct, aspects of this secular change. Firstly the binding of scientific knowledge *production* to capital-possessing, self-perpetuating institutions—universities, primarily—and the binding of the *produced* knowledge to other capital-possessing, self-perpetuating institutions: government agencies, industrial firms, and more recently a wider range of corporate entities. With some few exceptions, the former process—the binding of knowledge production to institutional bases—was effectively complete quite early in this century, and this circumstance was already emphasized by Max Weber in *Wissenschaft als Beruf.*[44] Yet a view of knowledge as production—and consequently as importantly dependent upon its material and institutional supports—was at variance with the reigning conception of science as a liberal pursuit, pertaining only to the man and not at all to his circumstances, and consequently was strongly resisted.[45] Thus many scientific journals continued to denote their contributors' institutional affiliations in disjoint or inspecific ways well into the middle of the twentieth century.[46]

Those journal editors and contributors alike would have insisted that there is, in any case, a great difference between an individual knowledge producer being dependent upon an institutional base for the performance of research—an undisputed fact ignored so far as possible—and the "binding" of the produced knowledge to a particular institution or interest—a circumstance deplored wherever it could not be disputed. But in those decades before and after World War II administrators of academic research had a better apprehension of the connections between these two forms of "binding."

Through the four decades of the Cold War the binding of knowledge to a national interest and its restriction to a national territory became the leading characteristic of a very large sector—depending on definition, even the largest sector—of the R&D industry in the United States,[47]

44. Weber 1919.

45. This fiction that knowledge production is independent of its institutional locus was a specifically modern perspective, to be distinguished from postmodernity on the one side and the Renaissance on the other side. In both the Renaissance (Forman 1973, 155; Biagioli 1993) and in postmodernity the sponsorship of the produced knowledge is one of its principal defining characteristics.

46. Gillmor (1986) has published data showing a saltation just after World War II in the fraction of papers in ionospheric research bearing indications of the authors' institutional affiliations, a jump from just a few percent to some 90 percent. However, this dramatic rise, which supports so well the point here made, is unfortunately largely an artifact, due to inconsistencies in the source from which the data were drawn.

47. Forman 1973 and Leslie 1992.

and a fortiori in the USSR. Although those nations' scientists—especially their exact and physical scientists—were deeply involved in classified research and embargoed technology, these features of knowledge production, lacking legitimacy in the scientists' modernist ideology, were bracketed out, compartmentalized, denied any acknowledged place in the scientists' picture of how science is done, in which picture the ideal was largely made to stand in for the real.[48]

Recently—very recently—this has begun to change as the modernist self-conception of a searcher after free and disinterested knowledge is exchanged for a postmodern acceptance of the legitimacy of proprietary, interested knowledges. "At issue," *now,* "is how academic science, primarily state funded, participates in a political economy that celebrates the market." Moreover, Sheila Slaughter continues, in the United States:

> As the presidents of elite research institutions moved segments of their universities closer to the market, they became effectively indistinguishable from chief executive officers of large corporations. They were concerned with maximizing the returns of any resources they could get for those segments of their institutions that were tied to the private sector.[49]

In this they have not only been oriented by the general shift in the political culture of the nation, but more specifically authorized and directed by new federal and state legislation and by new policies framed by their boards of trustees. Taking the University of Arizona as a case study, Slaughter and Gary Rhoades found that between 1969 and 1989

> [p]olicies and statutes moved from an ideology that defined the public interest as best served by shielding public entities from involvement in the market, to one that saw the public interest as best served by public organizations' involvement in commercial activities. Claims to the ownership and rewards of intellectual property shifted dramatically in that time, from faculty owning their products and time to complete ownership by the institution.[50]

This reorientation toward the market—which is, as such, a repudiation of the notion of "free" knowledge—together with the increasing orientation toward the particular problem, works powerfully to dissolve the scientist's attachment to his discipline, indeed to dissolve the disci-

48. Forman 1991, note 39; Forman 1996.

49. Slaughter 1993, 279, 296. Good work on this important question has been published by Eckert and Osietzki 1989, Krimsky, Ennis, and Weissman 1991, and, especially, Wright 1994.

50. Slaughter and Rhoades 1993, 287.

plines themselves and their disciplinary authority.[51] This is most clear perhaps in the growing acceptance of, even institutionalization of, secrecy.

In late modernity a sharp distinction was maintained between secret "classified" research (performed principally in the interest of "national security") and "unclassified" ("basic," "pure," and hence "free") research. A principal purpose of this distinction was to enable the scientists (and philosophers of science) to treat as exceptional all that transpired in that vast realm of classified research, while maintaining the authority of the scientific disciplines in the sphere of unclassified research. In postmodernity not only does this distinction between spheres fail, but so also does the effort to maintain it. Chu, for example, in submitting his first papers on high-temperature superconductivity, falsified crucial data in order that no one else be able to begin their research at the most advanced point which he had attained.[52] In X-ray crystallography, meanwhile, it has become accepted and formalized that the atomic coordinates will not be made available to other researchers even upon publication of the alleged crystal structure, but that the "discoverer" is *entitled* to hold them secret for one year in the U.S. and longer in Europe.[53]

In sum, the long-term historical process of binding of knowledge production to the interests of powerful institutions, which under the conditions of modernity operated under significant ideological constraints, has now, under those of postmodernity, advanced unhindered and extended in every possible direction.

Power and Policy Create a Convenient Reality

The binding of knowledge *production* to particular institutions with their own particular interests—formerly primarily institutions with a "military" purpose, but now increasingly market-oriented institutions—carries with it an implication, indeed an expectation, that the *produced* knowledge will "shape reality" in ways convenient to those powerful sponsoring institutions. That shaping of reality may be through technical devices extending their physical power and/or commercial returns, or it may "merely" be a definition of what the world contains and what is to be judged good about the world—advertising, in the widest

51. Gibbons et al. 1994, Ziman 1996.
52. Felt and Nowotny 1992.
53. Cohen 1995.

sense—which definition advances the sponsoring institution's perceived interests. " 'In the system I work in' "—Robert Bellah and collaborators are quoting a welfare supervisor—" 'our motto could be, "If you don't have to report it, it didn't happen." Appearances and regulations are all that count!' "[54] What is especially to be emphasized is that as "truth" recedes below the postmodern conceptual horizon, while instrumentalist programs of knowledge production spring up at every hand, the scientific disciplines are losing their authority as arbiters not merely of scientific conduct, but also of scientific knowledge, of the "shapes of reality" that are possible and impossible.

We have in "cold fusion" a striking example of this circumstance. Certainly the declining authority of the scientific disciplines is reflected in the "discoverers" covering themselves with a cloak of secrecy—"justified" by their intent to seek patents—and in their direct appeals to state and national legislatures for "authorization" of their research. But the most significant loss of authority of the scientific disciplines lies, rather, in their incapability of quashing this heresy. However much the physics discipline believes cold fusion to have been disproved, however much it seeks to ostracize and ridicule cold fusion research and researchers, it is simply unable to eradicate this research program. Science journalist Gary Taubes, placing himself emphatically in the side of "good science," wrote a thick book in which all who saw something in their experiments on cold fusion are treated as fools.[55] But at its end he has to admit that "[w]hat cold fusion had proven, nonetheless, was that the nonexistence of a phenomenon is by no means a fatal impediment to continued research. As long as financial support could be found, the research would continue."[56]

Here we enter, of course, on the most sensitive territory of postmodern doctrine, namely the linkage of power and truth. More particularly, at issue here is the reversal of the modernist reading of Bacon's "knowledge is power"—taking "is" as an implication, rather than an equivalence—to a postmodernist "power is knowledge," i.e., power includes the capability to create knowledge "in its own image."[57]

54. Bellah 1986, 126.

55. Taubes 1993.

56. Taubes 1993, 426.

57. This contention is most closely associated with Foucault, and it was taken by Lyotard as central in *The Postmodern Condition* (1984, 44–47). Thus Fekete (1987, 70) offers three alternative characterizations of "the nature of contemporary ('postmodern') culture," of which "[t]he first is the characterization of our times that follows from the work of Michel Foucault: an age dominated by the union of knowledge and power." Lyotard,

That circumstance to which the postmodern reading of "knowledge is power" points is by no means new; indeed it is far older than modernity itself, for the employment of the available knowledge producing capabilities to create a convenient reality is *in effect* the practice of all premodern, traditional societies and nearly all organized religions. In this, consequently, the critics of postmodernity who deplore it as a throwback to premodernity have a point.[58] Indeed, they could very well point to Galileo's inquisitors and Dostoyevsky's Grand Inquisitor. They are, however, more wrong than right, for in postmodernity this process of creating a convenient reality goes forward without that unambivalent confidence in the correctness of one's own beliefs, that conviction of the absolute validity of the ends which are to justify the means, that was so characteristic of the premodern, and recurred as aberration in modernity.

"Incommensurability" of Competing Programs for Knowledge Production

In *The Truth about Postmodernism,* Christopher Norris refers, impatiently, to "different (incommensurable) language-games," making it plain that such talk as Jean-François Lyotard's in *The Postmodern Condition* is commonplace in the postmodern parlance that Norris finds alarmingly widespread and imprudent.[59] Lyotard is indebted to Wittgenstein, of course. And a further source of the notion of incommensurability is Paul Feyerabend's writings in the sixties and seventies, where the concept appears in conjunction with an attack on modernist methodologic strictures and an advocacy of an "anything goes" instrumentalism that is remarkably far ahead of its still not quite postmodern time.[60] Nonetheless, the quantity of formal philosophical publication on the question whether such a thing as "incommensurablility" is possible or impossible, hence existent or nonexistent, points unquestionably to Kuhn's *Structure of Scientific Revolutions.*

My concern is not, of course, with the logical or ontological possibility or impossibility of "incommensurability," but with the popularity that

---

whose historical ignorance is abysmal, and Foucault, whom I will not presume to judge, both take this postmodern reading as the one transhistorical rule of Western society since the 17th century, at least. Thus the recognition of this (transcendent) fact appears as the fruit of postmodernism, but the fact as such has no special connection with postmodernity—which, indeed, Lyotard seems rather to hope might escape precisely this condition by virtue of its postmodern insights.

58. Norris 1993.

59. Norris 1993, 23; Lyotard 1984, 53–54.

60. See note 40, above.

the concept has attained, which is to say, the pertinence that it is widely perceived to have for contemporary experience just now, in the period and under the conditions of postmodernity. I hope, moreover, that the reader will see in the foregoing pages several perspectives on postmodern knowledge production and distribution from which "incommensurability" would appear advantageous to both *Kulturerzeuger* and *Kulturträger.* To wit: Overproduction of knowledge, and insuperability of the literature embodying it, must inevitably make "incommensurability" attractive as a strategy and an excuse for ignoring the greatest part of it. That this vast expansion of the knowledge-production industry is chiefly based upon the proliferation of instruments (i.e., tools) and of subdisciplines whose very existence originates with and depends upon said tool, itself suggests a sort of tool-constituted "incommensurability" with conceptual, practical, and sociological dimensions.[61]

Such a tool-constituted "incommensurability" is further reinforced when this instrumentalism of knowledge production is carried as instrumental rationality up to the ideological level. Then, as Herbert Marcuse saw so clearly: "One does not "believe" the statement of an operational concept but it justifies itself in action—in getting the job done, in selling and buying, *in refusal to listen to others.*"[62] And what indeed is this "refusal to listen to others," which I have underscored, but the operational definition of "incommensurability"?

Finally, standing in closest connection with the "refusal to listen to others"—indeed as Habermas would say, essentially a manifestation of it—is the disregard of disciplinary demands while harkening to the call of the market, or, otherwise put, the binding of knowledge production to institutions pursuing particular interests and using their power as institutions to create a convenient reality.

Thus the attractiveness of the concept of "incommensurability" under the prevailing postmodernist adaptation to the aggravated conditions of late-modern knowledge production is that it legitimates those conditions and the adaptation to them: "incommensurability" is then no longer a part of the problem of contemporary life but part of the resolution of that problem.

The question arises of course, how far can this process go? Obviously much, much further than was ever conceivable on the basis of modernist

61. Pickering 1995b, 187–189, 245–246. See also note 25, above.
62. Marcuse 1964, 103.

notions of knowledge.[63] Yet when we look at painting today we have reason to wonder (and worry): here is a cultural endeavor in which, a generation ago, practitioners and critics were unified by a moralistically sanctioned consensus as to the common goal of their efforts, and consequently paid the closest attention to each other's work. And today, in the words of Larry Rivers, one of those painters who "came afterward," "At this point no live artist is doing anything that will have the slightest influence on my work, nor will mine influence them, at least not my contemporaries."[64]

Postmodernity, Community, and Responsibility: Risk, Trust, and the Craving for Community

Beneath Rivers' braggadocio one can sense his regret, a common regret that underlies our contemporary search for "community." Our encounter with this word and theme at every turn, in every field and style of discourse, is the consequence of a diverse array of social and cultural problems and transformations. Postmodernity, as a failure of an accepted vision of progress, a failure of consensus regarding what is forward and what backward, what is up and what down, in social, political, and cultural development, amounts not merely to an explicit relativization of all judgements of value in these realms, but also to an implicit undermining of the traditional bases for the coherence and stability of modern culture, society, and polity. Under these circumstances, we are inclined to regard every "community"—i.e., every congeries of distinctive purposes, practices, and markers that succeeds in giving form and meaning to a collective life of any considerable number of human beings—as more than legitimate, as a positive value. To quote Bauman once again:

> No wonder that postmodernity, the age of contingency *für sich,* of self-conscious contingency, is also the age of community: of the lust for community, search for community, invention of community, imagin-

63. Thus on the basis of Kuhn's modernist axioms the coexistence, anyway *a la longue,* of even two, let alone many, incommensurable paradigms was logically precluded, for the existence of the scientific discipline depended on the uniqueness of the paradigm. Biagioli (1990, 207–208)—see note 39, above—approaches the question with postmodern pluralist axioms, which, however, he fails to recognize as such, and so misses the real basis of his disagreement with Kuhn.

64. As quoted from Rivers, *What Did I Do?,* with A. Weinstein (New York: Harper/Collins, 1992), in a review by R. Koenig, *New York Magazine,* Nov. 2, 91–92.

> ing community. . . . Community—ethnic, religious, political or otherwise—is thought of as the uncanny mixture of difference and company, as uniqueness that is not paid for with loneliness, as contingency with roots, as freedom with certainty; its image, its allurement are as incongruous as that world of universal ambivalence from which—one hopes—it would provide a shelter.[65]

To acknowledge such yearnings is not, of course, to fail to recognize that most of the innumerable deployments of "community" in contemporary discourse are modish, opportunistic, and disingenuous (as are also, we will shortly see, the deployments of "responsibility"). Indeed, to a very large extent, the shibboleth of "community" (and "responsibility") serves the conservative purpose of defanging criticism of the economic and political order. By downplaying differences of wealth and power—of interest, generally—it enables the underlying structures for the production and maintenance of such "goods" to continue out of sight and undisturbed. In this regard the call for community is an important contribution to postmodernity, regarded as an era in which "modern practice continues—now, however, devoid of the objective that once triggered it off."[66]

But if in recent decades in all industrial, democratic societies John Q. Public has become increasingly indifferent to disparities of wealth and power, he has also become increasingly alert to, and perturbed by, every putative threat to his own physical well-being, and especially to such "risks" arising out of the local, national, and global production and distribution of wealth and power.

Various sociologists have fixed upon risk as the typical preoccupation of contemporary society.[67] Among them is Ulrich Beck,[68] whose *Risikogesellschaft* (1986) is—or was a few years ago—"already one of the most influential European works of social analysis in the late twentieth century."[69] "The argument," wrote Beck, "is that, while in classical industrial society the 'logic' of wealth production dominates the 'logic' of

65. Bauman 1991, 246–247.

66. Bauman 1991, quoted above.

67. Luhmann's book on the *Soziologie des Risikos* was prompted by "the fact that present-day society is so much concerned with risk" (1993, viii). In the index to Giddens, *Modernity and Self-Identity* (1991), "risk" and "trust" have the largest number of entries, as do "risk" and "knowledge" in the index to Nico Stehr and Richard V. Erickson, eds., *The Culture and Power of Knowledge* (1992). Mary Douglas, *Risk and Blame* (1992, 15), asks "how to explain the new concern with risk?" Her answers show little understanding.

68. Beck 1992.

69. Lash and Wynne 1992, 1.

risk production, in the risk society this relationship is reversed."[70] To explain without Teutonic dialectic why this is so, Scott Lash and Brian Wynn[71]—underscored by Bauman[72]—point out that modernity, especially its late phases, has created "social dependency upon institutions and actors who may well be—and arguably are increasingly—alien, obscure and inaccessible."

So construed, the preoccupation with risk derives from a lack of trust: under the conditions of late-modern and postmodern life, to an ever-increasing extent we *must* trust experts and institutions, while the grounds for trust are continually eroding—specifically, we recognize ever more clearly the increasingly interested character and self-serving intents of the expert individuals and institutions on whom we depend to produce, purvey, and employ the knowledges indispensable to survival in postmodernity.[73] Otherwise stated, the decline in trust is the consequence of the general recognition that the acts of experts and institutions—indeed, those of all actors in contemporary society—are guided by a diminishing sense of *responsibility* toward those whose welfare depends upon their expert performance. It is through those ever more needful and ever less available desiderata, trust and responsibility—responsibility being the precondition for trust—that we can understand the craving for community in postmodernity. "Community" is precisely that social relation characterized by responsibility and, consequently, trust.[74]

Responsibility as Leading, and Most General, Normative Category

As Arie Rip has shown, "responsibility" is a term with a complex history and a complex network of meanings.[75] For the immediate purpose, however, it suffices to distinguish two broad uses of the term. The first, with no genuine connection to the concept of community, re-

70. Beck 1992, 12–13.

71. Lash and Wynne 1992, 4.

72. Bauman 1993, 203.

73. For the recent, widespread fixation upon "trust" as urgent social desideraturn, see Forman 1995b.

74. That "responsibility" trumps "trust" appeared quite clearly during the 1992 presidential campaign, where Bush, impugning Clinton's character, pushed "trust" as the principal issue. To which Clinton responded by charging that Bush "refused to assume a shred of responsibility" (*International Herald Tribune,* July 31, 1992, 3; as quoted by Warner 1993, 451).

75. Rip 1981.

fers to the accountability of an agent, official, or administrator for the efficient performance of assumed or assigned tasks. The second, properly communitarian, derives from the notion of responsiveness, and suggests, vaguely, attentiveness to the needs of others—of some wider or narrower class of "others." The former meaning predominated in modernity; the latter meaning is coming to predominate in postmodernity, as can be seen in the communitarian gloss commonly given today to the modernist, bureaucratic concept of responsibility. Thus, for example, the head of the Smithsonian Institution writes:

> To warrant continued success in attracting private resources, the Smithsonian bears a huge responsibility to the donor community. We are carefully identifying those needs that are best met through private resources and will use contemporary, yet appropriate, techniques to secure such funds.[76]

Indeed, some part of the great impact of "responsibility" as a demand of the day results from this left-right, one-two "punch," conflating its authoritarian with its antiauthoritarian meaning.

As we advance into postmodernity, "responsibility" is coming to play an enormously greater role than it had in modernity—to play *both* the role of *leading* normative category and the role of *most general* normative category. "If the key word for Lyotard in 1979 was 'legitimation,' perhaps the crucial word now is 'responsibility.'"[77] And so it appears in the thematic text introducing the exhibition "Science in American Life," which opened in the Smithsonian's National Museum of American History in the spring of 1994: "The challenge of the 21st century is to make responsible choices about science and technology." Similarly, MasterCard's public service message on U.S. National Public Radio in the summer of 1995 is "use your credit card responsibly." And every reader of today's newspapers knows that "personal responsibility is the catchword" of contemporary American politics.[78]

Carol Gilligan, bolstered by feminism's affirmation of the specifically female, was one of the earliest to put responsibility forward as leading normative category. Her observation that for *women* "the moral problem arises from conflicting responsibilities rather than from competing

76. Heyman 1994.

77. Jasper 1993, 2.

78. Robin Toner in *New York Times,* July 16, 1995, section 4, 1; likewise Michael Wines, *New York Times,* Sept. 10, 1994, section 1, 1. Such examples could be multiplied indefinitely.

rights,"[79] was not in itself new. What was new was her elevation of the values of community, commitment, and responsibility, i.e., a transvaluation of values placing a positive sign before the feminine apprehension of the moral problem, rather than the negative sign that previously, under masculine modernist axioms, had been assigned to the feminine failure to "understand" the primacy of individual rights.

Increasingly, now, rights are being subordinated to responsibilities. A "survey of recent work on citizenship theory" opens by noting that "there has been an explosion of interest in the concept of citizenship among political theorists."[80] After three paragraphs reviewing the changing temper of democratic polities since the late 1970s, the authors find "it is not surprising, then, that there should be increasing calls for a 'theory of citizenship' that focuses on the identity and conduct of individual citizens, including their responsibilities, loyalties, and roles," i.e., communitarian values, with no mention of rights. "Rights," we are told a few paragraphs further on, characterized "the postwar orthodoxy" but are now passé.

Such examples attest to responsibility's new role as postmodernity's leading normative category. No less significant is responsibility's new role as postmodernity's *most general* normative category, i.e., that category commonly appealed to in the widest range of ethical decisions. Indeed, in contemporary theory *and* in contemporary practice "responsible" is commonly taken as coextensive with "moral," as equivalent to "ethical." Bauman,[81] for example, leans heavily upon Emmanuel Levinas, particularly Levinas's thesis that ethics "does not supplement a preceding existential base; the very node of the subjective is knotted in ethics understood as responsibility."

Almost a decade ago in contributions to Paul Durbin's volume on *Technology and Responsibility,* Carl Mitcham pointed perceptively—and with a touch of pique perhaps?—

> To the central role the word ["responsibility"] will play in contemporary life, where "responsibility" has become a touchstone—if not cliche—in discussions of moral issues in art, politics, economics, business, religion, science, and technology. . . . The truth is that responsibility has become a general normative category.[82]

79. Gilligan 1982, 19.
80. Kymlicka and Norman, 1994.
81. Bauman 1992c, 43.
82. Mitcham 1987a, 3; Mitcham 1987b, 361.

Perhaps the best, the most authoritative, exemplification of this fact are the two volumes by Robert Bellah and collaborators diagnosing, and prescribing for, the cultural disease of contemporary America. *Habits of the Heart* (1986) and *The Good Society* (1992) are, to a great extent, a concerted attack on modernism—for its irresponsible individualism—and a plea for community. In the latter book we meet

> Marian Metzger [who] considers herself to be a *responsible* person—*responsible* in her personal life, *responsible* to the company she worked for, and *responsible* for improving the way she related to others . . . she . . . came up against the limits of her capacity for *responsible* action. . . . What she lacked was a way to think *responsibly* about the institutional forms that had brought about her quandry.[83]

Though the explicit emphases are mine, responsibility is implicitly the authors' most general, as well as leading, normative category.[84]

Modern Science as Flight from Responsibility

Recognizing the central and overriding importance attributed to responsibility not in America alone, but in all Western democratic societies in postmodernity, we must expect that these obligatory values will find some reflection in the attitudes of scientists toward themselves and their knowledge-producing activities. However, in order to appreciate how substantial a reorientation is involved in a scientist's affirmation of responsibility, it is first necessary to recognize that the traditional stance of the modern scientist, and the premodern as well, was one of radical irresponsibility.

By ancient and honorable tradition the ivory tower has been a place of retreat from individual moral responsibility, and only very rarely a support for its exercise. "I have been startled at how reluctant academics seem to be to treat ethical issues," said Clark Kerr looking back over a long career as professor of public administration and then chancellor of the Berkeley campus, and president of the University of California.

> As a young teacher at Berkeley [in the late 1930s], I was asked by a distinguished professor at the University of Chicago, who had a Quak-

83. Bellah et al. 1992, 22–23.

84. Bellah et al. 1986, Bellah et al. 1992. The truth is that the solution Bellah et al. offer (e.g., 1992, 40–41) to this quandary—how indeed to think and act responsibly in relation to the extremely powerful and intrinsically amoral institutions of our society—comes pretty close to fascism: institutions are for them beautiful things, embodiments of the highest ideals, etc.; in their index the word "bureaucracy" does not appear.

> erly concern about the academic study of ethics, to bring together influential members of the Berkeley teaching staff, particularly scientists, to talk with him about it. They were polite to him but not to me afterwards. They made it clear that this was not a subject which could hold any interest for scientists or scholars of any sort and that I should have known this; . . . that ethics was just a matter of personal taste and anything goes in matters of taste—with one extremely important exception: a commitment to scientific truth in the academic world.[85]

This commitment to a transcendent scientific truth overriding all moral considerations—and, in particular, overriding every responsibility—is, as Rip and Wim Smit point out, a prime example of Max Weber's "absolute ethic," the "ethic of ultimate ends."[86] In contrast with an "ethic of responsibility," which considers commitments and consequences in weighing incompatible interests and goals, an "absolute ethic," Weber wrote, "just does not *ask* for 'consequences.' "[87] Such is the ethos of pure science: "I think there is little more important to the individual scientist than that there should be opportunities for him to find and to record where his mind will take him, irrespective of consequences," said Percy Bridgman,[88] surely one of this century's most reflective scientists, just at the time of Clark Kerr's early, painful experience at Berkeley.

This self-conception of truth seeker/finder as elevated above all consideration of consequences was not elaborated by the modern scientist unaided. It has been effected and sustained by the culture of modernity. One can see this today in the remnants of the Western concept of genius to be found in any middlebrow organ, but very little any longer in highbrow journals. This romantic reworking of the Renaissance concept of immortal achievement[89] includes the notion that an individual capable of producing transcendent cultural goods is, ipso facto, placed beyond and above the reach of moral judgements.[90] With such broad cultural encouragement to flee responsibility, it is hardly surprising that "artists,

85. Kerr 1989, 139.

86. Arie Rip and Wim A. Smit, "Toward a responsible university: taking societal impacts of scientific research into account" (Paper presented at Seminaret i Vetenskapteori, University of Oslo, February 15, 1991).

87. Weber 1946, 120.

88. Bridgman 1938, 250.

89. Forman 1973.

90. A couple of examples at random in the *New York Times Book Review* (March 22, 1992, p. 29; March 8, 1992, p. 11): Andrew Sullivan, editor of the *New Republic,* reviewing a biography of the Australian novelist Patrick White, sees this undeniably mean and wretched character as "a vessel for extraordinary artistic and spiritual transcendence"; Walter Moore, author of a biography notorious for its revelations of Erwin Schroedinger's

writers and generally men and women whose genius forced them to keep the world at a certain distance"[91]—scientists among them—have taken the opportunity and run.

But the institutionalization of science in Western industrial societies in the last two centuries created two further determinants of, and supports for, an ideology of irresponsibility among scientists, academic and otherwise. The first of these is the modernist, disciplinary value of production as an end in itself, to which historian of mathematics Herbert Mehrtens and sociologist of mathematics Sal Restivo have each drawn attention. The overriding priority given to productivity by scientific disciplines had as inevitable consequence their refusal to allow consideration of the means by which high levels of productivity are maintained, whether those means be internal to the discipline or relate rather to the terms on which social support for its knowledge-producing activities is obtained. In this sense, as Mehrtens, especially, has emphasized, modern, discipline-directed science is "institutionalized irresponsibility."[92]

The third main determinant/support of a scientific ideology of radical irresponsibility became effective only after the Second World War had placed the scientists' claims to social support upon the basis of "national security." "National security" was not merely capable of, but insisted upon, maintaining an enormously larger scientific establishment than any previously known (and maintaining it at a much higher standard of scientific living).[93] Here now it became imperative for the scientists to fashion a self-image that allowed them to close their eyes to the real basis for the generous social support of their knowledge-producing activities and to maintain the illusion of personal and disciplinary autonomy.[94]

That new self-image, originating with the American physicists and then spreading to other disciplines, projected "fun" as the predominant feature and leading attraction of "doing science."[95] Its shibboleth, "phys-

---

irresponsible personal behaviors, reviewing David Cassidy's biography of Heisenberg, finds nothing unusual in the deficiencies of *this* physicist's personal character—"Scientists under all regimes tend to prefer expediency to morality"—but stresses, as his bottom line, that "[a]lthough Heisenberg's political life was far from heroic, his idealistic interpretation of physics will nonetheless endure."

91. The phrase, used by Hannah Arendt's biographer as descriptive of her, appears here as quoted by Thomas Söderqvist (1991, 151–152).

92. Mehrtens 1994. Cf. note 8, above.

93. Forman 1987.

94. Forman 1996.

95. Forman 1989.

ics is fun," was meant to trigger the fantasy of perfect autonomy: any eudemonic activity is an end in itself, and as such wholly autonomous—but also wholly irresponsible. Moreover, putting "fun" forward trumps any question of the end for which the scientist's knowledge-making is sustained, and thus supports from this side as well a stance of radical irresponsibility.[96]

In Postmodernity Scientists Embrace Responsibility

Though it would require much more space to develop the case, it is important here to state that the flight of the modern scientist from responsibility was integral with modernity generally[97]—modernity understood as "a gigantic exercise in abolishing individual responsibility."[98] The scientists were permitted their stance of irresponsible purity only because such a stance was compatible with the transcendence-oriented political, aesthetic, and cognitive ideologies of modernity, and, moreover, compatible with modernity's practice of constructing ever more elaborate—ostensibly more efficient—bureaucratic social machinery in which the individual is an expert but inconscient cog. Thus each of the three determinants of modern scientists' stance pointed out above—adherence to an "absolute ethic" of truth; overriding priority of discipline-directed productivity; and refusal to be fully cognizant of the grounds for society's support—is but an aspect of those more general modernist ideologies and practices.

What, then, if these supports for scientific irresponsibility should go, separately and collectively, by the board? But just that is the case in postmodernity, as we have seen in earlier sections of this chapter: the notion of a unique, universal, transcendent truth is now incredible; the scientific disciplines have lost much of their legitimacy, and the hierarchical ranking of them in scales of abstract-practical or pure-applied has lost nearly all its authority; with the failure of truth and the end of the primacy of disciplinary demands, a "space" is opened for moral judge-

96. This is the sense of Norbert Wiener's lament over the "degradation of the position of the scientist . . . to that of a morally irresponsible stooge in a science factory." Quite another dimension of irresponsibility in postwar science was stressed by various scientists caught up in the transition to 'big science.' To Melvin Schwartz, for example, it seemed that "in such a 'production-line' approach to scientific research '*nobody* need feel real responsibility for the accuracy of the results.'" Both quotations from Capshew and Rader 1992, 11.

97. Forman 1997.

98. Bauman 1992a, xxii.

ments, while, simultaneously, the need for an unambiguously clean conscience, which seemed so urgent in modernity, now appears hopelessly unrealistic.

Where previously the modernist recognition of the impossibility of any *purely rational* grounding of ethical norms—conjoined with the equally modernist refusal to accept the positive, pluralist consequences of that recognition—had long embargoed ethical discourse among philosophers and scientists, social and natural, the past two decades have been "a period that might be called 'the Great Expansion' . . . in ethics."[99] This "great expansion" has become a great explosion in consequence of various measures by the U.S. government mandating studies and instruction in bio-medical ethics.[100] And if, on top of all this, the common culture in which scientists too are immersed adopts "responsibility" as its leading and most general normative category, could scientists maintain—would scientists wish to maintain—their posture of irresponsibility? The answer is surely "no," and today one finds on every hand evidence of an ideological reorientation.

A point to be noted, with respect to physicists especially, as indicative of a much greater susceptibility to pressures from the cultural environment, is the self-inculpation which they now display in facing the recent decline in financial support for "curiosity-driven" research and in employment opportunities generally. Where even just a few years ago the American physicist saw any decrease in social support as manifestation of the failure by the general public to appreciate him at his true worth, it is today commonplace for physicists to accuse themselves of various faults, particularly being arrogant, and to hold themselves in large measure responsible for their current difficulties.[101]

Further it is important as a precondition, or at least concomitant, of the acceptance of moral considerations in directing their knowledge production that these physicists now also blame themselves for holding too tightly to their disciplinary orientation. "We physicists have become exceedingly conservative in our choices of research topics," the directors of the Princeton Materials Institute and of the Institute for Theoretical

99. Darwall et al. 1992, 123.

100. Thus 1989 in National Institutes of Health required that all institutions receiving its training grants provide instruction in research ethics to apprentice researchers, and more recently the funds for deciphering the human genome included a "set-aside" for ethical inquiries.

101. Striking evidence of this is to be found in the statements made in two "roundtable" discussions published in *Physics Today* (Coppersmith et al. 1994 and Byer et al. 1995).

Physics at the University of California, Santa Barbara, wrote recently in *Physics Today:*

> Large numbers of investigators tend to concentrate their efforts in a few well-established areas, while other topics—often the most interesting, interdisciplinary or otherwise unorthodox ones—remain relatively untouched . . . the term "strategic" should have an entirely positive meaning for us; there is no reason for it to have become a catchword symbolizing retreat from "pure" or "curiosity driven" research. Acting "strategically" should mean simply that more of us are working on projects that are interesting not just to ourselves but also to others, particularly areas outside our own specialties. In short, it should mean maximizing our impact on the world around us.[102]

Noteworthy about this admonition is not so much the allegations of conservativism—physicists have in fact become far less, not far more, conservative in their choices of research topics—but rather the antidisciplinary standard of progressivism that is here being applied. And this is perhaps truer still in physics paedagogy. Steven Strogatz, well known for his innovations at MIT, recently published a textbook based upon his course: *Nonlinear Dynamics and Chaos: With Applications to Physics, Biology, Chemistry, and Engineering.* The reviewer in *Physics Today* praised the work and noted that "the details associated with applications to lasers, pendula, fireflies, rabbits and sheep and foxes, superconductors, chemical reactions, love affairs, insect outbreaks and the coding of secret messages with chaos are self-contained."[103]

A yet more recent article in *Physics Today* provides striking evidence that the categories "moral" and "responsible" have indeed advanced to premiere positions in the rhetoric of this loosened discipline. Irwin Goodwin, the magazine's senior editor and a savvy analyst of physics' governmental relations, authored an account of a conference in honor of Hans Bethe's sixty years at Cornell University[104] titled "A tribute to a titan of physics"; it is subtitled "Physicists and friends celebrating Hans Bethe's scientific ingenuity and moral influence . . . ," and it begins, "Even at the age of 88 Hans Albrecht Bethe is one of the world's most resourceful and responsible physicists." The prominence of "moral" and "responsible" is the more significant in that, to judge by Goodwin's own

102. Eisenberger and Langer 1995.
103. Ronald F. Fox in *Physics Today,* vol. 48 (March 1995), 93–94.
104. Goodwin 1995.

report, these themes appeared only in a film about Bethe screened on that occasion and not at all in the viva voce presentations.

Another recent, striking example of the invocation of responsibility is a lecture, delivered in the spring of 1995 at the Livermore, California, "weapons" laboratory, by physicist Neal Lane, one year after assuming the directorship of the National Science Foundation.[105] "Responsibility" is laced through Lane's text, appearing eight times, once every four to five hundred words. ("Values"—usually in the morally more neutral combination "values and goals"—appeared fourteen times.)

Lane's uses of "responsibility," although highly inspecific, retain still remnants of the word's traditional meanings. But the breadth and dominance that "responsibility" has attained as normative category in the wider culture—where it is now commonly equated to "morality"—is also being reproduced in the culture of science. Prompted by the rash of "scientific misconduct" cases, and by the governmental concern about them, the U.S. National Academy of Sciences instituted a committee to say what good scientific conduct is and how it can be instilled.[106] The committee chose *Responsible Science* as the title of its report, and used the word repeatedly throughout. Although the report very carefully defines "scientific misconduct," the term "responsible" remains undefined, perhaps just because it is employed as equivalent to everything good.[107]

It should not, however, be supposed that because the word is, as used, empty of all specific content, its use is insignificant. On the contrary, such

105. Neal F. Lane, "The scientist in an age of unreason" (lecture delivered at Lawrence Livermore National Laboratory, May 18, 1995), typescript, 8 pp.

106. U.S. National Academy of Sciences 1992.

107. The report includes a second, supplementary, volume of commissioned background papers, none of which seeks to define responsibility. The committee, though it did not define "responsible," did state the "fundamental responsibility" of the "individual scientist," namely, "to ensure that their results are reproducible, and that their research is reported thoroughly enough so that results are reproducible"—an instrumentalist criterion that says nothing about the individual's conduct in research or toward others, and that, moreover, has rarely ever been fully met, and is increasingly deliberately unmet, as we have seen above. And while noting ironies, I cannot refrain from pointing out that spokesmen for the modernist resistance typically cannot themselves avoid appealing to postmodern normative categories. Thus Ernest Gellner (see note 1) said that "to pretend that the scientific revolution of the seventeenth century, and its eventual application in the later stage of the industrial revolution, have not transformed the world, but are merely changes from one culture to another, is simply an *irresponsible* affectation." And Gerald Holton, who gets the last word in the *New York Times*'s account of the May 31–June 2, 1995, "The Flight from Science and Reason" conference in New York (Browne 1995), calls American scientists to arms to defend "the *moral* authority of science." (Emphases are mine, of course.) For more postmodern-despite-themselves examples, see Forman 1995b.

use shows that all ethical questions are now being approached "from the side of responsibility," so to speak. The implicit meanings of the word remain powerfully at work. This becomes clearer when we take note of the frequent appeal these days to reasonably specific notions of responsibility even without use of the word. A revealing manifesto in this regard is Howard Georgi's chapter on "effective quantum field theories" in the imposing collection, *The New Physics.*[108] Even though his argumentation is largely traditional, his posture is "responsible"—though the word is not used. Thus in arguing against pursuing grand unification theories, Georgi says, "their only connection with reality is through cosmology. Cosmology is fun, but. . . ." Previously there would have been no "buts"; "fun" would have been trumps.

Conclusion

The foregoing exposition cannot be summarized in detail in any brief conclusion. Its principal burden is, however, that even as, in postmodernity, knowledge production becomes ever more closely integrated with the pursuit of "special interests," and fragmented for this purpose, the present acceptance of this state of affairs as real and inevitable has also opened up the possibility—which had remained closed in modernity—for moral considerations to play a role in science. We have, as Sam Schweber has said, given up our former "belief that rigid boundaries existed between the moral and the physical domains."[109]

> The scientific enterprise is now largely involved in the creation of novelty—in the design of objects that never existed before and in the creation of conceptual frameworks to understand the complexity and novelty that can emerge from the *known* foundations and ontologies. And precisely because we create those objects and representations we must assume moral responsibility for them.[110]

As important as these conclusions are for science, they are still more important for the historiography of science. In so saying I have in mind, of course, the obvious fact that any significant change in science is ipso facto also significant for us as its historians. Of far greater importance, however, are the implications of these conclusions relative to the distinction between "us" as historians and "them" as scientists, a distinction based fundamentally—however little we recognized it—upon the reality

108. Georgi 1989.
109. Schweber 1993a.
110. Schweber 1993b.

in modernity of that boundary between the physical and the moral.[111] In postmodernity, however—to quote the words of the Chairman of the Board of the Nobel Foundation from his opening address at the 1989 awards ceremony—, "[t]he steadily increasing demand for responsibility on the part alike of the researchers and the humanists is now bridging the gap between the two cultures."[112]

ACKNOWLEDGMENTS

This chapter is an elaboration of my oral presentation at the International Workshop on the Historiography of Contemporary Science, Technology, and Medicine in Göteborg, September 1994, and employs some material that appeared in the interim confirming my argument. For helpful criticism I thank Katherine Livingston, Herbert Mehrtens, Alan Morton, and Lorenza Sebesta.

REFERENCES

Ankersmit, Frank R. 1989. Historiography and postmodernism. *History and Theory* 28: 137–153.
Baird, Davis. 1993. Analytical chemistry and the "big" scientific instrumentation revolution. *Annals of Science* 50: 267–290.
Bauman, Zygmunt. 1987. *Legislators and Interpreters: On Modernity, Postmodernity, and Intellectuals.* Cambridge: Polity Press.
———. 1989. *Modernity and the Holocaust.* Cambridge: Polity Press.
———. 1991. *Modernity and Ambivalence.* Ithaca, N.Y.: Cornell University Press.
———. 1992a. *Intimations of Postmodernity.* London: Routledge.
———. 1992b. Survival as a social construct. *Theory, Culture, and Society* 9: 1–36.
———. 1992c. *Mortality, Immortality, and Other Life Strategies.* Stanford: Stanford University Press.
———. 1993. *Postmodern Ethics.* Oxford: Blackwell.
———. 1995. *Life in Fragments: Essays in Postmodern Morality.* Cambridge, Mass.: Blackwell.
Beck, Ulrich. 1992. *Risk Society: Towards a New Modernity.* London: Sage.
Bellah, Robert N., Richard Madsen, William Sullivan, Ann Swidler, and Steven Tipton. 1986. *Habits of the Heart: Individualism and Commitment in American Life.* New York: Harper and Row.
———. 1992. *The Good Society.* New York: Vintage Books.
Biagioli, Mario. 1990. The anthropology of incommensurability. *Studies in History and Philosophy of Science* 21: 183–209.

111. Forman 1991.
112. Nobel Foundation 1990, 7.

———. 1993. *Galileo, Courtier: The Practice of Science in the Culture of Absolutism.* Chicago: University of Chicago Press.
Bloch, Felix. 1952. W. W. Hansen. *Biographical Memoirs of the National Academy of Sciences* 27: 120–137.
Blume, Stuart. 1992. Whatever happened to string and sealing wax? In Robert Bud and Susan E. Cozzens, eds., *Invisible Connections: Instruments, Institutions, and Science.* Bellingham, Wash.: SPIE Optical Engineering Press.
Bridgman, Percy W. 1938. *The Intelligent Individual and Society.* New York: Macmillan.
Browne, Malcolm W. 1995. Scientists deplore flight from reason. *New York Times,* June 6: C1, C7.
Bud, Robert, and Susan E. Cozzens, eds. 1992. *Invisible Connections: Instruments, Institutions, and Science.* Bellingham, Wash.: SPIE Optical Engineering Press.
Byer, Robert L., et al. 1995. Roundtable: wither now our research universities? *Physics Today* 48: 42–51.
Cao, Tian Yu, and Silvan S. Schweber. 1993. The conceptual foundations and the philosophical aspects of renormalization theory. *Synthese* 97: 33–108.
Capshew, James H., and Karen A. Rader. 1992. Big Science: Price to the present. *Osiris* 7: 3–25.
Cohen, Jon. 1995. Share and share alike isn't always the rule in science. *Science* 268: 1715–18.
Collins, Randall. 1992. On the sociology of intellectual stagnation: the late twentieth century in perspective. In M. Featherstone, ed., *Cultural Theory and Cultural Change.* London and Newbury Park, Calif.: Sage.
Coppersmith, Susan N., Peter M. Eisenberger, Anthony M. Johnson, Neal F. Lane, Patricia McBride, Duncan T. Moore. Mark B. Myers, and Alvin Trivelpiece. 1994. Physics roundtable: reinventing our future. *Physics Today* 47: 30–39.
Darwall, Stephen, Allan Gibbard, and Peter Railton. 1992. Toward fin-de-siècle ethics: some trends. *Philosophical Review* 101: 115–189.
DeVorkin, David H. 1992. *Science with a Vengeance: The Military Origins of the Space Sciences in the American V-2 Era.* New York: Springer.
———. 1996. The military origins of the space sciences in the American V-2 era. In P. Forman and J. M. Sanchez-Ron, eds., *National Military Establishments and the Advancement of Science and Technology: Studies in Twentieth Century History.* Dordrecht: Kluwer.
Douglas, Mary. 1992. *Risk and Blame: Essays in Cultural Theory.* London: Routledge.
Dowler, Lawrence. 1993. Scholars, technology, and library resources. *Perspectives* (newsletter of the American Historical Association) April: 16–19.
Dupree, A. Hunter. 1986. *Science in the Federal Government.* Baltimore: Johns Hopkins University Press.
———. 1994. A knowledge policy for peace. *Technology in Society* 16: 289–300.
Eckert, Michael, and Maria Osietzki. 1989. *Wissenschaft für Macht und*

Markt: Kernforschung und Mikroelektronik in der Bundesrepublik Deutschland. Munich: Beck.

Edge, David. 1992. Mosaic array cameras in infrared astronomy. In Robert Bud and Susan E. Cozzens, eds., *Invisible Connections: Instruments, Institutions, and Science.* Bellingham, Wash.: SPIE Optical Engineering Press.

Eisenberger, Peter, and James S. Langer. 1995. Opinion: a case for strategic research. *Physics Today* 48: 78–80.

Fekete, John, ed. 1987. *Life after Postmodernism: Essays on Value and Culture.* New York: St. Martin's Press.

Felt, Ulrike, and Helga Nowotny. 1992. Striking gold in the 1990s: The discovery of high-temperature superconductivity and its impact on the science system. *Science, Technology, and Human Values* 17: 506–531.

Forman, Paul. 1973. Scientific internationalism. *Isis* 64: 151–180.

———. 1987. Behind quantum electronics: national security as basis for physical research in the United States, 1940–1960. *Historical Studies in the Physical Sciences* 18: 149–229.

———. 1989. Social niche and self-image of the American physicist. In M. De Maria, M. Grilli, and F. Sebastiani, eds., *The Restructuring of Physical Sciences in Europe and the United States, 1945–60.* Singapore: World Scientific.

———. 1991. Independence, not transcendence, for the historian of science. *Isis* 82: 71–86.

———. 1995a. "Swords into ploughshares": breaking new ground with radar hardware and technique in physical research after World War II. *Reviews of Modern Physics* 67: 397–455.

———. 1995b. Truth and objectivity. *Science* 269: 565–567, 707–710.

———. 1996. Into quantum electronics: the maser as "gadget" of Cold War America. In P. Forman and J. M. Sanchez-Ron, eds., *National Military Establishments and the Advancement of Science and Technology: Studies in Twentieth Century History.* Dordrecht: Kluwer.

———. 1997. Transcendence, or the flight from responsibility: modern science in postmodern perspective. In P. Galluzzi, ed., Proceedings of the conference Scienza & Potere. Florence, December 1994.

Fuller, Steve. 1992. Being there with Thomas Kuhn: a parable for postmodern times. *History and Theory* 31: 241–275.

Galison, Peter, and Bruce Hevly, eds. 1992. *Big Science: The Growth of Large-scale Research.* Stanford: Stanford University Press.

Gellner, Ernest. 1995. Anything goes. *Times Literary Supplement* June 16: 6–8.

Georgi, Howard M. 1989. Effective quantum field theories. In P. Davies, ed., *The New Physics.* Cambridge and New York: Cambridge University Press.

Gibbons, Michael, Camille Limoges, Helga Nowotny, Simon Schwartzman, Peter Scott, and Martin Trow. 1994. *The New Production of Knowledge: The Dynamics of Science and Research in Contemporary Societies.* London: Sage Publications.

Giddens, Anthony. 1990. *The Consequences of Modernity.* Stanford: Stanford University Press.

———. 1991. *Modernity and Self-Identity: Self and Society in the Late Modern Age.* Stanford: Stanford University Press.

Gilligan, Carol. 1982. *In a Different Voice: Psychological Theory and Women's Development.* Cambridge: Harvard University Press.

Gillmor, C. Stewart. 1986. Federal funding and knowledge growth in ionospheric physics, 1945–81. *Social Studies of Science* 16: 105–133.

Goodwin, Irwin. 1995. Bethe fest: a tribute to a titan of modern physics. *Physics Today* 48: 39–41.

Gornick, Vivian. 1983. *Women in Science.* New York: Simon and Schuster.

Greenberg, Daniel. 1967. *The Politics of Pure Science.* New York: New American Library.

Harwit, Martin. 1981. *Cosmic Discovery: The Search, Scope, and Heritage of Astronomy.* New York: Basic Books.

Heilbron, John, L. 1982. *Fin-de-siècle* physics. In C. G. Bernhard, Elisabeth Crawford, and Per Sörban, eds., *Science, Technology, and Society in the Time of Alfred Nobel.* Oxford: Pergamon.

Heller, Scott. 1994. At conference, conservative scholars lash out at attempts to "delegitimize science." *Chronicle of Higher Education* November 23: A18.

Heyman, I. Michael. 1994. Smithsonian perspectives. *Smithsonian* December: 12.

Hoyningen-Huene, Paul. 1993. *Reconstructing Scientific Revolutions: Thomas S. Kuhn's Philosophy of Science.* Chicago: University of Chicago Press.

Jasper, David. 1993. Introduction: religious thought and contemporary critical theory. In D. Jasper ed., *Postmodernism, Literature, and the Future of Theology.* New York: St. Martin's Press.

Kay, Lily E. 1988. Laboratory technology and biological knowledge: the Tiselius electrophoresis apparatus, 1930–1945. *History and Philosophy of the Life Sciences* 10: 51–72.

Kerr, Clark. 1989. The academic ethic and university teachers: a "disintegrating profession"? *Minerva* 27: 39–156.

Kevles, Daniel J. 1996. The assault on David Baltimore. *New Yorker* May 27: 94–109.

Kline, Morris. 1980. *Mathematics: The Loss of Certainty.* New York: Oxford University Press.

Krimsky, Sheldon, James G. Ennis, and Robert Weissman. 1991. Academic-corporate ties in biotechnology: a quantitative study. *Science, Technology, and Human Values* 16: 275–287.

Kuhn, Thomas S. 1991. The road since *Structure.* In Arthur Fine, Micky Forbes, and Linda Wessels, eds., *Proceedings of the 1990 Biennial Meeting of the Philosophy of Science Association.* East Lansing, Mich.: Philosophy of Science Association.

Kymlicka, Will, and Wayne Norman. 1994. Return of the citizen: a survey of recent work on citizenship theory. *Ethics* 104: 352–381.

Lash, Scott, and Brian Wynne. 1992. Introduction to Ulrich Beck, *Risk Society: Towards a New Modernity.* London and Newbury Park, Calif.: Sage.

Leslie, Stuart W. 1992. *The Cold War and American Science.* Cambridge: MIT Press.

Luhmann, Niklas. 1993. *Risk: A Sociological Theory.* Berlin and New York: W. de Gruyter.

Lyotard, J.-F. 1984. *The Postmodern Condition: A Report on Knowledge.* Minneapolis: University of Minnesota Press.

Marcuse, Herbert. 1964. *One-Dimensional Man: Studies in the Ideology of Advanced Industrial Society.* Boston: Beacon Press.

Margolis, Leonid B. 1992. Does American science need Russian humanitarian aid? *New Biologist* 4: 413–417.

Mehrtens, Herbert. 1990. *Moderne-Sprache-Mathematik: eine Geschichte des Streits um die Grundlagen der Disziplin und des Subjekts formaler Systeme.* Frankfurt: Suhrkamp Verlag.

———. 1994. Irresponsible purity: the political and moral structure of mathematical sciences in the National Socialist state. In M. Renneberg and M. Walker, eds., *Science, Technology, and National Socialism.* New York: Cambridge University Press.

Metropolis, Nicholas. 1992. The age of computing: a personal memoir. *Daedalus* 121: 119–130.

Mitcham, Carl. 1987a. Responsibility and technology: the expanding relationship. In P. T. Durbin, ed., *Philosophy and Technology, vol. 3: Technology and Responsibility.* Dordrecht: Reidel.

———. 1987b. Responsibility and technology: a select, annotated bibliography. In P. T. Durbin, ed., *Philosophy and Technology, vol. 3: Technology and Responsibility.* Dordrecht: Reidel.

Narin, Francis. 1994. Patent bibliometrics. *Scientometrics* 30: 147–155.

Narin, Francis, and E. Noma. 1985. Is technology becoming science? *Scientometrics* 7: 369–381.

Narin, Francis, and Dominic Olivastro. 1992. Status report: linkage between technology and science. *Research Policy* 21: 237–249.

Nobel Foundation. 1990. *The Nobel Prize.* Stockholm: Almqvist & Wicksell International.

Norris, Christopher. 1993. *The Truth about Postmodernism.* Oxford: Blackwell.

Pestre, Dominique. 1992. Les physicians dans les sociétiés occidentales de l'après-guerre: un mutation des pratiques techniques et des comportements sociaux et culturels. *Revue d'Histoire Moderne et Contemporaine* 39: 56–72.

Pickering, Andrew. 1989. Pragmatism in particle physics: scientific and military interests in the post-war United States. In F. A. L. James, ed., *The Development of the Laboratory: Essays on the Place of Experiment in Industrial Civilization.* London: Macmillan.

———. 1995a. Cyborg history and the World War II regime. *Perspectives on Science* 3: 1–48.

———. 1995b. *The Mangle of Practice: Time, Agency, and Science.* Chicago: University of Chicago Press.

Price, Derek J. de Solla. 1986. *Little Science, Big Science . . . and Beyond.* R. K. Merton and E. Garfield, eds. New York: Columbia University Press.

Remington, John A. 1988. Beyond big science in America: the binding of inquiry. *Social Studies of Science* 18: 45–72.
Restivo, Sal. 1990. The social roots of pure mathematics. In S. E. Cozzens and T. F. Gieryn, eds., *Theories of Science in Society*. Bloomington: Indiana University Press.
Rip, Arie. 1981. *Maatschappelijke verantwoordelijkheid van chemici.* Doctoral dissertation. University of Leiden. Privately printed.
Schilb, John. 1991. Cultural Studies, postmodernism, and composition. In P. Harkin and J. Schilb, eds., *Contending Words: Composition and Rhetoric in a Postmodern Age.* New York: Modern Language Association.
Schweber, S. S. 1986. The empiricist temper regnant: theoretical physics in the United States, 1920–1950. *Historical Studies in the Physical Sciences* 17: 55–88.
———. 1989. Some reflections on the history of particle physics in the 1950s. In L. M. Brown, Max Dresden, and Lillian Hoddeson, eds., *Pions to Quarks.* Cambridge: Cambridge University Press.
———. 1993a. [Review of W. Lanouette, *Leo Szilard*]. *Science* 261: 1461–1462.
———. 1993b. Physics, community, and the crisis in physical theory. *Physics Today* 46: 34–40.
Segerstraale, Ullica. 1990. The murky borderland between intuition and fraud. *International Journal of Applied Philosophy* 5: 11–20.
Shapin, Steven. 1991. "A scholar and a gentleman": the problematic identity of the scientific practitioner in early modern England. *History of Science* 29: 279–327.
———. 1994. *A Social History of Truth: Civility and Science in Seventeenth-Century England.* Chicago: University of Chicago Press.
Simon, Herbert A. 1978. Rationality as process and as product of thought. *American Economic Review* 68, no. 2: 1–16.
Slaughter, Sheila. 1993. Beyond basic science: research university presidents' narratives of science policy. *Science, Technology, and Human Values* 18: 278–302.
Slaughter, Sheila, and Gary Rhoades. 1993. Changes in intellectual property statutes and policies at a public university revising the terms of professional labor. *Higher Education* 26: 287–312.
Söderqvist, Thomas. 1991. Biography or ethnobiography or both? Embodied reflexivity and the deconstruction of knowledge-power. In F. Steier, ed., *Research and Reflexivity.* London: Sage.
Stehr, Nico, and Richard V. Erickson, eds. 1992. *The Culture and Power of Knowledge: Inquiries into Contemporary Societies.* Berlin and New York: W. de Gruyter.
Stine, Jeffrey K. 1992. Scientific instrumentation as an element of U.S. science policy: National Science Foundation support of chemistry instrumentation. In Robert Bud and Susan E. Cozzens, eds., *Invisible Connections: Instruments, Institutions, and Science.* Bellingham, Wash.: SPIE Optical Engineering Press.
Stone, Richard. 1995. Baltimore defends paper at center of misconduct case. *Science* 269: 157.

Taubes, Gary. 1993. *Bad Science: The Short Life and Very Hard Times of Cold Fusion.* New York: Random House.

Travis, John. 1993. Imanishi-Kari says her new data shows she was right. *Science* 260: 1073–1074.

U.S. National Academy of Sciences, National Academy of Engineering, and Institute of Medicine, Committee on Science, Engineering, arid Public Policy, Panel on Scientific Responsibility and the Conduct of Research. 1992. *Responsible Science: Ensuring the Integrity of the Research Process.* 2 vols. Washington: National Academy Press.

Warner, Daniel. 1993. An ethic of responsibility in international relations and the limits of responsibility/community. *Alternatives: Social Transformation and Humane Governance* 18: 431–452.

Weber, Max. 1946. *Max Weber: Essays in Sociology.* H. H. Gerth and C. W. Mills, eds. New York: Oxford University Press.

Wright, Susan. 1994. *Molecular Politics: Developing American and British Regulatory Policy for Genetic Engineering, 1972–1982.* Chicago: University of Chicago Press.

Ziman, John. 1996. Is science losing its objectivity? *Nature* 382: 751–54.

PART II

Science Conceived as a Production Process

3

The Simple Economics of Basic Scientific Research

Richard R. Nelson

I. Basic Economic Framework

Recently, orbiting evidence of un-American technological competition has focused attention on the role played by scientific research in our political economy. Since Sputnik it has become almost trite to argue that we are not spending as much on basic scientific research as we should. But, though dollar figures have been suggested, they have not been based on economic analysis of what is meant by "as much as we should." And, once that question is raised, another immediately comes to mind. Economists often argue that opportunities for private profit draw resources where society most desires them. Why, therefore, does not basic research draw more resources through private profit opportunity, if, in fact, we are not spending as much on basic scientific research as is "socially desirable"? In order to answer some of these questions, it seems useful to examine the simple economics of basic research. How much are we spending on basic research? How much should we be spending? Under what conditions will these figures tend to be different? Is basic research marked by these conditions? If so, what can we do to eliminate or reduce the discrepancy?

How much are we spending on basic research? In 1953, the latest date for which relatively sophisticated estimates are available, total expenditure on research and development was about $5.4 billion. Of that total, much more than half was for engineering development, much less than half for scientific research. Even less of the total, about $435 million in 1953, was spent on "basic research." All evidence indicates that since

Reprinted with permission of The University of Chicago Press. From *Journal of Political Economy* 67 (1959): 297–306.

1953 expenditure on research and development has increased markedly; $10 billion seems a reasonable estimate for 1957. Expenditure on basic research has also increased at a rapid rate, perhaps at a faster rate than total research and development expenditure. But basic-research expenditure today is probably under $1 billion, less than one-quarter of 1 per cent of gross national product.[1]

How much should we spend on basic research? Replacing the X_i of the familiar literature on welfare economics with "basic research" provides the theoretical answer. From a given expenditure on science we may expect a given flow, over time, of benefits that would not have been created had none of our resources been directed to basic research. This flow of benefits (properly discounted) may be defined as the social value of a given expenditure on basic research. However, if we allocate a given quantity of resources to science, this implies that we are not allocating these resources to other activities and, hence, that we are depriving ourselves of a flow of future benefits that we could have obtained had we directed these resources elsewhere. The discounted flow of benefits of which we deprive ourselves by allocating resources to basic research and not to other activities may be defined as the social cost of a given expenditure on basic research. The difference between social value and social cost is net social value, or social profit. The quantity of resources that a society should allocate to basic research is that quantity which maximizes social profit.

Under what conditions will private profit opportunities draw into basic research as great a quantity of resources as is socially desirable? Under what conditions will it not? If all sectors of the economy are perfectly competitive, if every business firm can collect from society through the market mechanism the full value of benefits it produces, and if social costs of each business are exclusively attached to the inputs which it purchases, then the allocation of resources among alternative uses generated by private-profit maximizing will be a socially optimal allocation of resources. But when the marginal value of a "good" to society exceeds the marginal value of the good to the individual who pays for it, the allocation of resources that maximizes private profits will not be optimal. For in these cases private-profit opportunities do not adequately reflect

1. National Science Foundation, *Basic Research: A National Resource* (Washington, D.C., 1957); *Science and Engineering in American Industry* (Washington, D.C., 1956); *Growth of Scientific Research in Industry—1945–1960* (Washington, D.C., 1957); *Federal Funds for Science—The Federal Research and Development Budget Fiscal Years 1956, 1957, and 1958* (Washington, D.C., 1957).

social benefit, and, in the absence of positive public policy, the competitive economy will tend to spend less on that good "than it should." Therefore, it is in the interests of society collectively to support production of that good.[2]

Society does, in fact, collectively support a large share of the economy's basic research. About 60 percent of our basic research work is performed by non-profit institutions, predominantly government and university laboratories. And a portion of the basic research performed in industrial laboratories is government financed. (This flow of funds is about equal to the flow of funds from industry to nonprofit laboratories in the form of grants and contracts for basic research.)[3] Much, though not all, of the government contribution to basic research is national defense–oriented. But defense-oriented expenditure aside, the American political economy certainly treats basic research as an activity that creates marginal social value in excess of that collectable on the free market.[4] Is this treatment justified? If so, since, in fact, society is collectively supporting much basic research and hence resources directed to basic research do exceed the quantity drawn by private profit opportunity, is present social policy adequate?

II. Scientific Research and Economic Value

What are the social benefits derived from the activity of science? It is sometimes argued that most of our great social and political problems would simply evaporate if all citizens had a scientific point of view and, hence, that the benefits derived from scientific research are only in small part reflected in the useful inventions generated by science, for science helps to make better citizens. And many scientists and philosophers take the point of view that the very activity of science—considered as the search for knowledge—is itself the highest social good and that any other benefits society might obtain are just by-products of the activity of science—social gravy. Dissents on both of these points are often

2. Of course, the resources supplied to the industry must be withdrawn from other industries which generate no external economies or less external economies. Although significant external economies are probably rare, they almost certainly exist in education and preventive medicine as well as in basic research. Though the burden of this paper is that more resources should be allocated to basic research, the argument is probably invalid if these resources are taken, for example, from education or preventive medicine.

3. National Science Foundation, *Basic Research: A National Resource.*

4. This paper will not consider the vital question of whether the Department of Defense is spending enough on defense-oriented basic research. It probably is not, but the analysis of the paper is independent of this.

sharp. The economist, after the usual perfunctory statement that he is fully aware that his definition does not capture everything, might define the benefits derived from the activity of science as the increase resulting from scientific research in the value of the output flow that the resources of a society can produce. In order to examine the extent to which a private firm can capture through the market the increased value of output resulting from the scientific research, in particular the basic scientific research, that it sponsors, it is necessary to examine the link between scientific research and the creation of something of economic value.

Scientific research may be defined as the human activity directed toward the advancement of knowledge, where knowledge is of two roughly separable sorts: facts or data observed in reproducible experiments (usually, but not always, quantitative data) and theories or relationships between facts (usually, but not always, equations). Of course, no strict line can be drawn between scientific research and all other human activities. Men have always experimented and observed, have always generalized and theorized; thus all men have always been, at least in a limited way, scientists. And knowledge has often (usually?) been acquired in activities in which pursuit of knowledge was of no, or negligible, importance. But even fuzzy definitions often have value. Scientific knowledge rests on reproducible experiments, but science is more than experimentation leading to new observations of facts which are believed observable by any other scientist undertaking the identical experiment. Science is most fruitful when it leads to ability to predict facts about phenomena without, or prior to, experimentation and observation. Scientific knowledge has economic value when the results of research can be used to predict the results of trying one or another alternative solutions to a practical problem.

Scientific research has increasingly been coupled to invention, where invention is defined as the human activity directed toward the creation of new and improved practical products and processes. But though many inventions occur as a result of a reasonably systematic effort to achieve a particular goal, many other inventions do not. They are a byproduct of activity directed in a quite different direction, often a scientific research project directed toward solving an unrelated problem. Mauve, the first analine dye, was discovered by W. H. Perkin while he was attempting to synthesize quinine, and calcium carbide and the acetylene gas that it produces were invented by a group attempting to develop a better way to extract aluminum from clay. And though many inventions are made possible by closely preceding advances in scientific knowledge, many others require little knowledge of science or occur long after the relevant scientific knowledge is available: scientific knowledge certainly had little to do with the development of such useful inventions as the safety razor and the zipper; scientists have long known that expanding gases absorb heat, thus cool whatever they contact, but the gas refrigerator is an invention of the twentieth century. But particularly in the institution of the industrial research laboratory, applied science and invention are closely linked, and inventions usually result from a systematic attack on a problem.

In the activity of invention, as in most goal-directed activities, the actor has a number of alternative paths among which he must choose. The greater his knowledge of the relevant fields, the more likely he will be eventually to find a satisfactory path, and the fewer the expected number of tried alternatives before a satisfactory one is found. Thus, the greater the underlying knowledge, the lower the expected cost of making any particular invention.

A rationally planned inventive effort will be undertaken only if the expected revenue of the invention exceeds the expected cost. In many instances the economic utility of a particular invention is so great that an inventive effort is economically rational, even though the underlying scientific knowledge is scanty and hence the expected cost of making the invention is great. Edison's attempt to develop an incandescent lamp and Goodyear's attempt to improve the characteristics of rubber are cases in point. In these cases, since there was little useful underlying scientific knowledge, the invention procedure was trial and error, the next trial being roughly—but only roughly—indicated by a very loose theory formulated as the research proceeded. But though the inventors knew that it would probably prove costly to achieve their objective, they believed that the gains, if they were successful, were sufficiently great to make the effort profitable.

But often, though the inventor believes that there is great demand for a particular invention, it is not rational for him to attempt the invention, given the state of scientific knowledge. Expected cost will exceed expected revenue unless additional scientific knowledge can be obtained. If the expected cost of acquiring the relevant scientific knowledge is low, an organization interested in making a particular invention may undertake an applied scientific research project. A profit-maximizing firm will undertake a research project to solve problems related to a development effort if the expected gains—for example, reduction in development costs, or improvement in the final developed product—exceed expected research costs and if total research and development cost is exceeded by the expected net value of the invention. To the extent that the results of applied research are predictable and relate only to a specific invention desired by a firm, and to the extent that the firm can collect through the market the full value of the invention to society, opportunities for private profit through applied research will just match social benefits of applied research, and the optimum quantity of a society's resources will tend to be thus directed.

However, by no means all scientific research is directed toward practical problem-solving, though the line between basic scientific research and applied scientific research is hard to draw. There is a continuous spectrum of scientific activity. Moving from the applied-science end of the spectrum to the basic-science end, the degree of uncertainty about the results of specific research projects increases, and the goals become less clearly defined and less closely tied to the solution of a specific practical problem or the creation of a practical object. The loose defining of goals at the basic research end of the spectrum is a very rational adaptation to the great uncertainties involved and permits a greater expected payoff from the research dollar than would be possible if goals were more closely defined. For commonly, not just sometimes, in the course of a research project unexpected possibilities not closely related to the original objectives appear, and concurrently it may become clear that the original objectives are unobtainable or will be far more difficult to achieve than originally expected. While the direction of an applied research project must be closely constrained by the practical problem which must be solved, the direction of a basic research project may change markedly, opportunistically, as research proceeds and new possibilities appear. Some of the most striking scientific breakthroughs have resulted from research projects started with quite different ends in mind.

Pasteur's discovery of the value of inoculation with weakened disease strains is one of the more famous cases in point, but what is important is that the case is so similar to many others. While studying chicken cholera, Pasteur accidentally inoculated a group of chickens with a weak culture. The chickens became ill but, instead of dying, recovered. Since Pasteur did not want to waste chickens, he later reinoculated these chickens with fresh culture—one that was strong enough to kill an ordinary chicken—but these chickens remained healthy. At this point Pasteur's attention shifted to this interesting and potentially very (socially) significant phenomenon, and his resulting work, of course, brought about a major medical advance.

Applied research is relatively unlikely to result in significant breakthroughs in scientific knowledge save by accident, for, if significant breakthroughs are needed before a particular practical problem can be solved, the expected costs of achieving this breakthrough by a direct research effort are likely to be extremely high; hence applied research on the problem will not be undertaken, and invention will not be attempted. It is basic research, not applied research, from which significant advances have usually resulted. It is seriously to be doubted whether X-ray analysis would ever have been discovered by any group of scientists who, at the turn of the century, decided to find a means for examining the inner organs of the body or the inner structure of metal castings. Radio communication was impossible prior to the work of Maxwell and Hertz. Maxwell's work was directed toward explaining and elaborating the work of Faraday. Hertz built his equipment to test empirically some implications of Maxwell's equations. Marconi's practical invention was a simple adaptation of the Hertzian equipment. It seems most unlikely that a group of scientists in the mid-nineteenth century, attempting to develop a better method of long-range communication, would have developed Maxwell's equations and radio or anything nearly so good.

The limitations of an applied-research project constrained to the solution of a specific practical problem, and the practical value of many research projects where the goal is simply knowledge, not the solution of a practical problem, is well illustrated by the development of hybrid corn. During the latter half of the nineteenth century several attempts were made to improve corn yields. Many of the researchers directed their attention, at one time or another, to the inbreeding of corn to obtain a predictable and profitable strain. But as corn plants were inbred, though they tended to breed true, they also tended to deteriorate in yield and in quality. For this reason, applied researchers attempting to improve corn dropped this seemingly unpromising approach. But George Harrison Shull, a geneticist working with corn plants and interested in pure breeds not for their economic value but for experiments in genetics, produced several corn strains that bred true and then crossed these strains. His project was motivated by a desire to further the science of genetics, but a result was high-yield, predictable hybrid corn.

III. Basic Research and Private Profit

It is clear that for significant advances in knowledge we must look primarily to basic research; the social gains we may expect from basic research are obvious. But basic research efforts are likely to generate substantial external economies. Private-profit opportunities alone are not likely to draw as large a quantity of resources into basic research as is socially desirable.

Significant advances in scientific knowledge, the types of advances that are likely to result from successful basic research projects, very often have practical value in many fields. Consider the range of advances resulting from Boyle's gas law or Maxwell's equations. On Gibb's law of phases rests the design of equipment in fields as diverse as petroleum refining, rubber vulcanization, nitrogen fixation, and metal-ore separation. Few firms operate in so wide a field of economic activity that they are able themselves to benefit directly from all the new technological possibilities opened by the results of a successful basic research effort. In order to capture the value of the new knowledge in fields which the firm is unwilling to enter, the firm must patent the practical applications and sell or lease the patents to firms in the industries affected.

But significant advances in scientific knowledge are often not directly and immediately applicable to the solutions of practical problems and hence do not quickly result in patents. Often the new knowledge is of greatest value as a key input of other research projects which, in turn, may yield results of practical and patentable value. For this reason scientists have long argued for free and wide communication of research results, and for this reason natural "laws" and facts are not patentable. Thus it is quite likely that a firm will be unable to capture through patent rights the full economic value created in a basic-research project that it sponsors.

A firm with a narrow technological base is likely to find research profitable only at the applied end of the spectrum, where research can be directed toward solution of problems facing the firm, and where the research results can be quickly and easily translated into patentable products and processes. Such a firm is likely to be able to capture only a small share of the social benefits created by a basic research program it sponsors. On the other hand, a firm producing a wide range of products resting on a broad technological base may well find it profitable to support research toward the basic science end of the spectrum.

A broad technological base insures that, whatever direction the path of research may take, the results are likely to be of value to the sponsoring firm. It is for this reason that firms which support research toward the basic-science end of the spectrum are firms that have their fingers in many pies. The big chemical companies producing a range of products as wide as the field of chemistry itself, the Bell Telephone Company, General Electric, and Eastman Kodak immediately come to mind. It is not just the size of the companies that makes it worthwhile for them to engage in basic research. Rather it is their broad underlying technological base, the wide range of products they produce or will be willing to produce if their research efforts open possibilities. (Eastman Kodak entered the vitamin business when a research project resulted in a new way to synthesize vitamin B.) Strangely enough, economists have tended to see little economic justification for giant firms not built on economies of scale. Yet it is the many-product giants, not the single-product giants, which have been most technologically dynamic, and, to the extent that we wish the private sector of the economy to support basic research, we must look to these firms.

The importance of a broad technological base as a factor permitting a company to engage profitably in basic research is clearly illustrated by Carothers' famous research project for Du Pont. Carothers' work in linear superpolymers began as an unrestricted foray into the unknown with no particular practical objective in view. But the research was in a new field of chemistry, and Du Pont believed that any new chemical breakthrough would probably be of value to the company. The very lack of a specific objective, the flexibility of the research project, was an important factor behind its success. In the course of research Carothers obtained some superpolymers which at high temperatures became viscous fluids and observed that filaments could be obtained from these materials if a rod were dipped in the molten polymer and then withdrawn. At this discovery the focus of the project shifted to these filaments. Nylon was the result, but at the start of the project Carothers could not possibly have known that his research would lead him to the development of a new fiber.

A wide technological base (usually involving a diversified set of products) does not imply a position of monopoly power in any or all of the product markets, nor does a monopoly position in a market imply a wide technological base. Focusing attention on market position, a business firm operating in a competitive environment will seldom find it profitable to engage in a research project which is not likely to result quickly in something patentable, even if the firm can predict the nature of the research results, unless the firm keeps tight secrecy. For if the results of research cannot be quickly patented and are not kept secret, other firms producing similar products using similar processes will be free to use the results as an input of a development program of their own, designed to achieve a similar patentable objective. If competing firms develop a patentable product first, or develop a competing product, these firms will in effect steal from the research sponsoring firm, through price and product competition, a large share of the social utility created by research. In fact, many companies engaging in research keep their research findings secret until the new knowledge is put to practical use and the results are patented.

Many industries have attempted to reconcile their need for new knowledge with the lack of incentives to individual private firms to produce that new knowledge by establishing co-operative industry research organizations. To the extent that an industry rests on a field of science that is likely to get little attention in the absence of sponsorship by the firms in the industry, it may be in the interests of all the firms that research in this field be pushed, though each firm would prefer the others to do the financial pushing. An industrial co-operative research laboratory may well develop under these conditions, supported by all or by a large number of the firms in the industry, and undertaking research likely to be applicable to the technology of the industry. The motivation for these co-operative laboratories is only in part the high cost of research. More importantly, these laboratories are motivated by the fact that most of the firms will gain from the results of relatively basic research in certain fields whether or not they pay for it; hence little research will be undertaken in the absence of co-operation.

The preceding argument has been focused on external economies that open a gap between marginal private and marginal social benefit from basic research. Two other factors, working in the same direction, must be mentioned, if not discussed. First, the long lag that very often occurs between the initiation of a basic research project and the creation of something of marketable value may cause firms much concerned with short run survival, little concerned with profits many years from now, to place less value on basic-research projects than does society, even in the absence of external economies. This is not to say that all firms have a greater time-discount factor than does society as a whole, but it can be argued that many firms do. Second, the very large variance of the profit probability distribution from a basic research project will tend to cause a risk-avoiding firm, without the economic resources to spread the risk by running a number of basic-research projects at once, to value a basic-research project at significantly less than its expected profitability and hence, even in the absence of external economies, at less than its social value.

IV. Is Current Social Policy Adequate?

It seems clear that, were the field of basic research left exclusively to private firms operating independently of each other and selling in competitive markets, profit incentives would not draw so large a quantity of resources to basic research as is socially desirable. But in fact basic research has not been the exclusive domain of private firms. Government and other non-profit institutions (principally universities) together spend more on basic research, and undertake more basic research in their own laboratories, than does industry. Since we are presently supporting collectively such a large share (more than half) of basic research, is it not possible that total basic-research expenditure (the sum of private and public efforts) equals or exceeds the social optimum? This is a tricky theoretical question. However, if basic research can be considered as a homogeneous commodity, like potato chips, and hence the public can be assumed to be indifferent between the research results produced in government or in industry laboratories; if the marginal cost of research output is assumed to be no greater in non-profit laboratories than in profit-oriented laboratories; and if industry laboratories are assumed to operate where marginal revenue equals marginal cost, then the fact that industry laboratories do basic research at all is itself evidence that we should increase our expenditure on basic research.

Public support of basic research has primarily been in the form of contracts let with private firms and in the establishment and support of a large number of non-profit laboratories. Save for the effects of tax laws (which apply to all business cost–incurring activities), public policy has not acted to *shift* the marginal cost curve of the basic-research industry. Public policy has resulted in shifts *along* the curve. Nor has public policy acted to drive marginal social utility to marginal private utility. External economies still exist at the margin. Clearly then, if industry laboratories are in profit-maximizing equilibrium, society would benefit from an increase in basic-research expenditure in industry laboratories, holding research efforts elsewhere constant, for the marginal social benefit of basic research in private laboratories exceeds marginal cost to the firm, which under our assumptions still equals alternative cost. But perhaps non-profit laboratories are spending too much on basic research—are operating beyond the point at which marginal cost equals marginal social benefit—and therefore it is desirable to reduce research expenditure in this sector. Given our assumptions, this cannot be. For, if marginal cost is no greater in non-profit laboratories than in industry laboratories, and society cannot distinguish between the fruits of research undertaken in the two kinds of laboratory—that is, if marginal social benefit is the same in the public and the private sector—and if it is socially desirable that expenditure on basic research be increased in industry laboratories, then it is also socially desirable that research expenditure be increased in non-profit laboratories. For if marginal social benefit exceeds marginal cost in industry laboratories, so does it in non-profit laboratories.

The assumptions on which the preceding argument is based rest but shakily on fact. Basic research certainly is not a homogeneous commodity. The types of knowledge generated, say, in an air-force project on high-speed gas flows, a Du Pont project on high polymer chemistry, or a Harvard project on solid-state physics are not perfectly substitutable. The knowledge generated will certainly be different, and in a reasonably predictable way. And, once the non-homogeneity of basic research is admitted, the concept of relative marginal cost becomes fuzzy. Thus one cannot make an airtight statement, based on welfare economics, that we are not spending as much on basic scientific research as we should. But I believe that the evidence certainly points in that direction.

V. Some Policy Implications

Though the profit motive may stimulate private industry to spend an amount on applied research reasonably close to the amount that is socially desirable, it is clear from the preceding analysis that under our present economic structure the social benefits of basic research are not adequately reflected in opportunities for private profit. Indeed, there is a basic contradiction between the conditions necessary for efficient basic research—few or no constraints on the direction of research with full and free dissemination of research results—and full appropriation of the gains from sponsoring basic research in a competitive economy.

This is not to say that some firms could not profitably increase their basic research effort. Some may presently be operating well to the left of their maximum profit point. But to the extent that we want our economy to remain competitive and want efficient use of basic-research funds, the laboratories of colleges, universities, and other non-profit institutions must perform a large share of our basic research if we are to put as much of our resources into basic research as we should. Although several laboratories of private industry have made significant contributions in the field of basic science, these contributions have been few and far between. If we advocate that basic research be increasingly undertaken by business and if we believe that business should be motivated by profit, we must accept the growth of large firms with a wide technological base, with virtual monopolies in several markets. If we do not want such an economic structure, then only to the extent that we think it desirable that private firms look to motives other than profit can we argue that industry laboratories should perform a significantly enlarged share of our basic research. In either case we undermine many of the economic arguments for a free-enterprise economy. If we want to maintain our enterprise economy, basic research must be a matter of conscious social policy.

This is not the place to suggest a menu of policies—the bill of fare offered in the National Science Foundation booklet on basic research lists some of the actions that might be considered. However, it does seem appropriate to suggest that public policy on basic research should recognize the following points:

1. The problem of getting enough resources to flow into basic research is basically the classical external-economy problem.[5] External economies result from two facts: first, that research results often are of little value to the firm that sponsors the research, though of great value to another firm, and, second, that research results often cannot be quickly patented. It therefore seems desirable to encourage the further growth of a "basic-research industry," a group of institutions that benefit from the results of almost any basic-research project they undertake. University laboratories should certainly continue to be a major part of this industry. However, an increasingly important role should probably be played by industry-oriented laboratories not owned by specific industries but doing research on contract for a diversified set of clients. Such laboratories would usually have at least one client who could benefit from almost any research breakthrough.

2. The incentives generated in a profit economy for firms to keep research findings secret produce results that are, in a static sense, economically inefficient. The use of existing knowledge by one firm in no way reduces the ability of another firm to use that same knowledge, though the incentive to do so may be reduced. The marginal social cost of using knowledge that already exists is zero. For maximum static eco-

5. The external economy aspect of basic research reacts back through the price system to undervalue pure scientists relative to engineers.

nomic efficiency, knowledge should be administered as a common pool, with free access to all who can use the knowledge. But, if scientific knowledge is thus administered, the incentives of private firms to create new knowledge will be reduced. This is another case in which static efficiency and dynamic efficiency may conflict. It is socially desirable that as much of our basic research effort as possible be undertaken in institutions interested in the quick publication of research results if marginal costs are comparable. In the absence of incentives to private firms to publish research results quickly (such incentives might be legislated) a dollar spent on basic research in a university laboratory is worth more to society than a dollar spent in an industry laboratory, again, if productivity is comparable.

3. If society places the brunt of the basic-research burden on universities, funds must be provided for this purpose. The current Department of Defense policies of letting huge applied research projects with universities should either be reconsidered or complemented with other policies designed to prevent the increased applied-research burden from drawing university facilities and scientists away from basic research. This is not to say that universities cannot effectively undertake applied research. Rather it is to say that their comparative advantage lies in basic research.

4

Economic Welfare and the Allocation of Resources for Invention

Kenneth J. Arrow

Invention is here interpreted broadly as the production of knowledge. From the viewpoint of welfare economics, the determination of optimal resource allocation for invention will depend on the technological characteristics of the invention process and the nature of the market for knowledge.

The classic question of welfare economics will be asked here: to what extent does perfect competition lead to an optimal allocation of resources? We know from years of patient refinement that competition insures the achievement of a Pareto optimum under certain hypotheses. The model usually assumes among other things, that (1) the utility functions of consumers and the transformation functions of producers are well-defined functions of the commodities in the economic system, and (2) the transformation functions do not display indivisibilities (more strictly, the transformation sets are convex). The second condition needs no comment. The first seems to be innocuous but in fact conceals two basic assumptions of the usual models. It prohibits uncertainty in the production relations and in the utility functions, and it requires that all the commodities relevant either to production or to the welfare of individuals be traded on the market. This will not be the case when a commodity for one reason or another cannot be made into private property.

We have then three of the classical reasons for the possible failure of perfect competition to achieve optimality in resource allocation: indivisibilities, inappropriability, and uncertainty. The first problem has been

Reprinted with permission of the National Bureau of Economic Research. From *The Rate and Direction of Inventive Activity* (Princeton: Princeton University Press, 1962), 609–26.

much studied in the literature under the heading of marginal cost pricing and the second under that of divergence between social and private benefit (or cost), but the theory of optimal allocation of resources under uncertainty has had much less attention. I will summarize what formal theory exists and then point to the critical notion of information, which arises only in the context of uncertainty. The economic characteristics of information as a commodity and, in particular, of invention as a process for the production of information are next examined. It is shown that all three of the reasons given above for a failure of the competitive system to achieve an optimal resource allocation hold in the case of invention. On theoretical grounds a number of considerations are adduced as to the likely biases in the misallocation and the implications for economic organization.[1]

Resource Allocation under Uncertainty

The role of the competitive system in allocating uncertainty seems to have received little systematic attention.[2] I will first sketch an ideal economy in which the allocation problem can be solved by competition and then indicate some of the devices in the real world which approximate this solution.

Suppose for simplicity that uncertainty occurs only in production relations. Producers have to make a decision on inputs at the present moment, but the outputs are not completely predictable from the inputs. We may formally describe the outputs as determined by the inputs and a "state of nature" which is unknown to the producers. Let us define a "commodity-option" as a commodity in the ordinary sense labeled with a state of nature. This definition is analogous to the differentiation of a given physical commodity according to date in capital theory or according to place in location theory. The production of a given commodity

1. For other analyses with similar points of view, see R. R. Nelson, "The Simple Economics of Basic Scientific Research," *Journal of Political Economy,* 1959, pp. 297–306 (chapter 3, this volume); and C. J. Hitch, "The Character of Research and Development in a Competitive Economy," The RAND Corporation, p. 1297, May 1958.

2. The first studies I am aware of are the papers of M. Allais and myself, both presented in 1952 to the Colloque International sur le Risque in Paris; see M. Allais, "Généralisation des théories de l'équilibre économique général et du rendement social au cas du risque," and K. J. Arrow, "Rôle des valeurs bousières pour la répartition la meilleure des risques," both in *Economètrie,* Colloques Internationaux du Centre National de la Recherche Scientifique, Vol. XL, Paris, Centre National de la Recherche Scientifique, 1953. Allais' paper has also appeared in *Econometrica,* 1953, pp. 269–290. The theory has received a very elegant generalization by G. Debreu in *Theory of Values,* New York, Wiley, 1959, Chap. VII.

under uncertainty can then be described as the production of a vector of commodity-options.

This description can be most easily exemplified by reference to agricultural production. The state of nature may be identified with the weather. Then, to any given set of inputs there corresponds a number of bushels of wheat if the rainfall is good and a different number if rainfall is bad. We can introduce intermediate conditions of rainfall in any number as alternative states of nature; we can increase the number of relevant variables which enter into the description of the state of nature, for example by adding temperature. By extension of this procedure, we can give a formal description of any kind of uncertainty in production.

Suppose—and this is the critical idealization of the economy—we have a market for all commodity-options. What is traded on each market are contracts in which the buyers pay an agreed sum and the sellers agree to deliver prescribed quantities of a given commodity *if* a certain state of nature prevails and nothing if that state of nature does not occur. For any given set of inputs, the firm knows its output under each state of nature and sells a corresponding quantity of commodity-options; its revenue is then completely determined. It may choose its inputs so as to maximize profits.

The income of consumers is derived from their sale of supplies, including labor, to firms and their receipt of profits, which are assumed completely distributed. They purchase commodity-options so as to maximize their expected utility given the budget restraint imposed by their incomes. An equilibrium is reached on all commodity-option markets, and this equilibrium has precisely the same Pareto-optimality properties as competitive equilibrium under certainty.

In particular, the markets for commodity-options in this ideal model serve the function of achieving an optimal allocation of risk bearing among the members of the economy. This allocation takes account of differences in both resources and tastes for risk bearing. Among other implications, risk bearing and production are separated economic functions. The use of inputs, including human talents, in their most productive mode is not inhibited by unwillingness or inability to bear risks by either firms or productive agents.

But the real economic system does not possess markets for commodity-options. To see what substitutes exist, let us first consider a model economy at the other extreme, in that no provisions for reallocating risk bearing exist. Each firm makes its input decisions; then outputs are produced as determined by the inputs and the state of nature. Prices are then set to clear the market. The prices that finally prevail will be a function of the state of nature.

The firm and its owners cannot relieve themselves of risk bearing in this model. Hence any unwillingness or inability to bear risks will give rise to a nonoptimal allocation of resources, in that there will be discrimination against risky enterprises as compared with the optimum. A preference for risk might give rise to misallocation in the opposite direction, but the limitations of financial resources are likely to make underinvestment in risky enterprises more likely than the opposite. The inability of individuals to buy protection against uncertainty similarly gives rise to a loss of welfare.

In fact, a number of institutional arrangements have arisen to mitigate the problem of assumption of risk. Suppose that each firm and individual in the economy could forecast perfectly what prices would be under each state of nature. Suppose further there were a lottery on the states of nature, so that before the state of nature is known any individual or firm may place bets. Then it can be seen that the effect from the viewpoint of any given individual or firm is the same as if there were markets for commodity-options of all types, since any commodity-option can be achieved by a combination of a bet on the appropriate state of nature and an intention to purchase or sell the commodity in question if the state of nature occurs.

References to lotteries and bets may smack of frivolity, but we need only think of insurance to appreciate that the shifting of risks through what are in effect bets on the state of nature is a highly significant phenomenon. If insurance were available against any conceivable event, it follows from the preceding discussion that optimal allocation would be achieved. Of course, insurance as customarily defined covers only a small range of events relevant to the economic world; much more important in shifting risks are securities, particularly common stocks and money. By shifting freely their proprietary interests among different firms, individuals can to a large extent bet on the different states of nature which favor firms differentially. This freedom to insure against many contingencies is enhanced by the alternatives of holding cash and going short.

Unfortunately, it is only too clear that the shifting of risks in the real world is incomplete. The great predominance of internal over external equity financing in industry is one illustration of the fact that securities do not completely fulfill their allocative role with respect to risks. There are a number of reasons why this should be so, but I will confine myself to one, of special significance with regard to invention. In insurance practice, reference is made to the moral factor as a limit to the possibilities of insurance. For example, a fire insurance policy cannot exceed in amount the value of the goods insured. From the purely actuarial standpoint, there is no reason for this limitation; the reason for the limit is that the insurance policy changes the incentives of the insured, in this case, creating an incentive for arson or at the very least for carelessness. The general principle is the difficulty of distinguishing between a state of nature and a decision by the insured. As a result, any insurance policy and in general any device for shifting risks can have the effect of dulling incentives. A fire insurance policy, even when limited in amount to the value of the goods covered, weakens the motivation for fire prevention. Thus, steps which improve the efficiency of the economy with respect to risk bearing may decrease its technical efficiency.

One device for mitigating the adverse incentive effects of insurance is coinsurance; the insurance extends only to part of the amount at risk for the insured. This device is used, for example, in coverage of medical risks. It clearly represents a compromise between incentive effects and allocation of risk bearing, sacrificing something in both directions.

Two exemplifications of the moral factor are of special relevance in regard to highly risky business activities, including invention. Success in such activities depends on an inextricable tangle of objective uncertainties and decisions of the entrepreneurs and is certainly uninsurable. On the other hand, such activities should be undertaken if the expected return exceeds the market rate of return, no matter what the variance is.[3] The existence of common stocks would seem to solve the allocation problem; any individual stockholder can reduce his risk by buying only a small part of the stock and diversifying his portfolio to achieve his own preferred risk level. But then again the actual managers no longer receive the full reward of their decisions; the shifting of risks is again accompanied by a weakening of incentives to efficiency. Substitute motivations, whether pecuniary, such as executive compensation and profit sharing, or nonpecuniary, such as prestige, may be found, but the dilemma of the moral factor can never be completely resolved.

A second example is the cost-plus contract in one of its various forms. When production costs on military items are highly uncertain, the military establishment will pay, not a fixed unit price, but the cost of production plus an amount which today is usually a fixed fee. Such a contract

3. The validity of this statement depends on some unstated assumptions, but the point to be made is unaffected by minor qualifications.

could be regarded as a combination of a fixed-price contract with an insurance against costs. The insurance premium could be regarded as the difference between the fixed price the government would be willing to pay and the fixed fee.

Cost-plus contracts are necessitated by the inability or unwillingness of firms to bear the risks. The government has superior risk bearing ability and so the burden is shifted to it. It is then enabled to buy from firms on the basis of their productive efficiency rather than their risk bearing ability, which may be only imperfectly correlated. But cost-plus contracts notoriously have their adverse allocative effects.[4]

This somewhat lengthy digression on the theory of risk bearing seemed necessitated by the paucity of literature on the subject. The main conclusions to be drawn are the following: (1) the economic system has devices for shifting risks, but they are limited and imperfect; hence, one would expect an underinvestment in risky activities; (2) it is undoubtedly worthwhile to enlarge the variety of such devices, but the moral factor creates a limit to their potential.

Information as a Commodity

Uncertainty usually creates a still more subtle problem in resource allocation; information becomes a commodity. Suppose that in one part of the economic system an observation has been made whose outcome, if known, would affect anyone's estimates of the probabilities of the different states of nature. Such observations arise out of research but they also arise in the daily course of economic life as a by-product of other economic activities. An entrepreneur will automatically acquire a knowledge of demand and production conditions in his field which is available to others only with special effort. Information will frequently have an economic value, in the sense that anyone possessing the information can make greater profits than would otherwise be the case.

It might be expected that information will be traded in, and of course to a considerable extent this is the case, as is illustrated by the numerous economic institutions for transmission of information, such as newspapers. But in many instances, the problem of an optimal allocation is sharply raised. The cost of transmitting a given body of information is frequently very low. If it were zero, then optimal allocation would obvi-

4. These remarks are not intended as a complete evaluation of cost-plus contracts. In particular, there are, to a certain extent, other incentives which mitigate the adverse effects on efficiency.

ously call for unlimited distribution of the information without cost. In fact, a given piece of information is by definition an indivisible commodity, and the classical problems of allocation in the presence of indivisibilities appear here. The owner of the information should not extract the economic value which is there, if optimal allocation is to be achieved; but he is a monopolist, to some small extent, and will seek to take advantage of this fact.

In the absence of special legal protection, the owner cannot, however, simply sell information on the open market. Any one purchaser can destroy the monopoly, since he can reproduce the information at little or no cost. Thus the only effective monopoly would be the use of the information by the original possessor. This, however, will not only be socially inefficient, but also may not be of much use to the owner of the information either, since he may not be able to exploit it as effectively as others.

With suitable legal measures, information may become an appropriable commodity. Then the monopoly power can indeed be exerted. However, no amount of legal protection can make a thoroughly appropriable commodity of something so intangible as information. The very use of the information in any productive way is bound to reveal it, at least in part. Mobility of personnel among firms provides a way of spreading information. Legally imposed property rights can provide only a partial barrier, since there are obviously enormous difficulties in defining in any sharp way an item of information and differentiating it from other similar-sounding items.

The demand for information also has uncomfortable properties. In the first place, the use of information is certainly subject to indivisibilities; the use of information about production possibilities, for example, need not depend on the rate of production. In the second place, there is a fundamental paradox in the determination of demand for information; its value for the purchaser is not known until he has the information, but then he has in effect acquired it without cost. Of course, if the seller can retain property rights in the use of the information, this would be no problem, but given incomplete appropriability, the potential buyer will base his decision to purchase information on less than optimal criteria. He may act, for example, on the average value of information in that class as revealed by past experience. If any particular item of information has differing values for different economic agents, this procedure will lead both to a nonoptimal purchase of information at any given price and also to a nonoptimal allocation of the information purchased.

It should be made clear that from the standpoint of efficiently distributing an existing stock of information, the difficulties of appropriation are an advantage, provided there are no costs of transmitting information, since then optimal allocation calls for free distribution. The chief point made here is the difficulty of creating a market for information if one should be desired for any reason.

It follows from the preceding discussion that costs of transmitting information create allocative difficulties which would be absent otherwise. Information should be transmitted at marginal cost, but then the demand difficulties raised above will exist. From the viewpoint of optimal allocation, the purchasing industry will be faced with the problems created by indivisibilities; and we still leave unsolved the problem of the purchaser's inability to judge in advance the value of the information he buys. There is a strong case for centralized decision making under these circumstances.

Invention as the Production of Information

The central economic fact about the processes of invention and research is that they are devoted to the production of information. By the very definition of information, invention must be a risky process, in that the output (information obtained) can never be predicted perfectly from the inputs. We can now apply the discussion of the preceding two sections.

Since it is a risky process, there is bound to be some discrimination against investment in inventive and research activities. In this field, especially, the moral factor will weigh heavily against any kind of insurance or equivalent form of risk bearing. Insurance against failure to develop a desired new product or process would surely very greatly weaken the incentives to succeed. The only way, within the private enterprise system, to minimize this problem is the conduct of research by large corporations with many projects going on, each small in scale compared with the net revenue of the corporation. Then the corporation acts as its own insurance company. But clearly this is only an imperfect solution.

The deeper problems of misallocation arise from the nature of the product. As we have seen, information is a commodity with peculiar attributes, particularly embarrassing for the achievement of optimal allocation. In the first place, any information obtained, say a new method of production, should, from the welfare point of view, be available free of charge (apart from the cost of transmitting information). This insures optimal utilization of the information but of course provides no incentive for investment in research. In an ideal socialist economy, the reward for invention would be completely separated from any charge to the users of the information.[5] In a free enterprise economy, inventive activity is supported by using the invention to create property rights; precisely to the extent that it is successful, there is an underutilization of the information. The property rights may be in the information itself, through patents and similar legal devices, or in the intangible assets of the firm if the information is retained by the firm and used only to increase its profits.

The first problem, then, is that in a free enterprise economy the profitability of invention requires a nonoptimal allocation of resources. But it may still be asked whether or not the allocation of resources to inventive activity is optimal. The discussion of the preceding section makes it clear that we would not expect this to be so; that, in fact, a downward bias in the amount of resources devoted to inventive activity is very likely. Whatever the price, the demand for information is less than optimal for two reasons: (1) since the price is positive and not at its optimal value of zero, the demand is bound to be below the optimal; (2) as seen before, at any given price, the very nature of information will lead to a lower demand than would be optimal.

As already remarked, the inventor will in any case have considerable difficulty in appropriating the information produced. Patent laws would have to be unimaginably complex and subtle to permit such appropriation on a large scale. Suppose, as the result of elaborate tests, some metal is discovered to have a desirable property, say resistance to high heat. Then of course every use of the metal for which this property is relevant would also use this information, and the user would be made to pay for it. But, even more, if another inventor is stimulated to examine chemically related metals for heat resistance, he is using the information already discovered and should pay for it in some measure; and any beneficiary of his discoveries should also pay. One would have to have elaborate distinctions of partial property rights of all degrees to make the system at all tolerable. In the interests of the possibility of enforcement, actual patent laws sharply restrict the range of appropriable information and thereby reduce the incentives to engage in inventive and research activities.

These last considerations bring into focus the interdependence of inventive activities, which reinforces the difficulties in achieving an optimal allocation of the results. Information is not only the product of inventive

5. This separation exists in the Soviet Union, accoring to N. M. Kaplan and R. H. Moorsteen of the RAND Corporation (verbal communication).

activity, it is also an input—in some sense, the major input apart from the talent of the inventor. The school of thought that emphasizes the determination of invention by the social climate as demonstrated by the simultaneity of inventions in effect emphasizes strongly the productive role of previous information in the creation of new information. While these interrelations do not create any new difficulties in principle, they intensify the previously established ones. To appropriate information for use as a basis for further research is much more difficult than to appropriate it for use in producing commodities; and the value of information for use in developing further information is much more conjectural than the value of its use in production and therefore much more likely to be underestimated. Consequently, if a price is charged for the information, the demand is even more likely to be suboptimal.

Thus basic research, the output of which is only used as an informational input into other inventive activities, is especially unlikely to be rewarded. In fact, it is likely to be of commercial value to the firm undertaking it only if other firms are prevented from using the information obtained. But such restriction on the transmittal of information will reduce the efficiency of inventive activity in general and will therefore reduce its quantity also. We may put the matter in terms of sequential decision making. The a priori probability distribution of the true state of nature is relatively flat to begin with. On the other hand, the successive a posteriori distributions after more and more studies have been conducted are more and more sharply peaked or concentrated in a more limited range, and we therefore have better and better information for deciding what the next step in research shall be. This implies that, at the beginning, the preferences among alternative possible lines of investigation are much less sharply defined than they are apt to be later on and suggests, at least, the importance of having a wide variety of studies to begin with, the less promising being gradually eliminated as information is accumulated.[6] At each stage the decisions about the next step should be based on all available information. This would require an unrestricted flow of information among different projects, which is incompatible with the complete decentralization of an ideal free enterprise system. When the production of information is important, the classic economic case in

6. The importance of parallel research developments in the case of uncertainty has been especially stressed by Burton H. Klein; see his "A Radical Proposal for R. and D.," *Fortune,* May 1958, p. 112 ff.; and Klein and W. H. Meckling, "Application of Operations Research to Development Decisions," *Operations Research,* 1958, pp. 352–363.

which the price system replaces the detailed spread of information is no longer completely applicable.

To sum up, we expect a free enterprise economy to underinvest in invention and research (as compared with an ideal) because it is risky, because the product can be appropriated only to a limited extent, and because of increasing returns in use. This underinvestment will be greater for more basic research. Further, to the extent that a firm succeeds in engrossing the economic value of its inventive activity, there will be an underutilization of that information as compared with an ideal allocation.

Competition, Monopoly, and the Incentive to Innovate

It may be useful to remark that an incentive to invent can exist even under perfect competition in the product markets though not, of course, in the "market" for the information contained in the invention. This is especially clear in the case of a cost-reducing invention. Provided only that suitable royalty payments can be demanded, an inventor can profit without disturbing the competitive nature of the industry. The situation for a new product invention is not very different; by charging a suitable royalty to a competitive industry, the inventor can receive a return equal to the monopoly profits.

I will examine here the incentives to invent for monopolistic and competitive markets, that is, I will compare the potential profits from an invention with the costs. The difficulty of appropriating the information will be ignored; the remaining problem is that of indivisibility in use, an inherent property of information. A competitive situation here will mean one in which the industry produces under competitive conditions, while the inventor can set an arbitrary royalty for the use of his invention. In the monopolistic situation, it will be assumed that only the monopoly itself can invent. Thus a monopoly is understood here to mean barriers to entry; a situation of temporary monopoly, due perhaps to a previous innovation, which does not prevent the entrance of new firms with innovations of their own, is to be regarded as more nearly competitive than monopolistic for the purpose of this analysis. It will be argued that the incentive to invent is less under monopolistic than under competitive conditions but even in the latter case it will be less than is socially desirable.

We will assume constant costs both before and after the invention, the unit costs being c before the invention and $c' < c$ afterward. The competitive price before invention will therefore be c. Let the corresponding demand be x_c. If r is the level of unit royalties, the competitive price after the invention will be $c' + r$, but this cannot of course be higher than c, since firms are always free to produce with the old methods.

It is assumed that both the demand and the marginal revenue curves are decreasing. Let $R(x)$ be the marginal revenue curve. Then the monopoly output before invention, x_m, would be defined by the equation

$$R(x_m) = c.$$

Similarly, the monopoly output after invention is defined by

$$R(x'_m) = c'.$$

Let the monopoly prices corresponding to outputs x_m and x'_m, respectively, be p_m and p'_m. Finally, let P and P' be the monopolist's profits before and after invention, respectively.

What is the optimal royalty level for the inventor in the competitive case? Let us suppose that he calculates p'_m, the optimal monopoly price which would obtain in the postinvention situation. If the cost reduction is sufficiently drastic that $p'_m < c$, then his most profitable policy is to set r so that the competitive price is p'_m, i.e., let

$$r = p'_m - c'.$$

In this case, the inventor's royalties are equal to the profits a monopolist would make under the same conditions, i.e., his incentive to invent will be P'.

Suppose, however, it turns out that $p'_m > c$. Since the sales price cannot exceed c, the inventor will set his royalties at

$$r = c - c'.$$

The competitive price will then be c, and the sales will remain at x_c. The inventor's incentive will then be $x_c(c - c')$.

The monopolist's incentive, on the other hand, is clearly $P' - P$. In the first of the two cases cited, the monopolist's incentive is obviously less than the inventor's incentive under competition, which is P', not $P' - P$. The preinvention monopoly power acts as a strong disincentive to further innovation.

The analysis is slightly more complicated in the second case. The monopolist's incentive, $P' - P$, is the change in revenue less the change in total cost of production, i.e.,

$$(4.1) \qquad P' - P = \int_{x_m}^{x'_m} R(x)dx - c'x'_m + cx_m.$$

Since the marginal revenue $R(x)$ is diminishing, it must always be less than $R(x_m) = c$ as x increases from x_m to x'_m, so that

$$(4.2) \qquad \int_{x_m}^{x'_m} R(x)dx < c(x'_m - x_m),$$

and

$$(4.3) \qquad P' - P < c(x'_m - x_m) - c'x'_m + cx_m = (c - c')x'_m.$$

In the case being considered, the postinvention monopoly price, p'_m, is greater than c. Hence, with a declining demand curve, $x'_m < x_c$. The above inequality shows that the monopolist's incentive is always less than the cost reduction on the postinvention monopoly output, which in this case is, in turn, less than the competitive output (both before and after invention). Since the inventor's incentive under competition is the cost reduction on the competitive output, it will again always exceed the monopolist's incentive.

It can be shown that, if we consider differing values of c', the difference between the two incentives increases as c' decreases, reaching its maximum of P (preinvention monopoly profits) for c' sufficiently large for the first case to hold. The ratio of the incentive under competition to that under monopoly, on the other hand, though always greater than 1, decreases steadily with c'. For c' very close to c (i.e., very minor inventions), the ratio of the two incentives is approximately x_c/x_m, i.e., the ratio of monopoly to competitive output.[7]

7. To sketch the proof of these statements quickly, note that, as c' varies, P is a constant. Hence, from the formula for $P' - P$, we see that

$$d(P' - P)/dc' = dP'/dc' = R(x'_m)(dx'_m/dc') - c'(dx'_m/dc') - x'_m = -x'_m,$$

since $R(x'_m) = c'$. Let $F(c')$ be the difference between the incentives to invent under competitive and under monopolistic conditions. In the case where $p'_m < c$, this difference is the constant P. Otherwise,

$$F(c') = x_c(c - c') - (P' - P),$$

so that

$$dF/dc' = x'_m - x_c.$$

For the case considered, we must have $x'_m < x_c$, as seen in the text. Hence, $dF/dc' \leq 0$, so that $F(c')$ increases as c' decreases.

Let $G(c')$ be the ratio of the incentive under competition to that under monopoly. If $p'_m < c$, then

$$G(c') = P'/(P' - P),$$

which clearly decreases as c' decreases. For $p'_m > c$, we have

The only ground for arguing that monopoly may create superior incentives to invent is that appropriability may be greater under monopoly than under competition. Whatever differences may exist in this direction must, of course, still be offset against the monopolist's disincentive created by his preinvention monopoly profits.

The incentive to invent in competitive circumstances may also be compared with the social benefit. It is necessary to distinguish between the realized social benefit and the potential social benefit, the latter being the benefit which would accrue under ideal conditions, which, in this case, means the sale of the product at postinvention cost, c'. Clearly, the potential social benefit always exceeds the realized social benefit. I will show that the realized social benefit, in turn, always equals or exceeds the competitive incentive to invent and, a fortiori, the monopolist's incentive.

Consider again the two cases discussed above. If the invention is sufficiently cost reducing so that $p'_m < c$, then there is a consumers' benefit, due to the lowering of price, which has not been appropriated by the inventor. If not, then the price is unchanged, so that the consumers' position is unchanged, and all benefits do go to the inventor. Since by assumption all the producers are making zero profits both before and after the invention, we see that the inventor obtains the entire realized social benefit of moderately cost reducing inventions but not of more radical inventions. Tentatively, this suggests a bias against major inventions, in the sense that an invention, part of whose costs could be paid for by lump-sum payments by consumers without making them worse off than before, may not be profitable at the maximum royalty payments that can be extracted by the inventor.

Alternative Forms of Economic Organization in Invention

The previous discussion leads to the conclusion that for optimal allocation to invention it would be necessary for the government or some other agency not governed by profit-and-loss criteria to finance research and invention. In fact, of course, this has always happened to a certain

---

$$G(c') = x_c\,(c - c')/(P' - P).$$

Then,

$$dG/dc' = [-\,(P' - P)x_c + x_c(c - c')x'_m]/(P' - P)^2.$$

Because of the upper bound for $P' - P$ established in the text, the numerator must be positive; the ratio decreases as c' decreases.

Finally, if we consider c' very close to c, $G(c')$ will be approximately equal to the ratio of the derivatives of the numerator and denominator (L'Hôpital's rule), which is x_c/x'_m, and which approaches x_c/x_m as c' approaches c.

extent. The bulk of basic research has been carried on outside the industrial system, in universities, in the government, and by private individuals. One must recognize here the importance of nonpecuniary incentives, both on the part of the investigators and on the part of the private individuals and governments that have supported research organizations and universities. In the latter, the complementarity between teaching and research is, from the point of view of the economy, something of a lucky accident. Research in some more applied fields, such as agriculture, medicine, and aeronautics, has consistently been regarded as an appropriate subject for government participation, and its role has been of great importance.

If the government and other nonprofit institutions are to compensate for the underallocation of resources to invention by private enterprise, two problems arise: how shall the amount of resources devoted to invention be determined, and how shall efficiency in their use be encouraged? These problems arise whenever the government finds it necessary to engage in economic activities because indivisibilities prevent the private economy from performing adequately (highways, bridges, reclamation projects, for example), but the determination of the relative magnitudes is even more difficult here. Formally, of course, resources should be devoted to invention until the expected marginal social benefit there equals the marginal social benefit in alternative uses, but in view of the presence of uncertainty, such calculations are even more difficult and tenuous than those for public works. Probably all that could be hoped for is the estimation of future rates of return from those in the past, with investment in invention being increased or decreased accordingly as some average rate of return over the past exceeded or fell short of the general rate of return. The difficulties of even ex post calculation of rates of return are formidable though possibly not insuperable.[8]

The problem of efficiency in the use of funds devoted to research is one that has been faced internally by firms in dealing with their own research departments. The rapid growth of military research and development has led to a large-scale development of contractual relations between producers and a buyer of invention and research. The problems encountered in assuring efficiency here are the same as those that would be met if the government were to enter upon the financing of invention

8. For an encouraging study of this type, see Z. Griliches, "Research Costs and Social Returns: Hybrid Corn and Related Innovations," *Journal of Political Economy,* 1958, pp. 419–431.

and research in civilian fields. The form of economic relation is very different from that in the usual markets. Payment is independent of product; it is governed by costs, though the net reward (the fixed fee) is independent of both. This arrangement seems to fly in the face of the principles for encouraging efficiency, and doubtless it does lead to abuses, but closer examination shows both mitigating factors and some explanation of its inevitability. In the first place, the awarding of new contracts will depend in part on past performance, so that incentives for efficiency are not completely lacking. In the second place, the relation between the two parties to the contract is something closer than a purely market relation. It is more like the sale of professional services, where the seller contracts to supply not so much a specific result as his best judgment. (The demand for such services also arises from uncertainty and the value of information.) In the third place, payment by results would involve great risks for the inventor, risks against which, as we have seen, he could hedge only in part.

There is clear need for further study of alternative methods of compensation. For example, some part of the contractual payment might depend on the degree of success in invention. But a more serious problem is the decision as to which contracts to let. One would need to examine the motivation underlying government decision making in this area. Hitch has argued that there are biases in governmental allocation, particularly against risky invention processes, and an excessive centralization, though the latter could be remedied by better policies.[9]

One can go further. There is really no need for the firm to be the fundamental unit of organization in invention; there is plenty of reason to suppose that individual talents count for a good deal more than the firm as an organization. If provision is made for the rental of necessary equipment, a much wider variety of research contracts with individuals as well as firms and with varying modes of payment, including incentives, could be arranged. Still other forms of organization, such as research institutes financed by industries, the government, and private philanthropy, could be made to play an even livelier role than they now do.

ACKNOWLEDGMENTS

I have benefitted greatly from the comments of my colleague, William Capron. I am also indebted to Richard R. Nelson, Edward Phelps, and Sidney Winter of the RAND Corporation for their helpful discussion.

9. Op. cit.

PART III

Science Conceived as a Problem of Information Processing

5

Note on the Theory of the Economy of Research

Charles Sanders Peirce

When a research is of a quantitative nature, the progress of it is marked by the diminution of the probable error. The results of non-quantitative researches also have an inexactitude or indeterminacy which is analogous to the probable error of quantitative determinations. To this inexactitude, although it be not numerically expressed, the term "probable error" may be conveniently extended.

The doctrine of Economy, in general, treats of the relations between utility and cost. That branch of it which relates to research considers the relations between the utility and the cost of diminishing the probable error of our knowledge. Its main problem is how with a given expenditure of money, time, and energy, to obtain the most valuable addition to our knowledge.

Let r denote the probable error of any result; and write $s = 1/r$. Let $\mathrm{U}r \cdot \mathrm{d}r$ denote the infinitesimal utility of any infinitesimal diminution, $\mathrm{d}r$, of r. Let $\mathrm{V}s \cdot \mathrm{d}s$ denote the infinitesimal cost of any infinitesimal increase, $\mathrm{d}s$, of s. The letters U and V are here used as functional symbols. Let subscript letters be attached to r, s, U, and V, to distinguish the different problems into which investigations are made. Then the total cost of any series of researches will be

$$\Sigma_i \int \mathrm{V}_i s_i \cdot \mathrm{d}s_i;$$

and their total utility will be

$$\Sigma_i \int \mathrm{U}_i r_i \cdot \mathrm{d}r_i.$$

Reprinted with permission of U.S. Government Printing Office. From *United States Cost Survey for the fiscal year ending June 1876* (Washington: USGPO, 1879).

The problem will be to make the second expression a maximum by varying the inferior limits of its integrations, on the condition that the first expression remains of constant value.

The functions U and V will be different for different researches. Let us consider their general and usual properties.

And first as to the relation between the exactitude of knowledge and its utility. The utility of knowledge consists in its capability of being combined with other knowledge so as to enable us to calculate how we should act. If the knowledge is uncertain we are obliged to do more than is really necessary in order to cover this uncertainty. And thus the utility of any increase of knowledge is measured by the amount of wasted effort it saves us, multiplied by the specific cost of that species of effort. Now we know from the theory of errors that the uncertainty in the calculated amount of effort necessary to be put forth, may be represented by an expression of the form

$$c\sqrt{a + r^2},$$

where a and c are constants. And, therefore, the differential coefficient of this multiplied by the specific cost of the effort in question, say h/c, gives

$$\mathrm{U}r = h\frac{r}{\sqrt{a + r^2}}.$$

When a is very small compared with r this becomes nearly constant, and in the reverse case it is nearly proportional to r. An analogous proposition must hold for non-quantitative research.

Let us next consider the relation between the exactitude of a result and the cost of attaining it. When we increase our exactitude by multiplying observations, the different observations being independent of one another as to their cost, we know from the theory of errors that $\int \mathrm{V}s \cdot \mathrm{d}s$ is proportional to and that consequently $\mathrm{V}s$ is proportional to s. If the costs of the different observations are not independent (which usually happens) the cost will not increase so fast relatively to the accuracy; but if the errors of the observations are not independent (which also usually happens) the cost will increase faster relatively to the accuracy; and these two perturbing influences may be supposed, in the long run, to balance one another. We may, therefore, take $\mathrm{V}s = ks$, where k represents the specific cost of the investigation.

We thus see that when an investigation is commenced, after the initial expenses are once paid, at little cost we improve our knowledge, and improvement then is especially valuable; but, as the investigation goes on, additions to our knowledge cost more and more and at the same time are of less and less worth. Thus, when Chemistry sprang into being, Dr. Wollaston with a few test tubes and phials on a tea-tray, was able to make new discoveries of the greatest moment. In our day, a thousand chemists, with the most elaborate appliances, are not able to reach results which are comparable, in interest, with those early ones. All the sciences exhibit the same phenomenon; and so does the course of life. At first, we learn very easily, and the interest of experience is very great; but it becomes harder and harder and less and less worth while, until we are glad to have done with life.

Let us now apply the expressions obtained for Ur and Vs to the economic problem of research. The question is, having certain means at our disposal, to which of two studies they should be applied. The general answer is that we should study that problem for which the economic urgency, or the ratio of the utility to the cost

$$\frac{\mathrm{U}r \cdot \mathrm{d}r}{\mathrm{V}s \cdot \mathrm{d}s} = r^2 \frac{\mathrm{U}r}{\mathrm{V}s} = \frac{h}{k} \cdot \frac{r^4}{\sqrt{a + r^2}}$$

is a maximum. When the investigation has been carried to a certain point this fraction will be reduced to the same value which it has for another research, and the two must then be carried on together, until finally we shall be carrying on at once researches into a great number of questions, with such relative energies as to keep the urgency-fraction of equal values for all of them. When new and promising problems arise, they should receive our attention to the exclusion of the old ones, until their urgency becomes no greater than that of others. It will be remarked that our ignorance of a question, is a consideration which has between three and four times the economic importance of either the specific value of the solution or the specific cost of the investigation, in deciding upon its urgency.

In order to solve an economical problem, we may use as variables

$$x = \int \mathrm{V}s \cdot \mathrm{d}s,$$

or the total cost of an inquiry, and

$$y = \frac{\mathrm{U}r \cdot \mathrm{d}r}{\mathrm{V}s \cdot \mathrm{d}s},$$

or the economic urgency. Then, C being the total amount we have to spend in certain researches, our equations will be

$$\mathrm{C} = x_1 + x_2 + x_3 + \text{etc.}$$
$$y_1 = y_2 = y_3 = \text{etc.}$$

Then, expressing each y in terms of x, we shall have as many equations as unknown quantities.

When we have to choose between two researches only, the solution may be represented graphically, as shown below.

From any point O_1 taken as an origin, draw the axis of abscissas O_1X_1, along which x_1, the total cost of the first investigation, is to be measured. Draw also the axis of ordinates O_1Y_1, along which y_1, the economic urgency of the first investigation, is to be measured. Draw the curve S_1T_1 to represent the relations of x_1 and y_1. Take, on the axis O_1X_1, a point O_2 such that O_1O_2 shall measure the total cost of the two investigations. Let x_2, the total cost of the second investigation, be measured on the same axis as x_1, but in the opposite direction. From O_2 draw the axis of ordinates O_2Y_2 parallel to O_1Y_1, and measure y_2, the economic urgency of the second investigation, along this axis. Draw the curve S_2T_2 to represent the relations of x_2 and y_2. Then, the two curves S_1T_1 and S_2T_2 will generally cut one another at one point, and only one, between the axes O_1Y_1 and O_2Y_2. From this point, say P, draw the ordinate PQ, and the abscissas O_1Q and O_2Q will measure the amounts which ought to be expended on the two inquiries.

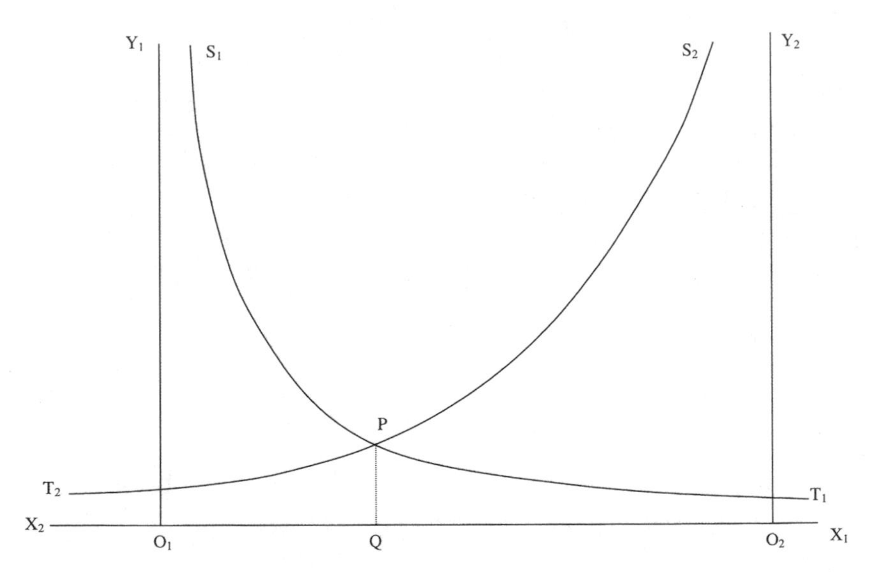

FIGURE 5.1. Peirce's Graphical Portrayal of the Economy of Research

According to the usual values of U and V, we shall have

$$y = \frac{1}{4} \cdot \frac{hk}{x\sqrt{ax^2 + \frac{1}{2}kx}}.$$

In this case, when there are two inquiries, the equation to determine x_1 will be a biquadratic. Two of its roots will be imaginary, one will give a negative value of either x_1 or x_2, and the fourth, which is the significant one, will give positive values of both.

Let us now consider the economic relations of different researches to one another, 1st, as alternative methods of reaching the same result, and 2nd, as contributing different premises to the same argument.

Suppose we have two different methods of determining the same quantity. Each of these methods is supposed to have an accidental probable error and a constant probable error so that the probable errors as derived from n observations in the two ways are

$$r_1 = \sqrt{R_1^2 + \frac{\rho_1^2}{n}} \text{ and } r_2 = \sqrt{R_2^2 + \frac{\rho_2^2}{n}}.$$

The probable error of their weighted mean is

$$\frac{1}{\sqrt{\frac{1}{r_1^2} + \frac{1}{r_2^2}}},$$

if their constant probable errors are known. The sole utility of any observation of either is to reduce the error of the weighted mean; hence,

$$Ur_1 = D_{r_1}(r_1^{-2} + r_2^{-2})^{-1/2} = (r_1^{-2} + r_2^{-2})^{-3/2} r_1^{-3}.$$

And as the cost is proportional to the number of observations

$$Vs_1 = k_1 \frac{1}{D_{n_1}s_1} = \frac{k_1}{D_{n_1}(R_1^2 + \rho_1^2 n_1^{-1})^{-1/2}} = \frac{2k_1 r_1^3 n_1^2}{\rho_1^2}.$$

Hence the urgency is (omitting a factor common to the values for the two methods)

$$r_1^2 \frac{Ur_1}{Vs_1} = \frac{1}{k_1\rho_1^2\left(1 + n_1 \frac{R_1^2}{\rho_1^2}\right)^2}.$$

And as the urgency of the two methods ought to be the same at the conclusion of the work we should have

$$\sqrt{k_1 \cdot \rho_1}\left(1 + n_1 \frac{R_1^2}{\rho_1^2}\right) = \sqrt{k_2 \cdot \rho_2}\left(1 + n_2 \frac{R_2^2}{\rho_2^2}\right),$$

which equation serves to determine the relative values of n_1 and n_2. We again perceive that the cost is the smallest consideration. The method which has the smallest accidental probable error is the one which is to be oftenest used in case only a small number of observations are made, but if a large number are taken the method with the larger accidental probable error is to be oftenest used, unless it has so much greater a probable error of the other as to countervail this consideration. If one of the two methods has only $(1/p)$th the accidental probable error of the other, but costs p^2 times as much, the rule should be to make the total cost of the two methods inversely proportional to the squares of their constant errors.

Let us now consider the case in which two quantities x_1 and x_2 are observed, the knowledge of which serves only to determine a certain function of them, y. In this case, the probable error of y is

$$\sqrt{\mathrm{D}_{x_1}y \cdot r_1^2 + \mathrm{D}_{x_2}y \cdot r_2^2},$$

and we shall have

$$\mathrm{U}r_1 = 2r_1 \frac{\mathrm{d}y}{\mathrm{d}x_1}.$$

$\mathrm{V}s_1$ will have the same value as before; but neglecting now the constant error, we may write

$$\mathrm{V}s_1 = 2k_1\rho_1 n_1^{1/2}.$$

Then the urgency (with omission of the common factor) is

$$\frac{\rho_1^2}{k_1 n_1^2} \cdot \frac{\mathrm{d}y}{\mathrm{d}x_1},$$

and as the two urgencies must be equal, we have

$$\frac{n_1}{n_1} = \frac{\rho_1}{\rho_2}\sqrt{\frac{k_2}{k_1}}\sqrt{\frac{\frac{\mathrm{d}y}{\mathrm{d}x_1}}{\frac{\mathrm{d}y}{\mathrm{d}x_2}}}.$$

The following is an example of the practical application of the theory of economy in research: Given a certain amount of time, which is to be expended in swinging a reversible pendulum, how much should be devoted to experiments with the heavy end up, and how much to those with the heavy end down?

Let T_d be the period of oscillation with heavy end down, T_u the same with heavy end up. Let h_d and h_u be the distances of the centre of mass from the points of support of the pendulum in the two positions. Then the object of the experiments is to ascertain a quantity proportional to

$$h_d T_d - h_u T_u.$$

Accordingly, if dT_d and dT_u are the probable errors of T_d and T_u, that of the quantity sought will be

$$\sqrt{h_d^2(dT_d^2) + h_u^2(dT_u^2)}.$$

We will suppose that it has been ascertained, by experiment, that the whole duration of the swinging being C, and the excess of the duration of the swinging with heavy end down over that with heavy end up being x, the probable error results are

$$dT_d = \sqrt{a + \left(b + \frac{c}{h_d^2}\right)\frac{1}{C + x}}$$

$$dT_u = \sqrt{a + \left(b + \frac{c}{h_u^2}\right)\frac{1}{C - x}}$$

where a, b, and c are constants. Then, the square of the probable error of the quantity sought will be

$$a(h_d^2 + h_u^2) + (bh_d^2 + c)\frac{1}{C + x} + (bh_u^2 + c)\frac{1}{C - x}.$$

The differential coefficient of this relatively to x is

$$-(bh_d^2 + c)\frac{1}{(C + x)^2} + (bh_u^2 + c)\frac{1}{(C - x)^2}.$$

Putting this equal to zero and solving, we find for the only significant root,

$$\frac{x}{C} = \frac{b(h_d^2 + h_u^2) + 2c}{b(h_d^2 - h_u^2)} - \sqrt{\left(\frac{b(h_d^2 + h_u^2) + 2c}{b(h_d^2 - h_u^2)}\right)^2 - 1}.$$

When b vanishes, x reduces to zero, and the pendulum should be swung equally long in the two positions. When c vanishes, as it would if the pendulum experiment were made absolutely free from certain disturbing influences, we have

$$\frac{x}{C} = \frac{h_d - h_u}{h_d + h_u},$$

so that the duration of an experiment ought to be proportional to the distance of the centre of mass from the point of support. This would be effected by beginning and ending the experiments in the two positions with the same amplitudes of oscillation.

It is to be remarked that the theory here given rests on the supposition that the object of the investigation is the ascertainment of truth. When an investigation is made for the purpose of attaining personal distinction, the economics of the problem are entirely different. But that seems to be well enough understood by those engaged in that sort of investigation.

6

Charles Sanders Peirce's Economy of Research

James R. Wible

Charles Sanders Peirce has authored an extraordinary "Note on the Theory of the Economy of Research" (1879; chapter 5 in this volume). The "Note" presents an economic model of research project selection in science. A case can be made that the "Note" was the first piece of modern scientific research in all of economics. This claim is based on the novelty of the method of argument, the graphical techniques, and the ratio of the marginal utilities found in the "Note." The "Note" is also significant for making economic factors a central part of a theory of scientific inference, something which contemporary economic methodologists and philosophers still have not done except for a few notable exceptions. And it has been used by philosopher Nicholas Rescher to interpret and criticize Karl Popper's notion of falsification. All of these contributions suggest that Peirce's "Note" may be of unusual interest to the economics profession.

> Economical science is particularly profitable to science; and that of all the branches of economy, the economy of research is perhaps the most profitable. (Peirce 1902b: 1038)

> The methodology of inductive practice is, in Peirce's view, pivotally dependent on the intelligent deployment of economic considerations from the very outset. (Rescher 1978a: 66)

More than a century ago, the brilliant founding thinker of American pragmatism, Charles Sanders Peirce, proposed an economic approach to scientific research project selection. Peirce's proposal is found in his

Reprinted with permission of Taylor & Francis Ltd., http://www.tandf.co.uk/journals.jec.htm. From *Journal of Economic Methodology* 1 (1994): 135–59.

piece titled "Note on the Theory of the Economy of Research" (1879; chapter 5 in this volume). Peirce's "Note" may be seen as inaugurating an economic approach to science, an economics of science. Peirce developed a cost-constrained utility model for choosing among alternative research projects. The purpose of this essay is to present and appraise Peirce's suggestion for an economics of research project selection in science. Peirce's proposal has been overlooked by the economics profession. Peirce's "Note" is significant for several reasons. It is the first piece to use economic analysis to address questions of scientific method. It may be the first truly scientific piece in all of economics. And it has been used by philosopher Nicholas Rescher to interpret and criticize Karl Popper's notion of falsification. All of these contributions suggest that Peirce's "Note" may be of unusual interest to the economics profession.

1. The Role of Economics in Peirce's Thought

Charles Sanders Peirce was the founder of pragmatism. Philosopher Philip Wiener writes that Peirce was "the most versatile, profound, and original philosopher that the United States has ever produced" (Wiener 1958: ix). Pragmatism was the distinctly American school of philosophy that sought to extend the experimental mindset of science to all of life.[1] Such a mindset was Peirce's prescription for making our thoughts and ideas clear.[2] Economics played a role in this process of clarifying ideas. Economic considerations were to be used in narrowing the number of hypotheses to be considered for actual testing. Peirce's pragmatism was expanded and reinterpreted by William James and John Dewey. But Peirce's economy of research was neglected and probably had little impact on the subsequent development of pragmatism. Other than founding pragmatism, Peirce made important contributions to mathematics, statistics, geodesy, symbolic logic, and scientific inference. These achievements are well known and widely recognized by contemporary scholars in these fields. But Peirce's economic contributions have been virtually ignored by economists. On the personal side, Peirce was a difficult man. He was something of a grating and eccentric personality.[3]

1. One presentation of the ideas in pragmatism is Wiener (1949). For an overview of the relevance of Peirce and pragmatism to controversial issues in economics see Kevin Hoover (1994).

2. The classic statement of Peirce's (1878 [1955]) pragmatism is his essay, "How to Make Our Ideas Clear."

3. Brent's (1993) biography documents that Peirce became an outcast in academia, government, and publishing.

Consequently, it was difficult for him to retain a university position. For most of his life, he worked for a government agency called the U.S. Coast Survey. Peirce's project for an economics of research is included in his annual report of 1876, which was printed and published as the "Note" in 1879 (Peirce 1879).

In the modern economics literature, the only place that I have seen Peirce's interest in economics referenced is in a compendium of essays documenting the emergence of mathematical economics. In Baumol and Goldfeld's (1968) *Precursors in Mathematical Economics: An Anthology,* the entry for Peirce is nothing more than a short letter.[4] Neither the "Note" nor other comments by Peirce on the economy of research or political economy are included. In the anthology, Baumol and Goldfield (1968) note that A. A. Cournot had published *Researches into the Mathematical Principles of the Theory of Wealth* in 1838.[5] It was the most advanced work in mathematical economics at that period of time. What is found in Peirce's letter are comments on Cournot. The letter was written to the astronomer Simon Newcomb (Peirce 1871a: 186–7). Newcomb also had an interest in political economy which resulted in a handbook many years later called *Principles of Political Economy* (1886).[6] Peirce's letter contains a few simple equations concerning profit maximization using calculus. In the letter, Peirce inferred that the results of profit maximization are dependent on whether there is unlimited competition or not. He concluded that "[t]his is all in Cournot."[7]

Peirce's interest in economics is much broader and more fundamental than what the "Note" by itself would suggest. His accomplishments in economics are mostly unknown in economics and philosophy. His interest in economics took three separate but complementary directions. First, Peirce had an abiding concern for creating a general mathematical political economy. Second, as part of his general evolutionism, Peirce recognized the economic function of the conscious mind and of scientific theories. And third, Peirce conceived of the economy of research as one of the major aspects of his theory of inference.

Peirce's interest in the political economy of his time vastly transcended the characterization of his thoughts as being merely interested in Cournot as portrayed in the Baumol and Goldfeld volume. For

4. A more complete edition of the correspondence with Newcomb is found in Eisele (1979b).

5. Cited in Baumol and Goldfeld (1968: 161–9).

6. My source is Eisele (1979b: 58).

7. Peirce (1871a) in Baumol and Goldfeld (1968: 187).

Peirce, Cournot's project for mathematically formulating economic ideas stimulated a vision for creating a mathematical political economy. The equations in Peirce's letter to Newcomb were intended to be a first step toward a mathematical political economy. Peirce had an enduring interest in creating a mathematical political economy. This aim is what drove his interest in mainstream political economy. Another economist who interested Peirce was David Ricardo.[8] What interested Peirce was the logical structure of the theory of rent in Ricardo's *Principles* (Peirce 1893 and 1897). Peirce thought that Ricardo's conceptualization of political economy made it more amenable to mathematical representation. Besides Ricardo, Cournot, and Newcomb, Peirce was aware of the economic contributions of Malthus, J. S. Mill, Charles Babbage, and possibly Irving Fisher.[9] Additionally, Peirce's father, Benjamin Peirce, a professor of mathematics at Harvard and the foremost American mathematician of his time, also had a keen interest in a mathematical political economy. Apparently, Peirce and his father collaborated in giving at least one public lecture on mathematical political economy in 1871.[10]

Besides mathematical political economy, the second major area of Peirce's interest in economics was the economic function of mind and scientific theories. A similar point of view can be found in John Dewey's philosophy.[11] In the context of an uncertain, evolutionary world, Peirce recognized the economic function of conscious thought and of scientific theories. To be effective, thoughts and theories must distil meaning and information and represent them in an efficacious way. Peirce believed science to be an area of human thought where the economic function of the mind reached its most general form. Peirce (1896: 48) referenced the writings of the physicist Ernst Mach, who developed this "instrumental" view of consciousness and science much more elaborately than Peirce. Mach's views on the economy of science and thought are found in two widely circulated late nineteenth-century works.[12]

Having briefly considered Peirce's concern for mathematical political economy and the economy of thought, the third aspect of his interest in economics can now be addressed. Of the three areas constituting Peirce's interest in economics, the one which he developed to the highest extent

8. Eisele (1979c) discusses Peirce's interest in Ricardo.
9. See Eisele (1979c and 1979d) and Peirce (1871b and 1882 [1986]).
10. Eisele (1979c: 253) and Fisch (1982: xxxv).
11. Wible (1984) discusses this aspect of Dewey's philosophy.
12. See Mach (1893 [1960] and 1893).

was his economy of research. Economic aspects of research were an explicit part of Peirce's theory of inference. The nature of scientific inference was a lifelong concern of Peirce's. The initial phase of inference was a process which Peirce called abduction. Abduction is the use of known facts to formulate as many alternative hypotheses as could be imagined by the investigator. By an economics of research, Peirce meant that an hypothesis which was less costly to investigate than others should be investigated first. This assumed a similar level of benefit among the hypotheses being compared. One can imagine researchers ranking alternative research projects by cost. Peirce would have the investigator choose the research project which was least costly. Such an economics of science implies that the scientist has some knowledge about the costs of doing research one way rather than another. This seems to suggest something like an economic abduction about the relative costs of alternative research methods as a supplement to the more basic abductive process involved in creating scientific hypotheses. The scientist would have general information about the cost of doing the investigation and should use this sort of economic knowledge in the selection of research topics.

Peirce's "Note on the Theory of the Economy of Research" is by far the most sophisticated contribution he has made to economics. The "Note" was written relatively early in his career. After writing the "Note," Peirce continued to make the economy of research an integral part of his theory of science. He repeatedly returned to this theme in his discussions of the nature of scientific inference. Several are worth quoting. In a comment written in 1896 titled "The Economy of Research," by the editors of the *Collected Papers,* Peirce restated the theory originally introduced in the "Note":

> There is a doctrine of the Economies of Research. One or two of its principles are easily made out. The value of knowledge is, for the purposes of science, in one sense absolute. It is not to be measured, it may be said, in money; on one sense that is true. But knowledge that leads to other knowledge is more valuable in proportion to the trouble it saves in the way of expenditure to get that other knowledge. Having a certain fund of energy, time, money, etc., all of which are merchantable articles to spend upon research, the question is how much is to be allowed to each investigation; and *for us* the value of that investigation is the amount of money it will pay us to spend upon it. *Relatively,* therefore, knowledge, even of a purely scientific kind, has a money value.

> This value increases with the fullness and precision of the information, but plainly it increases slower and slower as the knowledge becomes fuller and more precise. The cost of the information also increases with its fullness and accuracy, and increases faster and faster the more accurate it is. It therefore *may* be the case that it does not pay to get *any* information on a given subject; but, at any rate, it must be true that it does not pay (in any given state of science) to push the investigation beyond a certain point in fullness or precision.
>
> If we have a number of studies in which we are interested, we should commence with the most remunerative and carry that forward until it becomes not more than equally remunerative with the commencement of another; carry both forward at such rates that they are equally remunerative until each is no more remunerative than a third, and so on. (Peirce 1896: 49)

In 1901, Peirce again returned to his concern with economic issues of research. Peirce wrote a long essay approaching the length of a short monograph on the methodology of interpreting ancient manuscripts. This essay is titled "On the Logic of Drawing History from Ancient Documents Especially from Testimonies." In this essay Peirce considers again the nature of hypothetical inference. "Economical considerations" play a fundamental role in his conception of inference:

> Now economy, in general, depends upon three kinds of factors; cost; the value of the thing proposed, in itself; and its effect upon other projects. Under the head of cost, if a hypothesis can be put to the test of experiment with very little expense of any kind, that should be regarded as a recommendation for giving it precedence in the inductive procedure. For even if it be barely admissible for other reasons, still it may clear the ground to have disposed of it. In the beginning of the wonderful reasonings by which the cuneiform inscriptions were made legible, one or two hypotheses which were never considered likely were taken up and soon refuted with great advantage. (Peirce 1901: 754).

One of the last discussions of the economics of research by Peirce is contained in his 1902 grant application to the Carnegie Institution. The purpose of the request was to pull together many of his disparate writings into a coherent set of memoirs which would form a unified system of logic. The economy of research was to have an important role in the unified logic. At this point in his life, Peirce was destitute. He had lost his job with the Coast Survey. He was unable to find an academic position anywhere in the United States. Peirce survived with financial support orchestrated by his good friend William James and by his paid reviews and articles for various magazines and publications. In the grant application, proposed memoir number 28 was titled "On the Economics of Research." In this piece, Peirce restated many of the themes already introduced above:

> Many years ago I published a little paper on the Economy of Research, in which I considered this problem. Somebody furnished a fund to be expended upon research without restrictions. What sort of researches should it be expended upon? My answer, to which I still adhere, was this. Researches for which men have been trained, instruments procured, and a plant established, should be continued while these conditions subsist. But the new money should mainly go to opening up new fields; because new fields will probably be more profitable, and at any rate, will be profitable longer.
>
> I shall remark in the course of the memoir that economical science is particularly profitable to science; and that of all the branches of economy, the economy of research is perhaps the most profitable; that logical methodeutic and logic in general are specially valuable for science, costing little beyond the energies of the researcher, and helping the economy of every other science. (Peirce 1902b: 1038)

2. Peirce's Economic Model of Research Project Selection

In the "Note on the Theory of the Economy of Research," Peirce proposed an economic theory of research project selection in science.[13] The theory is much more rigorously developed than the intuitive summaries Peirce provided in later years. Its argumentation is so rigorous that it could be the first truly modern scientific piece in all of economics. Peirce's theory of the economy of science is premised on the idea that resources can be used to increase the precision of measurement.[14]

13. Presently there are three places where Peirce's "Note" has appeared. The "Note" was originally published in a report to the Coast Survey in 1879. This published report apparently was lost but then rediscovered in the 1960s. A reprinting of the Coast Survey version appeared in *Operations Research* in 1967 (Peirce 1879 [1967]: 642–8). A brief discussion of the Coast Survey version can be found in Cushen (1967). The "Note" previously appeared in the *Collected Papers of Charles Sanders Peirce,* vol. 7 (1958): 76–83, as edited by the Department of Philosophy at Harvard. The Harvard version is published from a handwritten manuscript in Widener library. More recently, the "Note" has been reprinted as part of a much more comprehensive edition of Peirce's papers than the *Collected Papers.* These are the *Writings of Charles S. Peirce: A Chronological Edition.* The "Note" appears in vol. 4 (1986): 72–8.

14. Brent's (1993) biography of Peirce documents that Peirce spent most of his professional life trying to measure differences in the earth's gravity at different geographical

The theory is based on the idea that resources can be expended to reduce probable error in science. Probable error is a nineteenth-century term dealing with the precision of a statistical estimator in scientific inquiry. It is a precursor to the notion of a confidence interval. The first paragraph of Peirce's "Note" introduces the notion of probable error and reads as follows:

> When a research is of a quantitative nature, the progress of it is marked by the diminution of the probable error. The results of nonquantitative researches also have an inexactitude or indeterminacy which is analogous to the probable error of quantitative determinations. To this inexactitude, although it be not numerically expressed, the term "probable error" may be conveniently extended. (Pierce 1879: 643)

The second paragraph of the "Note" restates the concern with probable error as an economic problem:

> The doctrine of Economy, in general, treats of the relations between utility and cost. That branch of it which relates to research considers the relations between the utility and the cost of diminishing the probable error of our knowledge. Its main problem is how with a given expenditure of money, time, and energy, to obtain the most valuable addition to our knowledge. (Peirce 1879: 643)

In the main body of the essay, Peirce focused on the allocation of additional resources to established, on-going research. Once the basic expenditures of a project have been made, Peirce maintained that additional expenditures would improve the accuracy of knowledge, but the benefits would begin to diminish:

> We thus see that when an investigation is commenced, after the initial expenses are once paid, at little cost we improve our knowledge, and improvement then is especially valuable; but, as the investigation goes on, additions to our knowledge cost more and more and at the same time are of less and less worth. . . . All the sciences exhibit the same phenomenon; and so does the course of life. (Peirce 1879: 644)

In "Economy of Research," Peirce followed a method of argument which any contemporary economist would recognize. Peirce created a

---

locations in Europe and North America. Peirce continually appealed for additional funding to support research at new locations and the construction of more accurate pendulums to reduce the inaccuracies introduced by the equipment. Peirce considered himself a first-rate gravimetric researcher and was well known in Europe for his skill.

mathematical model of the choices facing the researcher. The model anticipated many developments in modern microeconomics. Peirce theorized that the total utility of a series of research projects should be maximized subject to a limitation on total cost. He represented total cost and total utility as follows:

(1) $\Sigma_i \int V_i s_i \cdot ds_i$

(2) $\Sigma_i \int U_i r_i \cdot dr_i$

In these equations, V and U are functional symbols regarding the cost and utility of individual research projects and r is probable error. V is positively related to s, the reciprocal of probable error, and U is positively related to the "diminution" of probable error. As Peirce uses the notation, Vs today would be written as V(s) meaning that cost is a function of variable s and Ur would be written as U(r), meaning that U is a function of r. An index number i is used to designate each of the many alternative research projects under consideration. $V_1 s_1$ would represent the cost of the first research project, $V_2 s_2$ the cost of the second research project, and $V_i s_i$ the cost of the ith research project. The index number works in the same the research projects.

In the total cost function, equation (1), Peirce defines s to be the reciprocal of probable error, $s = 1/r$, and $V_i s_i$ to represent the cost of reducing probable error for any research project. One of the integrals represents the cost of reducing probable error for an individual research project. The total cost of reducing the probable error of all research projects is represented by the summation over the cost of all i research projects. In Peirce's theory, the cost of research rises as probable error is reduced. Peirce also speculates that cost, $V_i s_i$, is a linear, proportional function of s and that its integral is a quadratic function of s.

In the total utility function, equation (2), r represents the probable error of the research project and $U_i r_i$ is the utility resulting from the reduction in the probable error of the results. Again, just one of the integrals represents the utility of reducing probable error for an individual research project. The total utility of reducing the probable error of all research projects is represented by the summation over all of the research projects. Here Peirce also speculates about the functional form of the utility equation. He asserts that utility, $U_i r_i$, is a quadratic function of probable error.

Since Peirce's theory is developed in terms of a multiple-project model, the theory requires a decision criterion comparing one project to another. In order for the total utility of all research projects being considered to be maximized subject to a total budget or cost constraint, the relative value of one project must be compared to another. To solve this problem, Peirce created a ratio of the marginal utility of a project to its marginal cost. He called this ratio the "economic urgency" of the research project. Although he did not use the terms "marginal utility" and "marginal cost," Peirce wrote: "Let $Ur \cdot dr$ denote the infinitesimal utility of any infinitesimal diminution, dr, of r. Let $Vs \cdot ds$ denote the infinitesimal cost of any infinitesimal increase, ds, of s" (Peirce 1879: 643). In modern terminology, $Ur \cdot dr$ would be the marginal utility of reducing probable error and $Vs \cdot ds$ would be the marginal cost for decreasing the probable error of any one research project. Economic urgency for a single research project is denoted with the variable y as follows:

$$y = \frac{Ur \cdot dr}{Vs \cdot ds} \tag{3}$$

A ratio of economic urgency for every research project is designated as y_i. At this point, Peirce restates the maximization problem of the economy research with the following equations:

$$C = x_1 + x_2 + x_3 + \text{etc.} \tag{4}$$

$$y_1 = y_2 = y_3 = \text{etc.} \tag{5}$$

where C is the total amount that can be spent on all projects and x_i is the cost of an individual project. Equation (4) is an aggregate budget constraint for all projects undertaken by the researcher and equation (5) governs the optimal allocation of funds to each research project. It means that the marginal utility per dollar spent on reducing probable error in each research project should be equalized across all research projects in order to maximize the value of research from this series of scientific research projects. Peirce was writing before the development of modern price theory with indifference curve analysis and its technical vocabulary. Concerning the ratio of marginal utility to marginal cost, economic urgency, Peirce commented:

> When the investigation has been carried to a certain point this fraction will be reduced to the same value which it has for another research, and the two must then be carried on together, until finally we shall be carrying on at once researches into a great number of questions, with

> such relative energies as to keep the urgency-fraction of equal values for all of them. When new and promising problems arise, they should receive our attention to the exclusion of the old ones, until their urgency becomes no greater than that of others. It will be remarked that our ignorance of a question, is a consideration which has between three and four times the economic importance of either the specific value of the solution or the specific cost of the investigation, in deciding upon its urgency. (Peirce 1879: 645)

At a later point in the grant application to the Carnegie Institution, Peirce summarized the implications of the preceding model in the following way:

> Research must contrive to do business at a profit; by which I mean that it must produce more effective scientific energy than it expends. No doubt it already does so. But it would do well to become conscious of its economical position and contrive ways of living upon it. (Peirce 1902b: 1038)

Having developed his theory for the most general, multiple–research project case with mathematics, Peirce proceeded to simplify with the special case when two projects are under simultaneous consideration. The two-project case allows Peirce to graphically portray his theory of the economy of research. Peirce's illustration is reproduced as figure 5.1. The total cost of two projects is on the horizontal axis. The ratio of economic urgency, of marginal utility to marginal cost, is represented on the vertical axis by Y_1 and Y_2. The graph of the first research project is represented in the usual way by curve S_1T_1. The representation of the second project is reversed and read backwards from the X_2O_2 origin. Curve S_2T_2 represents the second project. For both projects, the ratio of marginal utility to marginal cost for reducing probable error decreases as more funds are expended on each project. At the intersection point of the two curves, P, the ratios for the two projects are equal. Peirce projected a line from P to the horizontal axis which established how additional funds should be allocated to the two projects. As Peirce portrayed it, greater additional funding should be allocated to the second project than the first.

3. The Notion of Probable Error and the Earliest Correct Usage of the Logic of Statistical Inference

The notion of probable error as Peirce used it is now no longer current in modern science. An understanding of probable error is central in understanding Peirce's analysis of the economics of research project selection both in the "Note" and in subsequent comments. As mentioned previously, the term "probable error" is a precursor of the notion of a confidence interval. Fortunately, Peirce's use of probable error has already been analyzed and interpreted in a work on the history of statistics. This work argues that Peirce was the first to correctly formulate the logic of modern scientific inference using probable error. If this is so, then the "Note" is a significant illustration of Peirce's path-breaking understanding of modern statistical inference.

A more elaborate sense of probable error can be gleaned from philosopher Ian Hacking's discussion of the concept in *The Taming of Chance.* According to Hacking (1990: 209), the standard use of the term "probable error" arose from astronomy. In an informal way, astronomers used probable error to mean that a measure of a position was within a given error of the true value half of the time. Or as Robert Dorfman has written: "It is a measure of the accuracy of an estimate equal to the range around the true value within which it has a 50–50 probability of lying."[15] We already know that Peirce corresponded with the astronomer Simon Newcomb and was employed by the Harvard Observatory and that Peirce's father was a consulting astronomer.[16] So it is not surprising that Peirce made use of a statistical concept developed by astronomers. Hacking describes Peirce's use of probable error in the following way:

> Peirce had a model for this kind of argument, based on the standard practices of astronomers, the "probable error." The probable error divides measurements into two equal classes. If the errors are Normally distributed, then in the long run half the measurements will err in excess of the probable error, and half will be more exact. But what does this amount to? (Hacking 1990: 209)

Relating probable error to more modern statistical terminology, Hacking answers the question just raised. Peirce's use of probable error apparently provides the core of the modern theory of confidence intervals:

> Peirce is original in understanding the logic of the situation. Readers familiar with the logic of statistical inference will have noticed that Peirce was providing the core of the rationale of the theory of confidence intervals and of hypothesis-testing advanced by Jerzy Neyman and E. S. Pearson in the 1930s, which is still, for many, the preferred

15. This statement is taken from private correspondence with Professor Dorfman. His comments sent me searching for a better understanding of the term "probable error."

16. See Fisch (1982: xxi–xxiii).

> route in statistics. As usual, I am unconcerned with Peirce the precursor. Neyman did not learn anything from Peirce. Still, there is a certain line of filiation. The first modern statement of the rationale of confidence intervals was given not by Neyman but by the Harvard statistician E. B. Wilson. Wilson had been a pupil of Peirce's cousin B. O. Peirce, and was a lifelong admirer of the family. He was one of the few readers of C. S. Peirce on errors of observation, and wrote a paper about it. He had the right perspective as regards predecessors. . . . But it appears that only Peirce, Wilson and then Neyman got clear about the logical principles of this type of reasoning. (Hacking 1990: 210).

Based on Hacking's appraisal, Peirce was the first to formulate the principles of statistical inference in a modern way with the idea of probable error. Probable error is a precursor to the more modern notion of a confidence interval. A confidence interval is a more explicit measure of the precision of a statistical estimator than the variance or calculated standard error of the estimator.[17] The confidence interval gives an interval of the parameter values that likely contain the true mean. A 95 percent confidence interval means that there is a 95 percent probability that the interval covers the true mean. It does not give the probability that the true mean is in the interval. The true mean is a constant. The probability is zero or one that the true mean is in the confidence interval. The endpoints of the confidence interval are random variables to which probability can apply. The true mean is not a random variable. Additionally, Peirce's probable error appears to be the statistical opposite of the confidence interval. Rather than giving the degree of our confidence in an estimator, probable error gives the opposite. Thus probable error is equivalent to one minus the confidence interval.[18] The probable error of a 95 percent confidence interval apparently would be 5 percent.

The precision of an estimator as captured in the confidence interval is multidimensional. It involves the width of the confidence interval, the percentage level of interpretation, and sample size. For example, for a

17. An estimator is a measure of a true parameter of a population under investigation. The true parameter is unknown unless the entire population is known. As an example consider the mean. A mean can be calculated from a random sample of the population and compared to the mean of the population. Thus sample mean is an estimator of the true mean of the population. What we want to know at this point is how accurate the sample mean is as an estimator of the true mean.

18. I am indebted to Jeff Sohl for suggesting this interpretation of probable error. It seems consistent with Hacking's (1990) interpretation of probable error in Peirce's writings. Kmenta (1971: 186–91) has a thorough explanation of the meaning of a confidence interval.

given sample size a 95 percent confidence interval would have wider endpoints than a 90 percent or 50 percent confidence interval. A 50 percent confidence interval would be quite small. With repeated sampling, the 50 percent confidence interval means that only half of the time would the interval contain the true value of the parameter being estimated. Thus a higher level of confidence means wider or less precise endpoints to the interval. If we want a high degree of confidence without widening the endpoints of the interval, then the sample size would need to be increased. Econometrician Jan Kmenta has expressed some of these matters in the following way:

> We should realize that the higher the level of confidence, the wider the corresponding confidence interval and, therefore, the less useful is the information about the precision of the estimator. . . . On the other hand, narrower confidence intervals will be associated with lower levels of confidence. . . . Obviously, given the level of confidence, a shorter interval is more desirable than a longer one. (Kmenta 1971: 187–8).

In his critical interpretation of the significance of Peirce's work on statistical inference, Hacking does not specifically discuss the "Note." This was unnecessary because Peirce had written frequently on both astronomy and statistical inference before he authored the "Note." The "Note" was not Peirce's first piece to use the notion of probable error.[19] Nor does Hacking discuss the economic aspects of statistical inference as Peirce did. Yet it is clear that Peirce continually conceived of statistical inference as having an important economic dimension that was crucial in the selection of those hypotheses which should be tested. It is also clear that Peirce's "Note" embodied the degree of sophistication regarding statistical inference that Hacking saw in Peirce's other writings.

Peirce's "Note" calls attention to the economic aspects of statistical sampling. These economic aspects of sampling are known to every well-trained statistician and econometrician. I suspect that the economic aspects of sampling are part of the common sense of graduate instruction in statistical inference.[20] With regard to statistical inference, Peirce's economy of research implies the following view of empirical research: that the value of scientific research depends on the degree of its precision, that the level of precision depends on sample size, and that sample

19. Peirce used the term "probable error" in a manuscript in 1861. See Peirce (1861: 70).

20. For example, Kmenta's (1971: 125–8) econometrics text provides a discussion of economic aspects of choosing the correct sample size.

size depends on costs of observation relative to available resources. None of this may be new to anyone experienced in the field of statistics. What is significant is that Peirce not only created a formal model of the economics of research and sampling, he also is the first to have correctly conceived of the logic of modern scientific inference.

4. Peirce's Note as the First Scientific Piece in Economics

Beyond the statistical and economic implications of Peirce's economy of research project selection, another significant claim regarding the novelty of Peirce's "Note" can be made. A case can be made for claiming that Peirce's "Note on the Theory of the Economy of Research" is the first modern scientific piece in all of economics. The general context for this claim is the Baumol and Goldfeld anthology of writings of the precursors of mathematical economics. Peirce's note can be directly compared with the contributions in the anthology. I believe that any economist who would read Peirce's essay would recognize it as the most modern and scientific of all of the early mathematical writings in economics. There are three specific aspects of Peirce's essay which make it an extremely innovative piece in economics compared to the mathematical economics of his time and in light of contributions which would come in the early part of the twentieth century. First, the method of argument which appears in the essay is novel. Second, the representation of the argument with a special two-case graph was extraordinary for its time. Third, Peirce's development of the economic implications of his theory with ratios of marginal utilities was nothing less than path breaking.[21]

The method of presenting scientific arguments in modern economics journals is known as the hypothetico-deductive method. This is the method for structuring the typical scientific article found in all of the major journals in economics. In brief, the hypothetico-deductive method requires the following: a clear analytical statement of a theory with mathematical equations or logical propositions that are independent of evidence or observation, the derivation of observable implications of the theory, and the testing of these implications in some appropriate man-

21. A search through recent articles regarding mathematical economics in the nineteenth century reveals that Peirce's contributions are unknown in economics except for the Baumol and Goldfeld volume. For example, Stigler (1972) does not include Peirce in his survey of the adoption of marginal utility theory in the late nineteenth century. Similarly, Peirce's contributions are not among those listed by Howey (1972) in recounting the origins of marginalism.

ner. In philosophy of science, the origin of this method is traced to logical positivism. Logical positivists created the method in response to the reconceptualization of mathematics and logic formulated in Bertrand Russell and A. N. Whitehead's *Principia Mathematica.*[22] The success of Russell and Whitehead in reducing mathematics to a small number of logical axioms convinced logical positivists of the possibility that analytical statements totally devoid of empirical content could actually be constructed. This belief in the possibility of a priori statements led to the formulation of the correspondence theory of truth.[23] In this view, theoretical statements were to be formulated as a priori analytical statements without empirical content. To be scientific, theoretical statements needed corresponding observational counterparts which could be used to determine the empirical content of the theory. While this view of truth has been much criticized for many decades, it still influences the way scientific arguments are presented. Everyone realizes that the logical positivists' conception of truth required but did not provide a solution to the problem of induction. In this regard the positivist program for science has failed. However, it did present a formal method of argument which still survives in most scientific disciplines.

What is extraordinary is that Peirce's essay exhibits the same structure of logical inference that the positivists created decades or perhaps a half century later. Peirce's "Note" is developed in exactly the manner of a modern theoretical journal article in economics. The sophistication of his "Note" rivals that of theoretical pieces authored during that innovative period which G. L. S. Shackle (1967) has called the years of high theory in economics during the 1920s and 1930s. Further evidence of the modernity of Peirce's method of scientific inference can be found in other writings of Peirce. It is not difficult to find an analysis of inference that is quite similar to the hypothetico-deductive method. One of these is an essay of Peirce's called "What is a leading principle?" In this piece, Peirce clearly maintained a logical theory of inference:

> Such a process is called an *inference;* the antecedent judgement is called the *premiss;* the consequent judgement, the *conclusion;* the habit of thought, which determined the passage from the one to the other (when formulated as a proposition), the *leading principle.* . . . A habit of inference may be formulated in a proposition which shall state that every proposition *c,* related in a given general way to any true proposition *p*

22. Suppe (1974: 12) discusses the impact of *Principia Mathematica* on positivism.

23. The correspondence theory of truth can be found in Suppe (1974: 12) and Feigl (1969: 17).

> is true. Such a proposition is called the *leading principle* of the class of inferences whose validity it implies. (Peirce 1880: 130 and 131)

In a peculiar way, the novelty of Peirce's method of inference is apparent in the opinion of Simon Newcomb. Newcomb did not fully comprehend Peirce's method of inference. In 1889, Peirce was still working for the Coast Survey and attempting to finish a report of the results of the pendulum experiments. Peirce sent his report to the superintendent who sent it to Newcomb for appraisal. Newcomb wrote a lengthy letter recommending rejection of the report as a publication of the Survey:

> A remarkable feature of the presentation is the inversion of the logical order throughout the whole paper. The system of the author seems to be to give first concluded results, then the method by which these results were obtained, then the formulae and principles on which these methods rest, then the derivation of these formulae, then the data on which the derivations rests and so on until the original observations are reached. The human mind cannot follow a course of such reasoning in this way, and the first thing to be done with the paper is to reconstruct it in logical order. (Newcomb letter of April 28, 1890, in Brent 1993: 197)

Although Peirce did not share the logical positivists conception of truth, his method of inference is quite similar to the one they created decades later. Peirce's "Note" clearly embodies the method of inference proposed by Peirce in his concept of a leading principle. Peirce started with a mathematical statement of the theory and derived implications from the statement of the theory. Unlike his gravimetric research, Peirce did not have experimental evidence to support his conclusions about the economics of research. Nevertheless, Peirce did believe that his own experience with experimental science gave some empirical validity to his call for an economics of research. And his method of drawing inferences in the "Note" is as modern as anything now being published in modern economics journals.

The second innovative aspect of Peirce's essay is the graph that he created for the two-case version of his economy of research. Graphs tend to be deleted from modern scientific articles in economics, but this was not always the case. And graphs are extremely useful in classroom instruction and in textbooks. The graph represented as figure 5.1 in Peirce's paper is a reproduction of the one from the published version of the essay published by the Coast Survey in 1879. While this may not seem significant and I easily could have glossed over it, it must be noted that Peirce had an extremely keen interest in graphical representations of scientific work. Peirce created a projection of the earth which was still in use for some navigational purposes during World War II.[24] Peirce also drew the graphs for his father's lecture on mathematical political economy in 1871. Peirce either actually drew or supervised preparation of the graph that is presented in the Coast Survey version of the "Note." Again in the context of the essays in the Baumol and Goldfeld volume, there is nothing which can compare with the representational sophistication of the graph which Peirce created in the "Note." Furthermore, although a similar graph can be found in Jevons's *Theory of Political Economy* (1871 [1957]: 97), Peirce's graphical representation is somewhat more advanced. In his figure, Jevons placed the marginal utility of two goods on the vertical axes and the quantities purchased on the horizontal axis. In contrast, Peirce put the ratio of the marginal utility per dollar of cost on the vertical axes and the dollars to be spent on the horizontal axis.

A third innovative aspect of Peirce's "Note" concerns the manner in which he developed utility theory. Again, I turn to an early work in mathematical economics in the Baumol and Goldfeld volume. In discussing the contribution of W. E. Johnson on the theory of utility curves which was written in 1913, Baumol and Goldfeld present criteria for assessing the theoretical advances represented in the piece. One criterion is particularly relevant to Peirce's "Note." One of the innovations was "an explicit analysis of utility which uses only the ratios of marginal utilities and hence is free from considerations of cardinal utility" (Baumol and Goldfeld 1968: 96). Again scrutiny of Peirce's "Note" with this criterion in mind suggests how thoroughly modern the "Note" is. In the "Note," Peirce made no mention of actually measuring utility. The implications of his theory have no dependence on cardinal utility. As Johnson did in 1913, the "Note," written as early as 1876, depended only on ordinal utility. The fundamental conclusions of Peirce's theory of the economy of research depend only on the ratios of the marginal utility of funds spent on rival research projects. This is the equation which I have designated as equation (5). A version of equation (5) appears in virtually every discussion of consumer theory in all of the principles texts I have ever encountered. Peirce appears to have been among the first to create this representation and formalization of utility theory. Again this suggests how truly innovative Peirce's "Note" in fact is.

24. See Eisele (1979e: 153).

5. Nicholas Rescher on Peirce's Economy of Research

Perhaps the only philosopher to take seriously Peirce's project regarding the economy of research is Nicholas Rescher (1976, 1978a, 1978b, 1989).[25] Rescher believes that the importance of Peirce's ideas on an economic theory of science have been overlooked. Rescher maintains that Peirce's theory is something like cost-benefit analysis:[26]

> Peirce proposed to construe the economic process at issue as the sort of balance of assets and liabilities that we today would call cost benefit analysis. On the side of benefits, he was prepared to consider a wide variety of factors: closeness of fit to data, explanatory value, novelty, simplicity, accuracy of detail, precision, parsimony, concordance with other accepted theories, even antecedent likelihood and intuitive appeal. But in the liability column, there sit those hard-faced factors of "the dismal science": time, effort, energy, and last but not least, crass old money. (Rescher 1978a: 69)

Behind Peirce's economy of research Rescher (1978a: 25 ff.) asserts is a geographic exploration model of science. This geographic model entails a finite physical world with a discernable and slowly changing nature. If studied long enough, scientific opinion would converge to a single view or theory of that world. According to Rescher (1978a: 26), with the geographic model Peirce thought of science in two stages: a preliminary stage of searching for the structure of qualitative relationships and a secondary stage of quantitative refinement. The preliminary stage was noncumulative but the secondary stage was a cumulative phase of increasing quantitative precision. During the first stage, there is disagreement regarding the content of a theory. In the second phase, given appropriate theoretical development, knowledge is increasingly refined.

25. There has been other research on economic aspects of science. Boland (1971) presents an economic analysis of conventionalism using welfare theory. I became aware of Boland's piece and Peirce's "Note" almost simultaneously while in graduate school in the 1970s. At that time, Boland's article lead me to attach a great deal of significance to Peirce's "Note." More recent works on the economics of science are Bartley (1990), Radnitzky (1987), and Wible (1991, 1992a, and 1992b).

26. Rescher's use of cost-benefit rhetoric and terminology stimulated a search of the cost-benefit literature. In searching that literature, a utility approach to research project evaluation was encountered in the work of F. M. Scherer (1965). What is so unusual about Scherer's model is that he chose a utility rather than a cost-benefit model for research project evaluation. Furthermore, it is astonishing to find that Scherer's model yields results that are nearly identical to those attained by Peirce almost a century earlier. Comparison of Peirce's and Scherer's models reveals the uncanny prescience of Peirce's model.

Scientific progress ultimately becomes a matter of exactness and detail. In the context of a geographic model of the phenomena being investigated, an economic approach makes a lot of sense. Additional resources could bring us closer to the truth. This is a situation ideally suited for economic analysis.

Rescher believes that the geographic exploration model of scientific progress which Peirce apparently assumed is flawed. Referring to the work of T. S. Kuhn (1970) and others who critiqued positivism, Rescher denies that all scientific progress can be understood as a process of cumulative accretion. The geographic model is misleading. Rescher believes that scientific progress is more a matter of changing the theoretical framework than filling in a "crossword puzzle" (Rescher 1978b: 29). On this narrower point, I am in agreement with Rescher.

Rescher's characterization of a geographic model of science behind Peirce's "Note" and his economics of science needs some amendment. This may be too restrictive an interpretation of Peirce. It is true that Peirce spent much of his professional scientific career attempting to measure the differential pull of the earth's gravity in various locations in the United States and Europe. It is also understandable how the number of locations and the quality of the instruments for measuring gravity would be affected by the availability of economic resources. Apparently, Peirce was continually involved with efforts to fund more observations and the purchase of more accurate pendulums to enhance the quality of his measurements. Peirce wanted to be known as one of the best scientists in the world in measuring the earth's gravity. But it is also quite apparent that the fundamental notion of statistical inference in the "Note," probable error, came from Peirce's extensive work in astronomy, not from geology. Furthermore, it is quite significant that Peirce continued to apply his economic vision of hypothesis formation outside the domains of both astronomy and geology. Peirce thought that economic dimensions of inquiry were just as relevant in other domains, such as the interpretation of ancient manuscripts. His most extensive elaboration of the economy of research is found in discussion of the logic of interpreting ancient manuscripts. Certainly this is an area more open to diversity than the fundamental measurements of the earth's gravity. Economic issues may need to be raised for research projects in more complex domains than geology and astronomy.

In other work, Rescher continues to expand on ideas concerning the economic dimensions of scientific research first encountered with Peirce.

What Rescher contributes is an awareness that economic concerns continuously affect science.[27] He believes that no conception of science can be complete if the economic dimensions of science are ignored. In his conclusion to his book *Cognitive Economy,* Rescher clearly portrays economics as central to an understanding of science:

> For inquiry—in science and elsewhere—is a human activity which, like any other, requires the expenditures of effort and energy in a way that endows the enterprise with an unavoidable economic dimension. Economic factors shape and condition our cognitive proceedings in so fundamental a way that they demand explicit attention. (Rescher 1989: 150)

6. Rescher's Peircean Interpretation of Karl Popper

Few economists may know of the significance of Peirce's philosophy and fewer still are aware of the economic aspects of his theory of science. Karl Popper is the philosopher of science typically recognized as having most influenced modern economics and methodology (Caldwell 1991). Popper's reputation in economics is due mostly to his idea of falsification. Yet there are connections between Peirce and Popper. Popper (1972b) has recognized the influence of Peirce in his adoption of an indeterminist interpretation of physics and Einstein's work. Both Popper and Peirce are evolutionary indeterminists. Beyond these similarities, there is another connection. Nicholas Rescher has applied the economics of research project selection from Peirce's work to Popper's widely influential notion of falsification. Rescher discusses falsification as one of several aspects of the problem of induction which can be resolved or enhanced in the context of an economic framework.[28] Rescher portrays Popper's methodology of falsification as an example of philosophy of science without economics. Like other philosophers, Popper makes no

27. In *Cognitive Economy,* Rescher (1989) also focuses on large-scale science that has become enormously vast and expensive: "This technological escalation has massive economic ramifications. The economics of scientific inquiry presents a picture of on-going cost escalation that is strongly reminiscent of an arms race. The technical escalation inherent in scientific research parallels that familiar arms race situation of inescapable technological obsolescence, as the opposition escalates to the next phase of sophistication" (Rescher 1989: 137).

28. Rescher (1978a: 72 ff.) maintains that an economic approach helps to deal with several disputes regarding aspects of induction. Among them are Carnap's total evidence requirement, Hempel's paradox of the ravens, Goodman's grue paradox, and Popper's notion of falsification.

mention of the economic dimensions of scientific research. Implicitly, the assumption seemingly would be that science is nearly costless so that economic factors can be ignored. Furthermore, no attention is given either to increasing costs or the opportunity costs faced by the researcher in attempting to falsify scientific theories.

According to Rescher, Popper presents an evolutionary methodology of inquiry in his *Objective Knowledge* published in 1972.[29] Popper portrays falsification in science as evolutionary competition among rival theories arbitrated by falsifying episodes and evidence generated by experiment and observation. Of course Popper was aware that falsification was no simple matter (Popper 1959: 86). More was required to reject a theory than disconfirming evidence. A better theory was required before one theory would be rejected for another (Popper 1965: 215–50). A better theory is the one which is most fit to survive the testing of both proponents and opponents. Between rival theories there is an evolutionary process of elimination through trial and error.

More specifically, Rescher summarizes Popper's methodology of falsification as containing three main elements (Rescher 1978a: 53): (1) a realization that the number of plausible hypotheses is infinite, (2) an awareness that science is a process of trial and error in eliminating alternative hypotheses, and (3) a sense that the process of eliminating hypotheses (falsification) is blind. The third element of Popper's methodology is the one which Rescher maintains is improved by reference to Peirce and economic considerations. Rescher does not believe that falsification is a "matter of blind, random groping" (Rescher 1978a: 53). Rescher asserts that Popper faces a real dilemma. If falsification operates among all conceivable or even proposed hypotheses, then there would never be sufficient time and resources for such a task. Popper's random and blind conception of the growth of scientific knowledge could not be carried out by humans with real resource constraints. Thus Rescher holds that Popper's theory of the growth of knowledge is incomplete because it does not explain the "rate and structure of scientific progress" (Rescher 1978a: 52).

In Rescher's (1978a: 65) view, the problem with Popperian falsification is underdetermination. There are an infinite number of potential

29. Recently another important evolutionary approach to science has appeared by David Hull (1988). His lengthy treatise *Science as a Process* has been reviewed and critiqued by Donoghue (1990), Oldroyd (1990), and Rosenberg (1992). Oldroyd (1990: 484) comments that Hull relies on invisible-hand-type arguments rather than those which depend on discredited altruistic norms to explain the behavior of scientists.

hypotheses and a finite body of empirical knowledge. Criteria of some sort must be used to select which hypotheses are going to be tested. Otherwise the task of falsification would be impossible. When Rescher (1978a: 51–63) applies economics to Popper's methodology for eliminating hypotheses, he maintains that it solves the underdetermination problem. Rescher maintains that there are just too many possible hypotheses to be considered with Popper's methodology. He believes there are almost an infinite number of hypotheses for scientific investigation. The researcher faces an impossible task. Faced with numerous hypothetical alternatives that could exhaust lifetimes, Rescher turns to economics. In the face of underdetermination, Rescher turns to Peirce and his economy of research. Rescher, taking his cue from Peirce's economy of research, believes that scientists should use economic criteria in deciding which hypotheses should be tested and subject to possible falsification. Using a cost-benefit metaphor, that hypothesis should be tested which offers the greatest net benefit to science. If two hypotheses are equal on grounds other than economics and if one involves much less time and resource expenditure than the other, then the least costly hypotheses should be tested. Rescher offers a general economic interpretation of his critique of Popper and falsification:

> The point is that our economically oriented approach is wholly undogmatic regarding generality preference. We replace Popper's purely logical concern for universality-for-its-own-sake with an economico-methodological concern for universality-relative-to-cost. If we take this economic line and do sensible decision making on the basis of the seemingly reasonable economic precept, "Maximize generality subject to the constraints of affordability," then our basic concern is one of cost-benefit analysis, seeking to optimize returns subject to resource-outlays. (Rescher 1978a: 83)

Judged from the vantage point of an economist, Rescher's critique of Popper and the Peircean economic corrective seem to make a lot of sense.[30] For an economist it is hard to imagine that science is unaffected by scarcity and incentives. Certainly the intellectual merits of the under-

30. Hands (1985) also considers economic aspects of Popper's philosophy of science. His conclusion in broad terms is analogous to Rescher's: "Popper's entire philosophy of science is simply *an application of the method of neoclassical economic theory.* The possibility of such an incredible conclusion only exemplifies the problems which await anyone interpreting Popper's writings on the topic" (Hands 1985: 85–6).

determination problem require evaluation from philosophers. However, without an economic dimension, Popper's method and those of other philosophers of science seem to be at such a high level of abstraction that they become dissociated from the conduct of real day-to-day science. Rescher's and Peirce's contributions make the fact that real science is done in an economic environment important for philosophy of science.[31]

All of this makes a great deal of sense to an economist. Rescher maintains that hypotheses should be tested to the maximum degree that resources allow. But there is one issue regarding Peirce and Popper that must be raised. The issue concerns the originality of the idea of falsification. Generally, the creation of the notion of falsification is attributed to Karl Popper. And I have no doubt that Popper should be credited with authorship in this case. Like so many of Peirce's intellectual contributions that were authored and then lost for so long, Peirce appears to have had an awareness of the essential concept of falsification. Peirce and his eccentric lifestyle must be given a great deal of the responsibility for obscuring many of his own contributions for so long. In the 1902 grant application to the Carnegie Institution, Peirce appears to have stumbled on to the idea of falsification:

> The true presuppositions of logic are merely *hopes;* and as such, when we consider their consequences collectively, we cannot condemn scepticism as to how far they may be borne out by facts. But when we come down to specific cases, these hopes are so completely justified that the smallest conflict with them suffices to condemn the doctrine that involves that conflict. This is one of the places where logic comes in contact with ethics. (Peirce 1902b: 1028–30)

Regarding this passage Brent (1993: 281) comments parenthetically, "compare Karl Popper's doctrine of falsification." And concerning another of Peirce's papers Brent remarks: "To read Popper or Carl Hempel on the logic of science after reading the 'Illustrations' shows how little has been added to the model first proposed by Peirce over a century ago" (Brent 1993: 117).

31. The use of economic models to study economic science raises some extraordinary issues. The most important issue concerns reflexivity or self-reference. The issues are summarized in a longer version of this essay, Wible (1992b), and by Hands (1994; chapter 19 in this volume).

7. Conclusions

Charles Sander Peirce's "Note on the Theory of the Economy of Research" is truly extraordinary. It is extremely unfortunate that it has been thoroughly neglected by the economics profession for more than a century. The "Note" presents an economic model of research project selection in science. A case can be made that the "Note" was the first piece of modern scientific research in all of economics. This claim is based on the novelty of the method of argument, the graphical techniques, and the ratio of the marginal utilities found in the "Note." The "Note" is also significant for making economic factors a central part of a theory of scientific inference, something which contemporary economic methodologists and philosophers still have not done except for a few notable exceptions. And the "Note" has also been used to criticize and reformulate the idea of falsification. Philosopher Nicholas Rescher has interpreted and extended Peirce's economic interpretation of science to Karl Popper's notion of falsification. Rescher argues that falsification needs to be supplemented with economic criteria. Together, the unprecedented contributions of the "Note," Peirce's subsequent comments on the economy of research, and his prescient ideas about scientific inference constitute a Peircean vision of science. His message is that science requires an explicit understanding of the economic factors of science. If this is so, then what the economics profession needs is a more explicit awareness of the economic aspects of economic science.

Peirce's essay may be pathbreaking for economics in a way that cannot be directly inferred from the "Note" or a more generalized Peircean vision of science. Except for similarities with a Marshallian perspective, a Peircean vision of economics is different than anything apparent in economics today. Mainstream economists in recent decades have tended to associate optimization and equilibrium theory with a Walrasian, mechanistic vision of economic activity. And the most prominent evolutionary economists of our time, the American institutionalists, have tended to reject optimization and equilibrium as relevant theoretical constructs. Like the institutionalists, Peirce rejected the mechanism that flourished in physics and in other disciplines during his lifetime. Peirce was an evolutionary indeterminist. Like the mainstream economists, Peirce saw optimization theory as a useful tool of economic analysis. For Peirce, optimization took place in the context of a more general evolutionary view of the world. Peirce's "Note" presented the core of a different vision of economics than either that of the present neoclassical mainstream or the American institutionalists. Peirce's vision of political economy was that of a mathematical theory assumed in the broader context of an evolutionary view of the world with an economy of research as an integral part of the conception of scientific inference. Reconstructing economics in a manner consistent with Peirce's evolutionism and his economics of the "Note" provides a new way of conceiving of economic science. A Peircean reconstruction of economics is one which surely will be worth pursuing.

ACKNOWLEDGMENTS

I would like to thank the Whittemore School for a grant in support of this research. I would like to thank Wade Hands, Robert Dorfman, Jeff Sohl, Paul Wendt, Avi Cohen, the students in my Economics of Science seminar, and anonymous referees for helpful comments.

REFERENCES

Bartley, W. W., III. 1990. *Unfathomed Knowledge, Unmeasured Wealth: On Universities and the Wealth of Nations.* La Salle, Ill.

Baumol, W. J., and Goldfeld, S. M., eds. 1968. *Precursors in Mathematical Economics: An Anthology.* London.

Boland, L. A. 1971. Methodology as an exercise in economic analysis. *Philosophy of Science* 38: 105–17.

Brent, Joseph. 1993. *Charles Sanders Peirce: A Life.* Bloomington, Ind.

Caldwell, Bruce J. 1991. Clarifying Popper. *Journal of Economic Literature* 29 (March): 1–33.

Cushen, W. Edward. 1967. C. S. Peirce on benefit-cost analysis of scientific activity. *Operations Research* 15: 641.

Donoghue, M. J. 1990. Sociology, selection, and success: a critique of David Hull's analysis of science and systematics. *Biology and Philosophy* 5: 459–72.

Eisele, Carolyn. 1979a. *Studies in the Mathematical Philosophy of Charles S. Peirce.* New York.

———. 1979b. The correspondence with Simon Newcomb. In Eisele 1979a.

———. 1979c. The mathematics of economics. In Eisele 1979a.

———. 1979d. Introductions to *The New Elements of Mathematics.* In Eisele 1979a.

———. 1979e. The problem of map projection. In Eisele 1979a.

Feigl, Herbert. 1969. The origin and spirit of logical positivism. In P. Achinstein and S. F. Barker (eds.), *The Legacy of Logical Positivism.* Baltimore.

Fisch, Max. 1982. The decisive year and its early consequences. Introduction. In E. C. Moore et al. (eds.), *Writings of Charles S. Peirce: A Chronological Edition.* 1867–1871. Bloomington, Ind.

Hacking, Ian. 1990. *The Taming of Chance.* Cambridge, England.

Hands, Douglas W. 1985. Karl Popper and economic methodology: a new look. *Economics and Philosophy* 1: 83–99.
———. 1994. The sociology of scientific knowledge: some thoughts on the possibilities. In Roger Backhouse (ed.), *New Directions in Economic Methodology.* London.
Hoover, Kevin D. 1994. Pragmatism, pragmaticism, and economic theory. In Roger Backhouse (ed.), *New Directions in Economic Methodology.* London.
Howey, Richard S. 1972. The origins of marginalism. *History of Political Economy* 2 (fall): 281–302.
Hull, David. 1988. *Science as a Process: An Evolutionary Account of the Social and Conceptual Development in Science.* Chicago.
Jevons, W. S. 1871 [1957]. *The Theory of Political Economy.* 5th ed. New York.
Johnson, W. E. 1913. The pure theory of utility curves. In Baumol and Goldfeld 1968.
Kmenta, Jan. 1971. *Elements of Econometrics.* New York.
Kuhn, Thomas. 1970. *The Structure of Scientific Revolutions.* 2d ed. Chicago.
Mach, Ernst. 1893 [1960]. The economy of science. In *The Science of Mechanics.* 6th American ed. LaSalle, Ill.
———. 1893. The economical nature of inquiry in physics. In *Popular Scientific Lectures* of Mach, trans. T. J. McCormack. 3d rev. ed. Chicago.
Oldroyd, David. 1990. David Hull's evolutionary model for the progress and process of science. *Biology and Philosophy* 5: 473–87.
Peirce, C. S. 1861 [1982]. [A treatise on metaphysics]. In *Writings of Charles S. Peirce: A Chronological Edition.* Vol. 1, 1857–1866, Max Fisch et al., eds. Bloomington, Ind.
———. 1871a [1982]. Letter to Simon Newcomb. In Baumol and Goldfeld 1968.
———. 1871b [1982] [Charles Babbage]. *Nation.* Vol. 13, November. In *Writings of Charles S. Peirce: A Chronological Edition.* Vol. 2, 1867–1871, E. C. Moore et al., eds. Bloomington, Ind.
———. 1878 [1955]. How to make our ideas clear. In *Philosophical Writings of Peirce.* New York.
———. 1879. Note on the theory of the economy of research. *United States Coast Survey* for the fiscal year ending June 1876. Washington, D.C. Reprinted in *Operations Research* 15: 642–8 (1967). Also reprinted in *The Collected Papers of Charles Sanders Peirce,* vol. 7, ed. A. W. Burks, Cambridge, 1958. And in *The Writings of Charles S. Peirce: A Chronological Edition,* vol. 4, 1879–1884, C. J. W. Kloesel, ed., 1986.
———. 1880 [1955]. What is a leading principle? In *Philosophical Writings of Peirce,* ed. J. Buchler. New York.
———. 1882 [1986]. Introductory lecture on the study of logic. In *Writings of Charles S. Peirce.* 1879–1884. C. J. W. Kloesel, ed. Bloomington, Ind.
———. 1893 [1960]. The logic of quantity. In *Collected Papers of Charles Sanders Peirce,* vol. 4, C. Hartshorne and P. Weiss, eds. Cambridge, Mass.
———. 1896 [1960]. The economy of research. In "Lessons from the history of science." *Collected Papers of Charles Sanders Peirce,* vol. 1, C. Hartshorne and P. Weiss, eds. Cambridge, Mass.

———. 1897 [1960]. Multitude and number. In *Collected Papers of Charles Sanders Peirce,* vol. 4, C. Hartshorne and P. Weiss, eds. Cambridge, Mass.

———. 1901 [1985] On the logic of drawing history from ancient documents especially from testimonies. In C. Eisele (ed.), *Historical Perspectives on Peirce's Logic of Science.* Berlin.

———. 1902a [1985]. Application for a grant [Carnegie Institution]. In C. Eisele (ed.), *Historical Perspectives on Peirce's Logic of Science.* Berlin.

———. 1902b [1985]. On the economics of research. Memoir no. 28 of Peirce 1902a.

Popper, Karl R. 1959. *The Logic of Scientific Discovery.* New York.

———. 1965. *Conjectures and Refutations: The Growth of Scientific Knowledge.* New York.

———. 1972a. *Objective Knowledge: An Evolutionary Approach.* Oxford.

———. 1972b. Of clouds and clocks: an approach to the problem of rationality and the freedom of man. In Popper 1972a.

Radnitzky, Gerard. 1987. Cost-benefit thinking in the methodology of research: the "economic approach" applied to key problems of the philosophy of science. In G. Radnitzky and P. Bernholz (eds.) *Economic Imperialism: The Economic Approach Applied Outside the Field of Economics.* New York.

Rescher, Nicholas. 1976. Peirce and the economy of research. *Philosophy of Science* 43: 71–98.

———. 1978a. *Peirce's Philosophy of Science.* Notre Dame.

———. 1978b. *Scientific Progress:* A *Philosophical Essay on the Economics of the Natural Sciences.* Oxford.

———. 1989. *Cognitive Economy: The Economic Dimension of the Theory of Knowledge.* Pittsburgh.

Rosenberg, Alex. 1992. Selection and science: critical notice of David Hull's *Science as a Process. Biology and Philosophy* 7: 217–28.

Scherer, Frederic M. 1965. Government research and development programs. In Robert Dorfman (ed.), *Measuring Benefits of Government Investments.* Washington, D.C.

Shackle, G. L. S. 1967. *The Years of High Theory.* Cambridge, England.

Stigler, G. J. 1972. The adoption of the marginal utility theory. *History of Political Economy* 4: 571–86.

Suppe, Frederick. 1974. *The Structure of Scientific Theories.* Urbana, Ill.

Wible, James R. 1984. The instrumentalisms of Dewey and Friedman. *Journal of Economic Issues* 23: 1049–70.

———. 1991. Maximization, replication, and the economic rationality of positive economic science. *Review of Political Economy* 3: 164–86.

———. 1992a. Fraud in science: an economic approach. *Philosophy of the Social Sciences* 22: 5–27.

———. 1992b. Cost-benefit analysis, utility theory, and economic aspects of Peirce's and Popper's conceptions of science. Department of Economics, University of New Hampshire.

Wiener, Philip P. 1949. *Evolution and the Founders of Pragmatism.* New York.

———. 1958. Introduction. In *Values in a Universe of Chance: Selected Writings of Charles S. Peirce.* New York.

7

Toward a New Economics of Science

Partha Dasgupta and Paul A. David

Science policy issues have recently joined technology issues in being acknowledged to have strategic importance for national "competitiveness" and "economic security." The economics literature addressed specifically to science and its interdependences with technological progress has been quite narrowly focused and has lacked an overarching conceptual framework to guide empirical studies and public policy discussions in this area. The emerging "new economics of science," described by this paper, offers a way to remedy these deficiencies. It makes use of insights from the theory of games of incomplete information to synthesize the classic approach of Arrow (chapter 4, this volume) and Nelson (chapter 3, this volume) in examining the implications of the characteristics of information for allocative efficiency in research activities, on the one hand, with the functionalist analysis of institutional structures, reward systems and behavioral norms of "open science" communities—associated with the sociology of science in the tradition of Merton—on the other.

An analysis is presented of the gross features of the institutions and norms distinguishing open science from other modes of organizing scientific research, which shows that the collegiate reputation-based reward system functions rather well in satisfying the requirement of social efficiency in increasing the stock of reliable knowledge. At a more fine-grain level of examination, however, the detailed workings of the system based on the pursuit of priority are found to cause numerous inefficiencies in the allocation of basic and applied scarce resources, both within given fields and programs and across time. Another major conclusion, arrived at in the context of examining policy measures and institutional

Reprinted with permission of Elsevier Science. From *Research Policy* 23 (1994): 487–521.

reforms proposed to promote knowledge transfers between university-based open science and commercial R&D, is that there are no economic forces that operate automatically to maintain dynamic efficiency in the interactions of these two (organizational) spheres. Ill-considered institutional experiences, which destroy their distinctive features if undertaken on a sufficient scale, may turn out to be very costly in terms of long-term economic performance.

1. Introduction and Motivation: Science, Economics, and Politics

To say that economic growth in the modern era has been grounded on the exploitation of scientific knowledge is to express a truism. To say that what goes on within the sphere of human activities identified as "The Republic of Science" has grown too important for the rest of society to leave alone is also something of a commonplace assertion.[1] Most of the industrial nations and many among the developing countries today acknowledge this, and virtually all societies in which modern science is practiced pay at least lip service to the belief that it is important to pursue some form of "science policy."

Indeed, in the West national science policies have tended to become more strongly interventionist and more explicitly committed to "planning and management." This is particularly so in the U.S. and Great Britain, where the remarkable post–World War II autonomy of Western science communities in setting their own basic research agendas has undergone significant erosion since the end of the 1960s to the fears of the public about scientific discoveries that may open the way to the creation of pernicious technologies, and the continuing calls for the exercise of greater "social responsibility" on the part of concerned scientists, have been added two more recent arguments for closer governmental direction and control over publicly supported science. One is the generic demand for the application of fiscal restraint to a noticeably large category of expenditures, arising more from obsession with controlling government budget deficits than from any significant real growth in the volume of public funds being devoted to civilian R&D.[2] The other, however,

1. For the origin of the term in quotes and its significance here, see Polanyi (1962; chapter 17, this volume) and note 5, below.

2. In the U.S., federal funding for civilian R&D in 1990 was only 3.7% larger, in constant (1982) dollar terms, than it had been in 1980, even though on a current-dollar basis this expenditure item had swelled by 56.3% during the same period, to reach the $23.1 billion mark. However, total U.S. expenditure for academic R&D, which in 1990 amounted to two-

reflects the more specific concerns of politicians and government administrators that the publicly supported parts of the scientific establishment were not addressing their efforts to the solution of national needs, and a growing conviction that the time had come to curtail the impulse of basic research scientists to "pursued knowledge for its own sake" in order to redirect researchers to work on "applied" projects that would bring more immediate discernable economic pay-offs.[3]

There is no little irony in the situation that consequently now presents itself. Supporters of political regimes that are ostensibly hostile to centralization and government planning in economic affairs, while jubilant over the comparative success of market capitalism and the manifest dysfunctions of socialist and communist systems, seem eager to subscribe to quite another position insofar as the management of science by the state is concerned; they view with deepening suspicion the largely autonomous, highly decentralized system for allocating resources that has been institutionalized and has flourished within "the Republic of Science."[4] Perhaps because demands for closer management control over

---

thirds of federally funded civilian R&D, had grown substantially more rapidly (in constant-dollar terms) than the latter during the 1980s—having increased by 69.3% in real terms between 1980 and 1990 (see National Science Board (1991), appendix tables 4-17 and 5-1).

3. In *Time Magazine* for 23 November 1992, Walter Massey, then director of the National Science Foundation, is quoted as saying: "The public hears that we're No. 1 in science, and they want to know why that fact isn't making our lives better. The one thing that works in this country doesn't seem to be paying off." The story goes on to point out that the shift under way in the U.S. government's stance towards science was not the product of the new administration. "Long before the [presidential] election, policymakers were concluding that they should assert more control over research by telling many scientists precisely what to work on. 'We've got to do some readjusting,' says Guyford Sever, co-chairman of a recent Carnegie Commission study on the future of American science." See Thompson (1992), p. 34.

4. To appreciate fully the irony in this, one need only read the following extract from a well-known critique of the outcome of the enormous growth of a science establishment supported by the State and large business corporations:

> as these developments have proceeded in a uncoordinated and haphazard manner, the result at the present day is a structure of appalling inefficiency both as to its internal organization and as to the means of application to the problems of production or welfare. If science is to be of full use to society it must first put its own house in order.

The author was not a businessman taking his turn at bat as a member of the Reagan-Bush "teams," nor a populist investigative congressman out to expose the abuses of academic research institutions, nor a science advisor or minister in the Thatcher government. Rather, the passage is quoted from *The Social Function of Science* (1939, p. xiii) by J. D. Bernal—a scientist and Marxist historian of science and technology, who took a leading role in the "social relations of science" movement in Britain during the 1930s!

government-funded science and engineering research to improve its social payoffs do seem discordant when emanating from circles that, in other contexts, are instinctively doubtful of the public sector's capability to allocate scarce resources efficiently, the idea of bringing the work of academic researchers into closer connection with market-oriented industrial R&D projects has lately been gaining a remarkable degree of support.

Thus, we in the West seem to have come full circle, and are approaching from a new direction the case for re-examining the institutions, organizational structures, and policies that constitute the mechanisms of resource allocation in that vital sphere of social activity we call Science. The popular press in the U.S. has moved beyond its earlier mode of sympathetically reporting new conquests on Vannevar Bush's "endless frontier" of science (see Nelkin 1987), along with speculations about the revolutionary consequences of such discoveries for our conception of the cosmos and the place of mankind within it, for the progress of industrial technology and health care, or for the sophistication and destructiveness of warfare. Instead, the newspapers increasingly divulge the more problematic aspects of the political economy of organized science, including some heated controversies over the criteria being used to select among alternative science "projects."

For example, has the U.S. Congress been successfully lobbied to spend excessive sums on enterprises in "big science," like the superconducting supercollider, at the sacrifice of a myriad undertakings in "little science"? Should we be trying to launch extensive, publicly funded programs like the effort to sequence the human genome as soon as techniques and instrumentation that permits such research become available? Or should some lines of investigation be deferred in favor of others? Should the availability of supplementary state and local government subsidies be made a deciding consideration when the National Science Board selects among groups of academic researchers (located at competing sites) those who are to receive the bulk of funds allocated for a particular scientific sub-field? Are public and private monies being channelled excessively into some fields of basic research, while other, potentially fruitful areas remain virtually drought stricken? Is the amount and distribution of public funding for university laboratories insufficient to permit them to keep pace with rising costs of state-of-the-art equipment, such as is made available to researchers in industry?

Still other news reports reflect growing recognition of the subtle and complex issues facing those who must frame science policies affecting the conduct of basic research in an environment of global economic competition. Why should any nation continue to devote a significant portion of its public expenditure to advancing scientific knowledge if, through the global networks of the international science community, those new discoveries soon will be made available to allies and rivals alike? How can one tell whether or not it is on balance advantageous for a particular country to have its scientists participating in a unrestricted exchange of information with colleagues working abroad? Is the U.S. failing to provide graduate (research) training for its nationals in the natural sciences, mathematics, and engineering, while at the same time subsidizing such training for too many foreign nationals? What is it about research universities that creates barriers impeding the easy and rapid "transfer" of new scientific knowledge to the sphere of commercial applications—even in the U.S., where cultural traditions of academic separation from the world of business affairs are far less strong than remains the case in western Europe? To what extent is it desirable to modify modern university institutions and operating rules to permit and encourage closer integration of academic and corporate research activities? Is it a legitimate cause for national concern when foreign corporations actively recruit university scientists to staff company-run basic research institutes set up in their immediate vicinity (as was the case when a computer science research institute under Japanese corporate management was located on the doorstep of the University of California in Berkeley)? Would there be no equivalent cause for concern were the business corporations American in ownership?[5]

It is thoroughly understandable why economists have remained much preoccupied with theoretical and empirical studies of the sources of technological innovation and its connections to productivity growth and improvements in economic well-being and national economic power. At the same time, in the present setting, it is a surprising and rather regrettable fact that elaboration of the economic analysis of technology was for some time allowed to run far ahead of the economics of science. In

5. For a sample of recent reporting of some of these issues, see: "Japanese Labs in U.S. Luring America's Computer Experts," *New York Times,* 11 November 1990; "Graduate Schools Fill with Foreigners," *New York Times,* 29 November 1990; "Foreign Graduate Students Know Hardship," *New York Times,* letters, 13 December 1990; William J. Broad, "Big Science—Is It Worth the Price?," *New York Times,* feature series, 27 May (introduction), 29 May (on superconducting supercollider), 5 June (human genome project), 10 June (NASA space station project), 19 June (NASA earth observing system project), 4 September (small-scale basic science funding impacts), 4 October (hot fusion project), 25 December 1990 (cutbacks funding for big science projects).

the following pages we undertake to report on recent analytical developments that may enable the laggard member of the pair to catch up, and which hold out the prospect that the two naturally interrelated areas of study can begin moving forward more swiftly in tandem. Our economic analysis of the organization of research within the spheres of science and technology emphasizes the point that we are dealing with an interrelated system, comprised of distinct activities that may reinforce and greatly enrich one another, but, furthermore, that it is a system that remains an intricate and rather delicate piece of social and institutional machinery whose constituent elements also may become badly misaligned. Indeed, there does not seem to be an adequate appreciation of the vulnerability of the science-technology systems in the West today, for all the frequency with which their importance to the modern economy and polity is acknowledged; nor of the basic features that are common to these variegated institutional and cultural structures, and which render all of them susceptible to destabilizing and potentially damaging experiments which may soon be embarked upon in the earnest hope of more fully mobilizing the respective national scientific research communities in the service of national economic security—the successor goal to military security—that is now being promoted under the euphemism of "competitiveness." Of course, the best preventative against blind and costly social experimentation that we can recommend is a prior investment in acquiring and disseminating deeper scientific understanding of the subject of concern. That, precisely, is our purpose in the present article.

2. The Old Economics of Basic Research, and the Emergence of a New Economics of Science

We must begin by acknowledging that it would hardly be fair to charge economists with having simply ignored problems of resource allocation in relation to science. Before there could be a new economics of the subject there must have been an "old" economics of science, and so there was.

. . .

In a pair of joint articles (Dasgupta and David, 1987, 1988) developed over the past several years we have tried, by synthesizing insights from economics and sociology, to carry the economics of science beyond the traditions of this older literature. A first goal was to provide an "explanation" for the prevalence of distinctive norms, customs, and institutions governing university science, on the one hand, and industrial R&D, on the other.[6] The phenomena that interested us naturally included those salient social arrangements that have occupied the attention of sociologists and philosophers of science: rules of priority and the role of validated priority claims in the reward structure of (academic) scientists, patenting and disclosure policies, and institutions associated with scientific communication and the functioning of a collegiate reputational reward system—ranging from "invisible colleges" to organized, professional academies in various disciplines. This explanatory approach, however, departs from that of the sociologists and philosophers.

A considerable part of the characteristic style of analysis in the new economics of science derives from its recognition of three features of the processes for the production, dissemination, and use of knowledge. First is the fact that certain crucial inputs, such as research effort, care, innate scientific talent, and the realization of elements of chance in the process of discovery, are very costly for outsiders to monitor, and in most cases are not even observable jointly by the "principal" who is sponsoring the inquiry (e.g., a private or public patron or employer) and the principal's "agent" (the researcher). Second is the observation that there often are significant aspects of indivisibility, and attendant fixed costs and "economies of scale," inherent in the underlying processes of knowledge production. Third is the point that the knowledge generated in research activities, rather than being inherently available to others, can be kept from the public domain should the researcher so choose; the characteristics of the reward system, along with the costs entailed, determining what information gets disclosed fully, what is disclosed partially, and what is kept secret. Given these structural conditions, and the long-standing observation that the production of knowledge is shot through with uncertainty,[7] it is quite in order to re-examine features of organized research activities from the perspective offered by the recently developed analysis of resource allocation under conditions of asymmetric information.

6. We associated university science with the world of "episteme" and the republic of science, whereas the latter we associated with the realm of technology.

7. For early efforts to quantify the uncertainty surrounding the duration, costs, and success of industrial research projects, see Mocking (1962) and Norris (1971). We are unaware of comparable studies based on the experience of university scientific researchers, but there is a presumption that, because the latter are less constrained to select problems where the feasibility of a solution is less clearly established, the research duration and costs associated with a successful outcome typically are even more difficult to predict with accuracy.

As will be illustrated by this paper, the new economics of science has the two-fold ambition of (1) exposing the underlying logic of the salient institutions of science, and (2) examining implications of those differentiating institutional features for the efficiency of economic resource allocation within this particular sphere of human action. To carry out this program we will be building upon the foundations laid down by the classic contributions in the sociology of science,[8] adding to the insights provided by the "old" economics of science some new ones that are drawn principally from the rapidly growing analytical literature that treats problems of behavior under incomplete and asymmetric information (including the economic theories of agency and optimal contract, or "mechanism design" theory), as well as issues in the dynamics of racing and waiting games.

3. Knowledge: Codified or Tacit? Public or Private?

As is the case in other specialized fields of inquiry, in this one some specific terminological conventions will be employed. A few initial definitions and distinctions will avoid later confusions especially since many of the terms we employ are given somewhat different meanings in ordinary language usage and in the scholarly literature on the intellectual and social organization of the sciences.

3.1 Knowledge, Information, and the Endogeneity of Tacitness

By the term "information" (following common practice in economics) we will mean knowledge reduced and converted into messages that can be easily communicated among decision agents; messages have "information content" when receipt causes some change of state in the recipient, or action.[9] Transformation of knowledge into information is, therefore, a necessary condition for the exchange of knowledge as a commodity. "Codification" of knowledge is a step in the process of reduction and conversion which renders the transmission, verification, storage, and

8. For example, Merton (1973), Polanyi (1951, 1962, 1966), Hagstrom (1965), Ziman (1968), Gaston (1970), Zuckerman and Merton (1971, 1972), Ravetz (1971), Cole and Cole (1973), Salomon (1973), Blume (1974), Mulkay (1977), Luhmann (1979), Nelkin (1987), Whitley (1984).

9. Shannon's well-known measure of the quantity of information (see Shannon and Weaver 1949) was formulated to serve the specific needs of communication theory, and has been found unsuitable for application in economics. On this see, e.g., Arrow (1984) and Marschak (1971).

reproduction of information all the less costly.[10] Yet, somewhat paradoxically, this transformation makes knowledge at once more of what is described (in the public finance literature) as a "non-rival" good, that is, a good which is infinitely expansible without loss of its intrinsic qualities, so that it can be possessed and used jointly by as many as care to do so.[11] Thus, codified scientific knowledge possesses the characteristics of a durable public good in that (i) it does not lose validity due to use or the passage of time per se, (ii) it can be enjoyed jointly, and (iii) costly measures must be taken to restrict access to those who do not have a "right" to use it.[12]

In contrast, tacit knowledge, as conceptualized by Polanyi (1966), refers to a fact of common perception that we all are often generally aware of certain objects without being focused on them.[13]

. . .

Going back to Polanyi's perceptual analogy, what gets brought into focus (and codified) and what remains in the background (as tacit knowledge) will be, for us, something to be explained endogenously by reference to the structure(s) of pecuniary and non-pecuniary rewards and costs facing the agents involved. Although the position of the boundary between the codified information and tacit knowledge in a specific field of scientific

10. This usage is related to, but does not carry the broader implications of the concept of "codification" in science as developed by Zuckerman and Merton (1972): "The consolidation of empirical knowledge into succinct and interdependent theoretical formulations." According to the latter, scientific fields differ in this regard, with those disciplines that remain in what Kuhn (1962) would call a "pre-paradigm state' being necessarily less "codified."

11. Were person A to give person B a piece of information concerning Q, that would not reduce the amount of information concerning Q that was retained by A (although, to be sure, the benefit to each would depend upon whether and in what manner B were to make use of the information). Romer (1990, 1993) has recently popularized this application of the term "non-rival" good, but see David (1993b) for the proposal of "infinite expansability" as an alternative term that is less confusing, in that it acknowledges the possibility of rivalries for possession of new information.

12. Even in the case of the codified knowledge the condition of strict non-excludability (namely, that once produced it is impossible to exclude anyone from benefitting from it), which is taken to be one of the hallmarks of a *pure* public good, does not hold: patents, copyrights, and trade secrecy are institutional devices for denying others beneficial access to information.

13. Whereas philosophers of science (e.g., Nagel 1961; Hempel 1966; Popper 1968) emphasized the epistemological bases for distinguishing between the scientific and other modes of human understanding, stressing formal methodological rules of "scientific procedure," Polanyi's (1966) insights concerning the ubiquitous role of tacit knowledge have led some modern sociologists of science (e.g., Latour and Woolgar 1979; Latour 1987) to study the practice of laboratory science as a "craft."

research may be shifted endogenously by economic considerations, the complementarity between the two forms of knowledge has important implications for the way research findings can be disseminated. We comment on these below in connection with recent discussions of measures to promote the transfer of university research results to industry for further development and commercial exploitation.

3.2 Science and Technology: Public and Private Knowledge

Although the foregoing discussion has referred to scientists and technologists in ways that suggest they are different kinds of knowledge-workers, the crucial distinction we seek to draw is between the social organizations of Science and Technology. By capitalizing the terms we signify that these labels are being used here to refer to the two sets of socio-political arrangements and their respective reward mechanisms affecting the allocation of resources for scientific research. For our purposes, what fundamentally distinguishes the two communities of researchers is not their methods of inquiry, nor the nature of the knowledge obtained, nor the sources of their financial support. To be sure, differentiations can be drawn along those lines, as already noted, but to our way of thinking much of the economics literature, like philosophy of science, has mistakenly focused upon science-technology distinctions that are epiphenomena of differences that lie at a deeper level. It is the nature of the goals accepted as legitimate within the two communities of researchers, the norms of behavior especially in regard to the disclosure of knowledge, and the features of the reward systems that constitute the fundamental structural differences between the pursuit of knowledge undertaken in the realm of Technology and the conduct of essentially the same inquiries under the auspices of the Republic of Science.

Loosely speaking, we associate the latter with the world of academic science, whereas Technology refers to the world of industrial and military research and development activities. What makes a knowledge-worker a "technologist" rather than a "scientist," in this usage, is not the particular cognitive skills or the content of his or her expertise. The same individual, we suppose, can be either, or both, within the course of a day. What matters is the socio-economic rule structures under which the research takes place, and, most importantly, what the researchers do with their findings: research undertaken with the intention of selling the fruits into secrecy belongs unambiguously to the realm of Technology. Secrecy, however, is more readily effected when knowledge is not codified in proprietary documents (e.g., blueprints, receipts for chemical syntheses) that can be purloined and published, but instead is retained in a tacit form. Training services that convey tacit information and access to contracting with the trained are commodities that can be, and are, exchanged for value by business organizations operating within the realm of Technology, and also by academic research organizations, although the enforceability of contracts tends to differ between the two cases, and the terms on which tacit information can be sold are correspondingly different.[14] None of this implies that profit-seeking business firms would not find it to their advantage on occasion to invest some resources in "basic" research, or organize research facilities in ways that emulated the open, cooperative environment characteristic of university campuses. Neither does it imply that academic scientists will never seek to benefit materially by patenting their inventions, or always refrain from rivalrously withholding research findings and methods from university-based researchers in their field.[15]

. . .

Although it is evident from the contradictory norms on which they are based that there is a tension between these two modes of economic and social organization, so that they do not "mix" easily, the two are not mutually exclusive ways to successfully organize the pursuit of scientific knowledge within the same society. Indeed, we will argue that in order to ensure a reasonably efficient allocation of resources in the production of knowledge, modern societies need to have both communities firmly in place, and attend to maintaining a synergetic equilibrium between them.

4. Priority and Adherence to the Norm of Disclosure in the Reward System of Science

It has long been recognized in the sociology of science that priority of discovery or development is the basis for legitimate reputation-

14. Of course, as we have acknowledged in the text above, epistemological and methodological differences exist between the two research communities. For example, it is widely believed that among applied industrial scientists, engineers, and like technologists the balance between codified and tacit knowledge ("practical knowhow") tilts more towards the latter than is the case among university researchers. Yet as we already have suggested, such a difference could well be a consequence of the different information disclosure rules and reward systems that obtain.

15. It is the identification of Weberian "ideal types" that endows the instances of deviance with interest and potential significance, as will be seen below in Section 4.2, for example.

building claims, and that an individual's reputation for "contributions" acknowledged within his or her collegiate reference groups is the fundamental "currency" in the reward structure that governs the community of academic scientists.[16] That scientists take intense interest in disputes over "priority," and spend much effort collectively in determining for what and to whom this "coin" shall be distributed, suggested to a functionalist like Merton (1973) the central role that competition for priority was playing in the organization of the Science community.[17] In the context of the reward system in science, the rule of priority serves two purposes at once: hastening discoveries, and hastening their disclosure. How it does that is seen readily enough.[18]

4.1 Priority and the Science Reward System

First, tying rewards to priority sets up a contest, a race, for scientific discoveries. Since a scientist's effort cannot in general be observed by outside monitors, payment cannot be based upon it. If funds were to be allocated for "effort," scientists like anyone else would be given an incentive to slack off while declaring that they were working hard. Nor can intention be the basis of payment, for intention cannot be observed publicly either. By contrast, performance, if disclosed, can be observed

16. See, for example, Merton (1973). See also Blume (1974) and Whitley (1984). Gaston (1970), Cole and Cole (1973), and Cole (1978) provide quantitative evidence supporting the view that science is an unusual institution in the sense that it comes close to achieving a reward system that is "universalistic," in Merton's sense, rather than particularistic, i.e., achievement-oriented rather than ascription-oriented. In a study of university physicists, Cole and Cole (1973) found that *quality* of published research was the most important determinant of recognition that came in the forms of honorific awards, appointments, and wide citation. While the relative weighing of quality and quantity of "contributions' have not been established across different fields of science, there is a presumption that where a dominant paradigm, or research program, reigns, there will be greater consensus concerning "quality." As fields of inquiry go through paradigm shifts, the quantity-quality weighing is likely to be disrupted, and it should not be supposed that stability in the relative weights will be maintained across fields.

17. See Lamb and Easton (1985, chap. 10) for a historical survey of priority disputes and races for priority. Scientists may be motivated to establish claims of "priority" because they seek fame through the attachment of their name to a discovery or hypothesis; because, as creative individuals, they need to secure the validation of their creation—in this case from an expert audience the feeling of having produced something new to the world and not just to the self (see Storer 1966)—; or because of material rewards like reputational standing among their scientific colleagues. For the purposes of the immediate argument, the precise nature of the underlying motivation does not really matter.

18. Material in this section draws upon our earlier papers (Dasgupta and David 1987, 1988).

and vetted publicly. So rewards can be based upon it; the greater the achievement, the larger the rewards—which may come, eventually if not immediately, in the form of salary increases, subsequent research grants, scientific prizes, eponymy, and, most generally, peer-group esteem.

A method of payment alternative to one based on priority would be a fixed fee for entering science, but this would dull the individual's incentive to work hard, since scientists could collect the fee irrespective of whether they produced anything of interest. So the reward has to be based in some way on achievement. However, it is often difficult to determine how far behind the winner the losers of a scientific race are when the winner announces his discovery. (Those who were left behind can merely copy the winner's results and claim that they were very nearly there). For this reason, it is not possible in general to award prizes on rank. Thus, unlike tennis tournaments, science does not pay big rewards to the runners-up. This suggests a system of payment which is compatible with individual incentives. It is one where, roughly speaking, the winner is awarded all that is to be dispersed by the community for the discovery. The rule of priority mimics this.

We have offered a rationale for the rule of priority among scientists involved in parallel research on the basis of what is publicly verifiable. Fortunately for society, there is congruence between this requirement and the relative social values of the outputs of parallel research teams. For note that among the discoveries (or inventions) made by rivals involved in parallel research only the first is worthwhile to society; there is no social value-added when the same discovery is made a second, third, or fourth time.[19]

There is, however, one immediately apparent difficulty about the rule of priority. If the losers of a scientific race were to receive absolutely nothing, the rule would place all the risks involved in the production of knowledge firmly on the shoulders of scientists. This cannot be an efficient system if scientists, like other mortals, are averse to taking risks which involve their survival and comforts. (One would expect individuals without private means to be particularly averse to absorbing all the risks.)[20] We conclude, then, that those who regularly engage in basic

19. By this we do not, of course, mean independent confirmations of a scientific discovery, which is a different matter altogether.

20. The analysis here is thus most appropriately interpreted to apply to arrangements for the patronage (or employment) of professional scientists, from either private or public sources, and does not bear on the pursuit of scientific knowledge by "amateurs."

science research need to be paid something regardless of the extent of their success in the scientific races they choose to enter. Otherwise many, if not most, scientists today would enter other professions.

All this suggests the desirability of a payment schedule which consists of something like a flat salary for entering science, supplemented by rewards to winners of scientific competitions, with the proviso that the better is the performance, the higher will be the reward. The flat-salary component of the public payment schedule acts as a drag on incentives to do research (for this reason it must not be all that high), but as we have seen, it is a socially necessary drag if there is to be Science.[21] Fortunately for the evolution of "academic science," it has been found possible to tailor the flat salary to a complementary, productive activity—teaching—and thereby reduce the wastage occasioned by the drag. Roughly speaking, a modern scientist is paid in the form of a fixed salary (e.g., for teaching, should he be in academia) and bonuses (e.g., promotions, scientific awards, and general recognition) for priority in discoveries and inventions.[22]

The second purpose the rule of priority serves is in eliciting public disclosure of new findings. Priority creates a privately-owned asset from the very act of relinquishing exclusive possession of the new knowledge. To put it dramatically, priority in science is the prize. Now, the public disclosure of new findings provides two additional social benefits. First, it widens the span of application in the search for new knowledge. It raises the social value of knowledge by lowering the chance that it will reside with persons and groups who lack the resources and ability to exploit it. Second, disclosure enables peer groups to screen and evaluate the new finding. The result is a new finding containing a smaller margin of error. The social value of "reliability" established by disclosure to

21. In addition, most academic institutions limit the potential inefficiency of paying flat-fees, unrelated to research productivity, by awarding tenure only after some extended period of trial during which research capabilities and motivation may be assessed (albeit with some error). Advancement to tenure, and the security represented by entitlement to future flat-fee payments, is thus a part of the performance-based-bonus feature contained in the young researcher's "contract."

22. Employment contracts in industry and government research establishments also exhibit the generic two-part structure of compensation described as characteristic in academic research institutions, although in the former cases the fixed portion has to be viewed as a payment corresponding to the option value of the proprietary rights to the knowledge about future inventions and discoveries that are relinquished as a condition of employment. It may be noted that these arrangements are not uniquely modern; the post-Renaissance system of noble patronage of mathematician-scientists also generated a two-part payment structure, as David (1991) points out.

the community of scientists is that users of new discoveries can thereby tolerate a higher degree of risk arising from other sources of incomplete knowledge and information.

There is a third beneficial consequence, stemming from the fact that for priority to matter the race must be run towards a goal that is widely recognized, either at the outset, or subsequently, as one worth achieving. The autonomous governance system that has characterized academic (and, indeed, much nonacademic) science in the West means that communities of scientific peers define what contributions to knowledge it is worth bothering to have arrived at before others. What effect does this have? It creates a cumulative, chain-linked impetus to the advance of knowledge, because what turns out generally to be appreciated is the disclosure of knowledge that aids (or is expected to aid) colleagues in the field in generating findings on the basis of which they can establish priority claims of their own.[23]

4.2 Priority and Secrecy in Science: Public Virtue and Private Vices

Of course, the reward system sets up an immediate tension between cooperative compliance with the norm of full disclosure (to assist oneself and colleagues in the communal search for knowledge), and the individualistic competitive urge to win priority races. This can engender neurotic anxieties on the part of researchers and "deviant" patterns of secretive behavior.[24] Conceptually one may want to distinguish between departures from the norm of disclosure that take the form of remaining taciturn until "a result" has been obtained and can be publicly announced, and incompleteness in disclosure—i.e., not revealing all that has been learned.

. . .

More generally, it can be seen that the boundary line between tacit and codified knowledge is not simply a question of epistemology; it is a matter, also, of economics, for it is determined endogenously by the costs and benefits of secrecy in relation to those of codification. One can see

23. The channelling of individual research efforts by means of the emergence of collegiate consensus as to which priority races are worth entering does certainly impart momentum to particular research programs, but, as will be seen in Section 5 below, it also has some undesirable effects on resource allocation within scientific disciplines.

24. The sociologist W. O. Hagstrom, in a 1967 study, found that 30–40% of American (university) scientists in some fields expressed concern over being "scooped' in their current work, to the extent that it could inhibit their willingness to discuss it freely with colleagues. See unpublished paper cited by Blume (1974, 38).

that accelerations in the progress of instrumentation and research techniques, made possible in part by synergisms and feedbacks between developments in the realms of Science and Technology, can raise the private marginal benefits of (a greater degree of) "tacitness" for researchers in Science. If the marginal costs of transferring tacit information to others, being largely the time of the researchers and their support staff, is constant or rising, there would be an understandable tendency for the boundary between private tacit knowledge and shared tacit knowledge to shift towards the former, which in turn would weaken the motivation to bear the marginal costs of codification for the purposes of public disclosure. At the same time, however, falling costs of information transmittal, deriving in large part from computer and telecommunications advances, have lately been encouraging a general social presumption favoring more circulation of timely information and a reduced degree of tacitness. One of the resultants of these conflicting forces would seem to be the emergence of more active sharing of intermediate results, via computer networks, among quasiprivate alliances of researchers. The advantages of pooling knowledge and swapping complementary techniques, being no less than formerly, and the costs of communication required for selective cooperation having fallen, this phenomenon is explicable by reference to the workings of self-interest. In other words, it is possible that cooperative behavior within a limited sphere can emerge and be sustained without requiring the prior socialization of researchers to conform, altruistically, to the norm of communalism.

We have here a rather straightforward instance in which insights from the theory of repeated games are applicable to explaining cooperative behavior among potentially rivalrous researchers. To give precision to the essential ideas, we may start with the simplified case of two researchers (or compact research teams) working towards the same scientific goal, which entails solving two subproblems. Suppose that each "team" has solved one of the problems. Once each gets the other solution, it will be a matter of writing up the result and sending it off for publication, the first to do so being awarded priority. We may further assume that the write-up time is determined by a random process, and that if both get both halves of the problem at the same moment, each will have the same (one half) probability of being the first of the pair to submit for publication. Whether the winner will be awarded priority will depend, however, on whether or not some other researchers also obtain the full solution and succeed in preempting its publication. The question is: should the first research team to learn any part of the solution follow the strategy (S) of sharing that information with the other one, or should they adopt the strategy (W) of withholding? If, without prior communication, they play the strategy pair (S, S) they can proceed immediately to the write-up stage; if they play (S, W) the second member of the pair will be able to proceed to the write-up, and the opposite will be true if they play (W, S). Should they both withhold (W, W), they must both spend further time working on the other problem. It is evident that if they are only going to be in this situation once, the rule of priority alone will induce each of them to withhold, and they will end up (collectively, if not individually) at a relative disadvantage vis-à-vis other researchers who are hurrying to publish. If nobody else has the full solution yet, society also will have been forced to wait needlessly, because each member of the pair has a dominant (private) strategy of withholding what it has discovered.

The game just described will be recognized to have the structure of a classic two-person "prisoners' dilemma," from which bad consequences can be anticipated.[25] It is well known, however, that an escape from the pessimal outcome (W, W) is possible in certain circumstances—namely, were this a game that was part of an open-ended sequence of such encounters (i.e., an infinitely repeated game), were the future not

25. It is not only in the fantasies of game theorists that bad things happen. Consider the following recent British newspaper account, appearing in *The Independent on Sunday* (31 October 1993):

> Several teams of scientists closing in on the discovery of a gene that causes breast cancer have abandoned collaboration in their intense rivalry to win the race. Secrecy—spiced with misinformation—has replaced the co-operation that once aided the efforts of geneticists in Britain, the United States, Canada and France; such are the rewards of coming first . . . Three years ago, when Mary-Claire King at the University of California at Berkeley placed the breast cancer gene somewhere on chromosome 17, scientific teams around the world formed a consortium to pool their resources in an effort to isolate it. They exchanged information regularly to identify regions of the chromosome that could be eliminated. But as the groups edged closer to identifying the gene they began to split apart, said Simon Smith, head of a Cambridge University research team funded by the Cancer Research Campaign. "Things have now gone quiet because none of us wants to give informtion to the others," he said. "In an ideal world we'd be talking to each other and not holding back information. But our work is judged on what is published. If we are always second, it's no good.

It is interesting to observe that the original consortium, or coalitional agreement, did not involve pre-commitments to joint authorship, presumably, because the number of participants was so large that internal monitoring of effort would be difficult, and because to do so would vitiate the point of a race for priority. Our two-person prisoners' dilemma game, above, abstracts from the possibility of forming sub-coalitions against the rest of the field, a consideration that will be developed in the text below.

discounted too heavily, and were the players to expect the other team members to remember, and punish on future occasions, their present refusal to cooperate.[26] However, the value in the future of developing and maintaining a good reputation for sharing has to be large to discipline the self-interested researcher into adhering to the sharing mode of behavior in the current period. If repetitive play comes to an end, or if the future is valued only slightly, cooperation will unravel from the distant terminal point in the game right back to its inception.

4.3 Culture: The Enforcement of Cooperative Rivalry and Collective Regulation in Science

Yet that is not the whole of the story. As there are other researchers in the picture, we should really be considering an n-agent game (again, where the agents may be individuals or small teams), involving the solution of an m-part problem, given $n \geq m$. Now the question of sharing information becomes one of sharing not only what you have learned yourself, but also what you have been told by others. It is obviously advantageous to belong to a coalition among whom information will be pooled, because that will give the coalition members a better chance of quickly acquiring all m parts of the puzzle and being the first to send it in for publication. On the other hand, if there are individuals who behave opportunistically by exchanging what they have learned from one group for information from people outside that group, but do not share everything they know within their group, they can expect to do still better in their current race for priority of publication. However, because others would see that such "double-dealing" will be a tempting strategy, cooperation will be unlikely to emerge unless double-dealers (who disclose what you tell them to third parties, but don't share their full knowledge with you) can be detected and punished. What is the form that retribution can take? Most straightforward will be punishment by exclusion from the circle of cooperators in the future; and even more

26. Indeed, there is a so-called folk theorem to that effect. For a non-technical introduction to the literature on the repeated prisoners' dilemma and its broader implications, see Axelrod (1984). The folk theorem of game theory holds that (if future payoffs are discounted by each player at a low rate) in the "super game" obtained by repeating a finite, two-person game indefinitely, any outcome that is individually rational can be implemented by a suitable choice among the multiplicity of Nash equilibria that exist. See Rubinstein (1979, 1980) and Fudenberg and Maskin (1984).

severely, not only from the circle that had been "betrayed" but also from any other such circle. This may be accomplished readily enough by publicizing "deviance" from the sharing norms of the group, thereby spoiling the deviators' reputation and destroying their acceptability among other groups.[27]

What, then, is the likelihood that this form of effective deterrence will be perceived and therefore induce cooperative behavior among self-interested individuals? If a coalition, i.e., "a research network" numbering g players ($g \leq n$), is large, identifying the source(s) of "leaks" of information and detecting instances of failure to share knowledge within it will be the more difficult. It is worth remarking that the power of a large group to punish the typical deviator from its norms by ostracism tends to be enhanced by the higher probability that all those individuals with whom potential deviators will find it valuable to associate are situated within the coalition.[28] In other words, the expected loss entailed in being an "outcast" is greater when there is only a fringe of outsiders with whom one can still associate. However, this consideration is offset by the greater difficulties the larger groups will encounter in detecting deviators. Smaller groups have an advantage on the latter count, and that advantage also enables them to compensate for their disadvantage on the former count. The more compelling the evidence that a particular individual had engaged in a "betrayal of trust," the more widely damaging will be the reputational consequences for the person thus charged. Hence, unambiguous detection and attribution of deviations (from recognized norms regarding the disclosure and non-disclosure of information) augment the deterrent power of the threat of ostracism that can be wielded by any group that remains small in relation to the total population of individuals with whom an excluded group-member could form new associations.

The foregoing suggests that small cooperative "networks" of information-sharing can be supported among researchers because cooperative behavior furthers their self-interest in the race for priority, and de-

27. See Greif (1989) and Milgrom et al. (1990) for analysis of repeated games of incomplete information that have this structure.

28. However, when there are inhomogeneities in communications that would tend to divide the coalition into tighter "sub-cliques," a grand coalition will be vulnerable to defections by some among its members. This seems to have been the situation of the breast cancer gene research consortium, which was formed from a number of pre-existing national research teams (as described above, in note 25).

nial of access to pools of shared information would place them at a severe disadvantage vis-à-vis competitors.[29]

. . .

Uncertainty in the outcome of research and the inevitable privacy of much relevant information, taken together, provide the basis for offering a rationale, or functionalist explanation, of much that is observed in the social organization and salient institutions of modern science. Sticking to this very gross level of observation, our discussion suggests that the distinctive institutional features and the reward system of Science does rather well in satisfying the requirement of social efficiency in the allocation of resources, but when one looks more closely at the detailed workings of this system, its many inherent inefficiencies begin to come into view. Taking these "fine-grain" inefficiencies together, the resulting malallocation of valuable resources may be far from negligible. The following two sections, therefore, will be devoted to examining some of their main manifestations and underlying causes.

5. Resource Allocation within Scientific Fields and Programs

Because the outcomes of research projects are uncertain, it is generally in society's interest to hold a portfolio of active projects which are run "in parallel" within a particular field, or under the auspices of any specific scientific program that is determined currently to be worth pursuing. Therefore, in and of itself, parallelism or a multiplicity of projects aiming at essentially the same result—isolation of a virus, or development of a vaccine, or development of superconducting ceramic filaments—does not imply waste.[30] Society should thus be prepared to

29. These "circles" or "networks," which informally facilitate the pooling of knowledge among distinct research entities on a restricted basis, can exist as exceptions to both the dominant mode of "private knowledge" characterizing Technology. Thus, von Hippel (1990) and others have described how firms in fact tacitly sanction covert exchanges of information (otherwise treated as proprietary and protected under the law of trade secrets) among their engineer-employees. Participants in these "information networks" who accepted money or remuneration other than in kind would most probably be dismissed and prosecuted for theft of trade secrets.

30. The exception to this rule is, of course, the set of circumstances where experimental facilities are indivisible and the fixed costs entailed are so large as to rule out the benefits of diversification. Under such conditions, which more or less fit the case of the superconducting supercollider project, it is desirable to pursue only one project within the program, if it is desirable to embark on the program at all. However, within such large and complex projects, typically, there will be many sub-projects that present opportunities to pursue several solutions in parallel.

tolerate multiple discoveries in the sense of Merton (1973).[31] Nevertheless, a legitimate question arises as to whether the rule of priority and the reward structure in academic science encourage a more than desirable degree of duplication of research efforts, leading both to too many projects being discontinued by those who perceive that they have lost a race for priority, and to an excessive probability that researchers will unknowingly "multiply" the findings of others. There are, in fact, a number of reasons why the incentive structure built around the rule of priority in Science is prone to cause wastage of resources in the form of excessive numbers of projects being launched in the same area, and an excessive correlation of research strategies among them. To identify these may suggest at least the broad lines along which remedial institutional adjustments and public policy interventions might usefully proceed.

6. The Timing of Research Programs within Science

In recent years the choice of the current mix of investment projects has been much discussed in the literature on social cost-benefit analysis; less so the timing of investments.[32] In almost all spheres of economic activity, most projects that are offered for appraisal are rejected. This is inevitable, but the fact that a project is not worth undertaking now does not mean that it will never be worth undertaking. Often enough, the right thing to do is to accept as a package a set of projects that are better when *sequenced* than when run simultaneously. Success in the first project might, for example, mean a reduction in the cost of running the second project, and so forth.

What is alluded to here is that there may be important positive spillovers across projects in the form of "learning effects." Quite aside from

31. The term "multiples" is ambiguous but, following Merton (1973, 364 ff.), its use among sociologists of science is not. Multiplicity connotes the occurrence of more than one research entity expressing essentially the same discovery or invention (including inventions of apparatus), and not that of a given research unit making more than one discovery. See Lamb and Easton (1985) for a recent treatment of the subject, which argues that the phenomenon of multiple discovery is inherent in the collective, evolutionary process through which scientific knowledge grows. In the present discussion, however, we are less concerned to account for what might be thought of as a "normal," or "background," level of multiplicity, and more with "excess multiplicity" created by certain features of the science-resource allocation mechanism.

32. On social portfolio choice and cost-benefit analysis, see e.g. Dasgupta et al. (1972), Little and Mirrlees (1974), Squire and Van der Taak (1975) and Lind (1982). David et al. (1992) seriously question the usefulness of applying the cost-benefit approach to public project evaluation in the case of basic science research projects.

the conceptual contributions that the codified findings of a research program in one field may make to accelerate research progress in another field, there is the question of spillovers affecting scientific equipment and skills, which often remain in the region of tacit knowledge but are nonetheless transferrable. Productivity gains in the performance of specified experimental tasks are likely to emerge as a by-product of the conduct of research through the development of superior instrumentation techniques (see, e.g., Moulton et al., 1990) including the development of generic computer software for performing data processing, storage, retrieval, and network transmission. The training of specialized technical staff and post-doctoral researchers, whose skills eventually become available to other projects, represents another important source of spillovers.

7. Policy Challenges: Maintaining Science and Technology in Dynamic Balance

The thrust of the analysis in the two immediately preceding sections has been to show that there are numerous features of the reward system and characteristic institutional structures of open science in the modern West that give rise to resource misallocations and static inefficiencies in the conduct of basic and applied research. Correspondingly, there may be a wide field here for economists specializing in contract theory and institutional mechanism design to familiarize themselves sufficiently well with the detailed internal workings of Science, as they have lately begun to do with regard to the realities of public regulatory bodies and procurement agencies. The problems in these two areas of public economics are not the same, of course, and the sources of the allocative inefficiencies to which we have pointed lie so close to the core of the collegiate reputation-based reward system of the open science system that, in a sense, they may be said to be intrinsic to it. Nevertheless, it is premature to declare that it is beyond the wit of economists to devise modifications of existing institutional procedures that would ameliorate some of the problems identified here.[33]

33. See, for example, Laffont and Tirole (1993) on the distinction between regulation and procurement (briefly, that the latter is a principal-agent relationship in which the principal is also the buyer of the commodity supplied, whereas regulation refers to situations where a firm acts as an agent of the government in supplying commodities to third-party purchasers). In its attention to the realities of the institutional environment and the informational, contractual, and political and administrative procedural constraints upon the public regulator (the principal), the "new regulatory economics" exhibits many points of kinship with the spirit of the analysis explored here. The key additional features with which the new econom-

Science, however, is not a self-contained system—and indeed, could not survive as such. Rather than risk suggesting that the agenda of the new economics of science is concerned exclusively, or even primarily, with the more static resource allocation issues internal to publicly supported research activities, it is now time for us to recognize that many of the most important challenges facing science policy-makers concern the dynamics of science-technology interactions—the disposition of research resources and the flows of information between the open science and proprietary science communities, and the consequences these will have for the improvement of economic welfare.[34]

8. Conclusion

The broad message emerging from recent advances in the economic analysis of science that have been reviewed here can be expressed in the following four propositions.

(1) Although the institutions and social norms governing the conduct of open science cannot be expected to yield an optimal allocation of research efforts, they are functionally quite well suited to the goal of maximizing the long-run growth of the stock of scientific knowledge—subject to the constraints on the resources that society at large is prepared to make available for that purpose.

(2) Those same institutions and social norms, however, are most ill suited to securing a maximal flow of economic rents from the existing stock of scientific knowledge by commercially exploiting its potential for technological implementations. The distinctively different set of institutional arrangements, and different modes of conduct on the part of researchers, that accordingly have been contrived for the latter (technological) purposes unfortunately leave unsolved the problem of securing the right amount of resources for the conduct of *open* science. Here, adequate public patronage is critical and warranted.

(3) The organization of research under the distinct rules and reward systems governing university scientists, on the one hand, and industry scientists and engineers, on the other, historically has permitted the evo-

ics of science has to deal are that the agents in question (the researchers) are supplying information products rather than conventional tangible goods and services, and have been assigned collective responsibilities for regulating many aspects of their activities.

34. On science-technology interactions and interdependences, see, e.g., the empirical studies in Grupp (1992), the survey in Freeman (1992) of modern formal institutions supporting science-based innovative activity, and the treatment of technological change as a dynamic system involving feedbacks between basic and applied research activities given in David (1993b).

lution of symbiotic relationships between those engaged in advancing science and those engaged in advancing technology. In the modern era, each community of knowledge seekers, and society at large, has benefited enormously thereby.

(4) The institutional machinery which has been performing these vital functions for our society is intricate, jerry-built in some parts, and possibly more fragile and sensitive to reductions in the level of funding for open science than often may be supposed. For all their importance to the modern economy and polity, the social mechanisms that allocate resources within the Republic of Science are still too little understood and remain vulnerable to destabilizing and potentially damaging experiments undertaken too casually in the pursuit of faster national economic growth or greater military security.

The foregoing propositions provide basic tenets to guide discussions of concrete problems and proposals that fall within the purview of decision-takers responsible for science and technology policies. Obviously, they are too general to have positive prescriptive value and are meant to be largely cautionary. If they are found to have some utility, it will reside not in instructing us what to conclude about this or that policy question, but rather that the economics of science can help frame better science and technology policies only insofar as it comes to grips with the logic and the performance of the specialized institutional structures that organize the production and distribution of that very peculiar asset: scientifically reliable knowledge.

ACKNOWLEDGMENTS

The authors are grateful for comments and suggestions on earlier drafts received from Richard Nelson, Laurence Rosenberg, Peter Temin, and Harriet Zuckerman; from Ashish Arora, Ed Steinmueller, and other members of the (fall quarter 1990) Technology, Organization and Productivity Workshop at Stanford University; from Chris Freeman, Keith Pavitt, and other participants in the SPRU Seminar at the University of Sussex (spring 1991); and from Alfonso Gambardella and other members of the IEFFE Seminar at the University of Bocconi, Milan, in April 1992. Weston Headley and Phillip Lim provided able research assistance in the early phases of this project. The present version has benefitted from the comments of two anonymous referees. This also is an appropriate place to acknowledge the financial support provided for this and related research by the Mellon Foundation Program on "Science and

Society," and (for P.A.D.) from the American Academy of Arts and Sciences, and the Information and Organization Program of the National Science Foundation (Division of Information, Robotics and Intelligent Systems, Grant IRI-8814179-02). The Center for Economic Policy Research (CEPR) of Stanford University provided administrative and other support for the research funded by those grants.

REFERENCES

Adrian, R. (Lord). 1992. *Do We Need Science Policies?* Proceedings of the American Philosophical Society. 136 (4) 526–532.

Arora, A. 1991. *The Transfer of Technological Know-How to Developing Countries: Technology Licensing, Tacit Knowledge, and the Acquisition of Technological Capability.* Unpublished PhD dissertation, Stanford University.

Arrow, K. J. 1962. Economic Welfare and the Allocation of Resources for Inventions. In: R. R. Nelson (Editor), *The Rate and Direction of Inventive Activity: Economic and Social Factors.* Princeton University Press, Princeton, NJ.

———. 1971. Political and Economic Evaluation of Social Effects and Externalities. In: M. Intrigator (Editor), *Frontiers of Quantitative Economics.* Contributions to *Economic Analysis.* North-Holland, Amsterdam.

———. 1974. *The Limits of Organization.* Norton, New York.

———. 1984. *The Economics of Information.* Harvard University Press, Cambridge.

Arrow, K. J., and W. M. Capron. 1959. Dynamic Shortages and Price Rises: The Engineer-Scientist Case. *Quarterly Journal of Economics* 73: 292.

Axelrod, R. 1984. *The Evolution of Cooperation.* Basic Books, New York.

Battaglini, A. O., and F. R. Monaco (Editors). 1991. *The University within the Research System—An International Comparison Handbook of the Law of Science: Comparative Studies.* Nomos Verlagsgesellschaft, Baden-Baden.

Baumol, W. J., and W. E. Oates. 1975. *The Theory of Environmental Policy: Externalities, Public Outlays, and the Quality of Life.* Prentice-Hall, Englewood Cliffs, NJ.

Bernal, J. D. 1939. *The Social Function of Science.* Routledge & Kegan Paul, London.

Blank, D. M., and G. J. Stigler. 1957. *The Demand and Supply of Scientific Personnel.* National Bureau of Economic Research, New York.

Blume, S. 1974. *Toward a Political Sociology of Science.* Free Press, New York.

———. 1987. The Theoretical Significance of Co-operative Research. In: S. Blume et al. *The Social Direction of the Public Sciences, Sociology of the Sciences Yearbook.* JAI Press, Greenwich, CT.

Blumenthal, D., et al. 1986. University-Industry Research Relationship in Biotechnology: Implications for the University. *Science* 232.

Boorstin, D. J. 1984. *The Discoverers: A History of Man's Search to Know His World and Himself.* Random House, New York.

Cohen, W. M., and D. A. Levinthal. 1989. Innovation and Learning: The Two Faces of R&D. *Economic Journal* 99 (397): 569–596.

Cole, J. R., and S. Cole. 1973. *Social Stratification in Science.* University of Chicago Press, Chicago.

Cole, S. 1978. Scientific Reward Systems: A Comparative Analysis, Research in Sociology of Knowledge. *Sciences and Art* 1: 167–190.

Cozzens, S. E. 1990. Autonomy and Power in Science. In: S. E. Cozzens and T. Gieryn (Editors), *Theories of Science in Society.* Indiana University Press, Bloomington.

Dasgupta, P. 1986. The Theory of Technological Competition. In: J. E. Stiglitz and F. Mathewson (Editors), *New Developments in the Analysis of Market Structures.* MIT Press, Cambridge, MA.

———. 1988. Patents, Priority, and Imitation, or, The Economics of Races and Waiting Games. *Economic Journal* 98.

Dasgupta, P., and P. A. David. 1987. Information Disclosure and the Economics of Science and Technology. In: G. Feiwel (Editor), *Arrow and the Ascent of Modern Economic Theory.* New York University Press, New York.

———. 1988. Priority, Secrecy, Patents, and the Economic Organization of Science and Technology. Center for Economic Policy Research Publication No. 127, Stanford University, March.

Dasgupta, P., and G. Heal. 1979. *Economic Theory and Exhaustible Resources.* Cambridge University Press, Cambridge.

Dasgupta, P., S. Marglin, and A. Sen. 1972. *Guidelines for Project Evaluation.* United Nations, New York.

Dasgupta, P., and E. Maskin. 1987. The Simple Economics of Research Portfolios. *Economic Journal* 97.

Dasgupta, P., and J. E. Stiglitz. 1980a. Market Structure and the Nature of Innovative Activity. *Economic Journal* 90.

Dasgupta, P., and J. E. Stiglitz. 1980b. Uncertainty, Industrial Structure, and the Speed of R&D. *Bell Journal of Economics* 11.

David, P. A. 1984. *The Perilous Economics of Modern Science.* Technological Innovation Program (TIP) Working Paper, Center for Economic Policy Research. Stanford University.

———. 1991. *Reputation and Agency in the Historical Emergence of the Institutions of "Open Science."* Center for Economic Policy Research Publication No. 261. Stanford University.

———. 1993a. Intellectual Property Institutions and the Panda's Thumb: Patents, Copyrights, and Trade Secrets in Economic Theory and History. In: M. B. Wallerstein, M. E. Mogee, and R. A. Schoen (Editors), *Global Dimensions of Intellectual Property Rights in Science and Technology.* National Academy Press, Washington, DC.

———. 1993b. Knowledge, Property, and the System Dynamics of Technological Change. In: L. Summes and S. Shah (Editors), *Proceedings of the World Bank Annual Conf. on Development Economics: 1992.* World Bank Press, Washington, DC.

David, P. A., D. C. Mowery, and W. E. Steinmueller. 1992. Analyzing the Economic Payoffs from Basic Research. *Economics of Innovation and New Technology* 2 (4): 73–90.

David, P. A., and W. E. Steinmuller. 1993. *University Goals, Institutional Mechanisms, and the "Industrial Transferability" of Research.* Center for Economic Policy Research. Proposal for an AAAS Study Project Conference. October.

Eisenberg, R. S. 1987. Proprietary Rights and the Norms of Science in Biotechnology Research. *Yale Law Journal* 97 (2): 177–231.

Freeman, C. 1992. Formal Scientific and Technical Institutions in the National System of Innovation. In: B.-A. Lundvall (Editor), *National Systems of Innovation.* Pinter, London.

Fudenberg, D., and E. Maskin. 1984. The Folk Theorem in Repeated Games with Discounting and with Incomplete Information. *Econometrics* 54: 533–554.

Fusfeld, H. I., and C. S. Haklisch (Editors). 1984. *University Industry Research Interactions.* Pergamon Press, New York.

Gaston, J. 1970. The Reward System in British Science. *American Sociological Review* 35: 718–730.

———. 1971. Secretiveness and Competition for Priority of Discovery in Physics. *Minerva* 9 (1).

Greif, A. 1989. Reputational and Coalitions in Medieval Trades. *Journal of Economic History* 49 (3): 857–882.

———. 1992, *Cultural Beliefs and the Organization of Society: A Historical and Theoretical Reflection on Collectivist and Individualist Societies.* Working Paper, Department of Economics. Stanford University.

Griliches, Z. 1957. Hybrid Corn: An Exploration in the Economics of Technological Change. *Econometrica* 25.

———. 1960. Hybrid Corn and the Economics of Innovation. *Science* July 29.

Grupp, H. (Editor). 1992. *Dynamics of Science-Based Innovation.* Springer, Berlin.

Hagstrom, W. O. 1965. *The Scientific Community.* Basic Books, New York.

Hempel, C. G. 1966. *The Philosophy of Natural Science.* Prentice-Hall, Englewood Cliffs, NJ.

Hoke, F. 1993a. Technology Transfer Boom Offers Scientist Rewards—And Challenges. *The Scientist* 17.

———. 1993b. Universities Reassess Social Role as Tech Transfer Activity Mounts. *The Scientist* 18.

Kuhlmann, S. 1991. The Research-Industry and University Industry Interface in Europe. Federal Republic of Germany, Research Memorandum and Frauhofer-Institut für Systemtechnik und Innovationsforschung, Karlsruhe.

Kuhn, T. S. 1962. *The Structure of Scientific Revolutions.* University of Chicago Press, Chicago, IL.

Laffont, J.-L., and J. Tirole. 1993. A Theory of Incentives in Procurement and Regulation. MIT Press, Cambridge, MA.

Lamb, D., and S. M. Easton. 1985. *Multiple Discovery: The Pattern of Scientific Progress.* Averbury Publishing, Trowbridge.

Latour, B. 1987. *Science in Action: How to Follow Scientists and Engineers Through Society.* Harvard University Press, Cambridge, MA.
Latour, B., and S. Woolgar. 1979. *Laboratory Life.* Sage, Beverly Hills, CA.
Lind, R. (Editor). 1982. *Discounting for Time and Risk in Energy Policy.* Johns Hopkins University Press, Baltimore MD.
Little, I. M. D., and J. A. Mirrlees. 1974. *Project Appraisal and Planning for Developing Countries.* Heinemann, London.
Luhmann, N. 1979. *Trust and Power.* Wiley, New York.
Mansfield, E. 1991. *Social Rate of Return from Academic Research.* University of Pennsylvania, Department of Economics Working Paper.
Marschak, J. 1971. The Economics of Information. In: M. Intilligator (Editor), *Frontiers of Quantitative Economics.* North-Holland, Amsterdam.
Meckling, W. H. 1962. Predictability of the Costs, Time, and Success of Development. In: R. Nelson (Editor), *The Rate and Direction of Incentive Activity.* Princeton University Press, Princeton, NJ.
Merton, R. K. 1957. Continuities in the Theory of Social Structure and Anomie. In: *Social Theory and Social Structure.* Free Press, New York.
———. 1973. Social Theory and Social Structure. In: N. W. Storer (Editor), *The Sociology of Science: Theoretical and Empirical Investigations.* University of Chicago Press, Chicago, IL.
Milgrom, P. R., C. North, and S. R. Weingast. 1990. The Role of Institutions in the Revival of Trade: The Law Merchant, Private Judges, and the Champagne Fairs. *Economics and Politics* 2 (1): 1–23.
Moulton, P. J. Young, and J. J. Eberhardt. 1990. *Technological Advances in Instrumentation: Impacts on R&D Productivity.* Battelle Institute Working Paper, Seattle, WA.
Mowery, D. 1983. Economic Theory and Government Technology Policy. *Policy Sciences* 16.
Mulkay, M. J. 1972. *The Social Process of Innovation: A Study in the Sociology of Science.* Macmillan, London.
———. 1977. Sociology of the Scientific Research Community. In: I. Spiegel-Rosing and D. de S. Price (Editors), *Science, Technology, and Society.* Sage, London.
Musgrave, R. A. 1968. *The Theory of Public Finance,* 2d ed. McGraw-Hill, New York.
Musgrave, R. A., and A. T. Peacock (Editors). 1958. *Classics in the Theory of Public Finance.* Macmillan, London and New York.
Nagel, E. 1961. *The Structure of Science.* Columbia University Press, New York.
National Academy of Sciences. 1989. *Information Technology and the Conduct of Research—The User's View.* Report of the Panel on Information Technology and the Conduct of Research. National Academy Press, Washington, DC.
National Science Board. 1987. *Science and Engineering Indicators—1987.* U.S.-G.P.O. (NSB87-1), Washington, DC.
———. 1989. *Science and Engineering Indicators—1989.* U.S.G.P.O. (NSB89-1), Washington, DC.

———. 1991. *Science and Engineering Indicators—1991.* U.S.G.P.O. (NSB91-1), Washington, DC.

Nelkin, D. 1987. *Selling Science: How the Press Covers Science and Technology.* W. H. Freeman, New York.

Nelson, R. R. 1959. The Simple Economics of Basic Scientific Research. *Journal of Political Economy* 67.

Nelson, R. 1990. *What is Public and What is Private About Technology?* University of California Center for Research in Management, CCC Working Paper No. 90-9. Berkeley, CA, September.

Norris, K. P. 1971. The Accuracy of Project Cost and Duration Estimates in Industrial R&D. *R&D Management* 2 (1): 25–36.

Pavitt, K. 1987. The Objectives of Technology Policy. *Science and Public Policy* 14: 182–188.

Peters, L. S. and H. I. Fusfeld. 1983. Current U.S. University/Industry Research Connections. In: *National Science Board, University-Industry Research Relationships, Selected Studies.* U.S.G.P.O., Washington, DC.

Pigou, A. C. 1932. *The Economics of Welfare.* MacMillan, London.

Polanyi, M. 1951. *The Logic of Liberty.* Routledge, London.

———. 1962. The Republic of Science: Its Political and Economic Theory. *Minerva* 1 (1): 54–73.

———.1966. *The Tacit Dimension.* Routledge & Kegan Paul, London.

Popper, K. R. 1968. *The Logic of Scientific Discovery,* 2d ed. Columbia University Press, New York.

Ravetz, J. R. 1971. *Scientific Knowledge and Its Social Problems.* Clarendon Press, Oxford.

Romer, P. M. 1990. Endogenous Technological Change. *Journal of Political Economy* 97: 71–102.

Romer, P. M. 1993. Two Strategies for Economic Development: Using Ideas vs. Producing Ideas. In: L. Summers and S. Shah (Editors), *Proceedings of the World Bank Annual Conference on Development Economics: 1992.* World Bank Press, Washington, DC.

Rosenberg, N. 1990. Why Do Companies Do Basic Research with Their Own Money? *Research Policy* 19: 165–174.

Rubinstein, A. 1979 Equilibrium in Supergames with Overtaking Criterion. *Journal of Economic Theory* 21: 1–9.

———. 1980. Strong Perfect Equilibria in Supergames. *International Journal of Game Theory* 9.

Salomon, J.-J. 1973. *Science and Politics.* Trans. Noel Lindsay. Macmillan, London.

Samuelson, P. 1954. The Pure Theory of Public Expenditure. *Review of Economics and Statistics* 36 (November).

Science. 1990. Data Sharing: A Declining Ethic? (Editorial). 248, 25 May 952–953.

Shannon, C. E., and W. Weaver. 1949: *The Mechanical Theory of Communication.* Free Press, Urbana, IL.

Squire, L., and H. Van der Taak. 1975. *Economic Analysis of Projects.* Johns Hopkins University Press, Baltimore, MD.

Stankiewics, R. 1986. *Academics and Entrepreneurs: Developing University-Industry Relations.* Frances Pinter, London.
Stiglitz, J. E. 1988. *The Economics of the Public Sector.* Norton, New York.
Storer, W. 1966. *The Social System of Science.* Holt, Rinehart & Winston, New York.
Thompson, D. 1992. Science's Big Shift. *Time Magazine* 23 November 34 ff.
Turner, S. P. 1990. Forms of Patronage. In: S. E. Cozzens and T. Gieryn (Editors), *Theories of Science in Society.* Indiana University Press, Bloomington, IN.
Vincenti, W. 1990a. *Engineering Knowledge, Type of Design, and Level of Hierarchy: Further Thoughts about What Engineers Know. . . .* Paper for the Eindhoven Conference on the History of Technology. Eindhoven, The Netherlands.
———. 1990b. *What Engineers Know and How they Know It: Analytical Studies from Aeronautical History.* John Hopkins University Press, Baltimore, MD.
Von Hippel, E. 1990. *The Sources of Innovation.* Oxford University Press, New York.
Wade, N. 1978. Guilleman and Schally. *Science* 200: 279–282, 411–415, 510–513.
Whitley, R. 1984. *The Intellectual and Social Organization of the Sciences.* Clarendon Press, Oxford.
Ziman, J. M. 1968. *Public Knowledge.* Cambridge University Press, London.
Zuckerman, H. A., and R. K. Merton. 1971. Patterns of Evaluation in Science: Institutionalization, Structure, and Functions of the Referee System. *Minerva* 9: 66–100.
———. 1972. Age, Ageing, and Age Structure in Science. In: M. W. Riley, M. Johnson, and A. Foner (Editors), *Ageing and Society.* Russell Sage, New York.

8

The Organization of Cognitive Labor

Philip Kitcher

1. Introduction

The general problem of social epistemology, as I conceive it, is to identify the properties of epistemically well-designed social systems, that is, to specify the conditions under which a group of individuals, operating according to various rules for modifying their individual practices, succeed, through their interactions, in generating a progressive sequence of consensus practices. According to this conception, social structures are viewed as relations among individuals: thus my departure from the tradition of epistemological theorizing remains relatively conservative.[1] . . . The general problem can be resolved into more specific instances. If people join the scientific community through the socialization process . . . ; if they form their divergent individual practices partly through interacting with nature and thinking by themselves, partly by borrowing from and lending to others; if what is transmitted to the next generation is shaped by their individual decisions and by their interactions, how will the whole system best work to promote a progressive sequence of consensus practices? Is it possible that the individual ways

 From *The Advancement of Science* (New York: Oxford University Press, 1993), parts of chapter 8.

1. More radical suggestions for socializing epistemology can be found in Fuller 1988, Rouse 1987, Longino 1990, and the writings of Bruno Latour. . . . A . . . line of objection, posed forcefully to me by Steven Shapin, is that the species of methodological individualism that I deploy in articulating my version of social epistemology cannot be sustained. For the moment, I shall rest content with challenging those who believe that my individualistic framework is too narrow to offer examples of social aspects of knowledge that cannot be accommodated within it.

of responding to nature matter far less than the coordination, cooperation, and competition that go on among the individuals?

In what follows, I shall focus on two clusters of problems. The first set is concerned with scientists' responses to others. The second considers the effects of individual efforts on communitywide belief. So, in the earlier sections of this chapter, I shall look at action *on* the individual. Later sections will study the effects that stem *from* the doings of the individuals who make up a scientific community.

In considering the first group of problems I conceive of trust in others as essential to scientific activity.[2] For an active researcher within a scientific community, there are important questions about the assignment of trust: Whom should one trust? When should one trust others more than oneself? When is it worth risking the errors that others might make? I shall explore some ways in which researchers might address these questions and consider the impact of their decisions on communitywide research.

The primary topic of the second group of questions is the distribution of effort within scientific communities. The framework of Chapter 3 depicts individuals with different individual practices pursuing diverse projects. To inquire about the structure of a well-ordered scientific community is to ask how these projects should be pursued so as best to promote the community project. How much division of effort is desirable? How can diversity be maintained in a scientific community? How should consensus be formed?

My general approach to both sets of questions is to draw on what we believe about the available possibilities for individual reasoning and the coordination of individual effort. Reflecting on the history of science and on current science, we can seek to identify recurrent problem situations that scientists face. At this point, there are two types of inquiry that are worth pursuing: first, we want to know what, given the range of possibilities, is the best approach to the problem situation in which we are interested; second, we should scrutinize which of the available combinations of individual decision procedures and sets of social relations would move

2. This point is elaborated extensively by Shapin in *A Social History of Truth* (1994). He is largely concerned with broader issues of trust than those that occupy me here, particularly the honesty of potential informants. As will become apparent, I shall be concerned more with assessing competence than evaluating honesty, but this should not be taken to imply that issues about sincerity are unimportant, either in Shapin's prime historical context (seventeenth-century Europe) or in the context of contemporary science. Ironically, many of the recent instances of scientific fraud reveal the extent to which scientists operate on trust.

the community closer to or further away from the optimal approach. More pedantically, given a problem situation, we seek

1. The optimum community response, A_{opt}, conceived as a distribution of the efforts of the individual members.

2. For each combination of individual decision rules and social relations, $\{D_i\}$, $\{R_j^s\}$, a representation of the success of A_{opt}.

Through the examination of a range of instances of recurrent problem situations, we can hope to understand the impacts that various systems of social and individual decisions have on the growth of science.

The inquiries just envisaged are often too ambitious. Optimal solutions to decision problems are usually only available in highly simplified situations. Much of this chapter is devoted to examining the kinds of equilibrium distributions of epistemic effort that would emerge under various types of social conditions and to exploring whether these distributions enable us to avoid certain types of cognitive disasters. Instead of thinking about how best to achieve a cognitive goal, we can consider whether one type of social arrangement avoids a pit into which another falls.

Although my analyses are very abstract, they ultimately connect with concrete issues about science policy. I do not find it troubling that questions of social epistemology might have practical consequences—enabling us, for example, to consider the merits of rival systems for awarding grants or refereeing scientific contributions. Philosophy of science should earn its way by trying to draw specific morals for the organization of scientific research. But I would caution against overinterpreting my results: although my analyses reveal neglected epistemic possibilities, they are far too idealized to enable us to be confident in reaching conclusions about practical strategies for (say) funding research. *Perhaps* that can come later.

In pursuing these problems, I shall employ an analytic idiom inspired by Bayesian decision theory, microeconomics, and population biology. The advantage of this idiom is that it enables me to formulate my problems with some precision, and that precision is important for both identifying consequences and disclosing previously hidden assumptions. Precision is bought at the cost of realism. My toy scientists do not behave like real scientists, and my toy communities are not real communities.

We can think of the problems that concern me as including those that would face a philosopher-monarch, interested in organizing the scientific work force so as to promote the collective achievement of significant truth. Science, of course, has no such benevolent dictator. In consequence, individual scientists face coordination problems. If we suppose that they internalize the (fictitious) monarch's values, how will they fare? If we assume instead that they are motivated in baser ways or that they are locked into systems of authority and deference, will they necessarily do worse than a society of unrelated individuals, each of whom is pure of heart?

The principal moral of my discussions is a cautionary reminder: do not think that you can identify very general features of scientific life—reliance on authority, competition, desire for credit—as epistemically good or bad. Much thinking about the growth of science is permeated by the thought that once scientists are shown to be motivated by various types of social concerns, something epistemically dreadful has been established. On the contrary, as I shall repeatedly emphasize, particular kinds of social arrangements make good epistemic use of the grubbiest motives.

Beyond this, I want to note that the details often matter. Although we shall see that competition is frequently helpful in enabling a community to achieve valuable cognitive diversity, there are also cases in which competition is impotent or in which it generates unwelcome homogeneity. Only close attention to the pressures actually present in scientific situations will teach us whether the social systems we employ do a good job of coordinating the efforts of individuals.

My investigations should be viewed as mapping a space within which identifications of the epistemically important characteristics of scientific communities can be made. They suggest obvious empirical questions: Can real scientists be construed as hybrid agents, torn between their epistemic and nonepistemic goals, and, if so, how much weight do they give to defending truth as opposed to receiving credit? How do scientists actually evaluate the chances that a proposal for modifying practice will succeed? I do not suppose that answering these types of questions will be easy. However, if my approach is right, then they are the kinds of questions that should be asked in developing a more substantive account of the growth of scientific knowledge in communities of scientists.

2. Authority

Reliance on authority affects all our cognitive lives, and, for present purposes, we can distinguish three ways in which it permeates the cognitive lives of scientists. First, there is the general epistemic dependence on the past that figures in everyone's early intellectual ontogeny. We absorb the lore of our predecessors through the teaching of parents and other authorities. Second, at the time of entry into the scientific community, novices endorse a communitywide conception of legitimate epistemic authority. Certain people are to be trusted to decide on certain issues, and the novice must accept whatever agreements they reach on those issues. Third, during the course of individual research, scientists interact with one another, adopting the claims made by *some* of their colleagues, investigating the proposals of others, ignoring the suggestions of yet others, when the claims, proposals, and suggestions in question go beyond what is agreed upon by the pertinent community.

. . . The overarching kinds of authority that fall under the first two types do not trap us in inevitable error. . . . We work our way free of the mistakes of earlier generations through further encounters with nature. But if this optimistic picture is to be sustained, it must be the case that the *third* type of attribution of authority does not interfere with the process of self-correction, but works constructively to further the community project. I shall be concerned with this type of attribution of authority, the differential assessment of peers.

To bring the problem into focus it is worth recalling some episodes from recent science. In the spring of 1989, two electrochemists, Stanley Pons and Martin Fleischmann, held a celebrated press conference, at which they announced the possibility of obtaining cold fusion on a tabletop. Prior to the announcement that claim was incredible: according to the consensus practice of nuclear physics, fusion cannot be achieved at room temperatures in the fashion that Pons and Fleischmann described. However, the payoff if Pons and Fleischmann were right would be enormous, and immediately after the press release laboratory telephones began to ring. Electrochemists knew and respected Pons and Fleischmann and, accordingly, took their apparently incredible claim seriously. Outsiders from physics had typically never heard of either Pons or Fleischmann. Many of the telephone calls they placed posed the same questions: Who are Pons and Fleischmann? Can they be trusted? When told of the high standing that Pons and Fleischmann had within the electrochemical community, the interested physicists began to consider the possibility that the outlandish finding might be right. So a significant expenditure of scientific effort was begun, as numerous physicists and chemists tried to replicate the Pons-Fleischmann experiment.

Contrast this situation with another. During the past twenty years, self-styled creation scientists have made periodic announcements about the co-presence of dinosaur and human tracks in the same strata and about the existence of human artifacts which, when dated by standard techniques, yield ages comparable to those of supposedly ancient rocks. Their pronouncements challenge paleontology just as Pons and Fleischmann questioned our understanding of nuclear fusion. But Duane Gish, Henry Morris, and their colleagues at the Institute for Creation Research do not inspire the same dedicated investigations. They catch the ear of the scientific community only when they are able to threaten, only when they have demonstrated an ability to influence legislators and publishers, making it necessary for scientists to divert time from profitable research to the enterprise of rebutting creationism.

The differences between the two cases are readily traced to differences in authority. Pons and Fleischmann had considerable authority among electrochemists, and they obtained authority among physicists because they had authority for some people who had authority for some physicists. Gish and Morris have no authority among paleontologists, nor do they have authority with anyone who has authority for paleontologists. However, though this contrast may throw into relief some basic features of the attribution and withholding of authority in science, more mundane illustrations are useful for bringing other points to our attention.

Consider anyscientist working in anylab. During the course of a day's work there will be numerous opportunities for relying on others. Some will be taken; others will not. Particular parts of the day's project will be assigned to technicians, graduate students, support staff. The scientist will typically perform other tasks herself. In addition, there may well be special opportunities to redesign some aspect of the course of research. Perhaps a journal arrives with an article which, if sound, would enable a time-consuming procedure to be abbreviated. Or a grant proposal, sent for review, may suggest an alternative sequence of experiments. Or a new catalogue may offer a novel version of a relevant piece of apparatus. All these opportunities, both those pursued or dismissed in the quotidian assignment of jobs and the more special chances, call for a decision by any scientist: Should I do this myself or rely on someone else? Can I trust *X* to do *A*? Is this procedure/instrument/technique reliable?

Understanding the role of authority in science requires us to probe these decisions. How should they be made? What kinds of individual decision rules and social relations facilitate the making of such decisions in ways that will profit the community to which anyscientist belongs? I shall start at the most general level, investigating the conditions under which cooperation among scientists, the relying on others or the borrowing from others, is worthwhile.

3. Cooperation

The most obvious advantages of deference to authority are that it enables individual scientists to pursue their epistemic projects more rapidly and makes feasible investigations that would be impossible for a single individual. Suppose that a scientist is dedicated to a particular inquiry: the scientist's overriding concern is to bring this inquiry to a conclusion by discovering the true answer to a particular question. In terms that I shall employ more systematically in later sections, this imagined scientist is a *pure epistemic agent,* one for whom the primary goal is to reach an epistemically valuable state. I further suppose that the scientist has total resources (time, energy, money) E and needs k items of information. Let the cost of acquiring each directly be C, the cost of acquiring each from an authority be c. The scientist's project is individually impossible but cooperatively feasible just in case:

(8.1) $kc < E < kC.$

We can assume that the investigation consists of a period in which the needed information is acquired followed by a period in which the scientist attempts to put the information to use. The chances of success in the project can be written as the product of two probabilities: the probability that the information acquired is correct and the probability of deployment of correct information at the second stage. The latter can be written as $F(E -$ acquisition costs), where F is nondecreasing and $F(x) = 0$ if $x < 0$.[3]

Consider the simplest case in which $k = 1$. You have two choices: you can do the work yourself or you can rely on authority. Which decision is preferable? It depends on two error rates, yours and that of your potential authority. Assume that borrowing is cheaper than doing the work yourself, $c < C$. Let your error rate be p, the potential authority's error rate q.[4] As noted, I suppose that your decision is motivated by the desire

3. Of course, the function F cannot be entirely arbitrary, but must meet the conditions on probability (thus, for example, taking only values in [0, 1]).

4. A cautionary note on notation. In the sections that follow, numerous probabilities and other parameters will be introduced. Given a limited lexicon, the same symbols have sometimes been assigned different interpretations in different sections. Although I have tried to maintain uniformity in notation for related problems, the reader should not assume

to bring this project to a correct conclusion. Then you should rely on authority if

(8.2) $(1 - p) \cdot F(E - C) < (1 - q) \cdot F(E - c)$.

This inequality is automatically satisfied if the potential authority is more expert at the relevant task than you are—$p > q$—for recall that $c < C$ and F is nondecreasing. But even if you are more reliable than the potential authority it may still be worth your while to take the risk of borrowing.

To see this, imagine that $F(x)$ is (a) 0 when $x < 0$, (b) mx when $0 < mx < 1$, (c) 1 when $mx > 1$ (where m is a constant representing the rate at which resources increase the probability of finding a solution). Assume also that $E - C > 0$ (it would be possible for you to do the work yourself and still have a chance of succeeding in the project), and that $m(E - c) < 1$ (you cannot have a probability > 1 of success, even if you borrow). Borrowing is epistemically preferable if

(8.3) $q < [(C - c) + p(E - c)]/(E - c)$.

Even when you are perfect ($p = 0$), if the costs of borrowing are negligible ($c = 0$), you can tolerate a maximum error rate of C/E in your potential authority, and C/E may be sizable if you would have to expend a lot of your resources on acquiring the information directly.[5]

The case in which the probability function F is linear is intermediate between two others. If a new piece of information would yield rapid initial returns, then, while it is still sometimes worth borrowing from others, the maximum tolerable error rate is decreased. By contrast, if you would have to expend considerable efforts, once the information is acquired, in learning how to use it most effectively in your project, then it will sometimes be worth borrowing even from extremely unreliable sources.

---

that symbols employed in sections dealing with very different topics have the same meanings.

5. As noted in the text, the discussion proceeds on the assumption that the project is relatively complex. If we relax this assumption, there are two subcases:

(a) The project is so straightforward that you can attain probability 1 of getting a solution whether you borrow or not; $m(E - C) > 1$ (a fortiori, $m(E - c) > 1$); unsurprisingly, the maximum tolerable error rate is now p; i.e., it is only worth borrowing from those who are more reliable than you.

(b) You can achieve probability 1 of succeeding if you borrow, but not if you do not; $m(E - c) > 1$, $m(E - C) < 1$; now the maximum tolerable error rate is $(1 - m(E - C)) + m(E - C)p$, it is easy to see that this value is greater than p, so that, here, it is worth borrowing from people who are less reliable than you are.

Up to this point, I have been assuming that the scientist's decision is dominated by a particular type of epistemic intention: the goal is that the individual scientist will solve the problem at hand. It is possible to imagine an even more epistemically devoted scientist, one who cares only that the problem be solved and who is prepared to pool resources with others in a richer cooperative effort. A scientist of this altruistic bent would engage in a different type of decision making, surveying the scientific community to ensure that her own efforts worked with those of others in advancing the community's understanding. When I consider cognitive diversity in later sections, I shall, in effect, be exploring the kinds of ideal distributions of effort that altruistic epistemically pure agents would aim to achieve. For the moment, however, I am interested in looking at the cooperative inclinations of a different type of agent, an *epistemically sullied* agent, one who is driven not only by a desire to solve the problem, but also by the quest for priority which Merton (1973) emphasizes.

Consider the predicament of an epistemically sullied scientist, X, engaged in a research project, when some peer, Y, announces a result that would, if correct, provide a way to simplify the investigation. X reasons as follows: "Suppose I make use of Y's result; there are two possibilities, Y is wrong or Y is right; if Y is wrong, then there is no chance of my solving the problem; if Y is right then I will have a probability $F(E - c)$ of arriving at a solution; my rivals, whose resources are equal to mine, who also borrow from Y will have an equal chance of solving the problem; if n of us borrow from Y and $N - n$ do not, then the expected number of problem solvers is $nF(E - c) + (N - n)(1 - p)F(E - C)$, so that, given that I produce a solution, my chances of being the first are $1/[nF(E - c) + (N - n)(1 - p)F(E - C)]$. Conversely, suppose that I do not borrow from Y; again, there are two possibilities, Y is wrong or Y is right; either way, I have a chance of getting the correct result if I set out to get the needed piece of information myself, and, given that I get the right result, then there's a chance of $F(E - C)$ that I will produce a correct solution to the whole problem; if Y is right, and n members of the community borrow from Y then the expected number of those who arrive at a correct solution is $nF(E - c) + (N - n)(1 - p)F(E - C)$ as before; if Y is wrong, then the expected number of competitors is $(N - n)(1 - p)F(E - C)$; either way, my chance of being the first solver varies inversely as the expected number of solvers."

This, I suggest, is eminently sensible reasoning for someone whose goal is to be the first solver of a scientific problem, a *scientific entrepreneur.* It involves a number of important assumptions. First, my imagined X makes no distinction of talent: all the N scientists who are engaged in trying to solve the problem are envisaged as equally likely to succeed. *One* way of relaxing this assumption would be to suppose that X reasons not about the actual community, but about a "virtual community" in which talented scientists are allowed to count more than once. (Imagine that a community of five sound but undistinguished scientists and one superstar is treated as of size ten, with the superstar being regarded as equivalent to five "regular" scientists.) Second, X takes the probability that he, *or any of the others,* will achieve the correct version of Y's result, if they do not borrow, to be the same value $1 - p$. In fact, X need merely assume that his own error rate is the average of those whom he takes to be in competition with him. Finally, X does not allow for any kind of partial credit: there is no payoff for being second or for showing that Y's result is indeed right.

At this stage I want to point to a moral that will become familiar later. From the community perspective, it is likely that sullied scientists will do better than the epistemically pure. This is because a pure community heads toward cognitive uniformity: either all the members find it worth borrowing from Y or they do not. By contrast in the sullied community, there are ample opportunities for division of cognitive labor. Some follow the strategy of aiming for a quick victory by borrowing from Y; others work independently. In this way, the sullied community hedges its bets.[6] That is, intuitively, a good thing. Later, I shall try to supply some arguments that would underwrite intuitions of this general type.

So far, I have assumed an unrealistic condition of symmetry among my imaginary scientists. All are supposed to have the same resources at their disposal, to assess the error rates of themselves and others in the same ways. By introducing asymmetries into the treatment of the entrepreneur's predicament, we make it easier for different members of the community to pursue different strategies. This is readily evident in instances in which, for one member of the community p is zero, while, for another, $p > q$. (Intuitively, one is well qualified to pursue the type of work that would lead to acquisition of the information Y promises, while

6. So presumably would a community of epistemic altruists, for they would adjust their behavior to what others were doing and attempt to maximize the community's epistemic utility.

the other is not expert in such matters.) The latter will find it profitable to borrow from Y, while the former will only borrow if the problem is sufficiently complex ($F(E - C)$ is sufficiently low).

There is a different type of asymmetry, one stemming from differences in resources. . . . We can expect those who have fewer resources sometimes to take far greater risks in borrowing information than would be justified by a sober epistemic assessment. If the scientific community is divided into those who have relatively large amounts of resources and those who have less, then we may expect that the latter group will include some members who pursue what seem to be rather unpromising ways of solving problems, while the resource-rich adopt more conservative strategies. As I have already intimated, it is likely that this type of cognitive diversity is no bad thing from the community perspective.

. . .

4. Attributing Authority

The discussion of the last section rested on the idea that scientists are able to assess the error rates of others, or, to put it more positively, that they are able to judge how reliable other members of the community are. How are such estimates made? How are authorities evaluated?

Sometimes, like instruments, potential authorities can be calibrated directly. We compare the output of an authority with our own opinions on topics where there is overlap. This is a temptingly simple suggestion, but I propose that, from the beginning, we think about authority in a broader way.

We are interested in a function $a(X, Y)$ that measures Y's authority for X, or, more exactly, X's assessment of the probability that what Y says will be true.[7] Here X is an individual scientist and Y may be another scientist, a research team, a journal, a series of scientific monographs, or some other composite entity on which scientists may potentially depend. (There are often serious questions about the entity whose authority is to be assessed: when an unreliable scientist publishes in a highly prestigious journal, known for its strict refereeing, the authority attributed to the article may be an amalgam of the authority of scientist and journal.) Attributions of authority are rarely uniform across topic. I shall assume,

7. More exactly, we are concerned with X's assessments of the probability that an arbitrary statement (belonging to a particular class—e.g., reports about a particular kind of topic) will be true, *given simply that Y produced or endorsed it*. . . .

in what follows, that the assignments of authority are relativized to a range of issues with respect to which the potential authority's deliverances can be assigned the same reliability.

An idealized treatment of authority that includes obvious social factors can proceed by breaking a scientist's credibility into two parts: there is *unearned* authority that stems from the scientist's social position (either within the community of scientists or in the wider society), the type of authority that arises from being associated with a major institution or from having been trained by a prominent figure; this contrasts with *earned* authority, that credibility assigned by reflection on the scientist's performances or through consideration of others' opinions of those performances. So I propose that we consider communities of scientists in which individuals evaluate one another in accordance with the equation

$$(8.4) \qquad a(X, Y) = w(X, Y)a_u(X, Y) + (1 - w(X, Y))a_e(X, Y)$$

where $w(X, Y)$ is a weight function, whose value indicates the relative importance X takes unearned authority to have in evaluating Y. I shall consider different possibilities for weighing unearned authority (different values of w) and different methods of computing earned authority (different measures of a_e).

Is this approach to authority complete? Perhaps the assessment of the reliability of others depends not just on such large social factors as prestige within the community but on the personal relations between evaluator and potential authority as well.[8] This could easily be incorporated by amending the basic equation 8.4 to

$$(8.5) \qquad a(X, Y) = w_1(X, Y)a_u(X, Y) + w_2(X, Y)a_p(X, Y) + w_3(X, Y)a_e(X, Y)$$

where $w_1 + w_2 + w_3 = 1$. This amendment would be useful in studying the ways in which differences in personal connections might be reflected in the pursuit of different lines of research in different subgroups. Initially, however, my principal concern will be to examine the balancing of considerations of track record (earned authority) against other forms of attribution, and, for these purposes, there is no loss in generality in

8. This point is clear in Shapin's studies of the importance of personal testimony about the probity of informants and would surely have to be treated if we were to respond to concerns about the honesty of others. Annedore Schulze has also emphasized to me the role of personal assessments of the character of others in the decision making of scientists working in small groups. I am indebted to both Shapin and Schulze for helpful discussion of these matters.

collapsing the a_u and a_p components. So I shall use the simpler version throughout the following sections. . . .

Before leaving the topic of authority, it is important to distinguish between authority and credit. A number of writers have emphasized the role that the search for credit plays in the conduct of scientific research (Latour and Woolgar 1979, chapter 5; Hull 1988). As I have already emphasized, authority is topic-relative. A scientist's credit, on the other hand, seems to be based on the *overall* assessment of that scientist's contributions. Thus it is very easy for a scientist to have high authority within a particular area, his area of expertise, but to have little credit. Scientific communities are full of respectable members whose deliverances about their assigned projects are dependable but who are viewed by their colleagues as pedestrian. In many instances, the authority structure of a community is that of an inverted pyramid: almost all those who have been trained, and who have survived their novitiate, have fairly high authority (with respect to the topics about which they make pronouncements). Nevertheless, the same communities can be sharply pyramidal in terms of credit, with a tiny fraction of the members aspiring to the highest levels of reputation (and concomitant resources). As I shall suggest in the next section, there is an intimate relation between credit and a particular type of authority, but the notions are not generally interchangeable.

. . .

9. The Entrepreneurial Predicament Revisited

After looking at some of the ways in which scientists might attribute authority to others, I now want to return to the approach to cooperation and the borrowing of others' ideas begun in Section 3. My first goal will be to understand how scientists would be expected to behave in circumstances in which the *perceptions* of others were given overriding importance. In Section 3, the focus was initially on epistemic values and pure agents who honored these values: scientists were supposed to plan their courses of action so as to maximize their chances of being right. Later, in setting up the entrepreneur's predicament, I supposed that the utilities were fixed by whether one is the *first* to be right. In integrating the behavior of scientists in cooperative competitive situations with the treatment of authority and credit, the first step will be to move further away from the purely epistemic. Let us now suppose that the utility of a course of action is determined by the *credit* it will bring, where credit is a function of the perceptions of others. What matters, then, is being *perceived* as the first to be right. The twin dangers are no longer being wrong and being pipped to the post, but being perceived to be wrong and being perceived to have been beaten.

It will be simplest to begin with the situation in which a scientist is working on a project without competition from others and to reconstruct the entrepreneur's predicament. The type of situation I shall consider is one which David Hull regards as central to scientific activity (Hull 1988). As he reminds us, there is too much published literature for scientists to engage in thorough search, and too little time for checking if scientists hope to make contributions of their own: "Scientists incorporate into their own work those findings that support it, usually without checking" (348). To obtain the credit that Hull (like other authors as diverse as Hagstrom and Latour) sees as one of the main motivations for devoting a large proportion of a human life to difficult, technical problems, scientists have to publish and market their ideas: "With the possibility of credit comes the possibility of blame. Scientists cannot spend very much time checking the work of others if they are to make contributions. They reserve checking for those findings that bear most closely on their own research, chiefly those that threaten it" (394). Hull thus offers us the picture of scientific entrepreneurs, operating within a credit economy.

As in Section 3, we suppose that *X* is engaged in some research project and that *Y* introduces something that could be used to reduce *X*'s question to some simpler issue, if it were accurate. Should *X* adopt *Y*'s innovation and pursue the apparently simpler project? Assuming that *X* has no competition with respect to this project and that the task is simply to receive credit by discovering a solution that is accepted by others, we can represent *X*'s situation as in figure 8.1.

This tree represents a simplified conception of the possible futures for a scientific entrepreneur whose possible strategies are to take over *Y*'s potentially helpful result or to ignore it and continue with the present course of research. The new version of the predicament differs from that of Section 3 in two respects: I have left out of consideration the pressures that may be set up by the presence of competitors who are working on the same project, and I have explicitly couched the outcomes in terms of *others' reactions* to what *X* does rather than in terms of *X*'s attaining a correct solution.

Writing the utility of O_i as U_i, I shall make the following assumptions:

(a) $u_1 = 0$ (Failure in the project has utility 0.)

(b) $u_{21} > 0$ (Unchallenged apparent solutions have positive utility.)

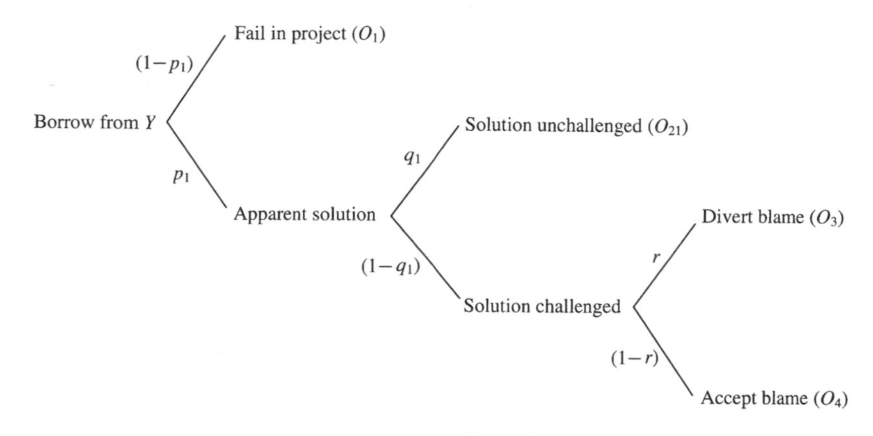

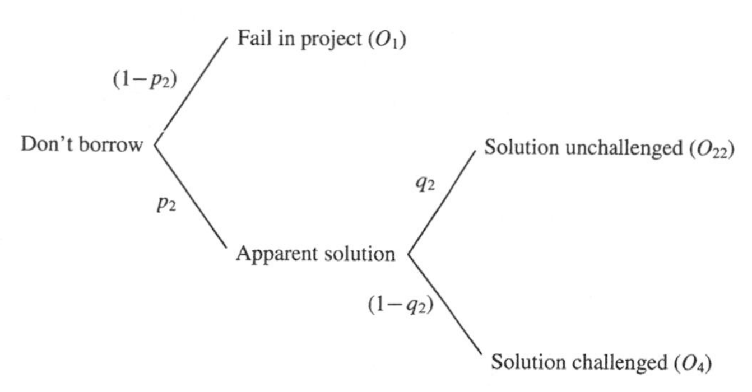

FIGURE 8.1. Decision Situation for Scientific Entrepreneurs

(c) $u_{22} > u_{21}$ (There is greater utility in achieving an apparent solution without Y's help than with Y's help.)
(d) $u_3 < u_{21}$ (The utility of achieving an unchallenged solution is greater than that of being challenged and successfully diverting blame.)
(e) $u_{32} > u_4$ (The utility of diverting blame is greater than that of receiving blame.)
(f) $u_4 < 0$ (Being blamed has negative utility.)
(g) $p_1 > p_2$ (Use of Y's idea would increase the chance of achieving an apparent solution).

. . . Many interesting scenarios satisfy these plausible conditions. Sometimes there is just the kind of pressure on scientists to borrow that Hull emphasizes. On other occasions, scientists engage in the type of careful scrutiny that Legend hypothesized. As we shall discover later (Section 12), if competition from rivals is not too severe, if Y's innovation promises increased probability of success, and if there is no benefit in building on a flawed result, the strategies of borrowing from *Y* and ignoring *Y* will be inferior to the strategy of checking *Y*'s proposal. Yet even when Legend delivers the right conclusions, they do not emerge for the traditionally accepted reasons.

. . .

10. Entrepreneurs, Authority, and Credit

I now want to consider the ways in which differences in authority affect the decision making of scientific entrepreneurs. . . . Ease of borrowing can be profoundly affected by considerations of relative rank. Both the traditional conception (according to which scientists only "borrow" results that they have independently checked) and Hull's emphasis on the acquisition of credit offer too uniform a picture of scientific behavior. Both capture something important that can be understood more exactly by probing the conditions of scientific decision making, seeing how all kinds of social and epistemic factors interact with one another.

A formal model excluded from this excerpt extends the analysis by allowing for effects of competition.[9] Competition situations will now be marked by interesting asymmetries: the chances of receiving credit as the first discoverer of a solution to a problem will depend on one's standing in the community. Scientific entrepreneurs thus have to evaluate not only their own authority and that of their potential sources, but also that of their competitors. In order to keep the analysis relatively simple, I deal only with the case in which *Y* announces a result that could be used by either *X* or *Z*, who are the only competitors on a project. Suppose that *X* has higher authority than *Y* and that *Y* has higher authority than *Z*. Technical Discussion 4 (Kitcher 1993, 329–34) examines the two-person game that *X* and *Z* play against one another. Social competition has appeared to promote welcome diversity within a scientific community. The present analysis shows that authority effects can weaken the competitive pressure. Where they do favor diversity, the asymmetries produced are different from those generated by differences in resources.

. . .

11. The Community Response to Innovation

I have been exploring ways in which individual scientists might attribute authority to others, and the effects of such attributions in a

9. See Technical Discussion 4 in Kitcher (1993, 329–34).

range of rather artificial examples. It is now time to return to the type of case that figured prominently in my initial discussion of reliance on authority. We imagine that someone in the scientific community announces a new finding that is at odds with current thinking. How should a well-designed community respond? What kinds of decision rules and social arrangements will lead scientists to distribute themselves so as to approximate the community optimum?

. . .

12. Individual Responses to Innovation

Let us now consider whether scientists, attributing authority in various ways and employing various decision rules, could work their way to the type of strategy that the community favors. Imagine, then, that you belong to a community of scientists and that some scientist, X, has announced a surprisingly heterodox finding. I shall suppose that you have two possible responses: you can ignore X and go on with your research project, or you can attempt to replicate X's result, deferring your current project. It is possible that there are conditions under which a third strategy, that of incorporating X's finding into your own research work without any antecedent check, might also become tempting—for reasons that we have explored in earlier sections—but I shall suppose that these do not obtain in the present case.

The strategy of ignoring X is basically a null option. By pursuing it you do not change your future well-being, so that the extra utility that accrues to you is 0. If you try to replicate, on the other hand, there are potential gains and losses. If you are the first to replicate then there will be some benefit u^*—possibly epistemic, possibly social, possibly mixed (the attainment of truth, the acquisition of status, the recognition of having discovered something important). If you are the first to show that X is wrong, then you will gain v^*, the utility accruing to the first refuter (again, this may be epistemic, social, or mixed). But, by diverting your effort from the ongoing research project, there will be costs of delay, d. On the basis of prior assessment of the authority of X and prior reasons for assigning probability 0 to X's challenging finding, you now suppose that there is a probability p that X is right. You are confident that, in any attempt at replication, you will judge correctly. However, to receive the benefits you have to be the *first* replicator or refuter. Your chances depend on the extent of the competition. Suppose that you conceive of all your colleagues as equally likely to win the race to replicate or refute. Then, if $n - 1$ join you in trying to replicate, the probability of your being the first is $1/n$. The expected extra utility of trying to replicate is therefore

$$pu^*/n + (1 - p)v^*/n - d.$$

Trying to replicate X is preferable just in case this quantity is positive; that is

(8.6) $p > (nd - v^*)/(u^* - v^*)$ if $u^* > v^*$ or

(8.7) $p < (v^* - nd)/(v^* - u^*)$ if $v^* > u.^*$

Assume that the new finding is sufficiently important that it would gain high credit for the first replicator and epistemic dividends for all researchers in the field, so that $u^* > v^*$. If you and all your colleagues decide in the same way, then at least one person will try to replicate X's result if

(8.8) $p > ((d - v^*)/(u^* - v^*))$.

(This is the condition under which, in a community all of whose members ignore the finding, it becomes preferable for a single person to switch to trying to replicate.) The community can reach this desirable outcome, even when the probability assigned to the finding after the announcement by X is low, provided that $d - v^*$ is small in comparison with $u - v^*$. This occurs trivially when the delay costs are smaller than the gains of first refutation: $d < v^*$. Surely many of the scientists who heard Pons and Fleischmann's claims about cold fusion were engaged in projects that could be delayed without serious cost. For them there were two significant questions: (i) Should a value greater than zero be given to the probability that the claim about cold fusion is correct? (ii) Would there be significant benefits from refuting Pons and Fleischmann if they were incorrect?

Both questions involve issues of authority. Assuming that the probability assigned to the experimental claim weighs the prior authority of the experimenters against the traditional considerations that underlie the antecedent view that there is zero probability of obtaining fusion on a table-top at room temperature, then, as before, we may write

(8.9) $p = z \cdot a(Y, PF)$

where Y is the scientist making the probability judgment and PF is the composite entity Pons-Fleischmann. To obtain a nonzero probability, all that is required is that nonzero weight be assigned to the prior authority of the experimenters and that they have nonzero prior authority. All that was needed from the telephone conversations *on this score* was the assurance that Pons and Fleischmann were reputable, not outsiders like Gish and Morris.

The second issue is more subtle. Each scientist could assess for herself the costs of delay in current research. The benefits of refuting a challenge to orthodoxy depend, however, on the authority of the challengers within the community. If nobody is prepared to defer to these challengers on this issue, then, supposing that their challenge is wrong, little or no epistemic damage will be done. Similarly, if the challengers are already seen as having low authority, exposing the flaws in the present challenge will not redound to the credit of the refuter. So v^* will be a nondecreasing function of the perceived authority of the challengers within the community.[10] Suppose, for the sake of simplicity, that

$$(8.10) \qquad v^* = V \cdot a^*(X)$$

where V is a constant, X is the challenger, and a^* is some averaging function over $a(Y, X)$ for the scientists Y who belong to the community. V will surely be large, for the benefits of exposing errors made by those to whom everybody assigns an authority of 1 are substantial. The crucial inequality for deciding to replicate is thus

$$(8.11) \qquad d < a^*(PF) \cdot V.$$

When telephone calls reveal that other scientists Y give a sizable value to $a(Y, PF)$, or when it is found that Pons and Fleischmann have high unearned authority, scientists whose delay costs are small will make the decision to investigate their finding.

Throughout earlier sections, we have faced the worry that appealing to authority could retard desirable changes in science, making it difficult

10. Michael DePaul pointed out to me the possibility that the utility of a successful refutation might depend not only on the perceived authority of the challengers but also on the number of people engaged in attempts to replicate. Thus, once the challenging finding is taken up, there is a possibility of bandwagon effects, even runaway bandwagons. This can easily be modeled by supposing that

$$v^* = V \cdot a^*(X)(1 + kn)$$

where v^*, V, a^* are as in the text, n is the number of scientists attempting to replicate/refute the challenging finding, and k is a constant. Since my first concern is with the conditions under which *someone* replicates, I am effectively looking at the decision problem for someone when $n = 0$, so that there is no difference between the approach in the text and that which would see utility as an increasing function of the number of other replicator/refuters. The problem of adjusting the number of potential replicators will be considered later.

for correct challenges to orthodoxy to be appreciated. Indeed, reliance on authority in assessing the credibility of third parties can lead scientists to take a jaundiced view of novel claims. I have explored a highly idealized model in which the community preference is for some efforts at replication, and I have suggested that a variety of epistemic or social forces could lead scientists to modify their research in ways that satisfy this *generic* preference. Moral: the impact of appeals to authority has to be evaluated by recognizing the other forces that come into play in scientific decision making—even when a high weight is given to unearned authority, even when calibration is indirect, a community is not doomed to stagnation if there are sufficient rewards (social or epistemic) for entrepreneurs and critics.

But this moral only considers the problem in a relatively gross way. If we have a community in which the new finding will be adopted into consensus practice if just one person succeeds in replicating it, then, as we have seen, it is quite likely that the pressures on individual scientists will lead someone to attempt replication. As yet, however, we have no assurance that, in communities with more complex consensus-forming mechanisms, *enough* member scientists will attempt replication, or, in general, that the distribution of potential replicators will bear any relationship to the community optimal distribution. Technical Discussion 6 (Kitcher 1993, 342–43) explores the much harder problem of how multiple replication might be achieved. It brings to light a serious difficulty. In a community in which findings are adopted into consensus practice if only one member succeeds in replicating, the first scientist knows that there is no danger that her labor will be wasted through the failure of anyone else to pursue the possibility of replication. However, if a single successful replication would not be sufficiently impressive to secure community decision, the potential replicators have to rely on the formation of a critical mass of size r. Thus part of their decision making must involve making judgments of the likely behavior of others, so that they can estimate the expected size of the group of replicators. We can imagine that this estimation is done through informal polling, and that the telephone calls that followed the Pons-Fleischmann announcement were also intended to discover whether a sufficiently large number of people took the alleged result seriously enough to devote their time to investigating it.

The picture that emerges from the preceding discussions of competition, cooperation, and authority is a complicated one. Despite the apparently harmful effects of deference to authority in inhibiting the reception of maverick ideas, the more extensive discussions of the past sections indicate ways in which authority structures within a scientific community might foster attention to novel, antecedently implausible, results. The exact relation between the actual distribution of effort and the community optimum will depend on the values of various parameters, but, even in communities in which intuitively worrying forms of authority play a major role in the assessments scientists make of their fellows, it is quite possible for those parameters to be set in ways that make for good epistemic design. To understand how well a particular community is likely to do, one cannot rest with simple descriptions of it as "authoritarian" or "individualistic." There is no substitute for looking at the types of models I have surveyed.

. . .

13. Division of Cognitive Labor

At various points in the previous sections I have suggested that there are advantages for a scientific community in cognitive diversity. Intuitively, a community that is prepared to hedge its bets when the situation is unclear is likely to do better than a community that moves quickly to a state of uniform opinion. Much of the rest of this chapter will be devoted to exploring this intuitive idea and trying to understand the kinds of social arrangements that might foster welcome diversity.

The problem is easily illustrated by many of the examples that have been considered in earlier chapters. There much was made of the idea that scientific debates are resolved through the public articulation and acceptance of a line of reasoning that takes considerable time to emerge, and that the working out of this line of reasoning depends crucially on the presence in the community of people who are prepared to work on and defend rival positions. So, for example, in the resolution of the Darwinian debate, and, even more, in the triumph of Lavoisier's new chemistry, it was important that opposing points of view were kept alive and that the objections they generated were used to refine the ultimately successful positions.[11] The example of "the great Devonian controversy" makes even more obvious the value of preserving rival approaches in

11. Kuhn (1962/70, 158–159) points out the importance for the scientific community of persistent opposition, and his (1977) advances the general idea of the value of cognitive diversity. Much of Feyerabend's earlier work is also devoted to celebrating the desirability of keeping minority approaches alive, using them to sharpen both the dominant views and one another (see Feyerabend 1963, 1965, 1970). The root idea goes back at least to Mill (1859).

situations of unclarity, for here the ultimate solution emerged from interaction between the previously dominant rivals. Even the much-decried hyperspecialization of science may play a valuable role.[12]

How, then, is cognitive diversity maintained? Given epistemic purity—a community of agents dedicated to modifying their practices so as to achieve purely epistemic ends—it is possible, even likely, that the outcome of separate decisions will be an epistemically homogeneous community. Even if differences between the alternatives are small, if one is slightly more developed, or more successful in overcoming difficulties, then those who value truth . . . will favor that. Consequently, in a community of clear-headed scientists, devoted to espousing epistemically virtuous individual practices, we may expect cognitive uniformity.[13]

. . .

24. Conclusion

As I noted in section 1, there are two main results of the foregoing analyses. First, motives often dismissed as beyond the pale of scientific decision making can, under a wide range of conditions, play a constructive role in the community's epistemic enterprise. Second, the details matter. The effect of various types of factors (authority, elitism), which we might have thought of as acting in a single fashion across scientific contexts, depends on features of the social situation and of the decision problem.

These results flow from a formal—but highly idealized—treatment of communities and of individual decision-making. What can we achieve by modeling scientific communities in so artificial a way? I have begun by employing an unrealistic model of human decision making (who can give precise probabilities? who has definite utility functions? who knows enough about others to be able to carry out, even in principle, the computations I have labored through?). I have proceeded by making numerous further assumptions in the interests of algebraic simplification (think of the many occasions on which I have regarded particular groups as homogeneous in some important respect). These are defects that I would

12. I am grateful to Kim Sterelny for this observation.

13. This will not occur if the scientists are altruistic. If they are prepared—and able—to adjust their cognitive commitments so as best to serve the community, then, of course, they will replicate the decision making that identifies the community optimum. Whether there are, in fact, any scientists so devoted to the community cognitive project as to waive any concerns for their own epistemic advancement in favor of adopting a minority approach ("because someone ought to do that") is a question I leave for the reader.

be happy to overcome. It is possible that a more realistic approach to cognition could be formulated with sufficient precision to enable us to achieve clear results about the outcomes of decisions made by the members of a community. It is possible that a more elegant general approach would expose the crudities of my analyses and offer us a more global perspective on the phenomena I have tried to study. Lacking these potential advances, the discussions of the previous sections should be viewed as first efforts, attempts to raise questions that I take to have been slighted within epistemology and to devise concepts and tools for solving them. I would be delighted were others to improve those concepts, or to transform the tools, rendering my treatment of the problems—but not the problems the themselves!—obsolete.

Finally, I want to address the puzzled reader who regards this chapter as part of a different book, one that does not contain its predecessors. . . . Fields of science, I have suggested, make progress in broadly cumulative fashion. As those fields progress, scientists come to employ cognitive strategies that are superior, when judged by versions of the external standard. Despite the existence of periods in which alternative ways of revising practice can be defended by equally good arguments, scientific communities typically resolve these indeterminacies by articulating a superior form of reasoning that can be used to support one of the rivals. . . .

All of these theses presuppose, more or less directly, that scientists' social involvement with one another does not interfere with the employment of epistemically virtuous individual reasoning. Defenders of Legend, holding a highly idealized picture of scientists, took it for granted that there could be no such interference. But we know that their vision of the scientist as pure seeker after truth is a myth. What are the consequences? Should we conclude that the growth of science is a process in which various kinds of social forces have shaped the doctrines that are accepted and the styles of reasoning that are prized?

The minimal contribution of this last chapter is to rebut the notion that one can infer directly from the existence of social pressures and nonepistemic motivations the conclusion that science does not advance in the fashion described earlier in this book. Philosophers who have studied Legend's critics have usually been haunted by the idea that the impotence of appeals to reason and evidence to resolve scientific controversies would create a vacuum in which unpleasant "social factors" would move the community in arbitrary directions. . . . This chapter reveals that the operation of social systems in ways that we might initially view as opposed to the growth of knowledge can be dependent on the use of complicated reasoning and can contribute to the community's attainment of its epistemic ends. The worry that Legend's heroes have feet of clay *and that, in consequence, science cannot have the progressive characteristics often attributed to it,* turns out to rest on a fallacy.

Imagine that we had started with a different image. From the beginning, let us suppose, we had conceived of scientists as ordinary people, subject to complex combinations of social pressures. Controversies, we might have believed, are typically settled by enrolling allies, and those who are most persuasive and recruit the most powerful followers win the day.[14] The present chapter attempts to show that, when the dynamics of communities who operate in this way is analyzed, the agents must be viewed as making complex epistemic evaluations. To turn the philosophers' fear inside out, the processes of enrolling allies depend critically on the potential allies' assessments of the consequences of the options open to them. Competitive and cooperative situations, as we have seen again and again, call for refined judgments about the merits of methods or of proposals for modifying practice. Thus we can say that a simple view of science as driven by "external" or "social" factors would create a vacuum *into which epistemic considerations would have to be introduced to explain how the social factors obtain their purchase on the individual actors.* The investigations of the past two chapters are thus complementary, the one showing how individuals reason and the other showing how their efforts, in admittedly artificial social contexts, combine to yield distributions of cognitive effort.

Yet, in the end, I want to claim more. *Part* of the epistemological task consists in responding to important skeptical concerns. Beyond that, epistemology should strive to formulate (fallible) claims about good reasoning, thus attempting to improve the ways in which we revise our practices. A rightly respected tradition has contributed much to one side of the meliorative epistemological project: thanks to the efforts of Locke and Hume, Kant, Whewell and Mill, Frege, Russell, and Carnap, we have a far clearer vision of good individual reasoning. The other facet of the meliorative project has been, as I have noted, almost completely neglected. Yet, just as it is important to uncover rules for the right direction of the individual mind, so too, it is necessary to understand how

14. Here I deliberately use the language of Latour 1987. However, it should be noted that Latour's position is far more subtle than the view I sketch, and he would be as unsatisfied as I am with my imagined polar opposite to Legend.

community strategies for advancing knowledge might be well or ill designed. So I conceive this chapter not simply as an attempt to complete my discussion of Legend and its critics, my case for objectivity without illusion, but as a first foray into epistemological terra incognita. I doubt that I have done more than scratch the surface of unfamiliar terrain, but I am confident that epistemological rewards await those who are prepared to dig more deeply.

REFERENCES

Carnap, Rudolf. 1951. *Logical Foundations of Probability.* Chicago, University of Chicago Press.

Feyerabend, Paul. 1963. How to be a Good Empiricist. *Proceedings of the Delaware Seminar.* New York, Interscience.

———. 1965. Problems of Empiricism. In R. Colodny (ed.), *Beyond the Edge of Certainty.* Englewood Cliffs, NJ, Prentice Hall.

———. 1970. Against Method. In *Minnesota Studies in the Philosophy of Science.* Minneapolis, University of Minnesota Press.

Frege, Gottlob. 1892. On Sense and Reference. In P. Geach and M. Black (eds.), 1952. *Translations from the Writings of Gottlob Frege.* Oxford, Blackwell.

Fuller, Steven. 1988. *Social Epistemology.* Bloomington, Indiana University Press.

Hagstrom, Warren. 1965. *The Scientific Community.* New York, Basic Books.

Hull, David. 1988. *Science as a Process.* Chicago, University of Chicago Press.

Kant, Immanuel. 1781/1968. *Critique of Pure Reason,* trans. Norman Kemp Smith. London, MacMillan.

Kitcher, Philip. 1993. *The Advancement of Science.* New York: Oxford University Press.

Kuhn, Thomas S. 1962/1970. *The Structure of Scientific Revolutions.* Chicago, University of Chicago Press.

———. 1977. *The Essential Tension.* Chicago, University of Chicago Press.

Latour, Bruno. 1987. *Science in Action.* Cambridge, MA, Harvard University Press.

———. 1988. *The Pasteurization of France.* Cambridge, MA, Harvard University Press.

———. 1989. "Postmodern"? No, Simply AModern: Steps Towards an Anthropology of Science. *Studies in the History and Philosophy of Science* 21: 145–171.

———. 1992. One More Turn after the Social Turn. In E. McMullin (ed.), *The Social Dimensions of Scientific Knowledge.* Notre Dame, University of Notre Dame Press.

Latour, Bruno, and Woolgar, Steve. 1979. *Laboratory Life.* London, Sage, (1986) reprint Princeton, Princeton University Press.

Longino, Helen. 1990. *Science as Social Knowledge.* Princeton, Princeton University Press.

Merton, Robert K. 1973. *The Sociology of Science.* Chicago, University of Chicago Press.

Mill, John Stuart. 1859. *On Liberty.* Indianapolis, Bobbs-Merrill.

Rouse, Joseph. 1987. *Knowledge and Power.* Ithaca, Cornell University Press.

Russell, E. B. 1916. *Form and Function.* Chicago, University of Chicago Press.

Shapin, Steven. 1994. *A Social History of Truth.* Chicago, University of Chicago Press.

PART IV

Science Conceived as an Economic Network of Limited Agents

9

From Science as an Economic Activity to Socioeconomics of Scientific Research

The Dynamics of Emergent and Consolidated Techno-economic Networks

Michel Callon

Talking of the economics of science, whether traditional or new, means agreeing that there is an activity—scientific research—and that it is the contribution or supposed contribution of that activity to the functioning of markets and the creation of wealth that the economics of science strives to describe. This point of view, which distinguishes between economics (a discipline) and the economy (a thing or activity) was recently challenged by certain sociologists of science. These sociologists applied to the social sciences an approach initially confined to the natural sciences and life sciences and showed the role of economics in the shaping of the economy (Callon 1998b, Power 1994). In this perspective, the formulation of economic theory is just a moment in a broader movement consisting of a series of comings and goings between the description of economic activities, the working out of analytical tools for explaining them, and their possible implementation in the form of instruments of calculation, material devices, and institutional rules that contribute to the restructuring of economic activities. This movement is never finished. The actors constantly invent new practices, set up new relations and interactions that flow over existing frames and force economists and sociologists to adjust their analytical tools.[1] Sometimes, after fierce controversy, this leads to the establishment of new frames and new rules of the game. In these conditions theoretical work is no longer reducible to the deciphering of an enigmatic reality beyond the observer. It is a stakeholder in a process of change, primarily of an institutional nature,

1. On the complementary notions of framing and overflowing see Callon 1998a. Zelizer (1998a, 1998b) has brillantly demonstrated that even in the case of money actors are able to subvert the existing frames.

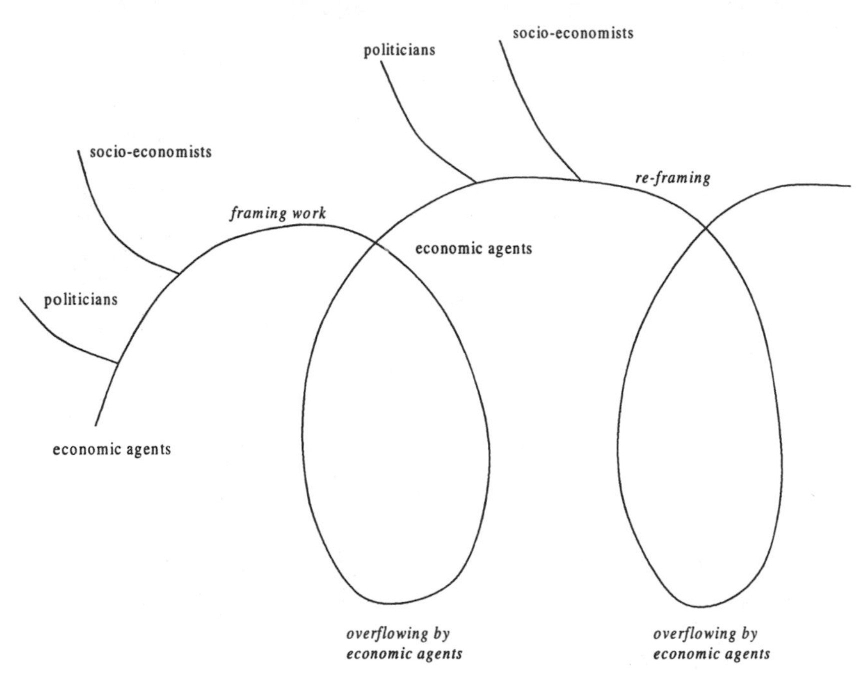

FIGURE 9.1.

in which the actors themselves play a key role, and in relation to which it must constantly take a stand. (See figure 9.1.)

The relevance of this approach is particularly striking when the economic status of science has to be elucidated. In the past few years there has been a proliferation of analyses highlighting a fairly radical change in the rules and conventions (or what might be termed institutional framing) governing scientific research, notably in relations with the economic market. Analyses emphasizing the existence of a distinct boundary between the scientific institution and economic markets were followed by detailed descriptions of relations, hybrid forms, and interactions. Different metaphors or concepts were proposed to describe what was presented as a new reality: triple helix (Etzkowitz and Leydesdorff 1997), interactive model (Kline and Rosenberg 1986), techno-economic networks (Callon 1992), networks of innovators (DeBresson and Amesse 1991), Mode 2 (Gibbons et al. 1994), boundary organizations (Guston 2000), and Knowledge Value Collective (Bozeman and Rogers forthcoming). Economic agents and public authorities have also helped to highlight the importance and efficiency of practices that, compared to existing frames, seem to be instances of overflowing (Caracostas and Muldor 1997).

These overflowings are not really new. Historians of science and industry note that such relations and interactions have been around for a long time. Examples are easy to find in the late nineteenth- and early twentieth-century chemical, electrical, or even metallurgical industries. It seems that the interactive model, or network model, was born with Western science. What it adds—and what is corroborated by many observers—is the establishment, in the aftermath of the Second World War, of an institutional frame largely related to physics, which asserted the autonomy of academic science (Pestre 1997, Mirowski and Sent, introduction to this volume) and the specific nature of its norms. In the next few decades this new framework, largely constructed by the actors themselves, found its theoretical justification in the analyses proposed by certain economists and sociologists. The theory of science as a public good (Nelson 1959, chapter 3 in this volume; Arrow 1962, chapter 4 in this volume), like Mertonian sociology of science, transformed into theoretical necessity what had been related essentially to historical contingency. By providing an unquestionable rhetorical resource to both politicians and academic researchers, as well as to industry, they helped simultaneously to bolster the establishment of Cold War institutions and the linear model associated with them. It is against this institutional background, raised to the level of a theoretical necessity, that networking appeared to some as a threat to economic prosperity and to others, by contrast, as the unleashing of creative and productive forces. This provides a particularly striking illustration of the contribution of theory to the formation of frames of economic action.

In these debates we could say, to simplify, that there are two contradictory, conflicting positions. Some plead for the restoration of scientific autonomy, which they consider to be threatened by these uncontrolled overflowings. They demonstrate, on the basis of theoretical arguments, the advantages of an open science that requires the reinforcement of existing rules (David and Foray 1996).[2] Others, by contrast, emphasize the radical nature of overflowing, which, they claim, can no longer be ignored or contained (Mowery 1999). According to them, new analytical tools must be constructed to account for the proliferation of links and interactions crossing over boundaries, and to identify the consequences so that new material and institutional frames can be discussed and de-

2. These rules or norms encourage researchers to disclose the results of their work.

cided. One of the essential points of the debate hinges on the rules of intellectual property rights and the aims of research carried out by non-profit organizations (universities, public institutions, government laboratories, etc.). Clearly, the issues here are not only theoretical, for they will ultimately impact on the arrangements opted for. Theory is directly geared to the social engineering of new institutional and material frames. In this case the entanglement between the prescriptive and the analytical dimensions of theoretical work is blatantly unveiled: the controversy is simultaneously scientific and political. This double component requires reflexivity from social scientists: they must show an interest not only in devising theoretical tools for identifying and explaining overflowing, they must also take into consideration if, and how, these tools should be implemented into institutional devices aimed at framing these overflowings.

In this contribution we consider as given the existence and development of overflowing, which multiplies interaction, exchange, cooperation, and even integration between scientific activity and economic markets. In previous work we proposed a number of concepts that could be used to study such networking and to describe its complexity and dynamics (Latour 1987, Callon 1992), but we need to go further than that. In the debate initiated by economists we have to take a stand, to voice our opinions—backed up by theoretical arguments—on the consequences of this hybridization and, depending on our assessment of it, on the institutional and material frames promoting or containing it. We are confronted with the performative part of our work and are engaged in the second part of the loop where theory becomes performative because it raises the question of how to reframe actors' (ir)resistible overflowing. This approach naturally contains numerous uncertainties and traps—which is why it may be useful to clarify our position.

Instead of defending at all costs the status quo, i.e., Cold War institutions, we have chosen to adopt a different perspective. Our starting point is the idea that Cold War institutions have had their day and that from now on they are becoming an obstacle to the contribution of science to economic development. In previous work we highlighted the importance of the role of scientific research performed in nonprofit organizations, as a constant source of technological diversification that thwarts the natural tendency of markets to render techno-economic trajectories irreversible (Callon 1994b). The question, once this argument has been accepted, is of course to imagine the conditions for the establishment of dynamics that make feasible this dual movement of exploration and exploitation, of the opening of new, as yet unprofitable, ways and of investing in existing and profitable ones.[3]

One way of answering this difficult question is to concentrate on the increasing number of configurations in which networks of relations are the longest and the most heterogeneous, and in which the actors are aware of the constant need to explore new fields of knowledge and simultaneously to invest in the systematic exploitation of those which are starting to be consolidated. Consequently, empirical cases that can serve as constant, even if implicit, references are sectors such as biotechnology or communication technology, characterized by the constant comings and goings between knowledge production, technological developments, and construction of demand, in close interaction with consumers. They constitute veritable social laboratories in which new arrangements and devices and original rules of the game are tried and argued. Finally, the framework of analysis proposed must respect the variety of modes of coordination and forms of action (avoiding, in particular, one of the usual types of reductionism that uses the economic metaphor for describing scientific production), while acknowledging the full importance of interactions and hybrid forms.[4]

Clearly, this project takes the opposite view to what some have proposed to call the new economics of science. It shares with the new economics of science its interest in the role and conditions of nonprofit research. However, instead of arguing and reinforcing the strict opposition between open science, based on Cold War institutions on the one hand and a market with clearly defined boundaries on the other, it aims to show the possibility of scientific research that is both autonomous and strongly connected to firms. One of the challenges of this analysis is to prove the economic and scientific viability of such configurations. Although overflowing challenges the dynamic equilibria associated with former configurations, it creates the possibility of new development regimes, of which the driving force needs to be identified.

To develop our analysis we shall start with three concepts defined and studied by the new economics of science. These concepts and their critique will enable us to introduce the two notions of emergent networks and consolidated networks. These will then be used to formulate

3. The theory of endogeneous growth, relying upon the public good assumption of basic science, has suggested the viability of the linear model. Unfortunately, as we will show, this assumption is unrealistic.

4. We chose to get rid of notions like market of ideas, scientific capital, credibility cycles, etc., which erase the differences between activities that are obviously different.

a very simple, obviously speculative model for demonstrating the existence of a dynamic regime compatible with the observed overflowing, while maintaining a radical differentiation between science and the market. The very possibility of this regime implies the establishment of new framing devices, already sketched out by actors themselves, of which a rough description could be useful to the design of public policies suited to this new regime. Thus, its potentially performative character will be taken into account in the construction of the model.

1. From the Circulation of Information to the Production and Transportation of Scientific Knowledge

Economic analysis of research activities must elucidate the economic status of science. Research can be seen as a productive process in which specific actors (researchers) transform inputs (knowledge, embodied skills, instruments, and materials) into outputs. This leads to the question of the nature of these outputs, whether they are comparable to goods and, if so, whether they can be commoditized. The conditions for market efficiency as a coordination mechanism should also be considered. In this section we first present the traditional approach with its three germane notions of rivalry, excludability, and generality. Then we show that these characteristics usually considered as intrinsic properties of science can better be seen as extrinsic properties depending on the state of the networks shaped by, and shaping, scientific research.

1.1 Rivalry, Excludability, and Generality as Intrinsic Properties of Science as Information

Following the work of Arrow and Nelson, the new economics of science attempts to bring a fresh approach to these issues, which, although diverse on the surface, is based on the same series of hypotheses and definitions. These should be briefly presented and discussed, for they will serve us in presenting emergent networks and consolidated networks.

The new economics of science is based on two founding hypotheses. Hypothesis H1 is related to the nature of scientific knowledge or, more particularly, the specific forms that it takes. It states that, from the perspective of economic analysis, scientific knowledge can be equated to information, which is “knowledge reduced and converted into messages that can be easily communicated among decision agents” (Dasgupta and

David 1994, 491; chapter 7 in this volume). This definition has two important elements. First, the idea of a message presupposes a material base that allows transmission. This message can be a written or oral statement or a series of statements. Equally, it might be inscribed in a human being, product, or machine. Second, whatever means are used to transmit it, the message is not considered to be information unless it leads to an action. The statement: "The structure of DNA is a double helix" is information only to the extent that it has utility for the recipient. In the same way, a laboratory that recruits a scientist or purchases an electronic microscope expects that this will open up new fields of research. Knowledge that has not been transformed into information is of no interest to the economist, because without utility it does not exist in a form that can be circulated and traded.

To the hypothesis that equates science with information and research activities with the production of information about the world, a second crucial hypothesis, H2, relating to the form of this information, is added. Leaving aside differences in terminology, the accepted distinction is between codified and embodied (tacit) knowledge.

A code, broadly defined, is a conventional system of rigorously structured signs and symbols, common to both emitter and receiver, that acts as a vehicle for communication. An archetypal example of codified knowledge is clearly a statement in ordinary language: "the structure of DNA is a double helix" or "the sun emits neutrinos." The available codes are numerous, ranging from the most esoteric mathematical formulas to the most simple digital sequences. They can also be classified according to their degree of disclosure: a message expressed in English will reach a wide audience, while encrypted messages using unique codes constructed at great expense (e.g., on the basis of certain properties of prime numbers) are understood only by a few who have access to the key.

Embodied knowledge brings together knowledge that is inscribed in the minds and bodies of individuals (researchers and technicians) or in instruments and machines. When they conduct an experiment, set out an argument, or interpret diagrams, scientists use a set of know-how, manual dexterity, and automatic mental modes of reasoning of which they are not necessarily aware. Scientific practice is an art: to solve an equation of partial derivatives or produce monoclonal antibodies, it is not sufficient to know how to do it, one must have done it oneself—in the same way that it is not sufficient to have read a manual, no matter how detailed, to be able to play the piano or ride a bicycle. The same is true of the use of instruments or experimental techniques (what Bachelard has called the phenomeno-technique) that crystallize knowledge and know-how that, even though not explicit in the form of plans of descriptions, is used by every researcher who applies them.

H1 treats scientific knowledge as information; H2 sees this information as having two distinct forms (codified/embodied). These two hypotheses taken together allow a description of the economic properties of scientific knowledge. Two variables are used for the analysis: rivalry/nonrivalry; and excludability/nonexcludability. They have a general application and lead back to the classic distinction between a public and private good.

A good is rival when actors who wish to consume or use it find themselves by necessity in competition with others: you can use the Renault Twingo with the registration number 3392 AZJ 78 or I can use it, but we cannot both use it at the same time for two different journeys. A good is nonrival in the opposite case: the fact that one actor uses or consumes it does not affect its utility for any other actor who also wishes to use or consume it. I can listen to a broadcast of a Gustav Leonhardt recital on Classic Radio or watch Green Bay Packers vs. New England Patriots in the Super Bowl without affecting the pleasure of other listeners or television viewers. More precisely, Dasgupta and David (1994) give the following definition of nonrivalry: "a good which is infinitely expansible without loss of its intrinsic qualities, so that it can be possessed and used by as many as care to do so" (493).[5] This intrinsic property presupposes that the production cost (or marginal costs) are zero or negligible. The proponents of the new economics of science agree to consider embodied knowledge as a rival good. A research director who recruits a scientist removes her or him from the labor market, for no other laboratory can employ her or him; a research team that monopolizes a unique telescope necessarily deprives colleagues of the possibility of developing alternative projects that also require its use. They also agree that codified knowledge is nonrival or infinitely expansible. According to them, to use a computer code, for example, or to introduce

5. Dasgupta and David's footnote is worth citing because it sheds light on one of the difficulties of this concept: "Were person A to give person B a piece of information concerning Q, that would not reduce the amount of information concerning Q that was retained by A (although, to be sure, the benefit to each would depend upon whether and in what manner B were to make use of the information)." This leads them to suggest "infinite expansibility" as an alternative term that is less confusing than nonrivalry, in that it acknowledges the possibility of rivalries for possession of new information (Dasgupta and David 1994, 493).

the law $F = ma$ or a Lorentz transformation in a demonstration in no way diminishes their utility for other potential users. By positing that codified knowledge is nonrival whereas embodied knowledge is rival, economic theory makes rivalry an intrinsic property determined entirely by the form of scientific knowledge.

The second variable that is useful for qualifying the economic status of knowledge relates to appropriable and nonappropriable goods. This characteristic is important. Indeed, for a good to be commercially traded, it is necessary not only for it to have a value or, if preferred, utility, but also for it to be able to be the object of property rights. A good is appropriable if it is possible and easy for the consumer or user to exclude all other potential users and consumers. When it is not possible to easily prevent access or use, a good is nonappropriable. The concept of excludability is sometimes preferred to that of appropriability. Rather than discussing excludability, it is more pertinent to consider another aspect: the cost of exclusion or appropriation. It is always possible to establish property rights for a given good. The variable, depending on the nature of the good considered, is the cost of the effective enforcement of these rights. In some cases it is very high, which leads to appropriation being costly, whereas in others it is, all things being equal, low, so that appropriation is easy. The cost of enforcing property rights for a given good is a characteristic of the good. Because it depends entirely on the nature of the good, it can be said to be intrinsic.

A car is easily appropriable. Given its physical characteristics, it is not overly complicated and, consequently, not overly costly, to prevent a third party from using it. Everyone agrees that the same is true of embodied knowledge. For example, it is not too difficult to control access to and use of scientific instruments such as a telescope, an electronic microscope, or an apparatus for determining the genetic sequence of rice. Neither is it difficult for an employer to oversee the movements and acts of a given researcher or technician. On the other hand, the new economics of science considers that it is more complicated, although not impossible, to exclude access to codified knowledge—as intellectual property rights demonstrate (patents, copyright, trade secrets). Such knowledge is more difficult to appropriate because its disclosure cannot be easily contained. Continuous progress in the means of reproduction and communication facilitate diffusion.

To SUM UP. For the new economics of science scientific knowledge is a good, because it can be equated to information. It can take two forms: codified or embodied information. This distinction is crucial because it affects the economic status of the knowledge under consideration. To simplify, codified knowledge demonstrates a high degree of nonrivalry and its appropriation cost is high. Conversely, embodied knowledge is rival and the cost of appropriation is lower. The former is a public good, the latter a private good.

Apart from minor differences, economists whose approach is based on the new economics of science not only share H1 and H2, they also agree that it is possible to distinguish between basic or fundamental research—characterized by the generality and universality of the laws it formulates—on the one hand and applied or technological research on the other. Thus, although their reasoning may be slightly different, the practical conclusion they reach is the same: basic research takes (by its very nature, according to Romer 1992) or should take (for reasons of efficiency and because embodied knowledge and codified knowledge are perfectly substitutable, according to David and Foray 1996) the form of codified statements. Consequently it is a public good. It follows that, for reasons of economic efficiency, this research must be organized according to the rules of academic science, as formulated in Mertonian-inspired sociology: profits are symbolic (recognition being linked to the priority of discovery) and disclosure is the norm. Observable overflowing, which leads to the privatization of basic science, therefore counteracts this efficiency. It has to be contained, which necessitates the reinforcement of Cold War institutional frames.

1.2. Replicating Scientific Knowledge: Rivalry, Excludability, and Generality as Network Properties

The analysis offered by the new economics of science has the distinct advantage of proposing a full range of concepts that render the study of research activities, i.e., knowledge production, accessible to economic analysis. As a result, the various institutional configurations, rules of operation, incentives, and rewards can, for instance, be accounted for and their efficiency evaluated (Stephan 1996, Stephan and Levin 1997).

It would nevertheless be a mistake simply to develop and expand the approach outlined. While the tools proposed can be used to describe, examine, and explain, in a general fashion, the various institutional configurations (academic research, market) in which research activities play a role, they are hardly adapted to an in-depth analysis of the dynamics of interactions between firms and laboratories. Indeed, they have nothing to say about the work that transforms scientific knowledge into commercial innovations. Economists stage a play with double-bound actors: they must interact but without crossing boundaries! The only relationships that these straitjacketed actors are allowed are those that emerge from the circulation of information. The conditions for their production are simply ignored. This is simple exchange economics, whereas one of the key points of science studies is that the circulation and production of knowledge are inseparable: they are one and the same process (Callon 1994a). Now, to obtain a representation showing the viability and economic efficiency of a configuration in which overflowing, that is, networking, is the rule, one has to be able to perform a microanalysis at the level of the agents and their cooperation and competition strategies, whether these actors are nonprofit laboratories or industrial firms.

The analytical framework that is now outlined takes on board the authors' main concepts, explored in the preceding section: codification/embodiment, rivalry, exclusivity, and generality. However, instead of associating these intrinsic properties with the notion of information, it relates them to the state of the networks whose construction and interaction are defined by the participating actors that produce, diffuse, and use scientific knowledge. The analysis is inspired by recent work in science studies and takes as its starting point work on the replication of knowledge. It thus highlights the fact that the extent of rivalry, appropriability, and generality of knowledge varies according to the structure of the networks in which they are produced.

Replicating Laboratories

At the heart of the new economics of science lies the assumption (in fact a commonsense hypothesis) that basic science can be reduced to codified or assimilated information and therefore is nonrival, is costly to appropriate, and has a high degree of generality. In its attempt to ensure the plausibility of this reduction, classic economic theory makes a strong hypothesis concerning knowledge diffusion: once produced, validated, and expressed in the form of statements, theoretical knowledge is supposed to be self-sufficient (being context free) and, when accessible, it is considered as usable everywhere. It might be difficult (and costly) for an agent to get access to these statements but once she or he acquires this codified information, she or he masters the whole scientific knowledge it is supposed to capture as well as its potential applications. Yet such a vision of the conditions of diffusion of (basic) scientific knowledge is wrong. In reality, it has been shown that knowledge is not simply disseminated or distributed but replicated at great expense because it is not statements that are duplicated but laboratories: consequently basic science can not be reduced to information.

Here it is necessary to look at the groundwork of H. Collins. By tracking the replication of experiments in several disciplines, this British sociologist has shown that the first reproduction of a finding or demonstration presupposes a lengthy process of collective learning, including the rigorous calibration of the instruments and competences embodied in human beings. Designing measurement apparatus, conducting experiments, interpreting results and diagrams, and developing an explanation are learned on the job through negotiation and discussion, trial and error, and, above all, direct example (Collins 1974, 1985, 1990, 1995; Aman and Knorr-Cetina 1988).

Thus, it has frequently been observed that to reproduce an experiment on a site other than that on which it was conducted for the first time (in order to validate the new finding and give it a context), it is necessary to transport instruments, technicians, and scientists (Cambrosio and Keating 1995). In pure emerging configurations, the movement of instruments themselves is problematic because it is not always easy to establish a clear frontier between the apparatus and what it is supposed to be measuring and because of the uncertainty on the testing procedures (Mallard 1998, MacKenzie 1990, Sims 1999). This demonstrates how misleading the common concept of information is in leaving aside the infrastructure, including metrological infrastructure, that facilitates its circulation and establishes its relevance. A statement in isolation is like a Boeing 747 without runways, radar, air traffic control, control towers, bus shuttles to conduct passengers to the airport, and ticket offices to make the reservations. A plane does not fly on its own. It is Air France that flies or, more precisely, Air France together with all the elements outlined above. In the same way, no research finding (scientific statements) contains its own meaning (Latour 1987). This is given to it extrinsically by the network of laboratories and competences within which it circulates. Once this infrastructure has been built, the concept of a network can be broadened to its logistics or metrological meaning and the term "information" will have some sense.

The construction of the universality of scientific knowledge looks more like a public works enterprise than like the miraculous conversion of minds convinced by sound arguments. Without reproduction of laboratories and without the infrastructure, training, and metrology this implies, statements would remain trapped in local settings and would soon be scrapped. They are not born universal but become universal; and their universality is networked, which means that it is valid only in the rare places that have been configured to cater for it.

Rivalry, Excludability, and Generality as Extrinsic Properties

If emerging science cannot be reduced to codified statements, even when it is what is commonly termed basic research, it cannot be considered nonrival. Nonrivalry or, if preferred, infinite expansibility implies in effect that the costs of reproduction are negligible relative to the costs of production. In reality, the reverse is true for knowledge based on an infrastructure that is still local. Emerging knowledge meets with a cold reception and often total incomprehension, not because there is a natural resistance to change or innovation but more simply because the laboratories have not been duplicated and the instruments, intellectual frameworks, and competences are not available. Mentioning resistance to change would be comparable to accusing the inhabitants of a remote island of being culturally or genetically hostile to Boeing 747s instead of recognizing that they have no airports, control towers, or maintenance systems. Once these investments have been made, they would take to the air like ducks to water. The more original the findings or, in other words, the scarcer the embodied competences that accompany them, the higher the cost of their reproduction because it cannot simply be amputated from all the other elements that form a whole.

Once the operational networks have been established and the calibration and standardization of instruments and knowledge carried out, i.e., once the configurations are consolidated, statements can circulate, supported by the ether they previously lacked. They then, and only then, become nonrival goods: anybody, at any point in the network, has the resources and competences available to use them freely or, at most, for the cost of a photocopy.

When findings emerge, they are not only rival but also easily appropriable. The good is not the statement but the set [statement + instruments + embodied competences]. At the outset, there is only one copy of this package and the first replication of the findings implies its total duplication in a way that is entirely comparable to the (re)production of a normal good such as an automobile or a lemon squeezer. On the basis of its physical characteristics, it is thus appropriable at a small cost. This is particularly true, for leaks are impossible since they presuppose the existence of a circulation space that does not yet exist. It is easy to take ownership of a good that nobody understands and that has no utility outside its place of production. Scientists the world over, when overcome with originality, routinely undergo the sobering experience of discovering that the problem is not so much protecting their discoveries as persuading a benevolent colleague to read what they have written and even—miracles do happen—show an interest in and an understanding of the content! To continue the aircraft metaphor, it cannot be hijacked unless it is flying and airports exist with which to negotiate a landing. In the case of emerging configurations, appropriability is guaranteed because, contrary to the traditional view, leaks and overflows are costly to organize.

Correspondingly, once competences and instruments have been duplicated in multiple copies and widely distributed, and the measurements have been calibrated and stabilized, the costs of appropriation of the knowledge produced, notably in the form of codified information, are high. The smallest leak can have considerable consequences for the responsible party, because everything is information: the smallest allusion or clue carries meaning and utility once competences are comparable. In the consolidated network configuration the concept of information is relevant, for each element (statements, calibrated instruments, embodied competences) is autonomous insofar as it can be reinserted, in each site, into the context that has given it meaning.

The concept of generality of knowledge can be subjected to the same analysis. Information, whatever its form, is not in itself characterized by a given degree of generality. Once again, the emerging/consolidated distinction is useful. The diffusion of the theory of relativity throughout the United Kingdom has been studied in some depth by Warwick. At first glance, what could be more universal and general than a theory that touches on the foundations of the universe? However, contrary to what is often thought, universality is an outcome and not a starting point. Diffusion cannot be explained by the intrinsic generality of the theory; the diffusion, i.e., the translation of statements from one place to another, gradually creates the generality. Warwick highlights the role of Cambridge mathematicians in the introduction of the theory of relativity into the United Kingdom between 1905 and 1911 (Warwick 1992–93). Their work was very different from that being undertaken by their German counterparts during the same period. The theory of relativity crossed the Channel not because of its universality but rather because it was considered to offer some interesting synergies with the groundwork of Lorentz and Lamour. It was only later, following the task of translation and integration, that a degree of convergence emerged between British and German physicists. This episode provides a good example: it shows that the field of application or the validity of a theory or statement is entirely dependent on the existence of relayers, translators, and mediators who, in pursuing their own interests, devote a lot of energy to integrating, adapting, and creating compatibility.[6] Without this work, generality would not arise.

In addition, the work necessary to the transport/translation of statements contributes to the frequent redefinition of the very content of knowledge. Not only is this costly in terms of time and resources; it is also productive in such a way that the distinction between the phase of creation and that of routine duplication is always arbitrary. Thus, the generality of information is the result of its being translated, in a costly and creative manner. This allows links to be established between sites that were, at the outset, unrelated and not even capable of communication. These translations, and the investment efforts that they imply, construct the equivalence, shape the network, design the relationships and interactions. Each point remains distinct from the others, but the passage from one point to another and the impact of one point on another—in short, what is called transmission—become possible.

The extent of the generality of knowledge is consequently built up gradually through successive investments that often require stubborn efforts and bold translations accompanying the establishment of the network. In emergent configurations the knowledge produced is local; in consolidated configurations its field of application is as wide as the number of points on the metrological network and as sound as the equivalence between these points. Nearly a century after the episode described by Warwick, Einsteinean physics is used throughout wide networks, including articles and textbooks that set out the statements, machinery, and equipment in which these statements are inscribed, and scientists and technicians trained to master and use the equations and basic concepts, etc. In this extended network, consolidated and aligned by the equivalencies produced, each element making up what has become the theory of relativity is available at any place for the various actors located at the various nodes of the network.

THIS ANALYTICAL framework demonstrates that the degrees of rivalry, appropriability, and generality of knowledge are strongly corre-

6. The mediators are numerous and active in direct proportion to the extent that the infrastructure of embodied knowledge is already soundly established.

lated to the state of the networks and the metrological infrastructures. If these are emergent, then knowledge is rival, appropriable, and specific. Once the complementary investments have been replicated in a number of places and adapted to each of these sites, it becomes nonrival, nonappropriable, and general, and is then a public good within the metrological networks so built. If nonrivalry, nonappropriability, and universality exist, they are not to be found in emerging science but rather in what Kuhn termed normal science or in what I prefer to call consolidated configurations. Compared to the standard theory, the about-face is complete. The Holy Grails of modern economics—nonrivalry, nonappropriability, and universality—are not given, but rather obtained at the price of costly investments. Their production should be taken into account in economic analysis.

2. Emergent Networks and Consolidated Networks as Framing Configurations

Once the link between the economic properties of scientific knowledge and the configuration of the network has been recognized, it remains necessary to characterize the modes of action or the types of agencies as framed by each of the configurations distinguished above. We shall consider, in turn, emergent configurations and consolidated configurations, with an emphasis on the former since they are usually underanalyzed by the social sciences.

2.1 Emergent Configurations

In an extreme case of an emerging network, actor A is committed to the production of knowledge that is local and will remain so as long as the replication and measurement work has not given rise to other localities capable of capturing this knowledge. Any attempt to describe and explain the basic replication process will look at those forces that push actor B, distinct from A, to start the replication game and to agree to the costly investments required.

The paradox of these first replications has been clearly formulated by Arrow, although from a slightly different perspective. Why should B be interested in the work of A, when she or he has no means of understanding the knowledge produced by A and is consequently totally incapable of assessing its utility? The paradox is even deeper than Arrow says: even if A transmits to B the knowledge produced, for example in the form of codified statements, B cannot assess their value because in an emergent configuration she or he has not (yet) invested in the necessary skills to understand them. This difficulty is accentuated by the fact that in an emerging network no actor is able to describe with any degree of precision the content of the knowledge that will be produced. Not only is there total uncertainty concerning the possibility of producing any given type of knowledge, but even the description of the content of knowledge can only be vague and general.

To describe these extremely ambiguous and paradoxical but nevertheless frequent situations, where actors decide to throw themselves into the production of emergent knowledge, Bruno Latour and I developed the concept of translation. This concept should not be seen as a theoretical one; it simply aims at describing those strategies developed by isolated actors when they try to survive and maintain their own identity. B can grasp the meaning of the knowledge produced by A only if A makes an effort to interest B. A translates for B: she or he attempts to convince B that a detour should be made, i.e., that it is in B's interest to pass by the competences produced by A. In these strategies of "interessement," it is clear that what is being negotiated is what A and B really want, i.e., their objectives and the programs that need to be designed to reach these objectives. Translation allows the establishment of a constantly negotiated equivalence between distant, immeasurable, and, to start with, uncertain and unstable interests (those of the knowledge producer and those of the potential user). It often leads to the production of new actors and interests. In its geometrical sense, translation relates to transport (of knowledge from one place to another) and in its linguistic sense, to an understanding between what at the outset are alien worlds (those of A and B) that are progressively reconfigured as they are connected. In the extreme case of completely emergent knowledge (A is the only agent to master it), there is no potential B to acquire it. A must make a serious attempt to interest, enroll, and ally with other actors in order to convince them of the utility of what she or he has produced.[7]

The story of French research into the electric car (VEL) is riddled with such translation strategies. For example, at the beginning of the 1960s the electrochemists at the French National Centre for Scientific

7. Contrary to what is frequently said by some Actor Network Theory (ANT) criticisms, an agent engaged in translation work cannot be assimilated to the optimizing, free-floating, unencumbered agent of neoclassical economics. To translate is to knit the first stitches of a web that does not still exist and that will maybe never exist; consequently translations are the very simple counterparts of the spatial and time persistence of actors: to translate is to exist.

Research (CNRS) persuaded their peers in industrial laboratories that they should work with them to study the physical properties of the interaction between platinum and hydrogen. This was thought to be a precondition for the construction of effective generators for electrical traction. This conviction developed at the same time as the interests, programs, and possible states of the world were constructed, negotiated, and elaborated in an evolutionary manner. Gradually, the programs consolidated the utility of the knowledge produced—or to be produced—to a point where this utility became visible and could be evaluated. Translation was deployed at several levels. In this case, it occurred at a very general level (VEL is the more effective weapon against pollution), but also at more specific, technical levels (the main problem was that of generators and for this it was necessary to make progress in electrocatalysis). It was the latter that rendered the program credible. At different levels, the actors participating and the nature of their interests differed: the range of utilities linked to the proposed research was wide, but each fed and reinforced the others. Because the electrochemists believed it would be possible and efficient to design cheap combustible batteries, policy makers and industrialists could attribute a utility to the research and raise the challenge of pollution. Conversely, because the VEL is a political challenge, the utility of basic research on the platinum-hydrogen interface was acceptable. This mechanism of the overlapping reinforcement of utilities, which is the outcome of the multiple connections and alignments performed by translation work, is general: it belongs to the dynamic of increasing returns. A lone researcher is powerless if peers and policy makers are indifferent. She or he cannot survive without devising projects and objectives capable of aggregating a wide range of heterogeneous actors and interests that are made dependent on one another (reduce pollution and energy consumption, develop a physics of interfaces, etc.). Such heterogeneous projects can therefore be interpreted as attempts to spawn emerging networks.

Little by little socio-technical networks emerge from such relationships, through progressive learning, iteration, negotiation, and adaptation. At each point in the network the same competences and knowledge become available. In this collective process, knowledge, competences, scientific artifacts, the identity of actors, their interests, projects, and expectations take shape and acquire consistency as interactions develop.[8]

8. Numerous studies highlight the variable geometry of actors during emergent phases. For a summary theoretical presentation: (Callon 1986, 1987; Latour 1987, 1992).

It is not possible in these configurations to talk of research programs (or even innovation programs) in the strict sense of the term, because the objectives are unstable and in a state of permanent revision, depending on the knowledge outputs and the alliances made. Research and innovation consist of trial and error, tinkering, and the cobbling together of homemade solutions that uncover the path to follow as they advance. It is even more difficult to speak of competing programs because competition presupposes a stable, highly structured and ordered world in which outcomes can be predicted and agreement already exists on the objectives to be attained.

It is to be underlined that it is in these tentative, occasionally successful, interactions that actors, confronted with others, learn to know what they want and what they are able to do. Shared objectives emerge through a series of successive approximations, and visible and clear demands are stabilized. Interests and demand for goods (knowledge, material goods, services) progressively replace the initial interests, the content of which can be defined only after the fact. Utility, whether expected or real, is dependent on the state and scope of the translations under way, of which the ultimate aim is to align logistic networks in which knowledge and machines circulate freely.

It can be said that in an emergent configuration, actor rationality is interactive. What an actor is, prefers, knows, and wants is the outcome of interactions into which she or he enters, and not the starting point for these interactions (Lane et al. 1996). Actors are condemned to interaction—i.e., translation and interessement strategies—if they do not wish to disappear, to be cut off from any possibility of pursuing their own action, deprived of any capacity to anticipate and know what others want, mastering knowledge that has no value other than local. To exist, to have intentions, projects, and interests, presupposes detours, discussions, and negotiations with at least somebody else who is also uncertain and unstable and will learn an equal amount from the interaction. This learning simultaneously and progressively shapes knowledge, know-how, programs, and identities. This is just another way of saying that it is necessary to act in order to understand and decide, and not the other way around. The actor as described by methodological individualism, one who has acquired information before acting and who acts with perfect knowledge of certain impacts, cannot exist in emergent phases. The actor builds her or his own identity through translation and being translated, she or he learns to understand and act on the world through interaction with it. If formal contracts are written, which of course sometimes occurs, their role and meaning are very different from those normally attributed to them. Contracts are only one mechanism among others that perform the joint construction of wills and knowledge, rather than representing the result of a compromise between two preexisting wills.

2.2 Consolidated Configurations

Consolidated (or aligned) networks are the other extreme. In this configuration competences have been replicated. Multiple sites exist at network nodes where the same instruments, knowledge, and know-how are available and exploitable. These give an identical or similar meaning and a perfectly predictable utility to each statement or technical device at any point in the network.

Consolidated networks are characterized by a strong similarity of actors involved. It is here that the contribution of the evolutionary theories of technological change is most evident. An economic actor, for example an industrial firm, is defined essentially in terms of its knowledge base (Cohendet and Llerena 1999). This base consists of codified knowledge but also, and importantly—evolutionary economists would argue—tacit or incorporated knowledge. In a consolidated configuration these knowledge bases (or competences) are largely similar.[9] This similarity explains the fact that in consolidated networks the actors, shaped by a common history, have the same view of the possible states of the world. Their expectations are also very much the same. Another consequence is the similarity of action programs conceived by the agents, especially R&D programs.

With the condition that institutional and material frames exist to ensure a certain degree of excludability for the knowledge produced, this similarity and proximity mean that competition prevails (I will return to these conditions below). This implies the definition and launching of R&D and innovation programs that aim to resolve similar problems and develop viable and acceptable solutions. In this highly ordered and stable world, programs can be defined ex ante because the possible states of the world are easy to identify: the available knowledge is stable and codified, the problems are formulated and recognized, the procedures to be implemented are well proved, and the actors are clearly identified along with their interests and preferences. In the most aligned networks,

9. Here we could suggest a difference with evolutionary economics, which emphasizes the irreducible singularity of each firm, a singularity that is embedded in its peculiar trajectory. Our analytical framework leads us to qualify this assertion: agents, who are strongly connected within the same network, are highly similar because they share the same history.

it is possible to give a probability for the occurrence of each of the possible worlds. What is important is that complete expectations can be formulated, i.e., the inventory of the states of the worlds and actions that permit their construction are known.

To summarize, this implies that the following are known and describable without too much ambiguity:

the current and future states of the world
the various actions that can be undertaken
the probable consequence of each of these actions (i.e., the states of the world they will help to bring into being).

Consequently, one can say that expectations are rational. Each actor has the same analytical capabilities, uses the same calculating tools as others, and is able to anticipate the behavior of each of her or his competitors.[10]

It follows that the R&D and innovation programs that precede and shape action are, to a certain degree, substitutes for one another because the objectives (i.e., the possible states of the world) and the competences to be used are similar. Each program (to which each actor can be equated) commits a certain amount of resources and establishes a link between the resources committed (embodied skills in researchers and technicians, instruments and equipment, etc.) and the objectives. Furthermore, it could be argued that the probability of achieving the objectives depends—to a large degree—on the amount of these resources. Each program is conducted in a linear fashion and consists of the succession of the three following sequences:

the formulation of problems and hypotheses
the search for one or several solutions
the setting out of the solutions; although they cannot be known in advance it is known that they do exist.

2.3 Characteristics of Emergent and Consolidated Configurations

The configurations discussed above are of course ideal types, purified forms that do not exist as such in reality. What we observe in the real world are hybrid evolving forms. But before proposing a model to account for network dynamics, we need to ensure that the dimensions

10. In stable configurations, actions conform closely to what is commonly called rational decision making. Actors have objectives, they acquire and process information, and possess a certain knowledge of possible actions and their outcomes. They can thus choose an action according to a calculation of the value of expected consequences.

used to characterize extreme forms are identical, for the progressive transformation and recombinations of configurations imply a continuous evolution along these dimensions. That is what table 9.1 enables us to confirm. It summarizes the main characteristics of the two extreme network configurations. Each configuration is described by the same three dimensions: the properties of scientific knowledge, the knowledge of future states of the world, and the modalities of action. In particular this table clearly unravels the ties between types of agencies and the state of the network, illustrating the framing process mentioned above.

2.4 Elements of the Dynamics

No (Inevitable) Network Life Cycle

These two configurations should not be considered as two independent and unrelated eventualities. One of the advantages of the proposed analysis is to show that they represent two snapshots of the same dynamic. Emergent configurations may give rise, through translation and successive alignments, to consolidated configurations. The reverse is also true. Aligned networks are liable to unravel and cede to emergent networks. One departure point would be, for example, a situation where actors, still prisoners of established networks, attempt to break out of the web of relationships in which they are locked. At this point the task of reconfiguration—made up of trial and error and learning—if successful, will result in the construction of a common space of competences and objectives through the transformation and progressive adaptation of actors, knowledge, and technical mechanisms. The story of Edison, as told by Hughes (1983), is a good illustration of the dynamics of these first moments. Numerous studies conducted by the Centre for the Sociology of Innovation (CSI) in the 1980s led to the design of tools and concepts to account for emergent situations and their progressive consolidation around what is now termed translation sociology or actor-network sociology (Akrich, Callon, and Latour 1988).[11] Evolution is not targeted once and for all: reconfigurations, for example rebirths, are possible, just as regressions and setbacks are foreseeable (MacKenzie and Spinardi 1995). As a result, intermediate configurations, neither completely stable nor completely emergent, are the most common.

Moreover, it would be a serious mistake to believe that there is a

11. Consolidated configurations are those that correspond to a socioeconomic world that is formed and predictable, the influence of technical and scientific consolidation being essential to understand and explain such developments.

TABLE 9.1. The Main Features of Emergent and Consolidated Networks

	Emergent Configurations	Consolidated Configurations
Knowledge	• statements + instruments + embodied skills	• statements are information because embodied competences are duplicated
	• nonsubstitutability between codified knowledge and embodied knowledge	• codified knowledge and embodied knowledge are relatively substitutable
	• knowledge is private: rivalry and excludability	• knowledge is public—i.e., nonrival, nonexclusive—in the networks in which it circulates
	• knowledge replication = laboratory replication	• knowledge replication = coding and replicating strings of symbols
	• local knowledge is generalized through successive and costly translations/transportations	• the degree of universality of knowledge is measured by the length of networks
States of the world	• list and identity of social and natural entities in constant reconfiguration	• list and identity of social and natural entities are known
	• states of the world revealed, ex post, through trials and interactions	• all states of the world are known ex ante and the probability of their occurrence can be calculated
	• uncertain and vague knowledge uses (this depends on the scope and state of the translations; they become standardized and stabilized along with the networks)	• uses of knowledge are predictable; they are more soundly established when there are multiple connections (principle of network externalities)
Modalities of action	• programs only exist ex post, as the outcome of action	• research and innovation programs (list of problems to solve and operations to accomplish to reach a solution) are defined ex ante and provide a framework for action (coordination)
	• mutual learning	• rational expectations
	• cooperation is an obligatory passage point for action, i.e., for translating identities and interests and for negotiating the content of knowledge	• cooperation is a strategy for cost and risk sharing or for consolidation of power positions

natural cycle that obliges a network to pass from an emergent to an aligned situation. Such cycles, like product cycles, are possible, as will be shown below,[12] but on no account should this be seen as a necessary and inevitable trajectory. As in a product cycle, it is possible to recommence at any stage. Dematurity is nearly as frequent as final decline (Abernathy and Clark 1985). This analytical framework supports the notion that the forces that push toward emergence or alignment act at every moment and decide on the prevalence, at a given time, of one configuration over the other.

This perspective goes beyond the normal opposition of static and dynamic analyses. It is more exact to consider that situations termed stable or consolidated are merely a moment in a particular dynamic process.

Emergent and Consolidated Networks Cross Institutional Boundaries

Emergent configurations and aligned configurations, as described and analysed, can both be found either in the marketplace or in the research world.

Examples have been given of emerging configurations for the production of scientific knowledge, considered as an activity in itself. Such situations are frequent—and possibly increasingly frequent—in markets where the good exchanged is a service. It is sufficient, and very common, for the content of the service, the conditions of its production and delivery, and the exact identity of its beneficiaries to progressively be negotiated from the process of its conception through to realization. In this case, the indeterminate nature of the service, which stabilizes only in parallel with translation strategies, is accompanied by the indeterminate nature of the knowledge to be deployed, also in a gestative stage (Gadrey and DeBandt 1994, Sundbo 1998).

Competition among (so-called) programs can be observed equally well in the world of academic science and in economic markets. For such competition to occur—which depends on the existence of aligned configurations—it is sufficient that incentives take a form that favors appropriation: patenting of the first arrival or the attribution of the discovery. Two examples will suffice.

—The duel between Guillemin and Schally brought into evidence

12. For a striking example of evolution from an emerging configuration to a consolidated one, see Granovetter and McGuire (1998).

two nearly identical programs to solve the same problem: isolation of the hormone that releases tyrotropine (Latour and Woolgar 1979, Wade 1981). The objective and problem to be solved were clearly identified. The only minimal difference concerned the choice of substances selected to extract and purify growth hormones. The probability of success, within a given time frame, could be estimated as a function of the resources allocated. The first past the post will win the Nobel Prize. Such a duel was possible only because the competences and instruments had been standardized and replicated and because the problem statement had attained a high degree of codification and recognition.

—Competition between pharmaceutical companies to isolate active molecules for the treatment of a given illness can be analyzed in the same terms. Each firm has competences that are closely related to those of others. Programs are similar and, in some cases, identical. The objectives are clearly identified, and procedures and experimental protocol defined and formalized in advance. It is even possible to break down the main program into parallel or sequential subprograms. At certain stages options, which are also predictable, appear. A subprogram will be chosen in relation to the results obtained. The more aware the firms are that to discover two effective molecules they must test several thousand, the easier it is to estimate the probability of success with given sums of investment (Bonazzi 1993).[13] In addition, they very often share the same tools for managing project portfolios. Competing firms fight to arrive first: the duel is comparable to that between Schelly and Guillemin. The only difference, which is minimal, is that the reward is not a Nobel Prize but a patent and monopoly rents that allow control.[14]

Basic Research and Fundamental Research: Toward a New Definition

One of the classifications most commonly accepted is that which distinguishes between basic research (which produces basic, codified knowledge with general validity) on the one hand and on the other applied research (which utilizes basic knowledge and develops it with specific knowledge, with a view to solving particular problems).

Network analysis escapes from this overly simple and widely criti-

13. The aluminum industry is another good example. Its development stretches over more than one century. Firms share the same knowledge basis; competition leads to cartels and to successive mergers (Le Roux 1998).

14. For different tentative modelings of these patent races see Reinganum (1989).

cized epistemology. The definitions I would like to propose are based upon the distinction between emergent and aligned knowledge. The first is orphan, it develops away from existing competences and networks. The second is trapped in already consolidated networks. If knowledge that aims to be general is termed basic, in both cases the knowledge produced can be basic. Indeed, according to the definition given, knowledge is all the more general insofar as it links (is linked to) numerous sites, i.e., numerous nodes of the network. This connection can be instantaneous (this is the case for the many ongoing projects in electronics or chemistry, where a finding is rapidly diffused because the metrological infrastructure is ready). Conversely, it can be the outcome of a costly achievement that has not yet reached its conclusion (this is still the case in numerous sectors of biology).

Two different concepts are necessary to distinguish between these two forms of basic research. I will define basic research as all activities that aim to produce general knowledge in aligned and extended networks. From this point of view, Guillemin's research probably belongs to basic research. I would suggest reserving the term "fundamental research" for research carried out in emergent configurations. Its capacity to produce basic knowledge depends on its further ability to participate in the establishment of long, heterogeneous, and consolidated networks.

In aligned networks, basic research constantly interacts with applied research. The Kline and Rosenberg model offers one of the most accurate pictures of these complex interactions. In fundamental research, the dynamic of applications can be described only as a consequence of the dynamics of the corresponding emerging networks.

3. A Model of Strategic Interactions between Laboratories and Firms

This analysis provides a new basis for the study of the conditions of efficiency of market coordination concerning the production, distribution, and application of scientific knowledge activities. It makes it possible to link the strategies of actors (firms and laboratories) with the interaction of knowledge and economy. To illustrate this, the main elements of a very simple model will be outlined.

This model should allow us to suggest the dynamic viability of an economy characterized by the existence of tension between emergent and consolidated networks. As in any model, and perhaps even more so in this one, the speculative nature is obvious. But these limits must be qualified by two observations. The first stems from the fact that this model—admittedly, in a stylized way—accounts for the mechanisms described by many authors who also tend to agree on their progressive generalization: intensified economic competition triggers product innovation that pushes firms to cooperate more closely with users and academic laboratories (Lundvall and Nielsen 1999). The second is the possibility, afforded by such a model, to identify the main issues around which conceptions of public policies and, more broadly, framing devices favoring the viability of these arrangements, could be organized.

3.1 A Two-Sector Model

The model posits the existence of two sectors. The first sector corresponds to consolidated configurations and the second to emergent configurations.

The main characteristics of sector 1 are given in column 2 of table 9.1. By convention, we call firms (Fi) the agents of this sector. Firms deliver goods (products or services) to the competitive markets. An essential driver of competition in such markets is the ability to renew the supply of goods and services. There is therefore a strong incentive to create and exploit novelty. Producing the same thing in the same way is not very rewarding, at least not in the long run. Funding new and more efficient methods of production and introducing new and more attractive goods into the market is necessary for survival in these highly competitive markets.

Moreover, in line with the most recent developments of evolutionary economics, firms present in these markets are considered to be knowledge based. Given the hypothesis of an aligned and consolidated network, their competence bases are broadly identical. Such firms demonstrate a high degree of similarity. In particular, their R&D and innovation programs are highly similar, as are their management tools. In accordance with table 9.1, they are consequently prone to develop rational expectations. This competition between very similar firms favors the emergence of two contradictory mechanisms.

The first mechanism leads to a narrowing of the technological options and to an increase in the consolidation of aligned networks. The main reason for this is to be found in increasing returns. To understand this mechanism the contributions of the economics of technical change are again invaluable. A firm tends to produce and innovate in fields that it masters because small marginal efforts are immediately and certainly profitable. Similarly, consumers prefer tried and tested knowledge. This conspiracy of supply and demand leads to technological lock-in, which feeds on an ever more intensive and profitable use of controlled knowledge and know-how. The more an industry progresses, the more it has an interest in not changing the trajectory: it becomes a prisoner of the competency trap that it has systematically and intensively developed. Incremental innovations are the normal outcomes of this trend.

The second mechanism has the opposite effect. The more the network expands and is consolidated, the more the competency bases of firms become similar and R&D and innovation programs substitutable. The competition ends up being ruinous for all the competitors. Provided that some conditions are met (see below), firms might engage in differentiation strategies. Each firm seeks to avoid confrontation and the potential risk of failure (winner-takes-all game) by building a unique demand and controlling it in such a way that it arrives at a monopoly situation. Among all possible differentiation strategies, the model favors customization strategies (made to measure on demand), which presuppose a high reactive capability if the lead is to be maintained. Micromarkets are thus built in strong cooperation with potential consumers, and are managed by each firm, which tries to take a lead over its competitors (Callon forthcoming). These differentiation strategies can be observed in many industries, notably in the service sector, and are powerful levers for reinvigorating emergent processes to explore new technological and productive combinations.

Consequently, market effects are potentially twofold. Whereas on the one hand the market presupposes a stable world that it helps to feed and reinforce, on the other hand, through its dynamic, it may undo previous investments, at least partially, to put new life in some types of emergent knowledge and reconfigurations.

We shall now consider the second sector (configuration 1 of table 9.1), characterized by emergent configurations. It brings together actors—that by convention we call laboratories (Li) (they may be something other than research centers as such)—which conduct fundamental research activities (in the sense given in section II.4).

To survive laboratories are doomed to devise and develop translation strategies to convince other actors (in this model it is irrelevant whether these are firms from the first sector or laboratories contemplating related research) to support and develop cooperation. The construction of these alliances presupposes lengthy negotiations and continuous relationships without which no adjustment is possible. Translation strategies oscillate between two extreme outcomes in which each laboratory, Li, attempts to find a compromise. The first outcome corresponds to the setting up of specific and narrow links between a small number of partners, an extreme configuration being that where the relationships of Li are limited to interesting one single Li or Fi. The second outcome, a mirror image of the first, corresponds to the maximal widening of the scope of alliances: Li allies with a large number of Lj and/or Fi. In reality, configurations are intermediary because these two strategic components are always combined. To interest and to translate require continuous and constant relationships that naturally restrict the scope of alliances. But risks of becoming trapped (to underscore the parallel with sector 1 we could talk of a trust trap, which is the counterpart of a competency trap) in a simple bilateral link push actors to diversify their translation strategies and to open the field of alliances. This propensity to broaden the scope of cooperation is more likely to be strong when numerous Li compete for enrolling and translating new allies.

3.2 Dynamics of the Model

The dynamic of the model is based on a fundamental strategic mechanism that can be captured in one sentence: the differentiation strategies of firms in their markets are stimulated by the existence of a wide range of possible cooperations with laboratories seeking on their own side for alliances.

Firm-Laboratory Cooperation as a Driver of Competition

Each firm attempts to capture demand to its own advantage by personalizing and coconstructing it with consumers. The main competitive advantage consists of being capable of rapidly reacting and deploying new resources to redesign the supply of products or services. This continuous reformulation of supply and demand, through which a firm captures a market from its competitors, occurs mainly through the setting up of cooperation and alliances with laboratories engaged in emerging configurations. The world of possible new options, tracks, trajectories, and diversification is thus enlarged. Firm Fi, in competition with firm Fj in the same market, will be more likely to differentiate its goods by committing itself to a policy of customization, if there is a possibility of an alliance with laboratories Li, where Li master competences and know-how that are sufficiently different from its own to allow the opening up of new technological options, but sufficiently similar to make communication possible at a reasonable cost. The existence of a reserve of cooperation and alliances with emergent actors is therefore seen as a threat.

Each firm has an interest in overtaking its competitors by sealing alliances with laboratories and thus avoiding the reconstruction of the end demand by a rival that was quicker to enroll the new scientific and technological resources.

This capacity that firms have to ally themselves with Li implies that they are organizations that we could agree to call dual. They have to be engaged in activities in which they simultaneously exploit existing trajectories and explore new ones. The temporary compromises reached establish a trade-off between openness, unexpected events, and new connections on the one hand and on the other closeness, known trajectories, and the reinforcement of links.[15] It is this creative tension that enables us to talk of learning organizations. Increasing competition is a major factor pushing firms toward learning and networking organizations. Such organizations are characterized by extensive delegation of responsibilities, cross-occupational working groups, quality circles, and the integration of different functions. In this model the firm is no longer reduced to a black box; its internal organization is taken into consideration and suggests that it be described as a network firm.

On their side laboratories, to continue to exist, are interested in these partnerships and thus contribute to the credibility of the threat. As seen above they are also caught in a form of competition forcing them to look for alliances, that is, to interest other agents in order to extend emergent networks. This twofold threat counteracts, on the sector 1 side, the tendency toward lock-in and the production of consolidated trajectories and, on the sector 2 side, trust trapping. The virtuous circle of differentiation begins to square the forces of alignment.

Just as dual firms are adapted to the establishment of relations with Li, so laboratories with diversified activities are the most prone to set up cooperation in order to extend their networks. Paradoxically, and despite the boom of laboratory studies in the eighties (Latour and Woolgar 1979, Knorr 1981, Lynch 1985), knowledge of the different forms and profiles of laboratories has made little progress. For want of better, we shall refer here to the model known as the "compass card" (Callon, Larédo, and Mustar 1997), or to concepts such as the Knowledge Value Alliance (Bozeman and Rogers forthcoming). The more a research collective develops diversified activities (production and circu-

15. This strategic conception is compatible with the view of the firm as a processor of knowledge, which manages a number of tensions between codified and embodied knowledge (Nonaka and Takeuchi 1995) or between different interests and logics.

lation of certified knowledge, participation in the production of competitive advantages, conception and delivery of collective goods, training and incorporated knowledge, expertise and public understanding of science, etc.), the more it is connected to diverse actors, and the more easily it will be able to manage basic research and fundamental research strategies by establishing trade-offs between the two.[16] Management of such research collectives implies the establishment of procedures and forms of organization that facilitate cross-fertilization and coordination between the different types of activity. Researchers managing such laboratories need to be real researcher-entrepreneurs (Laredo and Mustar 2000).

Framing Devices: The Role of the Public Powers

The proposed model suggests the dynamic viability of such a type of two-sector economy. The conditions of this viability are drawn largely from a body of empirical and theoretical work on the development of new forms of interaction between academic research and economic markets. In other words, the idea was to show that, through the invention of new practices, the actors define new forms of coordination that provide for a compromise between exploitation and exploration. Thus, an equilibrium is found between the renewal and the consolidation of the world of goods proposed to consumers.

But actors are not content merely to invent practices. They also strive to define and implement framing devices that enable these practices to emerge, to last and to spread. This is where a specific actor, absent until now, must come in, i.e., the public authorities. Governments might play a central role in regulating interaction between firms and laboratories. Their intervention contributes to the constitution of these relations. In other words, the state must not be seen as an external player; it is caught in this process of inventing and consolidating new types of practices and relations (Block 1994, Stark and Bruszt 1998). Here again, the proposed model aims at clarifying modes of intervention conceived and tested to guarantee the viability of these dynamics. In other words, what we wish to suggest is that the actors themselves set up devices that, by trial and error, can be used to reframe overflowing. The role of socioeconomists is to sort through these reframing devices, to identify the most effective ones, and to clarify the role of the state in their possible generalization.

16. Bozeman uses an empirical study to show that the complexity of relations does not prevent a laboratory from implementing a basic research strategy (Bozeman and Rogers forthcoming).

This work can only be preliminary and partial. It consists of illustrating the possibility of an approach, and not of producing an exhaustive inventory. One way of proceeding is to distinguish these tentative interventions according to their target: the consolidated sector, the emergent sector, and relationships between sectors. For each of these targets we give a few examples of intervention that, from our point of view, helps to frame the agents, that is, to set up devices, procedures, and decision-making tools that favor the dynamics of networking and bonding between the two sectors while ensuring their own autonomy.

The consolidated sector. We have seen that one of the motivations of research partnerships with laboratories in sector 2 was the development by firms of customization strategies, involving active customer participation. Modalities of intervention aimed at the stimulation of the coproduction of demands by firms and users are numerous.

A first set concerns the stimulation of contact between firms and consumers, so that the latter are closely associated with the conception of products and services intended for their use. Several empirical studies show that quality approaches constitute a powerful tool for managing this relationship. These include a range of measures that allow the expression of demand outside the market and, once visible, can direct a firm's strategy. More generally, devices that enable consumers to voice their demands and opinions (Hirschman 1970) favor this coproduction. We need to mention here a set of procedures that, on controversial subjects such as food safety, waste, or new drugs produced by biotechnology, involve concerned groups and laypeople in the discussion on the technical options and sometimes even on the choice of research objectives (Davison, Barns, and Schibeci 1997, Rabeharisoa and Callon 1998). The market is embedded in hybrid forums, public debates on technoscience, on the realm of products and services, and on the organization of markets. Everything that reinforces this embeddedness pushes firms to seek for new complementary assets and consequently to enter into partnership strategies with laboratories in the emergent sector.

More traditionally, property rights are another lever of choice, even if their potential is sometimes analyzed too simplistically. In a case where the patent races are symmetric (firms are identical and there is no competitive asymmetry) and where complete appropriation exists, players have an interest in developing different programs. Differentiation from the supply side is thus promoted. To favor the translation of this differentiation into a customization policy, industrial property could be conceived as being both strong and narrow. In this case, arriving second in the patent race means losing a large part of the investments plowed into R&D (even if certain intermediate results can be reused). Firms would be prompted to protect themselves by developing, and keeping control of, specific resources such as privileged relationships with their customers. This sometimes leads them to develop original lines of research and, in order to do so, to enter into partnerships with laboratories.

Interventions intended to favor certain forms of organization and governance can also contribute toward the framing of firms' behavior. The key concept here is the network firm. Many studies would be needed to clarify the different forms of state intervention and their effects. The work of Dobbin (1994) and of Gao (1998) proves, however, that such elucidation is possible.

The emergent sector. As regards sector 2 and interventions aimed at acting on the emergent strategies of laboratories, the government might have a role to play in supporting diversity, the very existence of which leads firms to depart from trajectories to which they were committed, in order to explore new tracks.

Financial support for those laboratories that leave aligned networks in order to experiment with new configurations is clearly one type of intervention. This support can take two extreme forms. The first is selective: the projects proposed by laboratories are examined by the funder against a criterion focused on the originality of the proposal. The second is automatic: funds are granted with the only condition being the development of emergent strategies (the amount and periodicity of these resources depend on the degree of originality of the research topics, which can only be assessed ex post). As has often been noted, one way of ensuring that such support is widespread is to involve researchers in teaching tasks. The more regular these tasks the more sure the support. Arrow is, to my knowledge, the first author to have developed this idea, which has had wide appeal. The recent empirical work of Geuna shows how valid this intuition is: the oldest European universities that have a solidly established tradition of teaching are those that are most effective at fundamental research (Geuna 1999). Teaching is not the only possible solution. If we revert to the compass card, each of the other forms of activity (expertise, etc.) may constitute, for both individuals and research collectives, a way of reducing risks associated with emergent strategies.

Moreover, a laboratory will undertake an emergent strategy only if there is some prospect of forming alliances and initiating cooperation, i.e., if translation strategies, which are always costly and difficult, are made easier. The basic mechanism is that of disclosure and circulation. In an emergent context knowledge is rival, appropriable, and local. As has been shown, interaction will be possible only if there are mechanisms to facilitate the knowledge breaking out of its isolation. The main measure is that which encourages publication—in the form of widely accessible articles—, conferences, or secondment of research scientists and technicians. Without such framing measures the visibility of emergent strategies is zero and no translation can be envisaged. In this case publication, it should be recalled, entails no risk of loss of knowledge, but is a signal on which to base the first contacts (Hicks 1995, Hicks and Katz 1997). Any measures aimed at multiplying personal interactions, especially those that develop in professional associations, or facilitate the mobility of persons, instruments, or materials, will also have a positive impact by creating bases for the establishment of relations.

Relationships between sectors. Let us consider as given a dynamic emerging sector (a large number of laboratories engaged in emergent strategies made visible by publication) and a consolidated sector where firms are strongly encouraged to differentiate and customize. If a virtuous circle is to be attained, it is necessary for actors from both sectors to be directly encouraged to form alliances and enter into cooperation. A few examples follow of public intervention that helps to constitute a framework favoring such alliances.

A firm that wishes to ally itself with a laboratory should at least be able to capture the signals emitted by it, i.e., it should have access to the laboratory's seminars and to the articles written by its researchers. Even better, it should be able to receive some of these scientists in its own research centers. This presupposes a sort of hybridization of firms, i.e., the integration by those firms of a core group of researchers playing the game of emergent configurations (publications, conferences, secondments), and thus becoming what Guston (2000) suggests calling "boundary organizations." For such hybrid firms to be mobilized in the dynamic of translation strategies, it is desirable that they be visible. Among other things, this implies that (and explains why) industrial firms publish in the best journals (publication is a credible signal of a willingness to cooperate and an indication of possible topics for such collaboration—see Hicks 1995 for a complete and cogent development of these points), employ quality research staff linked to, or coming from, public laboratories, and have researchers who participate in professional associations and public conferences where they have the opportunity of meeting representatives of laboratories.[17]

As far as the laboratories are concerned, a symmetrical approach can be adopted. A laboratory will try to increase the diversification of its activities if its funding is not guaranteed 100 percent by public sources, whether these are periodically reviewed or automatic. The laboratory will, in such cases, be obliged to find external resources and will naturally prospect for partners and for a mutual translation of interests. Pockets of basic research within a laboratory that has invested primarily in fundamental research, guarantee interessement strategies toward firms: by promoting the constitution of networks likely to be mobilized, they expose researchers to regular contact with enterprise. In addition, we need to mention all the incentives and procedures that consolidate and extend translation strategies developed by laboratories, for example the creation of firms by researchers from laboratories, which, as Mustar (1998) has shown, feed the existence and reinforce connections between the two sectors.

More generally, what is at stake is the constitution and development of a hybrid environment, a chain of intermediaries between the two sectors, between laboratories and firms. The biotechnology field and, in particular, that of genetic therapy is a good example of such a new social world (Powell and Brantley 1993). The emergence and growth of firms that carry out research upstream, such as the identification and localization of genes, epitomizes this evolution (Martin 1999). In this case a new type of boundary organization is emerging that can appropriately be called firm-laboratories. This emergence draws attention to two types of questions: first, of intellectual property rights, which are highly complex because they have to be reconciled both with protection as such and with the negotiating capacity they allow between agents in search of complementary assets (Bessy and Brousseau 1997); and second, of the financing of high-risk activities shrouded in uncertainty. The existence of a market like NASDAQ, based on more appropriate evaluations of profitability, constitutes a powerful mechanism for the survival of these firm-laboratories and one that is probably more effective than innova-

17. Moreover, any intervention aimed at facilitating the involvement of firms in fundamental research or their cooperation with academic laboratories (like CIFRE grants in France or CASE grants in the UK) contributes to the emergence of a virtuous cycle of collaboration between firms and laboratories.

tion fundings such as those distributed by ANVAR, a state agency in France.[18]

4. Concluding Remarks

In this paper we have tried to illustrate the relevance of an approach based on the explicit recognition of the performative dimension of economics in general and, more particularly, of economic sociology. The case selected, coordination between scientific research activities and economic markets, is from this point of view particularly interesting. The so-called Cold War institutions were put into place in the immediate aftermath of the Second World War and were both described and justified by economics and sociology. In this way the autonomy of academic research, with its modes of operation, and the linear model that describes the mode of diffusion of knowledge produced in the economic sphere, were reinforced and legitimized. Despite its strong institutionalization, this configuration did not manage entirely to frame the different agents and to contain overflowing. Thus, overflowing continued, proliferated, and, with the upsurge of certain sectors such as information technology and biotechnology, even became the rule. Existing frames consequently became restrictive and unsuitable. This led actors to conceive of new framing devices that favor the constitution and extension of complex networks of interaction. Faced with the dramatic increase in this overflowing, economists and sociologists have adopted two attitudes. Some, who place themselves under the flag of the new economics of science, have tried to prove the undesirable effects of this overflowing and, with reference to the work of Arrow and Nelson, have proposed, more or less explicitly, that we revert to Cold War institutions. One of their arguments is that without the existence of an open science, the social efficiency of research may well be undermined. Others stress and encourage the irresistible and irrevocable nature of this evolution, but to date no reasoned argument has been put forth to show the viability of this new configuration and provide elements of analysis that can both demonstrate its possible effectiveness and identify the framing devices supporting it.

In this text we have tried to progress in these two directions. Starting with concepts formulated by the new economics of science, we have

18. We should also consider procedures or devices (for example, stock options) intended to increase the capacity by such firm-laboratories to attract highly qualified engineers or researchers.

shown that the three characteristics making it possible to describe the economic status of the different forms of scientific knowledge are not intrinsic but extrinsic. More precisely, the degree of rivalry, exclusivity, and generality is closely correlated to the form and state of the networks concerned. On the basis of a set of empirical and theoretical studies, we have suggested distinguishing between two forms of network: emergent and consolidated. This distinction is at the base of a model that establishes the possibility of connections between fundamental research and the economic market, without denying the autonomy of these two types of activity. On the contrary, it is because these two sectors exist independently that the economic dynamic founded on the connection between them can be envisaged. At the same time, relying upon the observation and analysis of devices conceived and set up by actors to manage this dynamic productively, we have suggested ways to stimulate and simultaneously to frame this overflowing, and to transform it into legitimate strategies. All in all, we have shown that it is possible to reconcile these interactions with the maintenance of autonomous research. But instead of being defined in abstract terms by the general, codified nature of the knowledge produced, this research, which is also oriented toward the disclosure of its results, is defined by its capacity to spawn new networks and trajectories. In parallel, the model proposed, by extending the actors' experiments, indicates under what conditions such a dynamic can be maintained. Having contributed to the endless work of framing actors overflowing we have looped the loop presented in the introduction.

REFERENCES

Abernathy, W., and K. Clark. 1985. Innovation. Mapping the Winds of Creative Destruction. *Research Policy* 14: 3–22.

Akrich, M., M. Callon, and B. Latour. 1988. A quoi tient le succès des innovations? 1. L'art de l'intéressement; 2. Le choix des bons porte parole. *Annales des Mines, Gérer et Comprendre* 11: 4–17; 12: 14–29.

Aman, K., and K. Knorr-Cetina. 1988. Thinking through Talk. An Ethnographic Study of a Molecular Biology Laboratory. In L. Hargens, R. A. Jones, and A. Pickering, ed. *Knowledge and Society. Studies in the Sociology of Science Past and Present.* Greenwich, CT, and London: Jai Press.

Arrow, K. J. 1962. Economic Welfare and the Allocation of Resources for Inventions. In R. R. Nelson, ed. *The Rate and Direction of Inventive Activity. Economic and Social Factors.* Princeton: Princeton University Press.

Bessy, C., and E. Brousseau. 1997. Brevet, protection et diffusion des connaissances. Une relecture néoinstitutionnelle des propriétés de la règle de droit. *Revue d'Economie Industrielle* 79: 233–54.

Block, F. 1994. The Role of the State in the Economy. In N. J. Smelser and R. Swedberg, ed. *The Handbook of Economic Sociology.* Princeton: Princeton University Press.

Bonazzi, C. 1993. *R&D industrielle et concurrence. Le cas de l'agrochimie et de la pharmacie.* Working paper. Ecole des mines de Paris, CERNA.

Bozeman, B., and J. Rogers. Forthcoming. A Churn Model of Scientific Knowledge Value. The Non-economics of Science. *Research Policy.*

Callon, M. 1986. Some Elements for a Sociology of Translation. Domestication of the Scallops and the Fishermen of St. Brieuc Bay. In J. Law, ed. *Power, Action, and Belief. A New Sociology of Knowledge?* London and Boston: Routledge and Kegan.

———. 1987. Society in the Making. The Study of Technology as a Tool for Sociological Analysis. In W. Bijker, T. Hughes, and T. Pinch, ed. *New Directions in the Social Studies of Technology.* Cambridge: MIT Press.

———. 1992. The Dynamics of Techno-economic Networks. In R. Coombs, P. Saviotti, and V. Walsh, ed. *Technological Change and Company Strategies.* London: Academic Press Limited.

———. 1994a. Four Models for the Dynamics of Science. In S. Jasanoff, G. E. Markle, J. C. Petersen, and T. Pinch, ed. *Handbook of Science and Technology Studies.* London: Sage.

———. 1994b. Is Science a Public Good? *Science, Technology, and Human Values* 19 (4): 395–424.

———. 1998a. An Essay on Framing and Overflowing. Economic Externalities. In M. Callon, ed. *The Laws of the Markets.* Oxford: Blackwell.

———, ed. 1998b. *The Laws of the Markets.* London: Blackwell.

———. Forthcoming. The Economy of Qualities. *Economy and Society.*

Callon, M., P. Larédo, and P. Mustar, ed. 1997. *The Strategic Management of Research and Technology.* Paris: Economica International-Brookings Institution Press.

Cambrosio, A., and P. Keating. 1995. *Exquisite Specificity. The Monoclonal Antibody Revolution.* Oxford: Oxford University Press.

Caracostas, P., and U. Muldor. 1997. *La société, ultime frontière. Une vision européenne des politiques de recherche et d'innovation pour le XXIème siècle.* Luxembourg: Office des publications officielles des Communautés Européennes.

Cohendet, P., and P. Llerena. 1999. La conception de la firme comme processeur de connaissances. *Revue d'Economie Industrielle* 88: 211–36.

Collins, H. M. 1974. The TEA Set. Tacit Knowledge and Scientific Networks. *Science Studies* 4: 165–86.

———. 1985. *Changing Order. Replication and Induction in Scientific Practice.* London and Los Angeles: Sage.

———. 1990. *Artificial Experts. Social Knowledge and Intelligent Machine.* Cambridge: MIT Press.

———. 1995. Humans, Machines, and the Structure of Knowledge. *Stanford Humanities Review* 4 (2): 67–83.

Coriat, B., and O. Weinstein. 1995. *Les nouvelles théories de l'entreprise.* Paris: Le livre de poche.

Dasgupta, P., and P. David. 1994. Toward a New Economics of Science. *Research Policy* 23 (5): 487–521.

David, P., and D. Foray. 1996. Accessing and Expanding the Science and Technology Knowledge Base. *STI Review* 16: 13–68.

Davison, A., J. Barns, and R. Schibeci. 1997. Problematic Publics. A Critical Review of Survey of Public Attitudes to Biotechnology. *Science, Technology, and Human Values* 22 (3): 317–48.

DeBresson, C., and F. Amesse. 1991. Networks of Innovators. A Review and an Introduction to the Issue. *Research Policy* 20 (5): 363–80.

DiMaggio, P., and H. Louch. 1998. Socially Embedded Consumer Transactions. For What Kinds of Purchases Do People Most Often Use Networks? *American Sociological Review* 63 (October): 619–37.

Dobbin, F. 1994. *Forging Industrial Policy. The United States, Britain, and France in the Railway Age.* Cambridge: Cambridge University Press.

Etzkowitz, H., and L. Leydesdorff, ed. 1997. *Universities in the Global Knowledge Economy. A Triple Helix of Academic-Industry Relations.* Albany: State University of New York Press.

Gadrey, J., and J. DeBandt, ed. 1994. *Relations de services, marchés des services.* Paris: Editions du CNRS.

Gao, B. 1998. Efficiency, Culture, and Politics. The Transformation of Japanese Management in 1946–1966. In M. Callon, ed. *The Laws of the Markets.* Oxford: Blackwell.

Geuna, A. 1999. *The Economics of Knowledge Production. Funding and the Structure of University Research.* Cheltenham and Northampton, MA: Edward Elgar.

Gibbons, M., C. Limoges, H. Nowotny, S. Schwartzman, P. Scott, and M. Trow. 1994. *The New Production of Knowledge.* London and Thousand Oaks, CA: Sage Publications.

Granovetter, M., and P. McGuire. 1998. The Making of an Industry. Electricity in the United States. In M. Callon, ed. *The Laws of the Markets.* Oxford: Blackwell.

Guston, D. 2000. *Between Politics and Science. Assuring the Integrity and Productivity of Research.* Cambridge: Cambridge University Press.

Hicks, D. 1995. Published Papers, Tacit Competencies and Corporate Management of the Public/Private Character of Knowledge. *Industrial and Corporate Change* 4 (2): 401–24.

Hicks, D., and S. Katz. 1997. A National Research Network Viewed from an Industrial Perspective. *Revue d'Economie Industrielle* 79: 129–42.

Hirschman, A. 1970. *Exit, Voice. and Loyalty. Responses to Decline in Firms, Organization. and States.* Cambridge: Harvard University Press.

Hughes, T. P. 1983. *Networks of Power. Electric Supply Systems in the U.S., England, and Germany, 1880–1930.* Baltimore: John Hopkins University Press.

Kline, S., and N. Rosenberg. 1986. An Overview of Innovation. In R. Landau and N. Rosenberg, ed. *The Positive Sum Strategy. Harnessing Technology for Economic Growth.* Washington: National Academy Press.

Knorr, K. 1981. *The Manufacture of Knowledge. An Essay on the Constructivist and Contextual Nature of Science.* Oxford: Pergamon Press.

Lane, D., F. Malerba, R. Maxfield, and L. Orsenigo. 1996. Choice and Action. *Journal of Evolutionary Economics* 6: 43–76.

Laredo, P., and P. Mustar. 2000. Laboratory Activity Profiles. An Exploratory Approach. *Scientometrics* 47 (3): 515–39.

Latour, B. 1987. *Science in Action. How to Follow Scientists and Engineers through Society.* Cambridge: Harvard University Press.

———. 1992. *Aramis, or the Love of Technology.* Cambridge: Harvard University Press.

———. 1999. *Pandora's Hope. Essays on the Reality of Science Studies.* Cambridge: Harvard University Press.

Latour, B., and S. Woolgar. 1979. *Laboratory Life. The Construction of Scientific Facts.* Los Angeles: Sage.

Le Roux, M. 1998. *L'entreprise et la recherche. Un siècle de recherche industrielle à Péchiney.* Paris: Editions du CNRS.

Lundvall, B. A., and P. Nielsen. 1999. Competition and Transformation in the Learning Economy. *Revue d'Economie Industrielle* 88: 67–90.

Lynch, M. 1985. *Art and Artifact in Laboratory Science. A Study of Shop Work and Shop Talk in a Research Laboratory.* London: Routledge.

MacKenzie, D. 1990. *Inventing Accuracy. A Historical Sociology of Nuclear Missile Guidance System.* Cambridge: MIT Press.

MacKenzie, D., and G. Spinardi. 1995. Tacit Knowledge, Weapons Design, and the Uninvention of Nuclear Weapons. *American Journal of Sociology* 1 (July): 44–99.

Mallard, A. 1998. Compare, Standardize, and Settle Agreement. On Some Usual Metrological Problems. *Social Studies of Science* 28 (4): 571–602.

Martin, P. 1999. Gene as Drugs. The Social Shaping of Gene Therapy and the Reconstruction of Gene Disease. *Sociology of Health and Illness* 21 (5): 517–38.

Mowery, D. 1999. *The Changing Structure of the US National Innovation System.* Mimeo.

Mustar, P. 1998. Partnerships, Configurations, and Dynamics in the Creation and Development of SMEs by Researchers. *Industry and Higher Education* 12 (4): 217–21.

Nelson, R. 1959. The Simple Economics of Basic Scientific Research. *Journal of Political Economy* 67: 297–306.

Nonaka, I., and H. Takeuchi. 1995. *The Knowledge Creating Company.* New York: Oxford University Press.

Pestre, D. 1997. La production des savoirs entre académies et marchés. Une relecture historique du livre "The New Production of Knowledge." *Revue d'Economie Industrielle* 79: 163–74.

Powell, W. W., and P. Brantley. 1993. Competitive Cooperation in Biotechnology. Learning through Networks? In N. Nohria and R. Eccles, ed. *Networks and Organizations.* Cambridge: Harvard Business School Press.

Power, M., ed. 1994. *Science and Economic Calculation* (special issue of *Science in Context,* vol. 7, no. 3).

Rabeharisoa, V., and M. Callon. 1998. L'implication des malades dans les ac-

tivités de recherche soutenues par l'Association Française contre les Myopathies. *Sciences Sociales et Santé* 16 (3): 41–65.
Reinganum, J. 1989. The Timing of Innovation. Research, Development, and Diffusion. In R. Schmalensee and R. D. Willig, ed. *Handbook of Industrial Organization.* Amsterdam: North-Holland.
Romer, P. M. 1992. Two Strategies for Economic Development. Using Ideas and Producing Ideas. In *Proceedings of the World Bank Annual Conference on Development Economics.*
Sims, B. 1999. Concrete Practices. *Social Studies of Science* 29 (4): 483–518.
Stark, D., and L. Bruszt. 1998. *Postsocialist Pathways. Transforming Politics and Property in East Central Europe.* Cambridge: Cambridge University Press.
Stephan, P. 1996. The Economics of Science. *Journal of Economic Literature* 34 (September): 1199–235.
Stephan, P., and S. Levin. 1997. The Critical Importance of Careers in Collaborative Scientific Research. *Revue d'Economie Industrielle* 79: 45–62.
Sundbo, J. 1998. *The Organisation of Innovation in Services.* Frederiksberg: Roskilde University Press.
Wade, N. 1981. *The Nobel Duel.* New York: Anchor; Paris: Messinger.
Warwick, A. 1992–93. Cambridge Mathematics and Cavendish Physics: Cunningham, Campbell, and Einstein's Relativity 1905–1911. I: The Uses of History. *Studies in the History and Philosophy of Science* 23 (4): 625–56. II: Comparing Traditions in Cambridge Physics. *Studies in the History and Philosophy of Science* 24 (1): 1–25.
Zelizer, V. 1998a. How Do We Know Whether a Monetary Transaction Is a Gift, an Entitlement, or Compensation? In A. Ben-Ner and L. Butterman, ed. *Economics, Values, and Organization.* New York: Cambridge University Press.
———. 1998b. The Proliferation of Social Currencies. In M. Callon, ed. *The Laws of the Markets.* Oxford: Blackwell.

10

The Microeconomics of Academic Science

John Ziman

ABSTRACT: The macroeconomics of science is turning out to be problematic. Perhaps more attention should be paid to its microeconomics. Academic research can be represented as an interlocking system of "markets," some financial, some notional. Thus, universities compete for human and material resources to support specialized research entities. Research entities compete for research projects. Projects produce publications that compete for citations in the marketplace of ideas. Through their contributions to knowledge, researchers gain recognition and reputation, which they trade for academic posts or entrepreneurial profits. And so on.

Each of these "markets" has its own characteristic microeconomic rationale. One way of testing them for consistency is to imagine them totally commercialized. Such "thought experiments," although clearly unrealistic in practice, throw light on the macroeconomics of science and reveal many features that call for more sophisticated modes of economic analysis.

A Macroeconomic Monster

Ever since the discovery that science had a *politics,*[1] it has had to have an *economics.* Mostly this has been focused on the macro level. Numerous studies have been made of the overall expenditures of nations and firms on various broad categories of scientific activity. But as this volume shows, the macroeconomics of science is not really thriving. In the language of genetics, it is something of a *monster.* It is a mosaic of expenditures on heterogeneous activities by heterogeneous organizations to achieve heterogeneous ends. Even as an exercise in accountancy, it does not produce meaningful aggregates[2] that articulate into an organic whole.

THIS IS NOT to deny that scientific research is an economic activity. It is pursued with determination by professional experts working

1. Nelson (1968).
2. Irvine, Martin, and Isard (1990); OTA (1986).

together in large organizations, at no little cost, to produce a valuable product—reliable knowledge. But in its archetypical mode—*academic* science[3]—it is a complex social institution that does not fit neatly into the standard categories of either the private or public sector.[4] The typical unit of activity in academic science is a research entity of a dozen or so researchers, under the leadership of a person holding an appointment in a university or research institute. An *academic research entity* (ARE) thus has much more operational autonomy than a subunit of a bureaucratic hierarchy, but much less freedom than a small firm. It contributes consciously to a global enterprise, but not according to any plan. It is engaged in fierce competition with similar entities, but cooperates closely with them in various communal facilities. It profits notably from successful achievement, but not in direct financial terms—and so on.

NEVERTHELESS, IT is not altogether fanciful to represent scientific activity in quasi-economic terms.[5] In large part, of course, this is no more than an elaborate metaphor. Most of the "commodities" and "currencies" that one would see as being "traded" among researchers, AREs, academic institutions, industrial firms, research councils, etc. are purely notional. Even so, this is a metaphor that has often been invoked to explain particular features of the research process. It is frequently remarked, for example, that researchers make "contributions to knowledge" in exchange for "recognition,"[6] that scientists vie for employment in "the academic marketplace,"[7] and that research claims compete for acceptance in the public marketplace of scientific ideas.[8]

Indeed, as I showed in my original paper, the quasi-economic metaphor can be extended systematically to represent academic science as a whole as an interlocking system of just such "markets." In spite of its obvious deficiencies, this representation does provide some valuable insights into how academic science works, and especially how it has been affected by changing organizational and financial practices.[9] The question to be taken up here is whether the metaphor can be further elaborated into a more realistic microeconomic *model.* Each of the

3. Ziman (1996, 2000a).
4. Stephan (1996).
5. Ziman (1991).
6. Hagstrom (1965); Merton (1973).
7. Caplow and McGee (1958).
8. Ziman (1968).
9. Ziman (1994).

interlocking "markets" has its own characteristic features—monopoly, monopsony, speculative uncertainty, indeterminate quality, and so on. On the one hand, can more be understood about the research process by analyzing it in terms of conventional economic theory? On the other hand, do these notional "markets" articulate into an entity with properties typical of an element of a macroeconomic system?

These are, of course, vast questions, far beyond the scope of this paper. But one possible way of testing the rationality of this sort of approach is to ask what would happen if the notional commodities and currencies in each "market" were *monetarized.* Let me emphasize that this mental exercise is undertaken here simply as a thought experiment, not as a realistic practical proposal. Any suggestion that, say, basic research projects should be financed entirely by the private sector,[10] presumably in terms of their present-value discounted future earnings, would bring into play a much wider range of other considerations than can be discussed here. I am uneasily aware that just such fantasies are entertained in some political circles, and even unwisely put into practice. Not only would I not like to give such ideologies any further leverage, I would say, rather, that this study brings to the surface, in detail and overall, those aspects of science which are not amenable to economic analysis, in spite of what it contributes to the work, health, and happiness of humanity.

Institutional Frames

Notice, however, that this analysis concentrates on *academic* science—that is, the mode of scientific activity traditionally performed under "academic" conditions, as in universities. This is by no means synonymous with "basic," "pure," or "fundamental" research,[11] although it is typically directed toward the production of knowledge as such, rather than toward its perceived uses. As I argue at length elsewhere,[12] this can be considered a distinctive social formation, with characteristic practices, norms, social roles, and organizational structures. This is not to say that academic science is materially, vocationally, or epistemically self-sufficient, or that it is any more culturally isolated than other significant social formations such as "law," "government," or "business." But its long and extraordinarily productive history makes it the prime stereo-

10. Kealey (1996); David (1997).
11. Ziman (1998).
12. Ziman (2000a).

type of "science" as an established social *institution,* quite apart from its societal influence.

One of the primary observations in the "science of science" is that although scientific activity is almost universally homogeneous at the laboratory level, it is remarkably diverse at the macropolitical level. This has to be allowed for in general political, economic, and institutional accounts of academic science. For the sake of simplicity, I concentrate almost entirely on the research system in the UK, with which I am most familiar, and which does in fact exemplify most of the features of interest in a microeconomic analysis. Indeed, it is very likely that even in countries such as France and Germany, where academic science is organized along civil service lines, overt "market forces" will soon bring to the surface the quasi-economic features concealed in their scientific bureaucracies, just as they have, more brutally, in the countries of the former socialist bloc.

This specification of the institutional frame is necessary because the word "science" is customarily used to cover a very wide range of epistemic practices and organizational structures. But some of these, such as "industrial science"—i.e., research and development performed systematically for explicit utilitarian purposes—are so closely linked with normal commercial organizations that they are already adapted to conventional market environments. Despite its important economic role in society at large, academic science has traditionally claimed to be independent of direct market forces. It is thus particularly instructive, both for economics and for science studies, to discover quasi-economic processes at work in such an ideologically hostile environment.

It must be said, however, that the "academic" and "industrial" modes of knowledge production now seem to be merging into a "postacademic" form of science.[13] But this is no reason for not engaging in the present study. On the contrary, any new perspective on traditional academic practices and norms could be especially valuable at a time when these are undergoing radical change. Those of us—and that means all of us—engaged in this restructuring process do need to have some faint idea of what we are doing, in advocating, accepting, or resisting such changes.

At this point, readers not familiar with current work in the field of science studies[14] may need to be warned that this "Mertonian" (or

13. Gibbons, Limoges, Nowotny, Schwartzmann, Scott, and Trow (1994).

14. Jasanoff, Markle, Petersen, and Pinch (1995).

perhaps "neo-Mertonian") account of science is much out of fashion. Nevertheless, this naturalistic viewpoint is extraordinarily effective both in revealing the "internal" structure of science as a social formation and in differentiating this structure from that of other cognate formations such as corporate industry. I am not here challenging more general accounts of the relationship between "science" and "markets," such as that they are two distinct, independently self-organizing, but symbiotic social formations. Nor am I concerned with showing (surprise! surprise!) that the detailed behavioral patterns of the individuals involved in both classes of social activity are governed by just the same types of personal strategies and rationalized affects as other members of the general culture in which they are immersed. What I am interested in here is academic science as a *social institution,* that is, an orderly collective where people relate to one another according to mutually understood customs, usages, conventions, rules, etc. Never mind (which begs the whole sociological enterprise!) how or why such regulative principles are established and maintained. It is enough to say that the social order among academic scientists is largely sustained by their knowing, as individuals, how to behave toward other scientists and what they may correspondingly expect of others in return.

On the other side of my putative equation, the notion of a "market" is of a characteristic institutional form, exemplified as fully in a peasant village as on the New York Stock Exchange. Thus, one might describe it as a structure in which a number of individual participants perform voluntarily the roles of "buyers" and "sellers," set "prices," "bargain" with one another, make "purchases," and so on, according to the local conventions. That is to say, a "market" is more specific than the very general social processes of "exchange" and "competition" that it enables in practice, but is not confined to the realms of high finance or lowly commerce.

A "market" is thus as much a sociological as an economic concept. Of course economists have acquired an immense amount of detailed knowledge about conventional commercial markets, but much of this knowledge would seem to an outsider to be too elaborate to be relevant metaphorically. Indeed, this preliminary exploration of this theme with a relatively inexpert conceptual armamentarium may reveal aspects that a trained academic economist would dismiss as highly speculative or outside the formal canon of their discipline. Thus, at some points, in order to understand what was really happening in the scientific world, I found it helpful to refer to themes such as "evolutionary economics" that are still regarded with extreme suspicion within the neoclassical mainstream.

There is always the risk, of course, of "discovering" just the features one has set out to find. It is remarkable how much standard economistic jargon, ranging from "imperfect markets" and "rent-seeking" to "potlatch ceremonies," one can project onto the very different world of scientific research. Nevertheless, I certainly did not start with a list of such terms, hoping to find them exemplified, but asked myself in each case whether a "market" interpretation was genuinely valid, in the sense of contributing to our understanding of what was going on. Indeed, many of the items on such a list would not be checked. Academic science, for example, has no analogue of a "merger," and does not benefit organizationally from "economies of scale." Again, as a "naive amateur" one is not subject to the specialist's temptation of seeing everything through the glass of their profession, as if the market paradigm could be stretched to cover every aspect of life and thought. Academic scientists, for example, noticeably participate in collective modes of action that are almost inexplicable in neoclassical economic terms.[15]

How might this analysis benefit our understanding of science as a "knowledge production" system? The key point is that the marketlike mechanisms at various points in the academic research process facilitate the operation of the Mertonian norms, and thus keep the whole system open, flexible, progressive, relatively impartial, and self-critical. As I argue at length elsewhere,[16] the nature of the knowledge produced by this system is closely bound up with its social structure. But although the efficient functioning and gearing of its various "market" mechanisms is vital to the way that academic science performs overall as an epistemic institution, these mechanisms have no direct epistemological significance in themselves. This analysis makes no contribution to the bubbling brew of realism, relativism, empiricism, constructivism, rationalism, cognitivism, reductionism, objectivism, naturalism, empathic humanism, etc., etc. of contemporary philosophies of science.

Vendors and Customers, Commodities and Currencies

Schematically, a *market* is a social institution for the systematic exchange of *commodities* for *currencies* between *vendors* and *customers.* In the "perfect" form idealized by economists, numerous independent

15. Etzioni (1988).
16. Ziman (2000a).

vendors offer nearly interchangeable commodities to equally numerous customers, with complete freedom on both sides to enter into mutually satisfactory pairwise bargains, transacted in a stable common currency. Both vendors and customers are assumed to know all that is relevant about the intrinsic value of the commodities, especially in relation to the prospects for their supply and demand—and so on. As everybody knows, these conditions are very difficult to realize in practice, so that most commercial markets are highly "imperfect" in a theoretical sense.

Let me reemphasize that this concept of a market is not solely "economic." It also includes the institutional boundary conditions, the social conventions, the legal regulations, the politico-economic context, in which it is embedded. It requires, for example, a procedure for establishing initial property rights and for enforcing the contracts by which commodities change hands. In other words, the value put upon a commodity is relative to these circumstances, as well as to the respective needs of buyers and sellers at the point of sale.

Nevertheless, an essential characteristic of a "market" transaction is that it should be felt to be "fair" by both parties, in the sense that they have both come to put the same valuation on what is exchanged. In a barter system, this valuation is clearly a cultural construct. But the anthropologists must eventually give way to the economists, who have developed much more sophisticated ideas of what constitutes a "fair" exchange. In the end, however, economic theory would be vacuous unless exchange values could be generalized, and expressed, if only implicitly, in terms of a common currency. If we are going to move from a quasi-economic metaphor to a microeconomic model, we are bound to assume that the seller unconsciously has in mind a certain sum of *money* for which he would make the commodity available, and that the truly voluntary buyer would be willing to give much the same sum to obtain possession of it. Indeed, the supersession of barter by the introduction of money as a general medium of exchange is rightly regarded as a major socioeconomic advance, so why should scientific activity be exempted from its benefits?

As I have remarked, however, academic research in a country such as the UK does not conform to a simple market model, with a uniform category of "commodities" being bought and sold for a single universal "currency." It actually seems to involve at least half a dozen interconnected "market systems," each working on a different level or in a different metaphorical domain. In each system, "commodities" are being exchanged for "currencies" by "vendors" and "customers." But even when these "commodities" and "currencies" are not at all like the goods and services, the cash and credits, of real commercial markets, they still seem to generate social phenomena that can be understood in "economic" terms.

The National Institutional Market

The most general system of market relationships in UK academic science is what we might call the *institutional market* that has developed recently to modulate the funding of higher educational institutions. In effect, universities are now expected to make a living by selling educational and research services. For two decades, all quasi-academic research establishments in the public sector have been officially cast in the role of research *contractors,* competing directly for part of their funding from governmental and industrial *customers.* This customer-contractor relationship, for all its deficiencies in practice, is now fully institutionalized and has been extended throughout the higher educational system.

A regular market system has thus emerged, where the vendors are universities, quasi-academic, quasi-nongovernmental establishments and private sector "consultants," and the customers are research councils, government departments, charitable foundations, commercial firms and (increasingly) international organizations such as the European Community. The commodity is essentially a technical service—that is, highly specialized scientific research—and the currency, directly or indirectly, is simply hard cash.

Since the vendor institutions are now very largely sustained (at least in their research activities) by the income they thus gain, the competition in this market is very fierce. A great deal of effort is put into preparing tempting proposals, negotiating the finer details of contracts, and monitoring their performance. Even though much of this effort comes to nothing, economic theory tells us that it is not wasted, since it has the effect of motivating effort, allocating resources optimally and greatly improving the quality of the work that does, in fact, get funded. It would be absurd to suppose, however, that a market dealing in such intangible commodities as "education" and "research" could ever approximate "perfection" in the economists' sense. Certain areas of the national "institutional agora" are very far from level. Universities are constrained, moreover, by historical circumstances and traditional practices that make it difficult for them to respond quickly to market opportunities. These constraints include, for example:

traditional employment practices such as tenure, personal autonomy in teaching and research,[17] and barriers to interdisciplinary transfers;

administrative segmentation into quasi-independent departmental "cost centres" and research entities;

complex arrangements to separate the respective shares of the customer, the institution, and its academic employees in the intellectual property rights arising from a research contract;

nonnegotiable contractual conditions, such as the amount payable for indirect costs, fixed overheads, and academic staff time, set by the government, which controls and subdivides the total budget for basic research and is its sole customer.

Thus despite the monetarization of the institutional marketplace, universities are not yet fully adapted to direct commercial competition. But they are learning fast. Economic theory teaches us that market forces are inexorable. Having been kicked out of the public sector into the harsh world of private enterprise, universities are bound eventually to converge, in form and substance, to the other market players.

Nevertheless, the institutional marketplace is not yet perfectly competitive in the classical sense. Much of the public funding of UK universities is channeled through a block grant mechanism based on a centralized system of evaluation of the past research performance of academic departments—which depends very largely on how well they are already funded for research. In effect, the state intervenes in the research marketplace to reinforce success and eliminate weakness. This direct instantiation of the "Matthew effect"[18]—to him that hath shall be given—must surely increase preexisting inequalities of research competence—or esteem, which is not quite the same thing—resulting eventually in the concentration of all academic science in a "superleague" of research universities.

In other words, this process introduces a major nonlinear term, requiring a path-dependent microeconomic analysis.[19] Notice, moreover,

17. Ziman (1992).
18. Merton (1973).
19. Arthur (1994).

that this practice is entirely contrary to the optimum strategy for funding education—that is, giving extra support to the weaker institutions to bring them up to an acceptable national standard.[20] Failure to differentiate the goals of these two strategies is the cause of much confusion in the economics and politics of higher education.

Project Markets

A close scrutiny of the institutional market immediately reveals that institutions as such—whole universities—play a very limited role as active vendors of research services. Almost all the action takes place in numerous highly differentiated *project markets,* where research entities formulate proposals for specific research projects and apply for grants or contracts to undertake them.

Note, first of all, that it is now quite usual for such a project to be proposed as a joint enterprise of researchers in cognate departments from several different universities, regardless of the fact that these are supposed to be in open competition in the institutional market as a whole. This is one of the causes (or symptoms?) of the managerial fragmentation of universities, which often tend to be treated as bankers, or distant holding companies, by transinstitutional coalitions of their staff members competing in the project markets of their respective disciplines.

The fact is that individual academics and leaders of research entities have a much more direct and active interest in maximizing their operations in the project market of their specialty than in the prosperity of their university. Indeed, they are strongly motivated to sell their services as cheaply as possible, resenting the "additional" indirect costs that need to be charged to the customer if the institution is to remain solvent. This is why apparently arbitrary administrative variations in the incidence of these charges generate such discord and distortion in the institutional market.

The most serious feature of most project markets, however, is that they are near *monopsonies.* Most of the customers—typically, specialized project panels of funding bodies—are systematically differentiated in their interests, and even make administrative arrangements so as never to compete directly for the same commodity. This quasi-economic analogue of "product differentiation" is probably unavoidable in the

20. Ziman (1989).

case of the research councils, where all the funds eventually come through the same channel from the public purse, but there is also a tendency for charitable bodies to define specialized niches for themselves in the type of patronage they offer. Such a project panel may thus be almost the sole customer for academic research in this particular specialty and is not under any competitive pressure from other customers to optimize its choices.

Some of the imperfections of monopsony in project markets can be mediated administratively by making sure that customer groups such as research grant panels are widely representative of the potential "users" of the research results, that they are not dominated by a oligarchies of specialist vendors, and that all proposals are subject to independent peer review. But the only "market" solution to this problem is to facilitate the entry of further customers for basic research in the same specialty—for example, from the European Community, industry, and charitable foundations.

The conventions of "responsive mode" project funding do at least entitle research entities to have their proposals assessed on their intrinsic merits, and put the onus directly on customer committees to select wisely among them. From a microeconomic point of view, these should behave like the markets for original works of art, even up to the point of being strongly influenced by fads and fashions.

The trend toward "directed mode" funding could have much more profound consequences for specialized project markets. In effect, this means that the customers call for *tenders* to provide research services to meet stated specifications. The difficulty is that the most likely participants in the tendering process are precisely the research specialists who need to be consulted—often as an invited group—in drawing up these specifications, and who may thus gain "insider" advantages in the subsequent competition. In the end, this leads to a situation where research contractors become mere clients of the organizations that *commission* research projects from them—rather like the farmers that supply supermarkets.

Nevertheless, despite these imperfections, specialized project markets in academic science are highly competitive. Indeed, as a result of scientific and technological progress, they are always in a state of flux. AREs have to work hard, and innovate repeatedly, to keep ahead of the game, and many fall by the wayside in the process. In other words, they are in the situation of rather small firms competing for their share of a rapidly changing product market—essentially the conditions where the entry and exit of firms may require the extension of mainstream economic theory with ideas from *evolutionary economics.*[21]

Internal Institutional Markets

As we have seen, the specialized project markets and the national institutional markets mesh loosely through the administrative and accountancy machinery within individual universities. In principle, this linkage should be quite positive both ways, since research entities are legally and financially subordinate groups, such as departments or "centers," whose academic staff are mostly employees of the institution. In practice, for reasons already noted, this formal "command" system is usually very weak, and research entities enjoy (or suffer?) a great deal of independence in their scientific activities.

One response to this enduring characteristic of academic life is to deliberately encourage the managers of research entities to act as independent entrepreneurs *within their own institution.* Administrative mechanisms that force them to compete directly with one another for institutional resources, such as technical services and funds for new research equipment, are devised.

This may be an effective way of managing a "multiversity" with a diversity of missions, but it stretches the metaphor to describe it as an *internal market* system. Who are the "vendors" and "customers," and what "commodities" are they exchanging? From the point of view of *organization theory,* it might be expected to behave microeconomically as a *franchise* system, where the research entities are independent entrepreneurs in the external marketplace but have to pay rents to their parent institution in return for accommodation, office and technical facilities, senior personnel, etc. Above all, it provides them with access to venture capital without the serious risk of losing financial viability, which is the fate of the great majority of small, independent enterprises. But this is only one example of the many novel ways in which large firms nowadays manage the financial and administrative links with their subsidiaries.

Academic Job Markets

One of the reasons for the internal fragmentation of universities is that their most effective academic staff are often competing as individuals in national or international *job markets.* Each is a vendor of a highly

21. Nelson and Winter (1982).

prized commodity—their own research promise and grant-earning capacity. Institutional research performance depends vitally on assembling a staff of researchers whose individual quality is not always easy to spot and is always in short supply. Academic institutions at home and abroad are ready customers for the highest-quality products in this market, seeking out what they judge to be the best and offering the highest price they can afford in terms of a personal stipend and research facilities.

This is not a new phenomenon,[22] and not a novel analogy.[23] Indeed, the processes by which universities recruit their academic staff have become so complex and institutionalized that other social metaphors suggest themselves—the "slave market" of postdocs at the annual subject convention, the "patronage" dispensed by the "godfather" of the discipline, the "head hunting" for "distinguished professors" to adorn the faculty, and so on.

The academic job market thus plays a very important part in the institutional agora. The question remains, however, whether it can be understood in terms of conventional labor market economics.[24] For example, it is traditionally very price-inelastic. The attraction of a job depends much less on the stipend than on the standing of the institution, the facilities for research, the quality of other staff, etc. Academic scientists are also so specialized in their research interests and capabilities that they cannot really be treated as interchangeable commodities. Even basic employment statistics are uninformative. Thus, prognostications of imbalances of supply and demand in particular disciplines, such as computer science, are confounded by contradictions between the years required to train the necessary specialists and the actual capacity of individuals to retrain themselves to fill new niches.[25]

It may be, indeed, that academic jobs were traditionally traded more as *capital* than as labor. If, as was usual until quite recently, promising researchers are likely to gain early tenure, they have to be selected with all the care that a manufacturer gives to the purchase of a piece of capital equipment, such as a machine tool, which has got to perform well for another thirty years. But efforts to make academic employment more flexible—the abolition of tenure, pay differentials within academic grades, increasing numbers of researchers on short-term contracts, separate accountancy of teaching and research, and the facilitation of entre-

22. Ben-David (1971).
23. Caplow and McGee (1958).
24. Stephan (1996).
25. Ziman (1987).

preneurial ventures by academic staff—are transforming it into a conventional labor market, where human brain power can be purchased for longer or shorter periods for larger or smaller sums, like any other factor of production.

The Reputational Market

To understand the labor economics of traditional academic science, it is essential to make the connection with a much more metaphorical exchange system. Sociologists have emphasized[26] that the position of a vendor—that is, of an individual academic scientist—in an academic job market is closely related to his or her status in the corresponding *reputational market.* They argue further that scientific research is driven by a noncontractual but highly formalized social process, whereby scientists publicly present their "contributions to knowledge"—i.e., scientific papers reporting research results—in exchange for communal "recognition"—i.e., citations, employment, promotion, prizes, etc. In other words, it is a notional market where the vendors are individual researchers, the commodities are research results, the customers are "invisible colleges" of other researchers, and the currency is simply a public sign of personal esteem.

Although this style of functional sociology happens now to be out of fashion, the quasi-economic model is very instructive. It is no accident that Robert Merton named the traditional scientific norms to make up the acronym CUDOS: this is what the individual can expect to receive for obeying these norms in his or her social role. It is true that "recognition" is a very intangible currency, but nobody who has observed the ferocity with which scientists vie with one another for it will doubt its force as a psychological reality.

This exchange process is, of course, very indirect. It does not involve the negotiation of actual contracts between individual vendors and customers. Indeed, there is a strict taboo against any "trading" between those who seek esteem and those who award it. As Hagstrom pointed out,[27] the proper analogy is not with monetary commerce or even barter, but with a traditional system of *potlatch.* At periodic intervals, the tribe indulges in an orgy of public gift giving in which those who give away the most valuable goods receive the greatest honor. Even among scientists, the psychodynamics of such a system may be better understood through social anthropology than through economics.

26. Hagstrom (1965); Merton (1973).
27. Hagstrom (1965).

The validity of this analogy is not, however, the essential point. The job and reputational markets are very closely linked, and the former is not necessarily dominant. Many academic scientists still value, say, appointment or promotion to a full professorship more for its reputational implications than for its material returns. A highly cited paper is not only evidence of technically excellent research performance that will help win a more lucrative contract in the project market: it is a reputational asset that can be transformed into further symbols of public esteem such as membership of an academy or an honorary degree.

Nevertheless, here again, economic rationality is superseding communal convention. Scientific papers are treated as the quantifiable output of a "knowledge production process,"[28] arising from specific contracts for project funding or research employment. They are counted and evaluated to determine whether they constitute good "value for money"—real money that was actually paid over and may be renewed in the next round of project grants or institutional block grants. Their role as "public goods"—that is, freely available contributions to knowledge—is not given precedence, even by their authors, who may judge that they have more to gain by exploiting their discoveries commercially than by adding them to their individual lists of publications.

On the other side of the bargain, "reputation" is seen as an intangible asset that can be cashed in for something more solid than communal esteem. For example, the Nobel Prizes have recently been joined by a large number of other substantial monetary prizes for the high performers in various academic disciplines.[29] Wealthy universities are awarding well-paid "distinguished professorships" to eminent academics nearing retirement. Even researchers in full-time institutional posts are being permitted to gain lucrative consultancies and entrepreneurial profits, "in their spare time."

There is certainly much to be said for making the reputational status of scientists and scientific institutions more explicit and transparent, right across the board. Information on the past performance of research entities is an extremely important factor in project markets, and has significant managerial implications within universities. It is an axiom of economic theory that reliable information makes market competition more efficient. The expert judgment of peers is not necessarily unreliable, but

28. Gibbons, Limoges, Nowotny, Schwartzmann, Scott, and Trow (1994).
29. Stephan (1996).

it needs to be backed up with information on the actual contributions of researchers to the literature of their specialties.

Nevertheless, the commercialization of the reputational market does not produce conventional economic variables. Neither "contributions" nor "recognition" are additive over institutions. They derive from and are attributable to the work of individuals or small groups, and vary enormously in their intrinsic value. It is simply a fact of scientific activity that real progress is made through the contributions of a very small proportion of the competent researchers in a particular field.[30] Under these circumstances, an average over a large group is extremely misleading.

Indeed, academic science may be following commercial sport, the performing arts, the legal profession, and many other callings, toward a situation where there is an enormous disparity of income between the mass of typical practitioners and a few "stars." For example, the convention that a research discovery is attributed to the person who first published it is an obvious example of *winner-take-all economics,*[31] where such disparities are accepted as the norm. In a free market, one is bound to question the economic rationality of the tradition of not paying professional scientists the rate for the job, especially for such communal services as peer review. Perhaps one should go further, and negotiate with their personal agents the precise hourly rate payable to each reviewer for their invaluable expertise.

The Market for Research Claims

In the reputational market, "recognition" is exchanged for "contributions." But what is the value of a "contribution to knowledge"? What is the length of a piece of string? The typescript of a scientific paper submitted to a learned journal claims that the results that it reports are valid and significant. How is this claim tested, and perhaps transformed into an established item of scientific knowledge worthy of public recognition?

Research reports contribute nothing to science until they have been matched against one another in the worldwide *research claims market* of the scientific community.[32] They are refereed before publication, subjected to expert questioning at seminars and conferences, reviewed critically in survey articles, cited favorably or unfavorably in other papers

30. Ziman (1976).
31. Frank and Cook (1996).
32. Ziman (1968).

by other scientists, and tested by replication or experimental refutation. In effect, the literature of a research specialty is a public marketplace, where the pigs and potatoes produced by peasant farmers are prodded and weighed and shrewdly purchased by other peasants or by mercantile factors.

The research claims market in academic science is not driven by some abstract striving for truth. It is driven by competition between individual "vendors" seeking no more than public acceptance for their claims. It is a worldwide social institution, beyond the control of any one national scientific community. It is more global, more formalized, and more objective than the reputational market that it resembles. But its analytical power lies not in the economic analogy but in the *evolutionary* mechanism that underlies it. It invokes a *Darwinian* process of variation, selection, and reproduction[33] as the means for scientific progress. As Darwin perceived, the invisible hand is prodigal but precise. International competition for the validation of research claims is very harsh and very discriminating. A hundred flowers bloom for the one whose seed burgeons and bears fruit. A hundred research papers are published for the one that is still cited, still held to be valid, years later.

Indeed, even when "Lamarckian" factors such as intentional design are allowed for,[34] biological evolution[35] is probably a better metaphor for scientific and technological progress than the behavior of peasant markets. The explanation of phenomena such as the emergence of new levels of order out of systems near the edge of chaos requires very general mathematical and computational tools, such as complexity theory,[36] whose field of application is much wider than a traditional social science such as economics.

Intellectual Property Markets

Nevertheless, the "research claims market" is not altogether metaphorical. The knowledge generated by research can not only be traded notionally for "recognition": it can also be sold for cash. Basic science nowadays is seldom totally detached from its potential applications. Commercial firms are continually on the lookout for academic research findings that might be exploited technologically. From this point of view, all new scientific information is potentially "intellectual prop-

33. Campbell (1974).
34. Ziman (1999).
35. Callebaut (1993); Hooker (1995); Hull (1988); Nelson (1995); Ziman (2000b).
36. Cohen and Stewart (1994).

erty," with a legal owner empowered to demand payment for its use. There already exists an *intellectual property market,* where, so to speak, "futures" in the exploitability of current research results can be sold on a normal commercial basis.

In practice, this market is very imperfect.[37] Knowledge of a novel research finding is an intangible commodity, whose value depends entirely on the context in which it is transferred from the vendor to the customer. On the one hand, secret information can be bought and sold, but not publicly marketed. On the other hand, once it has been disclosed it becomes a public good, from which it is difficult to appropriate a rent. Only a small proportion of the public knowledge and private know-how that goes into a profitable technology can be protected by patents—and so on. Nevertheless, trade in patent rights is brisk, and many academic scientists and their institutions look forward to making a much better thing out of it than they used to.

From the viewpoint of the individual researcher, however, the *commercial* market in intellectual property competes uncomfortably with the *reputational* market, where the same research findings are subject to different "ownership" conventions and may have quite a different type of value. It is not even as if these alternative valuations were commensurable, since the commercial value may depend on keeping secret what only has reputational value if made public. There is an innate contradiction between a traditional academic career, where published research findings are accumulated, and eventually "capitalized" in the form of a prestigious, well-paid post and an entrepreneurial career, where the same skills are used to earn a much higher (if less secure) income by generating, exploiting, or selling research results.

Suppose then, in pursuit of economic rationality, that we resolve this contradiction by conflating the two markets. In principle, it would be perfectly feasible to monetarize the whole system of scientific publication and put it on an equal footing with the patent system. All it would require would be an amendment to the law of copyright to enable a fee to be charged for every published use of a previously published scientific idea—whether or not it had been cited or otherwise acknowledged. Free market theory would then insist that academic research entities could earn their keep directly by selling or renting out their intellectual products, whether as copyrights or patents.

This may sound somewhat fanciful. It would certainly have a very

37. Gibbons and Wittrock (1985).

serious epistemological effect, undermining the whole basis of academic science by putting scientific validity entirely at the disposal of financial interest. Yet it is consistent with the efforts of certain American officials, during the Cold War, to restrict the publication of *all* scientific work, however academic and basic, that might conceivably have a bearing on military technology. And it is just the direction being taken currently by commercial and governmental organizations seeking to patent sequences of the human genome as they are mapped.

Don't be frightened. Don't tell me all the practical problems. This is only a thought experiment. Nevertheless, it brings to the surface one of the major *economic* characteristics of all "intellectual property markets." They are highly *speculative.* Not only are estimates of the value of specific items of knowledge extremely conjectural: they are also highly *skewed.* There are immense variations in the amounts realized, practically or theoretically, out of academic research findings. In fact, 99 percent of what research scientists claim to have discovered turns out to be almost worthless commercially, and largely trivial scientifically. However assessed, the total value of the merchandise is large, but, like natural pearls in oysters, most of it is concentrated in only a tiny proportion of the goods on offer. And most pearl divers are wretchedly poor, as most scientists would be if they had to make a living out of what they individually discover. In the limit, winner-take-all economics gives way to lottery economics, where the inequalities are celebrated, but at least the risks are calculable.

The Noneconomic Dimensions to Be Reckoned With

It is clear that the market paradigm can be a fruitful source of insights into the behavior of academic scientists and their institutions. In certain circumstances they do perform roles analogous to the vending and purchasing of valuable commodities. This analogy also proves to be consistent with conventional economic reasoning, in that the idea of transforming those notional exchanges into normal commercial transactions does not reveal gross contradictions. Thus, the possibility of constructing microeconomic models of various stages of the research process is not ruled out on logical grounds.

Nevertheless, this expedition into an imagined ocean of monetarized research does not arrive at an overall representation of academic science that can be slotted into standard macroeconomic theory. If anything, it merely shows that the microeconomic mechanisms, genuine or metaphorical, inside the R&D black box are extremely diverse, and call into play a number of unorthodox modes of economic reasoning—a fact that was already obvious to the few economists who have looked at academic science with an understanding of its social structure.[38]

Thus, in spite of demonstrating the power of economic reasoning in the analysis of certain features of institutional mechanics, this study shows the inadvisability of trying to treat academic science as a social formation whose properties can be characterized entirely in conventional economic terms. I am not competent to enter into the debate about the general applicability of economic rationality to all forms of social action—in this case, whether *homo scientificus* can be proved, at heart, to be just a variant of *homo economicus*. But this study does suggest that the overall monetarization of academic science would effectively strangle a goose that has provided society with a wealth of golden eggs, and that we need a New Economics of Science that is consistent with what would normally be termed the *noneconomic* dimensions of this peculiar creature.

Here are some examples of features of scientific activity—many of them apparently contradictory—requiring much more sophisticated economic analysis.

Psychologically, scientific activity is strongly motivated by curiosity, obsession with puzzles, the exercise of technical virtuosity, and other intangible personal goods. But the pursuit of these goods is compatible with simultaneous or careerwise partitioned performance of other duties, such as teaching, technical management, etc.

Personally, scientists are highly individualistic, having been trained to be self-winding in originating and defending their own opinions, yet they are remarkably respectful of communal courtesies and responsibilities.

Interpersonally, scientists compete strenuously with one another for communal esteem and professional preferment, primarily as symbols of excellent performance, and yet are capable of combining for exquisitely elaborate collective projects.

Epistemologically, scientific knowledge evolves through the systematic, expert, unforced, but relentless, public criticism of freely published research claims.

Professionally, academic science is minutely subdivided into specialized research communities, each with its own distinctive repertoire of experts and expertise.

38. Dasgupta and David (1994; chapter 7 in this volume); Nelson (1959; chapter 3 in this volume); Stephan (1996).

Technically, scientific work is meticulously laborious, and increasingly involves the combined and coordinated efforts of a great many individuals, often from different specialties.

Sociologically, academic science is not so much a goal-oriented enterprise as a *culture*—or bundle of subcultures—characterized, structured, and sustained by a web of largely tacit social practices.

Managerially, knowledge-producing institutions such as universities are organizational frames within which academic scientists and their research entities are enabled to pursue creative scientific careers.

Critically, quality of performance is so variable, so decisive for successful achievement—and so difficult and uncertain to assess—that it is the dominant criterion in every selective process.

Dynamically, scientific progress is a typical evolutionary process—immensely beneficial in the long run but wildly profligate of failed variants in the short run—and therefore demands greater individual perseverance and collective patience than almost any other intentional human activity.

As scientists often say (like salesmen for their products): this is an exciting subject, raising numerous interesting questions on which a great deal more research will be needed!

REFERENCES

Arthur, W. B. 1994. *Increasing Returns and Path Dependency in the Economy.* Ann Arbor MI, U of Michigan Press.

Ben-David, J. 1971. *The Scientist's Role in Society.* Englewood Cliffs NJ, Prentice-Hall.

Callebaut, W. 1993. *Taking the Naturalistic Turn: How Real Philosophy of Science is Done.* Chicago IL, U of Chicago Press.

Campbell, D. T. 1974. Evolutionary epistemology. In *The Philosophy of Karl Popper,* Edited by Schilpp, P. A. La Salle IL, Open Court.

Caplow, T., & McGee, R. J. 1958. *The Academic Marketplace.* New York NY, Basic Books.

Cohen, J., & Stewart, I. 1994. *The Collapse of Chaos: Discovering Simplicity in a Complex World.* Harmondsworth, Penguin.

Dasgupta, P., & David, P. A. 1994. Towards a new economics of science. *Research Policy* 23: 487-521.

David, P. A. 1997. From market magic to calypso science policy. *Research Policy* 26: 229–55.

Etzioni, A. 1988. *The Moral Dimension: Towards a New Economics.* New York NY, The Free Press.

Frank, R. H., & Cook, P. J. 1996. *The-Winner-Take-All Society.* New York NY, The Free Press.

Gibbons, M., Limoges, C., Nowotny, H., Schwartzmann, S., Scott, P., & Trow, M. 1994. *The New Production of Knowledge.* London, Sage.

Gibbons, M., & Wittrock, B., eds. 1985. *Science as a Commodity: Threats to the Open Community of Scholars.* London, Longman.

Hagstrom, W. O. 1965. *The Scientific Community.* New York NY, Basic Books.

Hooker, C. A. 1995. *Reason, Regulation, and Realism: Towards a Regulatory Systems Theory of Reason and Evolutionary Epistemology.* Albany NY, SUNY Press.

Hull, D. L. 1988. *Science as a Process: An Evolutionary Account of the Social and Conceptual Development of Science.* Chicago IL, U of Chicago Press.

Irvine, J., Martin, B. R., & Isard, P. A. 1990. *Investing in the Future: An International Comparison of Governmental Funding of Academic and Related Research.* Aldershot, Edward Elgar.

Jasanoff, S., Markle, G. E., Petersen, J. C., & Pinch, T. J., eds. 1995. *Handbook of Science and Technology Studies.* Thousand Oaks CA, Sage.

Kealey, T. 1996. *The Economic Laws of Scientific Research.* London, Macmillan.

Merton, R. K. 1973. *The Sociology of Science.* Chicago IL, U of Chicago Press.

Nelson, R. R. 1959. The simple economics of basic scientific research. *Journal of Political Economy* 67: 297–306.

———. 1995. Recent evolutionary theorizing about economic change. *Journal of Economic Literature* 33: 48-90

Nelson, R. R., & Winter, S. 1982. *An evolutionary theory of economic change.* Cambridge MA, Belknap.

Nelson, W. R., ed. 1968. *The Politics of Science: Readings in Science, Technology, and Government.* New York NY, Oxford University Press.

OTA. 1986. *Research Funding as an Investment: Can We Measure the Returns?* Washington DC, Office of Technology Assessment.

Stephan, P. A. 1996. The economics of science. *Journal of Economic Literature* 34: 1199-1235.

Ziman, J. M. 1968. *Public Knowledge: The Social Dimension of Science.* Cambridge, Cambridge University Press.

———. 1976. *The Force of Knowledge.* Cambridge, Cambridge University Press.

———. 1981. *Puzzles, Problems, and Enigmas: Occasional Pieces on the Human Aspects of Science.* Cambridge, Cambridge University Press.

———. 1987. *Knowing Everything about Nothing: Specialization and Change in Scientific Careers.* Cambridge, Cambridge University Press.

———. 1989. Restructuring academic science. London, Science Policy Support Group.

———. 1991. Academic science as a system of markets. *Higher Education Quarterly* 45: 41–61. Republished in Ziman, J. M. 1995. *Of One Mind: The Collectivization of Science.* Woodbury NY, AIP Press.

———. 1992. Research: The next decade. In *Higher Education: Expansion and Reform,* edited by Miliband, D. London, Institute for Public Policy Research.
———. 1994. *Prometheus Bound: Science in a Dynamic Steady State.* Cambridge, Cambridge University Press.
———. 1996. Postacademic science: Constructing knowledge with networks and norms. *Science Studies* 9: 67–80.
———. 1998. Basically, it's purely academic. *Interdisciplinary Science Reviews* 23: 161–68.
———. 1999. The marriage of design and selection in the evolution of cultural artefacts. *Interdisciplinary Science Reviews* 24: 139–54.
———. 2000a. *Real Science: What It Is and What It Means.* Cambridge, Cambridge University Press.
———, ed. 2000b. *Technological Innovation as an Evolutionary Process.* Cambridge, Cambridge University Press.

11

A Formal Model of Theory Choice in Science

William A. Brock and Steven N. Durlauf

SUMMARY: Since the work of Thomas Kuhn, the role of social factors in the scientific enterprise has been a major concern in the philosophy and history of science. In particular, conformity effects among scientists have been used to question whether science naturally progresses over time. Using neoclassical economic reasoning, this paper develops a formal model of scientific theory choice which incorporates social factors. Our results demonstrate that the influence of social factors on scientific progress is more complex than previously thought. The patterns of theory choice predicted by the model seem consistent with historical episodes of theory change.

1. Introduction

> [S]tudents of the development of science, whether sociologists or philosophers, have alternatively been preoccupied with explaining consensus in science or with highlighting disagreement and divergence. Those contrasting approaches would be harmless if all they represented were differences of emphasis or interest. . . . What creates the tension is that neither approach has shown itself to have the explanatory resources for dealing with both . . . whatever success can be claimed by each of these models in explaining its own preferred problem is largely negated by its inability to grapple with the core problem of its rivals.
>
> Larry Laudan[1]

Since the classic work of Feyerabend (1975) and Kuhn (1970), analyses of scientific progress have been required to consider the socioeconomic environment in which scientific analysis is conducted. Whereas the predominant themes in the philosophy of science for most of the

 From *Economic Theory* 14 (1999): 113–30.

1. Laudan (1984, 2).

twentieth century have emphasized the development of normative criteria for the scientific process, and generally presumed from the positive perspective that scientists each possessed an identical desire to find "truth" as measured by these common criteria, recent trends both in the philosophy and history of science (Bloor 1976; Latour 1987) have been concerned with the social context in which research is conducted. This concern represents a challenge to standard accounts of science as a progressive enterprise, in which changes in accepted theories construct a trajectory towards greater verisimilitude (Popper 1972, 1976; Newton-Smith 1981), problem-solving capability (Laudan 1977, 1984), or other criterion or set of criteria for theory evaluation. The basic source of this concern is the belief that those motivations of scientists which are explicitly social, be they a desire for status, success, or conformity, can lead the evolution of science away from whatever criteria constitute the appropriate goals of science.

As our quotation from Laudan indicates, much of the debate concerning positive models of scientific change can be dichotomized between those who primarily focus on conditions which support consensus about goals and methods as the norm and those who primarily study conditions under which disagreement can occur. These differences tend to mirror the distinction between a vision of science as an activity conducted by disinterested truthseekers and one conducted by fallible and self-interested participants. A difficulty with these alternative assumptions is that they restrict the explanatory scope of their associated theories.

In this paper, we propose a model of theory choice which explicitly incorporates private as well as social influences on individual decision-making. We do this by employing a model based upon Brock and Durlauf (1997) which characterizes the behavior of binary choices in an interdependent population.[2] While the model is of course very stylized, it does provide a formalization of the sort of social environment in which theory choice occurs. In terms of antecedents, our work follows closely in the spirit of the pathbreaking study of Kitcher (1993; excerpted in chapter 8 in this volume) which formalizes science as a dynamic process in which individual scientists interact as purposeful actors.

2. See Blume (1993), Brock (1993), and Durlauf (1993) for related analyses of interaction structures and Oomes (1997) for the use of the basic framework we describe to study related issues.

At the outset, we wish to identify a number of assumptions which we employ although they are extremely controversial from the perspective of the philosophy of science. First, we assume that there is a unique scalar metric by which one theory can be judged as scientifically superior to another. Not only is there controversy between different criteria for scientific theory assessment, it seems clear that the perceived success of different theories depends upon distinct criteria (Putnam 1994). However, since our concern is with the interaction of "social" versus "scientific" criteria for theory acceptance, the problems with this assumption do not seem to be germane.

Second, we focus on a single type of social interaction—conformity effects, by which we mean the tendency of individual scientists to place greater weight on theories which others accept than otherwise. This assumption does not by itself imply the presence of nonscientific factors, since it is certainly possible that such weighting reflects incorporation of the scientific assessments of others. However, for our purposes, we interpret conformity exclusively as a nonscientific influence. Of greater concern is the possibility that this type of nonscientific influence loses some of the richness of the literature. In particular, we omit issues of financial support and ability to publish. Both of these factors introduce the issue of how a theory is able to generate evidential support, either directly through additional research or indirectly through altering the information set employed in research by others. While our intuition is that adding such factors would not qualitatively affect our results, this has not been formally shown and would certainly represent a useful extension of our model.

Our formal model possesses two especially interesting properties. First, our model shows how scientific consensus can rapidly emerge from a period of profound disagreement. In particular, we show how social interactions in the scientific community are an essential component of this process. Second, we demonstrate that social interactions do not necessarily represent, as is often assumed in the philosophy and (especially) sociology of science literatures, an impediment to the adoption of new and better theories over their entrenched predecessors. In fact, these social influences may actually accelerate the rate at which superior theories achieve a consensus.

Both of these features stem from the way in which scientific and nonscientific influences interact. These interactions produce nonlinear effects of the sort which frequently arise in the study of complex systems.

In this regard, we believe our analysis highlights the way in which formalization of arguments in the philosophy literature can lead to useful insights.

2. Framework

We consider the problem of theory choice between two theories, denoted as T_{-1} and T_1. Binary decisions of this type have been studied extensively in the economics literature. In particular, we consider a community of I scientists. Individual scientists are indexed by i; scientists i's theory choice at time t is $\omega_{i,t}$ with associated support $\{-1, 1\}$. The collection of theory choices by all scientists in the community is $\underset{\sim}{\omega}_t$. Finally, the vector of all decisions other than that of agent i is $\underset{\sim}{\omega}_{-i,t}$.

Each scientist is assumed to possess a way of assigning numerical valuations to the adoption of a particular theory, which we will refer to as "utility."[3] The utility a scientist receives from adoption of a particular theory is assumed to be measured by a function $V_{i,t}(\omega_{i,t})$.[4] Therefore, an individual scientist's theory choice is the solution to the maximization problem

$$(1) \qquad \max_{\omega_{i,t} \in \{-1,1\}} V_{i,t}(\omega_{i,t}).$$

In order to permit the analysis of the effects of private and social influences on theory choice, we place some restrictions on the utility function $V_{i,t}$.

In particular, we assume that the individual utility can be decomposed into three components

$$(2) \qquad V_{i,t}(\omega_{i,t}) = u_{i,t}(\omega_{i,t}) - E_{i,t} \sum_{j \neq i} J_{i,j,t}(\omega_{i,t} - \omega_{j,t})^2 + \epsilon(\omega_{i,t})$$

In this specification the three additive components refer to different types of utility. Specifically, $u_{i,t}(\omega_{i,t})$ represents deterministic private utility, $-E_{i,t} \sum_{j \neq i} J_{i,j,t}(\omega_{i,t} - \omega_{j,t})^2$ represents a general conformity effect, which we

3. Our use of the word "utility" to characterize the evaluation function for individuals carries no implications concerning the evaluative criteria of scientists. Also, notice that our use of a utility-maximizing framework has no implications for whether theory choice is active or passive, in the sense that while choice of goods in a grocery store is active, the determination of whether one is liberal or conservative might well be passive. In the passive case, our utility function determines which attribute is possessed by a scientist. Both active and passive elements to choice are presumably present in actual theory choice.

4. See Diamond (1988) for the use of a utility-maximizing framework to understand how scientists allocate time across theories, with resultant implications for the rationality of the scientific enterprise as a whole.

call deterministic social utility, and $\epsilon(\omega_{i,t})$ represents random private utility. Notice that the first and third components are idiosyncratic in the sense that they depend only on the individual's characteristics, whereas the second component is determined by the individual's characteristics as well as the (expected) choices of the rest of the population. We therefore interpret the first and third components as embodying the scientific judgments of each scientist and the second component as embodying the influence of social factors, namely conformity effects, on choice. While the individual-specific components to theory assessment could plausibly be argued to contain judgment factors which are nonscientific in nature, none of our results are fundamentally changed by this interpretation.

In formulating the individual scientist's decision problem this way, our analysis in some respect sidesteps Kuhn's (1970) argument that different scientific paradigms may be incommensurable due to different ontologies or epistemologies or Quine's (1951) related argument that theories are underdetermined by data. We do this as we require of scientific theories not that one can be interpreted in the ontology of another or that one theory can explain phenomena that another cannot, but rather that relative to whatever goals of science a community embraces, relative theory evaluation can be made. Such an assumption is commonplace in economic models of consumer choice in which individuals have preference orderings over different bundles of commodities, such as guns and butter, in which there is no intrinsic comparability between the individual commodities.[5]

Finally, the random utility components are assumed to be extreme-value distributed and independent across individuals, so that

$$\text{(3)} \qquad \text{Prob}(\epsilon(-1) - \epsilon(1) \leq z) = \frac{1}{1 + \exp(-\beta z)}; \beta > 0.$$

For our purposes, this assumption is used for analytical convenience as it allows us to explicitly characterize the probability measure of $\underset{\sim}{\omega}_t$. Anderson, dePalma, and Thisse (1992) provide some interpretations of and justifications for this functional form in the context of the theory

5. Laudan (1996), chapters 1–3, develops the related argument that while theories may be deductively underdetermined, theories may still be "ampliatively" determined, by which he means that nondeductive factors such as simplicity, explanatory scope, or relationship with other theories may nevertheless imply that a single theory dominates another. In the context of our utility specification, these ampliative factors are subsumed in the V function. See Hands (1994; chapter 19 in this volume) for a related discussion of the relationship between Laudan's approach to modelling scientists' decision-making and economic formulations of utility maximization.

of consumer choice. Observe that β indexes the degree of diversity of individual-specific theory evaluations in the community. Small values of β imply that there is wide diversity, whereas large values of β imply there is little diversity.

These assumptions are sufficient to imply (following formal arguments developed in Brock and Durlauf 1997) that at each t, scientist i's theory choice possesses the probability structure

$$\text{(4)} \qquad \text{Prob}(\omega_{i,t}|E_{i,t}(\underset{\sim}{\omega}_{i,t})) \sim \exp\left(\beta h_{i,t}\omega_{i,t} + \sum_{j \neq i} \beta J_{i,j,t}\omega_{i,t}E_{i,t}(\omega_{j,t})\right).^{6}$$

In this formulation $h_{i,t} = 1/2\,(u_{i,t}(1) - u_{i,t}(-1))$ and so provides a sufficient statistic for the private deterministic component of the comparative evaluation of the two theories. Note that incommensurability of the scientific evaluative criteria can explain differences across individuals in $h_{i,t}$, in the sense that different individuals may assign different scientific weights to theories as a result of differences of beliefs concerning favors such as which the phenomena are most important for a theory to explain or the mechanisms by which theories are evaluated.

Since the random components of individual utility functions are independent, the collection of theory choices is characterized by the joint probability structure

$$\text{(5)} \qquad \begin{aligned} \text{Prob}\Big(\underset{\sim}{\omega}_t|E_{1,t}(\underset{\sim}{\omega}_{-1,t})), \ldots, E_{1,t}(\underset{\sim}{\omega}_{-I,t})\Big) &= \prod_i \text{Prob}(\omega_{i,t}|E_{i,t}(\underset{\sim}{\omega}_{-i,t})) \\ &\sim \prod_i \exp\Big(\beta h_{i,t}\omega_{i,t} + \sum_{j \neq i} \beta J_{i,j,t}\omega_{i,t}E_{i,t}(\omega_{j,t})\Big)\Big) \end{aligned}$$

Our model of the evolution of theory choice will be complete once we specify the properties of β, $h_{i,t}$, $J_{i,j,t}$, *and* $E_{i,t}(\cdot)$.

3. Leading Case Analysis

i. Model Specification

We initially consider the case in which all dynamics are determined for fixed evaluation weights. Formally, this means that there exists a constant J such that

6. The term "$\sim$" means "is proportional to" and is employed to avoid the need to write cumbersome normalizations.

$$(6) \qquad J_{i,j,t} = \frac{J}{I-1}$$ [7]

and a constant h such that

$$(7) \qquad h_{i,t} = h$$

This means that any differences in the way in which members of the community make scientific evaluations of the competing theories are embedded in the $\epsilon_{i,t}(\underset{\sim}{\omega}_{i,t})$'s. Following Brock and Durlauf (1997), it is straightforward to show that

$$(8) \qquad \text{Prob}\left(\underset{\sim}{\omega}_t | E_{1,t}(\underset{\sim}{\omega}_{-1,t}), \ldots, E_{I,t}(\underset{\sim}{\omega}_{-I,t})\right) \sim \prod_i \exp\left(\beta h \underset{\sim}{\omega}_{i,t} + \beta J \omega_{i,t} m^e_{i,t}\right)$$

where

$$(9) \qquad m^e_{i,t} = \frac{1}{I-1} \sum_{j \neq i} E_{i,t}(\omega_{j,t})$$

The properties of this expression can be fully understood in two steps. First, assume that all scientists possess common expectations of the mean choice of others.

$$(10) \qquad m^e_{i,t} = m^e_t \qquad \forall i$$

Under this assumption, one can verify that the sample average choice level, $\overline{m}_{I,t}$, will, as the number of scientists becomes arbitrarily large, obey

$$(11) \qquad \lim_{I \Rightarrow \infty} \overline{m}_{I,t} = \tanh(\beta h + \beta J m^e_t)$$

Second, assume that the common expectation of the average choice is self-consistent in the sense that the common expected average choice level is one that is actually realized.

$$(12) \qquad m^e_t = \lim_{I \Rightarrow \infty} \overline{m}_{I,t}$$

Self-consistency, combined with equation (11), implies that the steady state behavior of the average choice level is any value m^* such that

$$(13) \qquad m^* = \tanh(\beta h + \beta J m^*)$$

The solution m^* to (13) depends only on β, h, and J.

This simple equation provides some insights into the interaction of

7. The $I - 1$ which appears in the denominator of this expression acts as a normalization factor.

private and social influences on theory choice. In particular, following Brock and Durlauf (1997), the following will hold:

Proposition 1. Existence of Multiple versus Unique Steady States

Under the assumption of the "leading case"

i. If $\beta J < 1$, there exists a unique solution to (13).

ii. If $\beta J > 1$, there exists a threshold H (which depends on βJ), such that

  a. for $|\beta h| < H$, there exist three roots to eq. (13), one of which has the same sign as h, and the others possessing opposite sign.

  b. For $|\beta h| > H$, there exists a unique root to eq. (13) with the same sign as h.

In the multiple steady state case, we will denote the three equilibria as m_-^*, m_m^*, and m_+^* in order to distinguish the extremal equilibria by sign.

Proposition 1 is interesting, as it provides a precise relationship between the strength of individual and social factors in determining whether the average theory choice of a scientific community is or is not a unique function of the set of scientific evaluations of the individual scientists. In particular, it illustrates that for any strength of conformity effects (as measured by J), there is a level of evidential support (as manifested by h) such that two communities of scientists will come to similar average conclusions on the relative merits of two theories. At the same time, for any level of evidentiary support h, there is some conformity level J such that social interaction can lead to a community consensus away from that theory which by scientific criteria is superior.

This feature has four implications for debates on the progressiveness of science. First, it suggests that the consequences of the introduction of social factors in the choice of scientific theories has no necessary implication for claims concerning the progressiveness of the evolution of theories.

Second, so long as there exist sufficiently decisive evidentiary differences between two theories (which in our model is equivalent to a sufficiently large value of $|h|$), strong social factors do not act to the detriment of progress in scientific models. Notice that this conclusion is not dependent upon any assumption about whether evidentiary support is theory neutral or not nor does it depend upon commensurability between theories (in terms of common or translatable definitions and objects of explanation) per se. The proposition only depends on the capacity of the scientific community to engage in relative theory evaluation.

Third, it suggests that the analysis of theory progression needs to distinguish between "local" and "global" progressiveness. What we mean is the following. The relationship between h and βJ means that so long as the relative scientific merits are large enough, social factors will never impinge on emergence of the consensus around the superior theory. However, for any theory and $\beta J > 1$, there will exist a local alternative theory (defined as one in which h is "small" but negative) which represents a dominant steady state choice for the community despite the presence of the superior alternative.

Fourth, the model suggests that the diversity of evaluative criteria and/or evaluative evidence within the population has important implications for the progressiveness of science. In particular, observe that a small enough β will, for any strength of social interactions, eliminate the multiplicity of steady states. Intuitively, a small β means that the dispersion of private theory evaluations is high, in the sense that a substantial fraction of individual theory choices is driven exclusively by private considerations, i.e., the relative scientific differences between the theories is so large that any conformity effect is overwhelmed. The presence of such "extreme" private beliefs in turn means that for the remaining members of the community, there will be insufficient capacity for consensus to allow for multiple steady states.

The qualitative results of the baseline model are robust to the assumption that all scientists agree on the relative scientific merits of the two theories. To see how the basic model can be generalized to account for fixed individual heterogeneity (as opposed to the heterogeneity associated with $\epsilon(\omega_{i,t})$), we simply reintroduce distinct values of $h_{i,t}$ for each scientist. In this case, for any common set of expectations, average theory choice will obey

$$\lim_{I \to \infty} \overline{m}_{I,t} = \int \tanh(\beta h_{i,t} + \beta J m_t^e) dF_{h_{i,t}} \tag{14}$$

where $dF_{h_{i,t}}$ denotes the probability measure characterizing $h_{i,t}$. Imposing self-consistency implies that the average choice level at t is any value m_t^* such that

$$m_t^* = \int \tanh(\beta h_{i,t} + \beta J m_t^*) dF_{h_{i,t}} \tag{14}$$

It is easy to see that when the $h_{i,t}$ values all possess the same sign, there will be a unique self-consistent solution when the magnitudes of the $|h_{i,t}|$ are sufficiently large.[8]

8. This follows immediately from the continuity and monotonicity of the tanh(·) function.

4. Dynamics

In order to analyze the dynamics of the baseline model, we make the following assumption:

$$m_t^e = m_{t-1} \tag{16}$$

Imposition of eq. (16) may be interpreted either as meaning that expectations of community beliefs are adaptive, so the expectation of the mean today is whatever transpired the previous period, or that theory choice at t is influenced by a desire to conform to average beliefs at $t - 1$. Note that since we assume that idiosyncratic $\epsilon(\cdot)$ terms are invariant (except for the choice of $\omega_{i,t}$), individual scientists do not flip opinions randomly once a steady state average has been achieved.

This assumption on expectations means that the dynamics of the model are described by the sequence of m_t values consistent with

$$m_t = \tanh(\beta h_t + \beta J m_{t-1}) \tag{17}$$

I. Stability

Brock and Durlauf (1997) verify that under these dynamics, the extremal equilibria are stable in the multiple equilibrium case, whereas the middle equilibrium is not.

Proposition 2. Stability of Steady State Mean Choice Levels

Under the assumptions of our "leading case,"

i. If eq. (13) possesses a unique root, that root must be locally stable.

ii. If eq. (13) possesses three roots, then the steady state mean choice levels m_-^* and m_+^* are locally stable whereas the steady state mean choice level m_m^* is locally unstable.

With reference to theory choice, Proposition 2 means that social factors can lead a community to make collective choices which on average differ from the choices which would be made due to purely "scientific" factors. Such choices will depend on the initial distribution of theory choices in the community. This establishes a type of path dependence in the evolution of theory choices.

II. Nonlinearity

This analysis of stability is incomplete in the sense that it treats the relative scientific content of the theories (h) as fixed and looks at the dynamics of the mean conditional on this. Of course, scientific theories evolve in the presence of changes in evidence and associated theories. Hence, we consider how the steady state average theory choice evolves in response to changes in h.

Given the multiple steady states which exist in the model, it is essential to distinguish between marginal and nonmarginal changes in h. A marginal change in h will alter the steady state mean according to

$$\frac{dm^*}{dh} = \frac{\beta(1 - \tanh^2(\beta h + \beta J m^*))}{1 - \beta J(1 - \tanh^2(\beta h + \beta J m^*))} \tag{18}$$

The sign of this function is of course positive. This function is highly nonlinear and means that the magnitude of the impact of changes in the evidentiary support of one theory versus another will depend on the current mean as well as the various behavior parameters of the model.

A second and more interesting type of nonlinear behavior can occur in the model when one considers a sequence of nonmarginal changes in h. Nonmarginal changes lead to the possibility that in addition to movement of a particular steady state, there may be a change in the existence of the steady state itself. This feature is illustrated in figure 11.1, in which

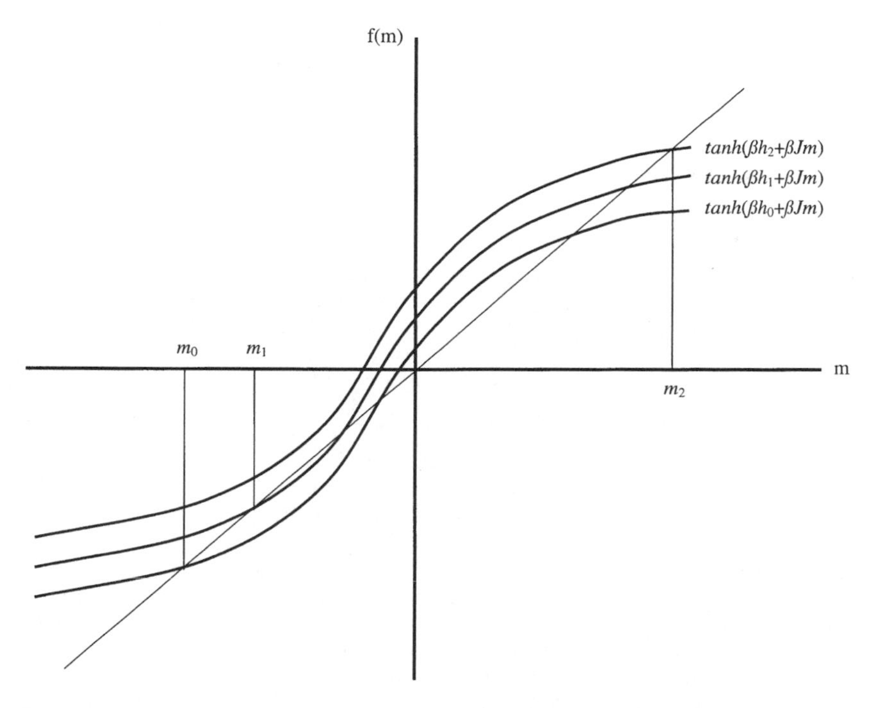

FIGURE 11.1. Evolution of Average Theory Choice

there is a set of equal-sized movements in h_t.[9] In this figure, it is assumed that the community of scientists is characterized by mean m_0, which corresponds to the case where social factors have caused community consensus to converge on the theory which the scientific valuation would have (on average) rejected. An increase in evidentiary support for theory 1, measured as a movement from h_0 to h_1, induces a shift in the average toward theory 1, although most scientists still choose theory -1. However, when the level of evidentiary support shifts the relative valuation from h_1 to h_2, the multiplicity of steady state outcomes disappears. The unique steady state of the model, m_2, possesses a sign consistent with that of h.[10]

The dynamics described in the figure match the empirical challenge described by Laudan at the beginning of this paper. Suppose that an accepted theory T_{-1} is challenged by an alternative theory T_1 which from the perspective of scientific criteria is superior, so that h_0 is positive. Social factors can lead to the continuing acceptance of T_{-1}. As evidence accumulates in favor of T_1, i.e., h_t increases, there will exist a point where a discontinuous change in the community occurs, corresponding to the elimination of the self-consistent steady state in support of the old theory. There is no conflict between an extended period of disagreement and a rapid emergence of theoretical consensus. Hence the "tension" between disagreement and consensus which Laudan describes is naturally addressed within our framework.[11]

An additional implication of this analysis is that decisive evidence in

9. We assume that the evolution of h_t is monotonic in favor of one theory. While monotonicity is not important for the subsequent analysis, the assumption that more and more evidence accumulates in favor of one theory (which by the evaluative criteria of community members is superior to the other) is essential. Kitcher (1993; chapter 8 in this volume) provides a model of the allocation of scientific resources which can naturally generate this assumption as an equilibrium in scientific effort. To be clear, there is no necessary reason to suppose that evidence in favor of a superior theory accumulates in this fashion, and certainly pathological episodes such as Lysenkoism indicate that the opposite could even happen. However, we regard this assumption as consistent with those scientific practices with which we are familiar and would certainly agree with Kitcher and others that the reward structure for scientific activity essentially achieves this over time.

10. This type of discontinuous behavior is common in models with positive interaction effects: see Brock and Durlauf (1997) for discussion and references.

11. Laudan (1996), chapter 13, argues that the emergence of consensus arises due to the evolution of new pieces of evidence which are able to satisfy scientists with different evaluative criteria from those who initially accept a theory. In our framework, this can be done by allowing the $h_{i,t}$s to differentially shift across time. Notice that Laudan's account does not include any conformity effects and so does not endogenously produce the threshold consensus formation found in our model.

a scientific controversy is contextual, i.e., the impact of an increment on the evolution of scientific consensus is determined by the distribution of relative weights at the time the new evidence is introduced in combination with the strength of the social interaction effect βJ. The most extreme version of this holds when an incremental piece of evidence eliminates two of the possible self-consistent equilibria, as illustrated in figure 11.1. Hence, the importance of Eddington's eclipse observations for the general acceptance of relativity was a function not just of the findings per se, but also of the totality of observational evidence and the ampliative virtues of the theory relative to Newton's. For example, Eddington's observations were interpreted by the scientific community against a background in which the aberrations in Mercury's orbit and the Michelson-Morley experiments were already known.[12]

III. How Does Diversity of Opinion Affect Scientific Consensus?

The introduction of heterogeneity in $h_{i,t}$ allows a straightforward analysis of the effects of an increase in the diversity of scientific evaluations of the relative merits of the two theories. We do this by introducing a mean-preserving spread in the $h_{i,t}$s. Suppose that

$$h_{i,t} = h_t + u_{i,t}$$

where $u_{i,t}$ is independently and identically distributed across scientists. When the variance of this shock, σ_i^2, is zero, the steady states of the model will correspond to those of eq. (13). When the variance is increased from zero, the effect on the average theory choice will depend on the slope of the $\tanh(\beta h_t + \beta J m_t^*)$, where recall that m_t^* is the initial self-consistent average. Assuming that we are at a locally stable steady state and given that these states, by Proposition 2, are those were the $\tanh(\beta h_t + \beta J m_t^*)$ function is locally concave, this implies that the population average will move towards the origin. Hence a mean-preserving spread in the relative theory evaluations will lead to a reduction of the average choice level in the direction of zero (i.e., towards half the community choosing each of the two theories). It is even possible for an increase in heterogeneity to cause the population to flip to a new steady state (and hence overshoot the origin) for reasons already

12. See Durlauf (1997) for a related discussion of how the interpretation of how the presence of particular alternative theories defines the context in which evidence evaluation can occur.

described in the discussion of the influence of β in the previous section. Taken together, the implications for the distribution of beliefs, as measured by the variance of $h_{i,t}$ and the level of β, indicate that increased heterogeneity of beliefs will act to mitigate the potential for social factors to cause a community to form a consensus around an inferior theory.[13]

IV. Do Social Factors Impede the Acceptance of More Successful Theories?

Our analysis of the progressiveness of science suggests that the role of social factors is more complex than is often recognized. From the analysis of stability, it is clear that social factors can impede the acceptance of a new, scientifically superior theory. This impedance is manifested by the existence of a negative-mean steady state.

However, once the mean choice level possesses the same mean as h, the presence of social factors will act to *increase* the degrees of consensus around the scientifically preferable theory.[14] This makes intuitive sense, since once the consensus of the community is centered on the superior theory, this consensus will influence its rapid acceptance. Notice that this holds even if there are multiple steady states in theory choice. In this case, social factors increase the absolute value of the extremal steady states, creating the possibility that, depending on which steady state is realized, social factors can lead to widespread consensus around either of the two theories.

These features are summarized in the following proposition, which assumes that h is positive.

Proposition 3. Interaction of Social Factors and Average Theory Choice

i. Conditional on the other parameters in the model, the greater the value of J, the greater the average theory choice in the model, when the average choice level and h possess the same sign.
ii. Conditional on the other parameters in the model, the greater the

13. See Kitcher (1993) for a complementary analysis in which heterogeneity diversifies the research output of a community, which in our notation will lead the distribution of $h_{i,t}$'s to evolve.

14. This is an immediate consequence of the monotonicity of the tanh(·) function and the assumption that $\beta J > 0$.

value of J, the lower the average theory choice in the model, when the average choice level and h possess the opposite sign.

5. Extensions

I. Leaders and Followers

Our baseline model assumes that each scientist possesses an identical weighting formula when considering theory choices of other members of his community. A natural modification, discussed and formalized in a very different context by Kitcher (1993; chapter 8 in this volume) is to decompose the community of scientists into two types, leaders and followers, which are distinguished by subscripts l and f, respectively. The average theory choice for the scientific community will then be a weighted average of the two groups. Letting $m_{l,t}$ and $m_{f,t}$ denote the means of the leader and follower groups, the overall mean m_t will be the population weighted average of the group specific means

$$m_t = \lambda m_{l,t} + (1 - \lambda)m_{f,t} \tag{19}$$

Leaders and followers may be distinguished with respect to the parameters h, β, and J. When each scientist possesses an h and β which depends on his status as a leader and follower as well as a separate social interaction weight which depends on the leader/follower combination involved in any pairwise interaction, the mean of the leaders and followers will be described by any joint solution to the pair of equations

$$m_{l,t} = \tanh(\beta_l h_{l,t} + \lambda\beta_l J_{l,l} m_{l,t} + (1- \lambda)\beta_l J_{l,f} m_{f,t}) \tag{20}$$

$$m_{f,t} = \tanh(\beta_f h_{f,t} + \lambda\beta_f J_{f,l} m_{l,t} + (1- \lambda)\beta_f J_{f,f} m_{f,t}) \tag{21}$$

One special case of the leader/follower model may be obtained from the further assumptions

$$J_{l,l} = \mathrm{J}_{l,f} = 0 \tag{22}$$

$$J_{f,l} > J_{f,f} > 0 \tag{23}$$

Together, these assumptions mean that leaders make decisions independent of a desire to conform whereas followers experience conformity effects which are especially strong relative to the average behavior of leaders. In terms of qualitative differences with the baseline model, there are several features worth noting. First, in this formulation, since the leaders of a discipline are assumed to make theory choice exclusively according to scientific merit whereas followers are influenced by social as well as scientific factors, the model can exhibit a complementary explanation to the disagreement/rapid consensus phenomena discussed in the previous section. In the leader/follower case, the development of scientific evidence supportive of theory 1 can, once it induces a sufficient consensus among leaders, generate a mean shift of followers towards the new theory. Of course, such dynamics only make sense if the leaders in the community do not have a vested interest in the continued acceptance of the old theory, which they have presumably helped to develop.

II. Schools of Thought

The leader/follower framework can be easily modified to account for schools of thought. In the simplest case, suppose that there are two schools of researchers, *A* and *B*. These schools are distinguished in that members of each school place different conformity weights on the theory choice of members of the same versus the alternative school. If each group weights internal and external conformity in the same way, the within-school means $m_{A,t}$ and $m_{B,t}$ can be expressed as the solutions to:

$$m_{A,t} = \tanh(\beta h_t + \beta J_s m_{A,t} + \beta J_D m_{B,t}) \tag{24}$$

$$m_{B,t} = \tanh(\beta h_t + \beta J_D m_{A,t} + \beta J_s m_{B,t}) \tag{25}$$

where J_s is the same-school interaction weight and J_D is the different-school interaction weight.

The important difference between this model and the previous one is the possibility that different schools of thought end up in different steady states, depending on the particular weights attached to the two communities. In the extreme case, suppose that $J_D = 0$. In this case, the communities will independently replicate our original model. So long as there exist multiple steady states, then it is possible for different communities to reach different conclusions based upon the same scientific evaluations.

III. Endogenous Evolution of h_t

A third modification of the basic framework concerns the evolution of evidentiary support distinguishing the two theories. One way to do this is to assume, for reasons ranging from relative ease of funding to the possibility that those who accept a particular theory are more likely to find evidence supportive of it than those who prefer another theory, that the rate of change in evidentiary support is a function of the percentage of individuals who accept it at a point in time. Formally, this would mean that

$$h_t - h_{t-1} = \phi(m_{t-1}). \tag{26}$$

We assume that $\phi(\cdot)$ is positive and monotonically increasing in m_{t-1} as we are again interested in the case where theory 1 dominates theory -1 in terms of comparative scientific evaluation, this suggests a second mechanism by which nonlinear theory dynamics can occur. When a consensus initially exists around a given theory, the introduction of a superior alternative will not necessarily produce a rapid accumulation of evidence in its favor. As evidence builds up in favor of one theory, this will feedback into the rate of change of the level of evidence, creating standard nonlinear dynamics.[15]

IV. Alternative Dynamics

Finally, our model can accommodate different dynamics from those implied by the expectation formation equation (16). In particular, we consider the case where the expected value of the average community choice at t equals a simple average of past realizations,

$$m_t^e = t^{-1} \sum_{j=1}^{t} m_{j-1}. \tag{27}$$

In this case, the sequence of expectations possesses a recursive structure in that

$$m_{t+1}^e = \frac{t}{t+1} m_t^e + \frac{1}{t+1} m_t = \frac{t}{t+1} m_t^e + \frac{1}{t+1} \tanh(\beta h_t + \beta J m_t^e). \tag{28}$$

In order to understand the limiting properties of this equation, we proceed as follows. First, straightforward manipulation of eq. (28) yields an adjustment equation for expectations of the form

$$m_{t+1}^e - m_t^e = \frac{1}{t+1} (\tanh(\beta h_t + \beta J m_t^e) - m_t^e) \tag{29}$$

Second, we decompose $\tanh(\beta h_t + \beta J m_t^e) - m_t^e$ as

$$\tanh(\beta h_t + \beta J m_t^e) - m_t^e = E_{t-1}(\tanh(\beta h_t + \beta J m_t^e) - m_t^e) + u_t \tag{30}$$

15. To give but one example, Holling and Sanderson (1996) provides an example of a mathematically similar story for ecological reasons.

where E_{t-1} denotes the conditional expectations operator given all information available by $t-1$ and u_t is (by construction) a martingale difference sequence. Together, we can rewrite the expectations adjustment process as

$$m^e_{t+1} - m^e_t = \frac{1}{t+1}\left(E_{t-1}(\tanh(\beta h_t + \beta J m^e_t) - m^e_t) + u_t\right). \tag{31}$$

This equation may be analyzed using results for a broad class of dynamic processes studied by Benaim and Hirsch (1994). In particular, they show that for a general process of the form

$$m^e_{t+1} - m^e_t = g_{t+1}\,(F(m^e_t) + u_t). \tag{32}$$

where g_{t+1} is a sequence of positive numbers decreasing to zero, u_t is a martingale difference sequence, the limiting behavior of this system is characterized by the differential equation

$$\frac{dm^e}{dt} = F(m^e) \tag{33}$$

In our case, which is an example of (32) when h_t is constant, the relevant differential equation is

$$\frac{dm^e}{dt} = \tanh(\beta h + \beta J m^e_t) - m^e \tag{34}$$

Hence, Benaim and Hirsch (1994) implies that with probability one, under the expectations process eq. (27), the average population choice will converge to the one of the locally stable steady states described by Proposition 2.

How does the evolution of h_t influence this conclusion? The evolution of h_t introduces a way in which path dependence in theory choice is influenced by the evolution of evidence. Suppose that h_t evolves towards some fixed $\bar{h}$, achieving this limit at some finite T. The specific sample path of h_t will determine (along with the sample path of u_t) which of the steady state solutions is realized. The evidentiary noise associated with the deviations of h_t from $\bar{h}$ can thus mean that either of the deterministic steady states is reached with positive probability.

Finally, as shown in Benaim and Hirsch (1994) it is possible to retain the qualitative features of this analysis if eq. (27) is replaced with an alternative averaging mechanism which assigns greater weight to more recent observations.

6. Conclusions

This paper has attempted to contribute to the analysis of the sociology of science by constructing a very stylized model of community theory choice. Despite the many simplifications, the model was shown to be able to replicate an important nonlinearity in the evolution of scientific paradigms—the movement of a community of scientists from a period of extended disagreement to a period of rapid consensus formation. This feature was shown to naturally fall out of a framework which assumes that theory choice is *comparative,* in that the community is choosing between a pair of alternative theories. In addition, the model suggests that the influence of social factors in scientific theory choice may assist rather than hinder scientific progress, at least as measured by progressiveness in the evolution of theories. While we do not establish any presumption that social factors facilitate rather than impede scientific progress, our analysis does indicate that any such conclusions will depend upon a range of specific details of community interactions. Hence we are skeptical that there is any generic empirical regularity to be found in the role of social factors in the evolution of scientific theories; rather, careful case-by-case historical studies need to be conducted.

This analysis, of course, does not speak to the question of how alternative theories emerge for consideration the first place. In Popper's language, we have discussed the community-level logic of discovery. Any conclusions we draw on the possible facilitating role of conformity effects on movement towards superior theories is strictly conditional on the available theory options. Given the viability of scientifically inferior local theory alternatives at any point in time, it is possible that evolution of theory choices could lead to long-run path dependence in scientific knowledge.[16] Extension of our model to endogenize the evolution of theory components is an important complement to the current analysis.

ACKNOWLEDGMENTS

The Vilas Trust, Romnes Trust, and National Science Foundation have generously provided financial support. This paper was written while Durlauf was visiting the Santa Fe Institute, whose support and hospital-

16. The failure to explicitly consider whether and how small local differences between theories at a point can lead to global differences over time is, in our opinion, a serious weakness of attempts to argue that social factors have long-term effects on the development of science. For example, none of Bloor's (1976) conjectured alternative mathematics notions is ever shown by him to possess any substantive implications for the growth of mathematical knowledge per se.

ity he gratefully acknowledges. We thank Kim Sau Chung and participants at the Notre Dame Conference on the New Economics of Science for comments on an earlier draft.

REFERENCES

Anderson, S., dePalma, A., Thisse, J. F. 1992. *Discrete choice theory of product differentiation.* Cambridge, MIT Press.

Benaim, M., Hirsch, M. 1994. *Dynamics of Morse-Smale urn processes.* Mimeo. Department of Mathematics, University of California at Berkeley.

Bloor, D. 1976. *Knowledge and social imagery.* London, Routledge and Kegan Paul.

Blume, L. 1993. The statistical mechanics of strategic interaction. *Games and Economic Behavior* 5: 387–424.

Brock, W. 1993. Pathways to randomness in the economy: emergent nonlinearity and chaos in economics and finance. *Estudios Economicos* 8: 3–55.

Brock, W., Durlauf, S. 1997. *Discrete choice with social interactions.* Mimeo. Department of Economics, University of Wisconsin at Madison.

Chalmers, A. 1994. *What is this thing called science?* Indianapolis, Hackett.

Diamond, A. 1988. Science as a rational enterprise. *Theory and Decision* 24: 147–167.

Durlauf, S. 1993. Nonergodic economic growth. *Review of Economic Studies* 60: 349–366.

Durlauf, S. 1997. Limits to science or limits to epistemology? *Complexity* 2: 31–37.

Feyerbend, P. 1975. *Against method.* London, Verso.

Feynman, R. 1965. *The character of physical law.* Cambridge, MIT Press.

Friedman, M. 1953. The methodology of positive economics. In *Essays in positive economics.* Chicago, University of Chicago Press.

Hands, D. W. 1994. Blurred boundaries: recent changes in the relationship between economics and philosophy of science. *Studies in the History and Philosophy of Science* 20: 751–772.

Holling, C., Sanderson, S. 1996. Dynamics of (dis)harmony in ecological and social system. In Hanna, S. (ed.), *In rights to nature.* Washington, DC, Island Press.

Kitcher, P. 1993. *The advancement of science.* Oxford, Oxford University Press.

Kuhn, T. 1970. *The structure of scientific revolutions,* revised edn. Chicago, University of Chicago Press.

Latour, B. 1987. *Science in action.* Cambridge, Harvard University Press.

Laudan, L. 1977. *Progress and its problems.* Berkeley, University of California Press.

———. 1984. *Science and values.* Berkeley, University of California Press.

———. 1996. *Beyond positivism and relativism.* Boulder, Westview Press.

Newton-Smith, W. 1981. *The rationality of science.* London, Routledge and Kegan Paul.

Oomes, N. 1997. *Market failures in the economics of science.* Mimeo. Department of Economics, University of Wisconsin.

Popper, K. 1972. *Objective knowledge.* London, Oxford University Press.

———. 1976. *Conjectures and refutations.* London, Routledge and Kegan Paul.

Putnam, H. 1994. The diversity of the sciences. In *Words and life.* Cambridge, Harvard University Press.

Quine, W. V. O. 1951. Two dogmas of empiricism. In *From a logical point of view.* Cambridge, Harvard University Press.

12

Scientists as Agents

Stephen Turner

A large proportion of the time and effort of scientists is spent in activities that have no obvious place in the traditional model of "basic science discovery leading to application in a marketable product." Some of this time is spent on basic science that does not lead to application in a marketable product; this can be assimilated, though very problematically, to the traditional model by regarding it as failed effort, or the production of a different kind of good, such as generally available scientific knowledge that can be seen as a "public good." But an overwhelming proportion of the time and effort of scientists is spent on a series of activities that fits neither the core model nor its variations: writing grant proposals, negotiating revisions of proposals, evaluating proposals; evaluating other scientists for promotions or appointments, writing and reading letters on their behalf; evaluating students and postdocs, in ways ranging from grading students in classes to making admissions and funding decisions about them; reading, as a peer-reviewer, articles, notes, abstracts, and the like submitted for publication or to conferences, and evaluating, as an editor, the comments made by referees; evaluating other scientists for prizes or awards, membership in honorific bodies; serving as a consultant and evaluating proposals, scientists, or ideas for firms; performing site visits on behalf of funding agencies, accreditation agencies, and the like.

These evaluative activities are tremendously expensive. If one calculated the time that scientists spend on them at the same rate as consulting fees, the costs would be astronomical. But they are inseparable from the way science is presently conducted. Although obviously some of these activities can be internalized to firms, and costs reduced, it is not clear that it would be possible for there to be "science" anything like the science we are familiar with if all of these activities *were* internalized to firms producing for the market. Firms, indeed, rely on the existence of these activities in many ways, from obtaining financing to gaining acceptance for their products, and this has been the case at least since the era in the middle of the nineteenth century in which chemical fertilizers were introduced.

In this chapter I will attempt to say something about the economic character of the "evaluative" part of science, the part that is most completely hidden by usual accounting methods. Yet despite the fact that the costs are hidden, there is no question that these activities involve transactions. The transaction that occurs when a scientist agrees to spend an afternoon evaluating the work of a colleague at another university who is applying for promotion is sometimes a cash transaction: some universities pay a small honorarium for this service, and scientists are paid consulting fees by firms for professional opinions of various evaluative kinds that are substantively identical to those that they give within a university as part of their job, for "free" in connection with colleagues, or as part of their membership on an editorial board, as a journal reviewer, or for a nominal fee as part of a grant-reviewing team. So for the most part the "payments" are implicit. And we lack a coherent understanding of what goes on in these transactions: why they exist at all, what the larger functions they serve are, and what sorts of choices, markets, and competition is involved in them. These transactions, markets, and competitions are the subject of this chapter.

Getting a Grip on the Economics of Evaluation

The costs of the these "evaluative" activities are borne in peculiar ways that have the effect of hiding them: it costs about $50,000 to evaluate and editorially prepare an article for a major medical journal, including the costs of editing and the operation of the editorial office, but not including the cost in time of peer reviewing. These costs are paid through journal revenues, which are largely the result of sales to libraries, and partly through advertising and sometimes membership dues of professional associations. So libraries contribute to the support of this expensive machinery, as do peers, who donate their valuable time, which is in turn paid for by universities and research institutions. If the journal is subscribed to by practitioners, costs are borne ultimately by customers, in this case patients and the third parties that pay on the behalf of patients. The time of scientist professors is paid for by universities, and the journals are paid for out of library budgets. These costs are in turn borne by funding agencies in the form of indirect costs, by students in the form of tuition, and so forth. So the payments are indirect and much of what makes the journal valuable—the contributions of the authors, editorial board, and peer reviewers—are part of a system of implicit payments whose character is difficult to grasp in terms of standard economic models, but nevertheless is quantitatively enormous if we consider the time and effort that these activities involve.

The peculiarly indirect and complex structure of actual payments involving science makes it very difficult for economic analysis to deal directly with flows, transactions, and choices. And since science is paid for in significant part by a complex system of state subsidies, it is difficult to value it. It is even difficult to say what the portion is. The difficulties are evident in the structure of expenditures of universities themselves, which operate through an elaborate system of cross-subsidies. Tuition is paid at one price for undergraduates and for large categories, such as graduate students, without regard for differentials in actual costs of instruction. There are consequently huge cross-subsidies, most of which cannot be determined from an examination of university budgets. The internal economy of universities is itself complex and often purposely mysterious. For example, university managers construct elaborate strategies to deal with fact that there are restrictions on different kinds of funds, and that much of the management of funds consists in using funds collected under one pretext for other purposes. Donations from aged alumni, for example, rarely support daily operations and are often for things that the university could do without. But there are ways of making assets acquired with restrictions, such as buildings, produce funds that the university can employ for other purposes. Thus a dormitory donated by an alumnus can be used to extract fees that can then be used for other purposes. Similarly, of course, for tuition fees.

One of these means has a great deal to do with science: "indirect cost" payments for grants are for the most part payments for assets the university not only owns but in many cases can raise funds (and thus double bill for), such as buildings, library assets, and so on (Turner 1999c). So universities can construct strategies or simply benefit from past donations in the form of money that is unrestricted in its uses and only very indirectly connected to present costs. Indeed, in some universities, particularly state universities where there are legislative restrictions on the use of particular funds, there is more or less open trading in types of funds, such as funds that may be spent on equipment or maintenance as distinct from funds that can be spent on alcoholic beverages, whose exchange value varies in the way that the exchange rates of different currencies vary. Needless to say, these strategies are not reflected in the "books" in any easily accessible way, and it is even very difficult to ask such apparently simple questions of managerial accounting as "which departments do we make money on and which do we lose money on?"[1] This means that it is also very difficult to estimate the total expenditures that universities make on science, and consequently the share of non-grant funding in the total cost of science, and indeed the total cost of science itself. If we are to include the costs of evaluation, the subsidization of scientific publications by library budgets, and so forth, the picture would certainly look different.

There are other ways in which the economic processes in this domain can be approached. One can begin piecemeal with studies of such things as the value of a Nobel Prize in raising money for a biotechnology venture, for example, and such studies certainly point to the profound economic significance of evaluations in science. But because of the peculiar history of science studies, there is another option: one can reinterpret prior findings and theoretical efforts from other fields, notably sociology, in economic terms, in order to get a general understanding of the processes, an understanding that can subsequently be tested against piecemeal studies. Within this strategy there are a number of different possible options, and the approach I will take here reflects a set of choices in the face of these difficulties that needs to be explained in advance.

The first choice is theoretical. There is no question of simply "applying" market models to the activities I will consider. But I do not propose, or see the need for, market analogies, such as the notion of a marketplace of ideas or such notions as public goods. The main difficulty with studying and conceptualizing the vast body of activity I have de-

1. An interesting internal study of this question at the University of Chicago done in 1971, when the university, responding to undergraduate student activism, considered converting to an entirely graduate institution, attempted to do this; the problematic character of the assumptions that the author, economist and dean David Gale Johnson, was forced to make indicate something about the difficulties here, but also point the way to the possibilities of empirical analysis. Interestingly, at this time, which in retrospect is now considered a period of prosperity for the sciences (though shortly after the beginning of a sharp decrease in the rate of growth of funding), the sciences were not money makers for the university. This hints that as competition for funding became more intense and more universities competed, the sciences required even greater subsidies, and perhaps even that the increases in tuition in the following period ultimately, through the complex cascade of cross-subsidization that is characteristic of universities, actually wound up supporting science.

scribed above results from the weakness of conventional models in cases in which the usual simplifications cannot be matched up with economically measurable quantities. The activities themselves, however, are not merely analogous to markets. They are genuinely economic activities, which have real (if almost always implicit and indirect) costs and consequences, and they are engaged in both rationally and self-consciously strategically in a setting in which choices are made that have direct consequences for who gets what and how. Choices are made by agencies that dispense and receive real funds and sign real contracts. The choices that are made greatly affect the decisions of the scientist participants in these activities about how to choose what to do, and what to invest time and effort in.

This approach contrasts sharply with some other approaches. Uskali Mäki in a recent (1997) paper comments on his disappointment with Coase's paper on the marketplace of ideas (1974), which, Mäki observes, is actually about publications such as newspapers (and their costs) rather than about ideas and knowledge.[2] As my comments suggest, I think that basic facts about the costs of activities in science such as the costs of journal articles have been ignored, perhaps because of the difficulties of understanding the web of implicit and indirect transactions involved. And I suspect that the basic facts about the costs of tangible and tangibly economic activities provides a necessary, and perhaps sufficient, place to start. I leave the construction of marketplace analogies and the kinds of analogies needed to apply notions like "public goods" to others, but my bias is that they are unneeded and, in the case of the evaluative activities I will consider here, actually serve to obscure the way in which scientists and the evaluative institutions of science create value. And in the end I think an answer to *this* question suggests a transformation of the way in which science has been understood, both by economists and by the community of science studies scholars.

My second choice has already been alluded to. My approach to the economic structure of these evaluative activities will itself be somewhat indirect. The activities can be interpreted in noneconomic terms, but the noneconomic terms point to and illuminate their economic structure. My core argument will be that there is a clear and an important element that can be understood in terms of, and resonates with, though it does

2. Actually it is about the First Amendment, and the strange contrast between attitudes toward the regulation of newspapers and similar sources of opinion and the regulation of other products.

not precisely match, some familiar economic conceptions associated with principal-agent theory. Some of the similarities are obvious. Clearly scientists act and use science and have their science used in chronic situations of information asymmetry, or to put it somewhat differently, in situations where the users are at best only partly competent to judge the validity of the science that they use, and are also characteristically not even competent to fully understand the science that they use. This situation of chronic lack of ability to fully understand makes the notion of "information," i.e., something that might not be known by one party to a transaction would be transparently understandable to them, of very problematic applicability in these settings. Nevertheless, asymmetries are obviously central to the activity of evaluation in science, and so is overcoming them.

The application of these concepts is obviously not easy even in the paradigm cases of principal-agent theory, and is substantially more difficult in the complex system of mutual monitoring and agency risk spreading that I will discuss. Thus while the general similarity between the problem of principal-agent theory and the problematic of monitoring in science is undeniable, because of the implicit character of transactions in science, there is no straightforward way to "apply" these notions. But there are both actual transactions and implicit transactions with significant actual costs and detectable pricelike variations throughout this system. For example, in some respects the incredibly expensive and time-consuming processes of evaluation discussed above are surrogates for the kinds of monitoring expenses employed in the regulation and approval of drugs by regulatory agencies, which require expensive research conducted according to standardized protocols. But in another sense, these monitoring processes are inseparable from the evaluation system of science and dependent upon it: the research must be conducted by qualified researchers, i.e., those who have subjected themselves to many of the monitoring activities discussed above by becoming Ph.D. scientists. In addition, drug companies are eager to have peer-reviewed researchers publish material related to drug testing.

To ask why this should be so is to enter directly into the complexities of concern to this chapter. A simple, but central, issue is the obsessive concern by the participants in science over pecking orders, particularly with respect to such arcane questions as the value of publishing in one journal over another or the much more extreme difference of, say, an abstract published in an obscure proceedings of a meeting and distributed and printed by a drug company. Are these obsessions irrational?

The facts of science—the "informational contents"—are literally the same in each case: nothing about the "application of basic science" or the public goods model distinguishes the kinds of publication. But drug companies and researchers invest very heavily in these distinctions, with real money and costly time and effort. Are they irrational? Or do these approaches to the economics of science simply fail to understand what these activities and choices are about?

Norms and Counternorms: A Typical Principal-Agent Structure

We may begin with a simple observation. In one of the most famous papers ever written on science, "A Note on Science and Democracy," Merton described four "norms" of modern science: universalism, disinterestedness, "communism," and organized skepticism. Merton inferred these norms, we are told, from the practice of scientists (1942; cf. 1973). Three of the four relate directly to evaluative processes, and the fourth, "communism," turns out to be even more directly relevant to the concerns of this paper.

Merton's analysis has been controversial, but the controversy is itself revealing. One of Merton's critics, Ian Mitroff, argued that the "norms of science" as formulated by Merton were essentially backwards (1974). What scientists actually valued was exactly the opposite of, for example, disinterestedness. They valued passionate dedication to the goals of a project and fervent beliefs in the ideas that were being developed or the methods that were being used. Something similar could be said about the other norms. Yet Merton anticipated much of this criticism, and noted especially that, despite the norm of community possession of knowledge, there were intense priority disputes, and argued that the peculiarity of priority disputes is that they were not matters of egoism alone, but were themselves normative in character. The signpost of their normative character is the fact that disinterested observers often enter into the discussion of priority disputes to secure the norm (Merton 1973, 292). Michael Mulkay (1976) treated the Mertonian norms as an "ideology" in the pejorative sense of a story told to the public (though he conceded that this was also a story that could be employed within science by scientists, i.e., a political mythology), but also argued that, in fact, there are no strongly institutionalized norms of this sort in science. The mythological character of the "norms" turns out to be a useful clue.

All norms exist to restrain impulses of some sort, so there would be no "norms" unless there was something valued, desired, or sought for them to be "counter" to. However, there is a special class of human relationships in which self-denying norms are especially prevalent, and it is this class of relationships that is the subject of principal-agent theory. The theory concentrates on those relationships in which we rely on someone whose interests conflict with ours, and on the costs of dealing with the problem of conflicts of interest. A good paradigm case of this is the relationship between a client and a lawyer. The client trusts the lawyer to exert himself on behalf of the client. However, the client, not being a lawyer, is not in a position to effectively judge whether the lawyer is properly representing the client or giving the client adequate legal council and advice. This is *why* trust is required in this relationship. Not only is the client suffering from a deficiency in information or inability to make judgments, but the lawyer is a person with interests as well which the lawyer can advance, potentially, by cheating the client. Another case of an agency relationship is the relationship between a client and a stockbroker. The stockbroker benefits, as the lawyer might, by doing commission work for the client. The stockbroker also advises the client on what work needs to be done. Similarly the lawyer advises a client not only about what legal steps to take but benefits from the client's decision to take those legal steps that are necessarily costly to the client and beneficial to the lawyer who is paid for carrying out those steps.

The term "trust" has a certain seductiveness, but it is a seductiveness born of the illusion that something is being explained by it. Unfortunately, in the writings of more than a few commentators, trust functions as an unmoved mover, or as a kind of mystery that produces results without having a cause. Although principal-agent theory is not the only approach to trust, it deals with the costs of routine systems of trust that are impersonal and rely on the existence of, or beliefs about, the incentives under which people operate. Incentives means tangible, if sometimes indirect and difficult-to-measure, costs. Just as love is free, but roses are not, the cognitive respect we give science is not purchased by science directly, but is maintained through activities that do have costs.

In the cases of lawyers and stockbrokers, matters are simple: the self-denying normative structure of the codes and rules that these two professions operate under is clear. A stockbroker should be aggressive and motivated by money. A stockbroker who is not would be unlikely to do what he needed to do to keep your business. Similarly, a lawyer needs to be aggressive and motivated by money in order to do anything on your behalf. Yet, at the same time, both lawyer and stockbroker must substitute the client's interest for their own interests. The "norms" of legal representation and stockbroking are characteristically statements of the absolute subordination of the stockbroker's and lawyer's interests to the client's interests. The "counternorm" is that the stockbroker and lawyer should be tenacious and highly motivated, and motivated by the fact that someone is paying them for these services. Even lawyers, when they hire other lawyers, want the shark to be an entirely altruistic shark who puts their interests before the shark's own interests in every respect. And we know that there are some mechanisms for punishing a lawyer who violates the rules, and that indeed the bar association disbars people, uses its dues to support actions involving the punishment of lawyers, and so forth.

At first glance this seems to have little to do with science. Sharks in science are not sharks on behalf of the interests of a client. But science is also about ambition, score keeping, playing by particular rules of the game, and being a shark in debunking false claims. Scientists do all of these things in the context of a large and complex set of institutions, a large number of which are devoted to regulation, which in one fashion or another involve people acting authoritatively on behalf of science. Science is thus "political" in a mundane sense. As I have suggested, scientists make authoritative *decisions* in the name of others, such as decisions on behalf of organizations or collective bodies or in the name of science itself, and have such decisions made about them. Many of these are "gatekeeping" decisions, and indeed the business of gatekeeping is perhaps the primary means of exercising authority in science.

The making of evaluative decisions and the exercise of authority or advisory authority is a pervasive fact of scientific life: in directing the work of subordinates, in asking funding bodies for resources, and the like. But this political "decision-making" character of science is also a largely undiscussed fact—whether by commentators on science, philosophers of science, or sociologists of science.[3] The reason for this neglect, in part, is that these decisions occur under a particular theory or ideology: the idea that the scientists making the decisions are operating neutrally or meritocratically, and that the public role of science itself is neutral. Science is thus mundanely political, but its overtly political features are conceived to be unpolitical. This line of reasoning has the false implication that the activities are somehow inessential to science, or that it

3. There is a good but small literature on peer review (cf. Chubin and Hackett 1990) and on ethical issues (La Follette 1992).

is transparent that there are merely administrative, and that they work so effectively because scientists do in fact always agree on merit.

The problem of concern to me in the rest of this discussion is the problem of what this "regulatory" activity means. My concern is not to debunk the "meritocracy" argument so much as to explain the mystery that it produces—the mystery of why, if merit is so transparent, there are so many and such expensive mechanisms for assuring it. There is a sociological literature, to be discussed below, that treats it as a matter of the maintenance of "cognitive authority," and this is certainly part of the story. But reading institutions politically is not the same as explaining why they exist, how they came to exist, and why they take the form they do. The fact of decision making on behalf of "science" is the feature of these institutions that is the topic of the norm-counternorm structure, and it is this activity—judging and evaluating—that produces many costs and incentives, such as the incentive of publication in a prestigious journal or winning a Nobel Prize that is the proximate goal of many scientists' efforts. But the activity itself deserves some consideration, for it is unusual, especially when compared to the usual subjects of principal-agent theory.

Representation, Bonding, and Membership

All collective or political activities depend on some model—an "ideology," in the nonpejorative sense—of representation (Pitkin 1989). As I have noted, one peculiarity of decision making in science is the idea that any scientist, or at least the most accomplished scientists, can truly represent or speak for "science" as a judging or evaluative body, without being elected or otherwise "chosen." Thus scientists engaged in peer review are understood to represent not their personal or economic interests but to be impartial judges evaluating from the point of view of scientific opinion at its best. Although in practice it is accepted that scientists who have made accepted research contributions to a particular area are the most competent to judge new work in that area, it is also believed that they "speak for science" rather than for themselves when acting as a representative.

Why do these arrangements exist at all? As we have seen, this is an ambiguous question, in that one might choose to interpret decision-making processes in accordance with the political mythology of science and say that they are part of the administration of things rather than the governance of people. Some scarce resources—jobs, research opportunities, and so forth—do exist, and allocating these requires some administrative expense. It might be thought that mere administration—at least if the mythology were true—would be a matter whose costs could be minimized, delegated to scientists who function essentially as clerks, leaving talented scientists for more important work. Obviously that is not how these processes work. And as difficult as it might be to estimate the costs, it is clear that the costs go very far beyond administrative expense, and even the administrative expenses are largely connected with a degree of evaluative scrutiny that occurs in these cases. So the problem of the economic function of these expensive activities is intact. Why are they performed in the way they are?

Central to what follows is the relation between *who* evaluates and the value of the evaluation—something that the "transparent merit" model necessarily denies, and that conflicts with at least some understandings of Merton's "norm" of universality. It is obvious that some evaluations are more valuable than others because the evaluators are more valuable as evaluators, the distinction is more competitive, or the journal more prestigious. The pursuit of more valuable evaluations or certifications has real costs—the costs of investing the time and effort to publish in an especially competitive journal, for example. My thesis here is that there is a market for evaluators and evaluations, and that the existence of this market is critical to understanding science. The market is bilateral; that is to say the evaluators compete with one another—how and why will become clear shortly—and the scientists who are being evaluated compete as well, but also make complex market choices between the forms of evaluation available to them. The market is competitive on both sides. But there are also potential problems with these markets that can make them seriously defective, and it is a matter of policy interest, and interest to scientists, to consider the effects of the workings of these markets, and to alter institutional arrangements in response to these defects.

The markets are about agency relationships, and to understand them it is necessary to understand something about the complexities of the agency structures that occur in science. When scientists exert discipline, reward, exclude, and accept, they act as "agents" or representatives, and when it is done on behalf of a journal or professional society, the society itself is the political embodiment of a collectivity. The simplest agency structure is "representation." The decision maker is an agent of a collectivity, either an actual one, such as the American College of Obstetrics and Gynecology, which is in the business of certification (quite literally) for practitioners, has a publication, and selects scientific articles through the actions of editors who act as representatives, or a hypothetical one, such as the "scientific community" or "science" itself. The agency relationship has costs for the members and for the agents, though the costs may be difficult to estimate and the transactions themselves may be implicit.

Science differs from medical practice in a decisive way. In medical settings (like the law and stockbroking) there are typically single certifiers, such as the American College of Obstetrics and Gynecology, or the bar association. In these monopolistic cases, there are costs, benefits, enforcement, and so forth, but no markets in which certifying organizations compete. In science, matters are considerably more complex. The basic logic of the relationship is nevertheless the same: getting a Ph.D. or submitting a paper to a journal is a transaction in which someone pays for a form of certification. The difference is this: in medicine or law, one either passes one's exams and is certified or one fails. In science, the individual pursuing a career can choose between different forms of certification, add different forms of certification to one's CV, pay different "prices" for different forms of certification and thus develop and act on a strategy for accumulating forms of certification. These certifications are then valued in various marketlike transactions—such as employment decisions—by buyers who themselves can think and act strategically about the market of certifications.

What function does this system of agency relations perform? There are various possible ways to answer this question, but the most straightforward and obvious one is this. The agents or evaluators by the act of evaluation assume risks. The economic function of the activity is to spread risks—risks that arise from wrong answers, scientific error, bad scientists, and so forth. The process here has many imperfect analogues in more familiar economic activities: cosigning for a loan, bonding, signaling, and so forth. My preference is for the notion of bonding: an act in which an agent pays for an assurance (in this case an assurance by other scientists) that the agent will act in accordance with a principal's interests in a situation of information asymmetry (cf. Jensen and Meckling 1976). But the analogy is imperfect in several respects. The asymmetry is not simply a matter of "information" and the assurances take various forms, such as assurances about minimal competence as a scientist, or minimal adequacy sufficient for publication in the case of a journal article. In each case—for example when an academic program awards a degree or a journal accepts an article, the program or journal assumes a risk that its assurances of adequacy will be found out to be false, and the consequence of error is damage to "reputation," which translates into a loss of the value of future assurances of the same type. This feature is central—and for this reason, and for convenience, I will retain the term "bonding."

The term seems merely to be about reputation. Reputation is a deceptively simple notion, however. There is a sense in which reputation is contrasted with reality, and thus that there is some sort of falsity or illusoriness of reputation. But in the case of the bonding that happens in science, this is potentially misleading, for it suggests that no real value is created by the activities of evaluation. Something like this suggestion appears in what Merton called the Matthew effect (1973), by which scientists whose achievements are recognized in various ways "accumulate advantage" so that a scientist who has gone to the right schools, published in the right journals, and won the right prizes is more likely to have his achievements cited. The implied distinction is between the intrinsic value of the science done by the scientist and the increased impact that results from advantage. But if we think of the process of accumulating advantage in terms of bonding, it becomes clear that at each point of accumulation something has actively been done, at a cost, to create value through reducing risks, specifically by distributing risks to people other than the scientist accumulating the advantages. So the total value of the "product" in question, the science, is not only the ideas, the intrinsic value, but the guarantees that come along with it, in the form of risk-bearing actions taken by editors, hiring departments, and prize givers, each of whom has put the value of their journal, department, or prize at risk by their actions. The accumulation of advantage is thus like the accumulation of cosigners to a loan. So where Merton, operating in terms of notions about merit, and concerned to make the argument that science proceeds successfully and without external interference in terms of merit, finds it puzzling and problematic that advantage accumulates, the "bonding" approach finds it understandable.

Why is there so much "bonding" in science? One of the concerns of principal-agent theory is the problem of adverse selection, and there is a sense in which one can see the web of evaluation activities that is so important to science itself as a means of avoiding a large number of potential adverse selection problems. Scientists themselves need to select ideas to pursue, to believe, and to take seriously. Academic departments and businesses employing scientists need to make hiring decisions.

These decisions are almost always made in the face of asymmetries, and where there are risks of making bad choices. "Bonders" provide ways of spreading these risks. To put this very simply, because in science only a few can seriously evaluate an idea, the rest must rely on their certifications of the idea, and thus certification takes on a massively important role.

A somewhat more puzzling question is why are there so many *forms* of bonding in science. The system, incarnated in the CV, is one in which single distinctions do not suffice. Not only does a scientist seek many kinds of certifications, they are sought from, and available from, a wide variety of sources. Understood as a total market or system, it is complex in the following ways: most of the certifications are indirect with respect to matters of truth. What is judged is the minimal adequacy or interest of a journal article or Ph.D. dissertation. The effect is that a scientist acquires various "certifications" from a variety of sources, that these certifications overlap, sometimes involving the personal quality of the scientist, sometimes of a particular article, sometimes, as in the case of prizes, for a "discovery" or "contribution" that spans many years and consists of many items. The fact that they overlap amounts to the building in of redundancy. An established scientist will have passed through many tests, of which the CV is the archaeological record. The existence of all this redundancy is a relatively recent phenomenon in the history of science, and it deserves its own discussion, which I will provide shortly. It must suffice to say that there is obviously a market here, with both demand for bonding and an incentive to supply it. So far I have focused on the demand side of this relationship. I now turn to supply.

Bonding and Value

What I have said here about bonding suggests that assuming risks is a major activity of science, that one might expect that it will show up in the form of transactions, and points out that it does, for example in the phenomenon of accumulated advantage. But this account assumes that there is something of value already there for the transactions to be about, and in a sense this is the central puzzle of the economics of science: what is the value of knowledge? Several points ought to be made about scientific knowledge that bear on this question. First, scientific knowledge is "embodied knowledge," in the sense that it has little or no value as pure information. One needs to know what to do with it, and this requires an investment in scientific education.

This simple fact is critical to understanding the way that scientists solve the problem that any possessor of embodied knowledge faces: how to convert one's knowledge into money. One way is to employ it in the production of things that can be bought and sold—this is the model around which much discussion of science has revolved, and it produces the familiar puzzle of the value of basic science, and such solutions as the idea of science as a "public good." Patent law is a kind of extension of the notion of product that creates an artificial right to produce something that can itself be bought and sold but is valuable only if it either produces something or prevents something from being produced. But much of science cannot be converted into such a right simply because there are no products with sufficiently high market value that patenting makes any sense, and the patent process is limited to things that it makes sense to create rights about. But there are other options for scientists to convert their embodied knowledge into money.[4] One widely employed option is consulting. Here there is no mystery about the existence of a market, the existence of competition, and the existence of market pricing processes.

A primary way in which knowledge can be converted into money is through the transmission of that knowledge to others, at a price. Teaching is a primary source of income, or a large source of income, for the possessors of many kinds of knowledge, such as musicians. Musicians are also paid for performances, and this is a useful model for understanding the value of scientific performances. Musicians, however, are not, for the most part, their own judges. There is a public audience for music. Science, however, is different. There is a public audience for science, to be sure, but it is an audience that assesses science in very general ways, and through its products, but does little to distinguish among scientists, and to the extent that it does, is inclined to grant much greater credit

4. Elsewhere I have argued that it was fatally misleading for Polanyi to have used the word "knowledge" with respect to these competencies or skills (Turner 1994). As I have repeatedly pointed out since, "knowledge" used here is an analogical notion but one in which the distinctive characteristics of knowledge are absent precisely because of the nature of the stuff that is being captured by the analogy (1999a, 1999b). Philosophers sometimes think of knowledge as justified true belief, for example, but the stuff that is being analogized is not, except analogically ("implicitly"), beliefs, and by definition not warranted nor justified in the usual public sense of this term, and consequently not the sorts of things one can say are true. I have elsewhere suggested that the best one can do with these notions is to talk about the habits that one acquires by successfully going through certain kinds of public performances (Turner 1994). It is these habits that I have in mind with the problematic phrase "embodied knowledge." The issues here are primarily of importance to social theory and philosophy, but it is worth at least alluding to them to indicate that there is no easy way of eliminating them in favor of some simple substitute notion.

to the kinds of scientists who write for the public than scientists themselves do.

If we invert the usual way of thinking about the relationship between these performances and judging and think of them on the analogy of the kind of music teacher who charges, and raises, tuition on the basis of successful performances, we have a model for the role of scientific publication, as well as a very good explanation for scientists' concerns about having their successes credited to them. Performances, or discoveries, are a source of demand for the other things that the scientist can do, such as teach and make judgments. This close link suggests an alternative view of Merton's norm of "communism." There is no need for musicians to have an "ethic" of public performance, since performance is, in effect, a condition of their making money through tuition. One can of course, as a musical nonexpert, judge the products of musical tuition, so the link is not so close. Still, performing demonstrates one's possession of the embodied knowledge that one charges tuition to transmit to students.

In science, unlike art, production and assessment are closely related, and indeed, since developing hypotheses and testing them almost invariably have implications for the validity of previous hypotheses, there is a sense in which all science is authentication. Yet the activity of criticism and production is separated in various ways, in that there is a moment of pure thought, so to speak, in which hypotheses are formulated; and other moments when hypotheses are, sometimes very expensively, tested; and another when scientists assess the work of other scientists by means such as peer review or simply reading and criticizing the works of other scientists. Finally, with art, there is an object whose value is increased by the efforts of art historians. In science, matters are much more obscure. People are bonded; ideas are bonded; some bonding, such as degree granting, is the sort of thing one pays for, some, such as publication in a famous journal, is not.

One might say, however, that this pattern is characteristic of the old economics of science, in which a particular kind of competition for talent, between university departments of physics, chemistry, and the like, was central. Whether there has been, or might be, a radical change as a result of present changes, such as the corporatization of bioscience, is a question I will take up in the conclusion.

The Market for Bonders

The problem of judgment of peers is that these judgments themselves require bonding because there are known conflicts of interest that arise in judgments of competitors and judgments of work in the same specialty. Evaluators may be biased in judging the general significance of work in a particular area that resembles theirs, and, if they are competing for fame and fortune, they have a general interest in seeing work that cites and extends theirs being funded. By the same token, they may have an interest in impeding the research of a competitor. Competition of exactly the same kind exists between "bonders" as such. Journals compete, prizes compete, and departmental programs and universities compete. Departments compete with corporations in the hiring of scientists. And there are various incentives to the creation of novel bonding mechanisms, such as official journals and scientific societies.

The potential risk, from the point of view of the journal or the department, is the risk of bonding too much, or the wrong things, so that credibility as a bonder (and consequently the value of implicit "certification") is diminished on the one hand. The more demanding the standards, the more prestigious the certification might be. But if the standards are different in such a way that there are well-recognized achievements that fail according to these standards then the value of certification is likely to diminish generally, for the users of certification will be unable to say when the certification is useful and when it is not. Overly generous certification obviously risks the same effect, especially if it is potentially biased, for example in the direction of allowing dangerous or useless therapies. The market, in short, demands a certain uniformity and punishes those who deviate from it, though the uniformity is in effect a moving equilibrium. But the "market" is not closed: the rise of new forms of scientific activity may make previously important certifications peripheral or worthless.

Central to what I have argued here is that there are different kinds of bonding—from journal gatekeeping, to grant giving, to degree granting and award giving—that are redundant and overlapping. By this I mean that no single mechanism is sovereign or final. It must be noted that it is part of the founding mythology of modern science that no person is authoritative in science. The experience of the prosecution of Galileo by the church was formative: scientific bodies were generally reluctant to certify scientific truths, to avoid the risks of error. If we recognize the lack of sovereignty in the agency activities of scientists as bonding representatives, it becomes obvious that under normal historical cirsumstances—that is to say circumstances of polyarchy rather than a Stalinist uniformity—there will always be a variety of overlapping communities, consumer groups, and forms of certification that will sometimes agree and sometimes fail to agree, sometimes produce equivalent bonding results and sometimes not.[5]

The model I have constructed here points toward some way of characterizing the intentions of the parties to at least some of these processes. Buyers of science have various purposes, such as being cured or annihilating their enemies by the use of nuclear weapons. The consumers have a need for certified knowledge because it is cheaper for them to accept certification than to find out for themselves. But there may be competing certifiers or bonders. In the simplest case there is a kind of equilibrium of bonding. The "results" of science are thus standardized so that one can choose between the scientifically bonded and the unbonded. In situations of scientific change, the situation will typically be more complex. Users may be buying, and happy with the results of, scientific work that is not universally certified or bonded by scientists but only by some small faction. This may persist indefinitely, though to retain the credibility of bonding as such, scientists have an interest in giving some sort of final judgment on particular claims, especially if those claims have market appeal and are bonded by competing bonders. To fail to recognize a genuine new achievement is to become less valuable as a bonder. And it is perhaps for this reason that bonding agencies seldom subject themselves to the risk of making direct and unqualified endorsements of scientific claims. In the end, the majority does not rule, but rather the credibility market and its demands rule.

It is this market that produces the distinctive norms of science described by Merton. "Disinterestedness" is just a norm of bonding as such. A journal or prize that was known to be biased would lose some of its value for bonding. "Organized skepticism" is not so much a norm as a description of a situation in which participants recognize the inadequacy of the means they possess to judge the claims of science—the very situa-

5. Latour's actor network theory (1987), from the point of view of this chapter, is in effect a description of the connections between parts of the process of the rise and fall of scientific ideas, but it is not an explanatory model. It works, to put it very simply, by a kind of inverted behaviorism. Rather than denying intentionality to human beings and explaining them and their interactions as though they were purely causal phenomena, as behaviorism attempted to do, actor network theory inverts this explanation and grants a kind of quasi intentionality to all sorts of objects, including the subject matter of science, making networks into quasi-intentional conglomerations. Nevertheless, this actually explains nothing because the relevant intentions are not themselves explained by anything. What is right about the Latourian model is the denial of a point of epistemic sovereignty, such as "the scientific community." What is lost is any account of the intentions themselves, and how they relate to one another to produce the particular social forms that arise in science.

tion that produces the demand for bonding. The term "organized" is an important qualification: there are actual means of bonding, such as degree granting and the like, between which the skeptic chooses. "Communism" or publicity is another matter entirely, and I think one on which, as I have suggested, Merton was simply in error. There are advantages to keeping trade secrets if they are secrets one can exploit. There is a patent mechanism for preserving the secrets' use and protecting it from copiers. Because science can be produced as a public performance or, sometimes, kept secret and used as a product, scientists have a choice: the choice is dictated by their interests and those of their funders, and does not always lead to making scientific results public. "Universalism" is a descriptive feature of the market and another way of expressing the political mythology of representation. If by some bizarre circumstance, scientists were confined to something like a sovereign buyer such as Stalin, it would be rational to be nonuniversalistic, as indeed Lysenko and his associates were, and of course Hitler's scientists were as well. Is this, or Islamic science, some sort of normative aberration? In the sense that granting epistemic sovereignty to any particular tradition or agent is a violation of the basic political mythology of science that is the foundation of its notion of representation, it certainly is. But universalism does not demand universal agreement, just the agreement of representatives speaking properly for science.

The New Situation of Science

The basic argument of this chapter has been simple. Bonding is an economically essential feature of science that is a result of the high cost of assessing alternatives personally. Bonding, certifying, accepting, using, and so forth are real acts, acts that are not at all cost free. In the end, however, the overall benefits are very close to those conferred by the rule of law and the reduction of transaction costs that the rule of law enables. The real scientific revolution is the revolution that substantially reduced those costs by the emergence of an effective bonding market, itself the product of incentives, notably incentives rooted in the imperative of making money out of the possession of knowledge.

Scientists and nonscientists alike rely on, or treat as information, things that scientists say without having any knowledge of the opinions of this handful of certifying figures, and the certifying figures are themselves certified and recertified by the many and highly redundant indirect means discussed above. Instead, there are many certifying mechanisms of the sort I have described that the user does rely on, such as the fact of acceptance in a journal, the granting of research funding and support of the project, the high academic position of the person reporting the finding, and so forth. The striking development from the kind of gemeinschaft of early science is that science is now able to proceed effectively by accepting the certifications and recognizing the certifying mechanisms of other groups in science. And, unlike face-to-face mechanisms, these mechanisms of certification can be extended in such a way as to permit an enormously complex division of labor in which no one's understanding extends very far. If we can trust the market for bonding to produce adequate bonders, we can trust the results of science where there are such markets. In short, it is the market for bonding in which we place our trust, and on which our trust in science properly rests.

From the point of view of this chapter, the basic question raised by the corporatization of science, and especially biotechnology, is whether the system described here is historically doomed—bound to an era of academic competition that perhaps has already passed its peak and is being transformed into something else entirely. The usual way of thinking about what it might turn into is a world that is indirectly dominated by the demands of investors, of the source of the funds necessary to produce science. Since these funds far outstrip the funds available through teaching, it stands to reason that they will become—as government funding became—the locus of competition. But government funding of science was in effect an extension of the academic system, in which academic prestige governed peer review and if anything simply made it more powerful. So the new question is this: is the autonomy of science in the sense of scientific control over the standards of science compromised by these new funding sources?

To begin to address this question it is perhaps useful to discuss some ways in which the academic-governmental system itself has sometimes gone wrong or been accused of going wrong. It is well known, for example, that there are critics of the HIV-AIDS connection who argue that the huge AIDS research establishment is committed to a hypothesis and refuses to fund other approaches. High-energy physics research facilities and telescopes, notoriously, are scarce resources: opportunities to do research are allocated by like-minded committees, with the effect that the possibilities of research are limited. When there are no alternative sources of funding, these committees are effectively monopoly buyers, and "market" competition disappears in favor of science by decision. Nuclear power researchers failed to preserve their independence from the industry (Morone and Woodhouse 1989). In many disciplines, notably economics and sociology, it has been argued that the power of certain top journals, together with the peculiarities of competition for space in those journals, has served to elevate particular styles of research and exclude others, thus distorting the standards of the disciplines and the disciplines themselves. In each of these cases there is a dominant player whose conduct dictates the conduct of others, including the standards of evaluation.

These problem cases raise some serious questions about whether the arrangement I have described in this chapter, in which a scientist's acting as an agent for science provides certification or what I have called bonding independently, competitively, and in different but overlapping ways, is vulnerable to a shift in the weights of forms of market power of the sort that corporate science represents. In its most advanced form, the form familiar to us, it created a powerful internal market and set of incentives that propelled the development of science, reduced the risk of bad science. But the "system" I have described is an oddity in the history of science. University science and the research university came to their position of dominance in science only during the twentieth century. The idea of scientific authorship and the system in which a scientist acting as an agent certified the science of others, which is so characteristic of twentieth-century science, was very weakly developed.[6]

It would be ironic, but hardly surprising, if the success of science governed by this evaluative market were to lead to the demise of the system itself. The system depends on mutual monitoring through implicit agency transactions. But corporations do not participate in this process. They don't benefit by editing journals, and the journals would probably not be trusted if they did. Yet the problems of monitoring and evaluation that this loose system solved do not vanish. The "information asymmetries" that always exist in science still exist. Investors may need, or come to need, more and different types of monitoring and information. Conflicts of interest would still arise over evaluations. And the fact that cash has already invaded the "old" system of bonding in the form of payments

6. I am grateful to David Stodolsky for bringing the fascinating work of D. A. Kronick (1988) to my attention. Kronick discusses the evolution of authorship, and the previous use of anonymity and collective publication, and the demise of these alternatives to the present system. Kronick, however, uses the language of Merton, and thus runs together two things that from the point of view of the present paper ought to be sharply distinguished: the collective nature of science and the act of publication of particular anonymous contributions by particular scientific collectivities. The latter I take to be a case of collective endorsement that effaces the contributor. Kronick seems to think of it as a microcosmic instantiation of the collective nature of science itself.

for evaluation, suggests that a surrogate for the old system of cashless "credit" is in the process of emerging. This will almost certainly be a system that is less "public" than academic evaluation was, and in which academic science and its scheme of competitive incentives is marginalized and made ineffectual, and its web of indirect devices replaced by more direct forms of monitoring and assessment. The real irony is that this replacement will almost certainly, and justly, command less respect by the consuming public. The strong resistance to genetically modified foodstuffs in much of the world testifies to the suspicion in which corporate science will likely be held. Replacing the old evaluative system may thus kill the goose that laid the golden eggs.

ACKNOWLEDGMENTS

The writing of this chapter was supported by a grant from the National Science Foundation Ethics and Values Studies Program (SBR-6515279). Revisions were done at the Swedish Collegium for Advanced Studies in the Social Sciences and also with the support of a grant from the NSF Science and Technology Studies Program (SBR-9810900). I would like to acknowledge the suggestions of several commentators on earlier versions of this paper, especially including Phil Mirowski, George Alter, and Richard Swedberg.

REFERENCES

Chubin, Daryl E., and Edward J. Hackett. 1990. *Peerless science: Peer review and U.S. science policy.* Albany: State University of New York Press.

Coase, R. H. 1974. The economics of the First Amendment: The market for goods and the market for ideas. *American Economic Review* 64:384–91.

Jensen, Michael C., and William H. Meckling. 1976. Theory of the firm: Managerial behavior, agency costs, and ownership structure. *Journal of Financial Economics* 3:305–360.

Kronick, D. A. 1988. Anonymity and identity: Editorial policy in the early science journal. *Library Quarterly* 58:221–237.

La Follette, Marcel C. 1992. *Stealing into print: Fraud, plagiarism, and misconduct in scientific publishing.* Berkeley and Los Angeles: University of California Press.

Latour, Bruno. 1987. *Science in action: How to follow scientists and engineers through society.* Cambridge: Harvard University Press.

Mäki, Uskali. 1997. Free market economics of economics: Problems of consistency and reflexivity. Paper presented to the New Economics of Science Conference, Notre Dame, Indiana.

Merton, Robert. 1942. A note on science and democracy. *Journal of Legal and Political Sociology* 1:15–26.

———. 1973. *The sociology of science.* Chicago: University of Chicago Press.
Mitroff, Ian. 1974. *The subjective side of science: A philosophical inquiry into the psychology of Apollo moon scientists.* Amsterdam: Elsevier.
Morone, Joseph G., and Edward Woodhouse. 1989. *The demise of nuclear energy? Lessons for democratic control.* New Haven: Yale University Press.
Mulkay, M. J. 1976. Norms and ideology in science. *Social Science Information* 15:637–656.
Pitkin, Hannah. 1989. Representation. In Terence Ball, James Farr, and Russell Hanson, eds., *Political innovation and conceptual change.* Cambridge: Cambridge University Press.
Turner, Stephen. 1999a. Searle's social reality. *History and Theory* 38:211–231.
———. 1999b. Practice in real time. *Studies in the History and Philosophy of Science* 30:149–156.
———. 1999c. Universities and the regulation of scientific morals. In John M. Braxton, ed., *Perspectives on scholarly misconduct in the sciences.* Columbus: Ohio State University Press.
———. 1994. *The social theory of practices: Tradition, tacit knowledge, and presuppositions.* Chicago: University of Chicago Press.

PART V

Contours of the Globalized Privatization Regime

13

Making British Universities Accountable

In the Public Interest?

Shaun P. Hargreaves Heap

1. Introduction

Knowledge is not a commodity that is well suited to regulation via the market, primarily because its production is fraught with uncertainty and once produced it tends to disseminate freely (see Arrow 1962, chapter 4 in this volume; and Dasgupta and David 1994, chapter 7 in this volume). As a result there is a strong argument for some kind of public intervention. Two obvious examples are the patent system and the public funding of research in universities. Unfortunately the optimal design of such interventions is difficult for precisely the same reasons that make knowledge ill suited to regulation by the market. In the case of publicly funded research in universities, the difficulties arise because the output of research cannot be easily given a market value and it is difficult to monitor research effort, with the result that university researchers are liable to pursue their own idiosyncratic research or other interests. The latter, the agency problem, has come to the fore in recent times with the "informational turn" in economics (see Mirowski and Sent, introduction to this volume).

The traditional response to the agency problem in the UK has been to rely on the norms of academic life and the institution of competition for research grants. However, since 1986 the UK government has introduced a new mechanism of control: the Research Assessment Exercise (RAE). Under this arrangement, each university department's research is assessed every four years on a scale of 1 to 5, and public funding of the department for research is based on this ranking until the next RAE. This paper is concerned with how the RAE process for allocating research funding has affected research in the UK. Since there have been notable changes in the funding of science in many countries (Slaughter and Rhoades [1996, chapter 1 in this volume] see these changes in part as a consequence of the end of the Cold War), this paper can be viewed as a case study in what is a more general "rationalisation" of science.

I begin in section 2 with a sketch of the RAE and its role in the resource allocation process. Section 3 focuses on how the RAE process has influenced the level and type of research activity, using the reports of two independent consultants, commissioned by the Higher Education Funding Council for England (HEFCE) and a survey of faculty at the University of East Anglia (UEA). Section 4 assesses how these changes have contributed to creating value for money in these public expenditures. There has been a clear increase in research activity, and in this sense there has been success with respect to the agency problem. But it is not always clear that the type of research has also changed in ways that contribute to creating value for money. The latter is worrying not only because it detracts from the benefits of greater research activity, but also because some of these changes seem to have been unrecognized by one of the independent consultants. One of the key sources of concern here is the way that the RAE seems to have encouraged research to take on many of the attributes of research in the private sector.

This tendency connects with recent arguments in the literature on the economics of science concerning how the distinctive norms of "science" or university research may be giving way to those of "technology" or commerce (see Dasgupta and David 1994, chapter 7 in this volume; Slaughter and Rhoades 1996, chapter 1 in this volume; and Stephan 1996). This is worrying because it means university research is less likely to fill the gaps where markets fail in the production of knowledge. Indeed the claim on the public purse is bound to be weakened if university research increasingly replicates rather than complements the type of research that is undertaken in the private sector. Or to express this more starkly and paradoxically: the very action of making universities more accountable has actually led universities to give a less good account of themselves with respect to compensating for the market failures that are likely in the production of knowledge. It seems natural in these circumstances to seek ways of reinforcing the distinctive academic norms of science. I turn to this issue in section 5, where I argue that what seems to be a rather peculiar sense of norm that has emerged in the economics of science literature is liable to prove deeply misleading on how this might be achieved and I suggest an alternative course of action.

2. The RAE System

Research in UK universities is publicly funded through two sources: research councils that give awards for specific projects (through some form of competition); and the Higher Education Funding Councils (the HEFCs) that give block grants to universities for the activities of teaching and research. There are now separate HEFCs for England, Wales, Scotland, and Northern Ireland; formerly there was a single funding authority for the UK: the UFC, and before that the UGC. In 1986 the UGC introduced a mechanism of formula funding for the allocation of the block grant, and as part of this exercise it launched the first RAE. A small part of the research element of the block grant was allocated using these assessments in 1986. Further RAEs have been conducted in 1989, 1992, and 1996, and since 1992 the bulk of the research component of the block grant has been allocated using these assessments. For instance, this element (referred to as QR) now accounts for 94 percent of the research funds that are distributed by HEFCE.

The form of the assessment and the way that a ranking affects funding has changed over time in several respects. Here is a snapshot of the system around the time of the last RAE. Each department filled in a form for this RAE by 1 June 1996 detailing each assessed person's four publications submitted for evaluation purposes, the department's research grant and contract income, and numbers of postgraduate students from January 1992 to March 1996; an essay explaining the research strategy of the department, its successes, and its future plans (maximum length of six pages) was also submitted. The department could choose which people to forward for the assessment; to be eligible, they had to be in their current position on 31 March. The department chose the people knowing that the HEFCE research funding would depend on the grade received in the RAE and on a volume indicator (and each full-time academic forwarded for assessment would count as 1 in the volume indicator). Thus each department faced a potential trade-off between the number of people generating research income and their per capita level of support, which was based on the RAE grade.

An integer scale of 1 to 5 was used in grading departments, except that the 1996 assessment level "3" was subdivided into 3a and 3b and panels were given the option of awarding a 5* to some departments. The panels were selected through consultation with the appropriate professional associations and research councils. For example in economics, there were three parts to the process. The HEFCs drew on the advice of the 1992 chair (Tony Atkinson) of the panel to appoint the 1996 chair (David Hendry). The Committee of Heads of University Departments of Economics invited nominations from their own membership and forwarded the five names with most nominations to the chair of the panel. The chair accepted this list and drawing on the advice of other learned and professional bodies and in consultation with the chair of the Committee of Heads of Departments in Economics, he appointed four further members (details from Lee and Harley 1997).

The translation of a grade on the 1 to 5 scale into per capita funding was based between 1992 and 1996 on the following relativities in England: 5/4 rated = 4/3; 5/3 rated = 2; 5/2 rated = 4; and 1-rated departments got nothing. In 1997, the relativities were sharpened using an amended scale. Departments rated 2 or 1 received nothing, and the relativities became: 5*/5 rated = 1.2; 5*/4 rated = 1.8; 5*/3a rated = 2.7; and 5*/3b rated = 4.05. These relativities together with the volume indicators and a global allocation of research funds to a subject area (informally known as its "pot of gold" or more recently in HEFCE documents as the subject's "quanta") enables the HEFCs to distribute each subject's pot of gold to the various departments in that field. The origin of the global sum available to each subject area for this purpose is somewhat obscure (and perhaps usefully lost in the institutional metamorphosis from UGC through UFC to the HEFCs today), but apparently when the UGC first began this exercise in 1986 it relied on a desk diary exercise conducted in the late 1960s regarding how academics in different disciplines spent their time. In 1997 the system changed. The base value for a subject quanta was determined by the number of research active staff times a subject weighting that depends on which of the three groups (laboratory based, part–laboratory based, and mainly library based) the subject falls into and that should reflect the relative cost of doing research in these broad areas. The quanta are then adjusted to include a "policy factor" that will reflect "judgements on national needs and international standing."

The motives behind the introduction of the RAE and formula funding do not seem to be in dispute. The purpose was to make universities accountable for their use of public money. This was seen by the UGC as an imperative in the 1980s because of perceived political pressure. The origins of this pressure need no real explanation. They were partly political in the broad sense that the Thatcherite revolution involved everyone learning that "there's no such thing as a free lunch"; they were also partly political in a narrow sense that, with the incorporation of the old polytechnics into the university system, it seemed unfair to give all the research funds to the old universities just because they were older than the ex-polytechnics; and they were also partly narrowly economic in the sense that once universities had become accountable it would be easier to control public expenditure, both in the aggregate and in its allocation between uses. As a result the HEFCs now typically list the principles used in allocating research funds to demonstrate their accountability. For instance HEFCE (1996) lists several guiding principles and the objectives that lie behind them. The first one, "Quality and selectivity," is particularly helpful for the purposes of this paper because it establishes explicitly the perceived part of the RAEs in making universities accountable.

> a. Quality and selectivity. To make best possible use of public money, the Council allocates funds selectively according to international and national standards of excellence in research as assessed through RAEs. (4)

Thus the RAEs are a tool for making the "best possible use of public money." There are other guiding principles that reflect other objectives such as avoiding duplication of other funding agencies, encouraging research potential, the development of new and interdisciplinary areas, and meeting national needs for research. These are perhaps best understood as offering an insight into how the "best possible use" of funds will be judged. None though is explicitly connected to the RAE exercise and the allocation of QR funds. Although it is clear how meeting something like national needs can be incorporated directly into the RAE allocations through the proposal to adjust subject quantas by the policy factor. Indeed, it is perhaps worth noting that there was a general concern during this period to make research contribute to the economic and social welfare of the country, reflected for instance in research council statements from this period and in the creation of the Technology Foresight Programme.

3. The Effects of the RAE

The broad institutional facts of the allocation that has resulted from the RAE process are well known. The distribution of the HEFCs' research funds is concentrated on a few institutions. The top institutions have the lion's share of the allocated funds. In part this is accounted for by the fact that they also have the lion's share of the research active staff in British universities. In addition the top institutions have a disproportionate number of the best researchers (as judged by the RAEs) with the result that their share of QR funds is disproportionate to their share of staff. For example, in England, in 1994–95 the top five universities received 29 percent of QR and contained 21 percent of the research active staff, the top ten received 44 percent of QR and contained 36 percent of the research active staff, while the bottom 50 percent of universities received only 5 percent of QR and accounted for 13 percent of active researchers (see Segal, Quince and Wicksteed 1996, hereafter "SQW"). In this section, I shall focus on the way that the RAEs have influenced the level of research activity and the type of research conducted in English universities.

1. The Level of Research Activity

The most obvious effect of the RAE is the boost to research activity. This is evident from the two independent assessments of the exercise performed by consultants on behalf of HEFCE (see McNay 1996 and SQW). It is also reflected in answers to a survey of faculty conducted in my own university. The two consultants (and they agree with experience at UEA) find three aspects to this increase.

First, active researchers now spend more time on research. This is readily understandable as institutions and individuals have responded to the incentives that now exist to do research. Thus at an institutional level it is typical to find that some extra teaching has been allocated to the nonresearch active staff to release time for the research active and that there is better central organisation and support for research. At an individual level, people now perceive even stronger incentives to conduct research because there is a closer connection with promotion and salary and so allocate more time to research at the expense of leisure and teaching time.

Second, the largest increase in research activity has been among the most able researchers (in an RAE sense). This effect arises in two ways: the most highly rated departments (which have on average the most able researchers in RAE terms) have received a disproportionate amount of the research funds and so need to spend less time on average on other income-generating activities like teaching; and within departments, the most able researchers have typically received most relief from other duties. Thus both among and within departments there has been a tendency for the increase in research efforts to be concentrated on the most able researchers. These two effects arise as a result of the interaction between the national system of selective allocation and the internal resource allocation decisions at the university level, both with regard to the funding of departments, and at the department level, with respect to individual teaching and administrative loads.

In fact, it seems from the consultants' reports that there has sometimes been an interesting difference in approach at these two levels of university decision making. At the level of university allocations among departments, there have been various responses to the new incentives created by the RAE process of research funding. Some have closed or reduced the size of poorly performing departments and switched resources into better-performing ones (i.e., they have reinforced the national system of selective allocations), while others have attempted to build up the poor performers through explicit or implicit reallocations from the earnings of the better-performing departments (i.e., they have mitigated the effects of the national system of selectivity). Doubtless, some of the differences in response reflect differences in objectives. Some universities wish to see strength across the board while others seek excellence in a few areas. Some wish to maximize income while others want high RAE ratings, and so on. The decision is also complicated because changing the activity level of different departments can affect not only research income but also teaching income, and in ways that are typically not well understood because the details of HEFC rules in this respect are often opaque. Nevertheless, even when universities have decided to build up poor departments, neither consultant found that these reallocations amount to more than a partial offset to the systemic bias of the national funding system toward the high-rated departments. As a result, decision makers at the departmental level face a similar set of incentives to those making decisions at the university level; and they must make similar strategic decisions over how best to allocate their research funds (i.e., between researchers of different ability). Here, however, the consultants find much less diversity of approach as most departments have opted to boost disproportionately the time of their most able researchers.

Third, there has been an increase in the number of research active staff in most subjects (economics is one of the exceptions). All universities seem to have engaged in staff restructuring programs where nonresearchers have been encouraged to take either early retirement or extra teaching and administrative duties. When early retirements have resulted, departments have often replaced the individual with someone who is research active, thus boosting numbers in the aggregate. This boost is clear, but also probably underrecorded in the national data on numbers included in the RAEs, as the early returns to the RAE probably included a number of marginally research active individuals and as departments have learned that a tail of poor performers can significantly affect the overall rating of a department, these marginal researchers have been dropped. In other words, the criteria for identifying the research active used by departments when entering people in the RAE have become more restrictive over time, thus masking in these statistics the tendency for the numbers of the research active to increase.

Crude statistics convey some of this change. For instance, nationally there has been increase in the average grade received by all departments (McNay 1996). Likewise, 81 percent of the sample at UEA thought the volume of research activity had increased, while only 3 percent thought it had decreased in their units. To use biology as an illustration, the numbers of active researchers increased from 1,382 in 1992 to 1,546 in 1996. Anecdotally, one respondent to the questionnaire at UEA remarked that the RAE had made "everyone aware that research is one of a university's products . . . previously research was regarded more like a leisure activity which faculty did or did not do as they chose."

2. Changes in the Type of Research

SQW discuss several possible changes in the type of research. On the matter of new initiatives and interdisciplinary research, they judge that the impacts have been neutral essentially "because QR has been concentrated in the larger and stronger universities and these are often best placed to assume the risk in launching new initiatives. They also have the breadth of skills required for interdisciplinary research" (p. iv). This is, of course, true, but it overlooks the powerful encouragement to disciplinary research that comes from being judged by a panel of peers from within the discipline. Indeed, the typical way in which these panels have been constituted is with specialists from the discipline, and when confronted with interdisciplinary work, they have sought the advice of the panel of the other discipline involved in the research. This is hardly a process that is likely to reward interdisciplinary work when the point of that work is precisely to transcend the boundaries of each discipline. It is more likely to produce a spurious agreement between the panels of the respective disciplines that interdisciplinary research is less valuable. Indeed, McNay (1996, 24) is much less sanguine with respect to the fate of interdisciplinary research, noting that 46 percent of department heads and 20 percent of staff in his survey report that the disciplinary base of assessment is a hindrance to the development of interdisciplinary research.

With respect to the quality of research and its orientation, SQW's view is most clearly expressed when discussing how greater selectivity, concentrating more funds on the top universities, might affect research outcomes:

> We would therefore expect these HEIs [the top higher-education institutions] to produce more high quality research. We would however, note that the tendency is for such HEIs to focus on research council funds and basic research and to become relatively less involved with industry. (44)

The downside is, they note, the impact on low-rated departments. These come in two types according to the report. One does research similar to the top-rated ones and here "the higher outputs we would expect from the stronger HEIs would more than compensate for decreases elsewhere in the system" (44). The other has developed applied, specialist niche research expertise, and while there has been some help from HEFC funding, this has not usually been a key to their initiation and so is unlikely to be affected. Thus, as far as quality is concerned, they judge that selectivity has raised the average quality of research without obviously damaging the national balance between basic and applied research.

What underpins this analysis are two beliefs about research in universities; and both are in different ways potentially misleading. The first is that selectivity concentrates research staff at large departments and as there are economies of scale, this raises quality. On the basis of the consultants' own evidence it is a moot point whether selectivity has concentrated research staff in this way. Furthermore, since the publication of SQW's report, the results of the 1996 exercise have been published and it is possible to see how research staff numbers changed after the 1992 assessment and funding arrangement. Tables 13.1 and 13.2 reproduce the figures for average size and average change in numbers between 1992 and 1996 by RAE rating in 1992 for selected subjects in the sciences and arts respectively. Only one subject, chemistry, is close to fitting the SQW model; the other subjects have seen growth in numbers at most RAE rating levels and the largest growth (both proportionately and in terms of absolute numbers for the discipline as a whole) is often in moderately rated departments (i.e., the 3s and 4s).

Suppose, however, for the sake of further examination of the SQW argument, selectivity did lead to an increase in the size of the most highly rated departments; then there is the question of whether there are econ-

TABLE 13.1. Faculty Numbers in Selected Science Subjects

Subject	RAE Score	Number of Departments	Average Size, 1992	Change in Size, 1992–96
Biology	5	8	41.9	5.4
	4	13	28.9	4.1
	3	16	18.7	4.0
	2	18	14.8	−1.8
	1	4	7.6	0.3
Chemistry	5	7	38.2	5.6
	4	13	29.8	2.4
	3	10	22.1	−0.2
	2	10	11.4	0.8
	1	11	9.6	−6.3
Computing	5	8	30.5	4.4
	4	15	19.4	2.1
	3	4	27.0	−2.0
	2	3	14.0	4.0
	1	3	13.3	−12.0
Environmental science	5	4	23.6	4.6
	4	2	19.5	14.4
	3	3	12.0	−1.0
	2	10	9.9	0.5
	1	6	5.7	−4.5

omies of scale. The evidence offered by the SQW report is "a strong correlation between RAE grades and size" (42). This evidence is presented in a figure where total number of research active staff in an institution are plotted against the average RAE grade. This seems a strange way to test for economies of scale, not the least because economies of scale are usually thought to be subject specific. Or to put this slightly differently, few people have argued that increasing the numbers of humanities staff raises the research productivity of those in the sciences. In fact, tables 13.1 and 13.2 indicate that there is a similar correlation at the subject level as average size does seem to rise with 1992 RAE grade. (Although there are exceptions, such as media studies. Also, in 1996 the correlation weakened in several subjects, notably economics, where it ceased to be significant.) The deeper problem with such correlations, however, is that they are not a good test of economies of scale. Since highly rated departments are better funded for research and so have the opportunity to hire more staff, one would expect higher-rated departments to be larger for this reason as well as because of any putative economies of scale. Indeed, as I have just indicated, the logic of the

TABLE 13.2. Faculty Numbers in Selected Arts Subjects

Subject	RAE Score	Number of Departments	Average Size, 1992	Change in Size, 1992–96
English	5	5	39.5	2.8
	4	13	21.1	1.8
	3	14	17.0	3.2
	2	10	8.8	2.7
	1	12	5.3	2.3
History	5	5	27.3	−0.6
	4	20	25.5	3.3
	3	17	13.2	0.9
	2	18	6.4	2.4
	1	8	3.8	1.7
Sociology	5	5	26.0	5.3
	4	10	14.0	5.2
	3	17	11.3	3.1
	2	14	10.4	1.8
	1	10	8.2	−1.2
Economics	5	10	35.7	0.3
	4	9	18.6	−2.4
	3	14	15.3	−1.3
	2	9	10.7	−2.7
	1	4	11.0	−7.0
Media studies	5	3	5.8	4.0
	4	2	4.0	3.8
	3	6	7.7	5.4
	2	9	5.0	2.9
	1	4	4.3	3.1

SQW argument depends on this reaction by departments. Thus, insofar as the SQW argument begins to make sense, it means that the putative correlation demonstrating economies of scale correlation could result from the national system of selective funding.[1]

The second belief is that there is a diversity of research among British universities with those enjoying high public funding concentrating on basic research and having little connection with industry, while those institutions that have received little public funding have tended to rely more on industry and government funding for applied research. The evidence for this diversity given in the report is financial: the fact that high-

1. This is a matter of some consequence because HEFCE (1996) makes plain, on the basis of the review conducted by SQW, that selectivity should be "no less than now" (7). The new relativities have now been announced and they have broadly maintained the relativity between 4s and 5s, but they have reduced funds to 3s and 2s; thus, selectivity has increased.

QR institutions rely on research council external funding while low-QR institutions rely on high industry and government external funding. Doubtless this picture is broadly correct, but it fails to capture some of the finer-grained changes that seem to be occurring in QR-funded research in universities. For example, in the survey UEA faculty perceived, on balance, that the RAE had encouraged research with quick payoffs, research similar to that found in commercial laboratories and directly relevant to the economy, and that it had discouraged research with uncertain outcomes and research on basic or fundamental issues; the surveyed faculty also believed that it has discouraged the free flow of information during early stages of research. Table 13.3 sets out the details of these survey results for UEA as a whole. There were very few differences in responses between faculty from different subject areas. One is recorded in the table for the question regarding the free flow of information, where the balance of opinion in faculty from science and professional subjects points more strongly to a reduction. The importance of these changes will be discussed in the next section.

The surveys by McNay (1996) also reveal some of the same fine-grained changes. For example, McNay finds agreement with the propositions that the RAE has inhibited the establishment of "new research areas because it takes some time to establish a high grade of reputation"; that the RAE "has encouraged a more conservative approach to re-

TABLE 13.3. UEA Questionnaire Responses

Question	Has Decreased	Has Not Changed	Has Increased
Volume of research	3.2	15.5	81.3
Projects with quick payoffs	2.0	32.0	66.0
Influence of peers	3.9	59.1	37.0
Early publication	1.9	46.1	52.0
Free flow of information (free flow for	27.3	57.3	15.4
sciences and professionals)	34.3	54.3	11.4
Competition	0.0	16.7	83.3
Planning research	1.3	15.8	82.9
Relevance to economy and society	5.3	68.7	26.0
Similarity to private research	5.4	77.7	17.0
Uncertain projects	49.7	43.0	7.3
Basic projects	35.1	52.3	12.6

NOTE: Percentage responding that "time spent on *the question area* [e.g., uncertain projects] has decreased/increased as a result of the RAEs" or "*the question area* [e.g., volume of research] has decreased/incresed as a result of RAEs." The questionnaire was sent to all faculty (440) in October 1996, and there was a response rate of over 40 percent.

search with a reduction in open ended, speculative, 'blue skies,' and curiosity driven work"; and that "staff have published prematurely." However, he also finds agreement with the proposition that "research has moved away from the applied end of the spectrum towards more basic, pure research"; and this would appear to contradict the finding in the UEA study. There is, though, a matter of interpretation here. While the UEA study found some support for such a proposition, there were many more who held the contrary view (i.e., research was less focused on basic/fundamental issues). As McNay merely reports on those agreeing with the view and not those disagreeing, it is difficult to decide whether there really is a difference in results here. Since his report on the average score on his scale for this question suggests that there must have been significant disagreement, it seems doubtful that his result really does support his assessment, at least as far as all research is concerned.

4. An Assessment of the RAE

Has the introduction of the RAE improved the value for money that the public receives from research in British universities? This is the question that I now turn to; and rather than making any pretense at precision, I shall draw up a rough balance sheet.

Research Activity

On the positive side of the ledger, there is the increase in research activity that has been achieved without any significant increase in real public expenditure. There are more research active individuals, and each research active individual spends more time on research than before. In this way the RAEs have had a considerable impact on the agency problem with respect to the block grant for research, and this is a considerable bonus. However, the value of research depends on other factors as well as the sum of individual research activity. For example, it depends on the type of research undertaken and on the distribution of this greater activity across individuals of differing research skills.

Type of Research

With respect to the type of research, there are at least two aspects that are worth considering in connection with the RAE.

Orientation toward Privatelike Research

Research in all subject areas appears on balance to have become more short-term in orientation, less occupied on uncertain projects, and less concerned with basic or fundamental issues. Of course, it is possible that too much time was spent in the economy as a whole on research on fundamental, long-run, and uncertain issues in the past and so this change may have helped restore a better balance. However, if this had been the case, then this is not the best way of achieving a better balance. The point is that these changes in research signal a shift in the type of research undertaken toward that found in the private sector and this undermines the distinctive character of publicly funded research and with this, its claim on the public purse. The public funding of research is usually justified because there are gaps in private provision, particularly in those areas of research where outcomes are uncertain and long-term and refer to basic or fundamental issues and where the free flow of information at an early stage is important. This is because these are areas where either there are major externalities, as in fundamental research, or where markets are thought to suffer from distortions, as in the case of short-termism, or where markets do not encourage the right incentives, as in the case of sharing information. If it was thought that too much of this kind of basic/uncertain/etc. research was being undertaken, then the way to rectify the problem would be to switch resources from public research to private research; it would not be well served by encouraging publicly funded research institutes to mimic that done in the private sector. This mimicking simply solves one putative problem by creating another. In fact, since there is no apparent evidence that the balance was wrong in the first place, the net effect of this change in the type of research conducted in the public sector has in all likelihood been negative as it means public research is now less likely to fill the gaps created by market failures.

In some respects, this broad conclusion is hardly unexpected. It is one of the lessons of central planning that performance indicators encourage agents to concentrate on those aspects of an activity that are rewarded, and when these performance indicators fail to capture all aspects of the activity agents will concentrate on only some aspects of the activity to the detriment of others. The experience of higher education in this respect mirrors the problems, albeit perhaps less dramatically than the famous oversized nails or extremely heavy machinery found in the old Soviet Union. Again the distortions that emerged in universities are highly predictable. With a short time period between assessments, the increased focus on projects that are less uncertain and offer quick returns is not surprising. Likewise, if departments feel that competition has increased between them, it is not surprising to find that the free flow of information between them has decreased.

Discipline Mix

The type of research undertaken in the public sector is also affected by the distribution of research activity across different disciplines. Here it is difficult to judge how the RAE has changed matters because there are no easy points of comparison on the distribution of research activity by subject area before the RAEs. However, it is difficult to believe that the anomalies in subject areas' pots of gold that became apparent after 1992 have not encouraged a growth in activity in those areas that are particularly well funded (and there is some evidence to support this, since geography's research active staff grew in the UK by 30 percent and those of chemistry and biology both increased by around 10 percent, while those of economics and physics both declined). This may or may not be a good thing; all that can be said for sure is that if it is a good thing, that is pure serendipity since it is an unintended consequence of the quasi-historical method for determining subject pots of gold.

Distribution of Research across Individuals/Departments

With respect to the distribution of research activity across individuals, the selective aspect of the rewards for research under the RAE system appears to have encouraged the greatest growth of research activity among those who are deemed the best researchers. Does such a change represent an increase in value for money? In other words, would a smaller or larger change in the research time of the more able have produced greater research value? To get some idea of what is involved in answering this question, I shall begin by supposing that there is an individual production function for research value by individual i (R_i) that depends only on the skill of the researcher (s_i) and the time spent on research (t_i), as in (1).

(1) $R_i = f(s_i, t_i)$ for individual i

Let us also suppose that this individual production function exhibits constant returns to scale. The distribution of skills across individuals is given and we wish to consider whether a reallocation of time among individuals of different quality raises the overall value of research. Whether such a change improves the overall value of research will depend on the respective marginal products of those losing and those gaining research time, since an individual's marginal product with respect to time is an increasing function of the skill-to-time ratio (s/t). Assume individual i loses time and individual j gains time, then there will be an improvement if

(2) $s_i/t_i < s_j/t_j$

To get some idea of the possible size of these ratios at the moment, I need to make some further assumptions. I shall suppose that the RAE panels rate individuals in a department on a similar 5-point scale plus 5* (there is some evidence that this is what they do) and that the department as a whole is given a rating based on the average score for the department (where 5* counts as a 6 for this purpose). Further I shall suppose that the average person receives an average amount of teaching and administrative relief for the department as a whole, based on the research funding received by the department, with the result that average research time depends directly on the research funding of the department. With these assumptions, the ratios of the time given to relief for the average person in variously ranked departments is given by the ratio of their per capita research funding. That is, up to 1997,

(3) $t_5/t_4 = 5/3; t_5/t_3 = 2; t_5/t_2 = 4,$

where the subscripts indicate ratings. Now consider a change in the time of a 5-rated person that comes at the expense of a 2-rated person. This change in time would raise the value of research if the ratio of the skill levels for these two people exceeded 4 (i.e., $s_5/s_2 > 4$, where s_5 and s_2 refer respectively to the skill level of a 5-rated and a 2-rated person). Is this inequality likely to hold? We do not observe skill levels directly, but we do observe research outputs and if the skill ratio did exceed 4, then this would mean that the value of the 5-rated researcher before the transfer would have to be more than 4 times that of the 2-rated person (this follows from the assumption of constant returns to scale as the time inputs of the 5-rated person are 4 times those of the 2-rated person and the skill inputs of the 5-rated person are more than 4 times those of the 2-rated one). Similar ratios apply for other marginal transfers of time between differently rated people (i.e., a transfer of time from a 3-rated person to a 5-rated person would raise research value if the current value of a 5-rated person was more than double that of a 2-rated person, and so on).

Of course, such reflection is likely to be useful only for those in a particular discipline as they will know something of the research produced by differently rated people. Another more public way of getting at such a calculation is through a comparison of the Scottish universities with those of England. Scotland operated a different set of relativities for differently rated departments up to 1997:

$$(4) \qquad t_5^s/t_4^s = 1.4; t_5^s/t_3^s = 1.96; t_5^s/t_2^s = 2.74.$$

Hence if we make the same assumptions about the allocation of research times in Scotland, then the English system should have produced a bigger set of differences in the allocations of time to people of different abilities than Scotland. If we further assume that distribution of time across skill levels was roughly the same in Scotland and England in 1992, then any change in the research outputs of the two systems between 1992 and 1996 is liable to reflect the differences in the reallocations of research times between individuals/departments in the two countries. Of course, there may be other reasons for change in research output (e.g., changes in the average level of skill and average time spent on research). Nevertheless a comparison may be instructive. Table 13.4 contains some comparisons in these outputs as measured by the RAEs.

Naturally care is required in interpreting any of these comparisons. I have made heroic assumptions about how a national system of selectivity might be translated into a departmental and individual selective allocation of time. They have the effect of causing the national system of selectivity to be mirrored at the level of the department and the individual. In reality, the impact of the national system is bound to be more diverse and complicated because individual universities and departments have differing objectives and probably perceive the constraints that they operate under differently. Likewise, I have assumed that a change in national relativities is reflected directly in a change in the relativities of time spent on research when the situation is undoubtedly more complicated. The provision of time for research does not necessarily lead to the same expenditure of time on research and one of the effects of a system of relativities is to encourage the use of research time on research. These are all complications that also underline the difficulties of HEFCE in using the lever of relativities to influence in predictable and precise ways the value of research. Nevertheless, it would be difficult to argue on the basis of the evidence in table 13.4 that English universities have achieved better value for money from their more extreme relativities than the ones used in Scotland.

Table 13.4. Comparison of Changes in England and Scotland

Subject	RAE Score	England				Scotland			
		Number of Departments		Number of Faculty		Number of Departments		Number of Faculty	
		1992	1996	1996	1992	1992	1996	1992	1996
Biology	5 or 5*	8	15	335	553	1	2	70	106
	4	13	16	376	534	2	3	60	86
	3 or 3a,b	16	18	298	326	4	3	108	93
	2	18	6	266	91	3	2	42	13
	1	14	8	106	41	3	0	40	0
Chemistry	5 or 5*	7	10	268	424	0	1	0	38
	4	13	11	387	323	2	2	49	54
	3 or 3a,b	10	13	221	255	3	3	91	69
	2	10	8	114	80	2	0	29	0
	1	11	6	105	30	3	2	26	5
Computing	5 or 5*	8	13	244	376	2	3	95	125
	4	15	17	291	322	2	3	13	29
	3 or 3a,b	16	17	248	274	4	4	63	45
	2	18	13	210	172	3	1	31	9
	1	15	6	89	6	1	1	2	12
Economics	5 or 5*	10	13	357	409	0	0	0	0
	4	9	11	168	192	3	7	56	98
	3 or 3a,b	14	11	214	108	5	1	50	9
	2	9	3	28	0	1	1	11	7
	1	4	0	28	0	1	1	11	7

Cost of the RAE

The final negative feature of the RAE system is the cost. There are no estimates of this, but they arise at several points. There is the cost to HEFCE of running the exercise. There is the cost of faculty time spent on putting the submission together, planning research, etc.; and there is the cost of time spent on administration. These costs are not inconsiderable as anyone who experienced the football-like, fevered transfer activity before the 31 March deadline will attest. McNay (1996) is inclined to play down this aspect of things, noting that the turnover affected only a small percentage of staff, but I think that simple statistics of this sort miss both the ripple effect of a few high-profile transfers and the time spent by institutions defending themselves from transfer bids, often successfully with the result that they do not show up in turnover statistics. In addition, there is the possible cost, familiar from the analysis of winner-take-all markets (see Frank and Cook 1995), that comes in the form of higher per capita research costs at top-ranked departments as competition for scarce research talent bids up the relative salaries of the most productive researchers. There is some evidence of this effect from UEA salaries (assuming professors are the most research productive): between 1992–93 and 1997–98, the average professorial salary rose by 9 percent, while that of senior lecturers rose by 7.2 percent and and that of lecturers by 6.5 percent. Nevertheless, insofar as the costs are either faculty time or the cost of faculty, then these losses have been swamped as overall assessment faculty research activity has increased.

5. The Norms of "Science"

From the perspective of the economics of science literature, the most disturbing feature of this balance sheet is probably the suggestion that the type and practice of research have become more like those found in the private sector. This literature often distinguishes the norms of "science" or academic life from those found in commerce or the pursuit of "technology," and it is concerned that the new emphasis on making publicly funded research relevant to economic and social well-being may be undermining the distinctive norms of "science," particularly those with respect to the free flow of information regarding research findings (see Dasgupta and David 1994, chapter 7 in this volume; Slaughter and Rhoades 1996, chapter 1 in this volume; and Stephan 1996). It seems from the UEA survey that there has been a reduction in the free flow of information (which is probably more pronounced than the figures in table 13.3 suggest, as a number of those citing an increase in flow also appended a comment noting that this was due to the Internet). And this result is hardly surprising since people also feel competition has increased between departments and research is increasingly on private sector–like topics.

If this is so, and it remains the case that market failure in the generation of knowledge creates the scope for productive publicly funded research to fill some of these gaps, then the question is what might be done to repair the distinctive norms of "science." Toward this end, I argue in this section that we need a different model of norms from that which is found in the economics of science literature.

In that literature so far as I can judge, and this may be unfair, norms are treated as behavioral regularities that arise from the incentives created by the institutions. In the case of academic research the two key institutions are the "priority of publication" reward system coupled with a basic salary that comes, in effect, from teaching activities. The putative virtue of these institutions is that the one encourages the free flow of information while the other compensates for the risk and so does not mean that risky and difficult problems are not addressed. This seems a strange and misleading sense of norm. Norms are, of course, regularities in behavior, but it is a peculiarity of some economists to treat a norm as arising from an instrumental calculation when most social scientists have traditionally cast norms as something that comes into play when there are gaps in the advice that instrumental calculation offers (see Hargreaves Heap 1989, 1999a).

Let me use an example from Dasgupta and David (1994 , chapter 7 in this volume) to bring out what is at issue here. They discuss the "norm" of disclosure. The strength of their analysis is precisely that it examines the circumstances under which a priority-based reward system will lead to disclosure among instrumentally rational scientists. Modeling the disclosure decision as the cooperative move in a repeated prisoners' dilemma (or free rider) interaction, they argue that with indefinite repetition (or as they put it, infinite repetition and discounting) a cooperative equilibrium can be sustained. "Does this imply that the normative content of Merton's communalistic norm of disclosure is really redundant . . . ?," they ask (since cooperation has arisen from instrumental calculation). "Not at all" they reply:

> For it can be shown that networks of cooperative information sharing will be more likely to form spontaneously if the potential participants

> start by expecting others to cooperate . . . and cooperative patterns of behavior will be sustained for longer if participants have reason to expect refusals to cooperate will be encountered only in retaliation for transgressions on their part. (504)

There are two ways of reading the role assigned to norms here. Both involve norms in equilibrium selection; and that seems to me to be the right general track for blending norms with instrumental calculation, as it is well known (from the various folk theorems) that while cooperation can emerge in repeated play of these games between instrumentally rational agents, it is a feature only of some of the many equilibria that can arise in these games. So one needs something in the situation that is going to explain equilibrium selection, and "norms" in a sociological sense (as an influence on behavior that is distinct from instrumental calculation) are an obvious candidate.[2] Where the two ways differ is in the mechanism linking "norms" in this sociological sense to equilibrium selection. One regards agents as boundedly (instrumentally) rational and sets the repeated play in an evolutionary setting. This is the setting suggested in the text by the reference to the "spontaneous" generation of cooperation; but it is not made explicit, so I hesitate to draw this inference. On this account, norms predispose agents initially to use a strategy like tit-for-tat and this then spreads through the population. This gives norms an important historical role, but it does not allow norms to play a conscious role in explaining behavior because agents are boundedly rational.[3] In this sense, it concedes very little to the typical noneconomist's understanding of how norms influence behavior. Norms are no more than behavioral predispositions/expectations that push what would otherwise be an open-ended evolutionary process in one direction rather than another.

The alternative way of introducing norms is to say that they provide reasons for agents to select through their choice of strategies the cooperative equilibrium. They can be thought to do this in two ways. One is

2. Actually, I have some doubts over whether punishment strategies are viable in these types of interactions, partially because it is difficult to control disclosure so that only a subgroup receives information and partially because it is difficult to detect when someone has transgressed by not disclosing. This is, in a sense, a separate issue. Nevertheless, the thought lends support to the idea that the norm of disclosure in academic life owes something to a more substantial and distinctive sense of norms and norm-guided behavior than this account will allow.

3. Schelling's (1963) ideas of salience and focal points, which have been more recently developed by Sugden (1995), are another way of introducing norms in this same spirit as subconscious bits of grit that aid decision making. Also see Greif (1994).

that norms provide a different kind of reason for action from the usual instrumental reason studied by economists, which agents draw upon when instrumental reason is stymied (as in cases of equilibrium selection). Alternatively, norms can be construed as changing the payoffs from actions so that in this instance the interaction no longer corresponds to the prisoners' dilemma or free-rider game (e.g., see Rabin 1993). In this way cooperation emerges from instrumental calculation, and there is no reason to expand the model of rationality as such. I have a preference between these two ways of norms giving agents reasons to act cooperatively, but that is not an important issue for the purpose of this argument. The important point is that both ways introduce norms into the conscious process of decision making. This seems right because people do not just follow a norm of disclosure in some unconscious way.

Now consider how each approach to norms would tackle the question of how to change or repair the norms of science. The first approach would need to disturb the payoffs to action so that boundedly instrumentally rational agents would start to change their rules of thumb as part of the evolutionary process governing behavior. Of course, evolutionary process can throw up strange results, but the natural thing to do would be to change the incentives in such a way as to encourage the desired behavioral patterns. This is also the response of those who are not inclined to see any role for norms in explaining behavior and so can be plausibly associated with the typical economist's approach to the problem. After all, it is economic common sense to conclude that if people are not behaving in the way desired then you need to change the incentives.

The second approach is skeptical about the *general* value of such a prescription because it views instrumental and normative reason as two different types of motivation and insofar as policy is fine tuned to encourage instrumental reason into areas where it has hitherto not operated, it will often drive out normative reason and this can have perverse effects. This is the message from the psychological literature on "extrinsic" and "intrinsic" reason (see Deci 1975), which has been tellingly applied to a variety of economic settings by Frey (1997). This literature argues that people seek "sufficient" reason for their action with the result that an increase in one type of reason (say, the "extrinsic" kind that comes from an increased material incentive to undertake some action) can induce people to revise downward their assessment of the other type of reason for taking the action (i.e., its perceived "intrinsic" value). There are differences in terms here that come from differences in discipline, but the translation between extrinsic and instrumental and intrinsic and normative is not difficult to make and the experience of the RAEs seems to illustrate precisely how one kind of motivation can crowd out another. The norms of science have retreated because the RAE makes people now think instrumentally about their actions in ways that they did not previously.[4]

Of course, this diminution of the role of norms may or may not be a good thing depending on the character of those norms, but in general it means substituting a rather inferior mechanism of control. The point here is simply that control through instrumental incentives always involves costly monitoring and is prone to throw up the anomalies found under any central planning regime. Hence, in the case of the prisoners' dilemma, it is always cheaper when people do not want to cheat in the first place than it is to have a system of incentives and monitors that make people decide on instrumental grounds not to cheat. Or more specifically in this case, it is simpler if people think that the advance of science is a good worth pursuing rather than making the advance of science the upshot of people pursuing their various interests under a specific set of incentives. This may be conceded, but it leaves open the question of how to change behavior if not through instrumental leverage when norms need changing (as they might well have in the case of the UK before the RAE) or are in need of repair (as now).

This is a key question, but this is not the place to address it. It serves my purpose to establish this as an important question: first because the question naturally arises from a more plausible account of norms and second because normative control, if it is possible, is liable to be more efficient than the rigmarole of the RAE and its like. There is, however, space for one brief observation.

The encouragement of norms need not mean ignoring incentives altogether. Instead it may entail retargeting incentives for the purpose of selecting people to a group rather than influencing behavior directly. Thus one way of encouraging the development of a norm within a group is to select those to the group who are predisposed to the norm, and as Brennan (1996) argues this can be done through the use of incentives. For example, with respect to the agency problem in universities, it may be more sensible to devise a reward structure that selects those to the academy who value research (and so will do research because it has "in-

4. On this point, it is no coincidence that the earnings of academics are or were surprisingly flat (see Stephan 1996) because the norms of science could thrive only when behavior is underdetermined by instrumental calculation (see Hargreaves Heap 1999a, 1999b, for a fuller discussion of the relation between instrumental and normative reason).

trinsic" value) rather than constructing one that encourages those who are in the academy to do research (i.e., because it has "extrinsic" value). His particular proposal is to make part of the remuneration package of an academic a research budget, arguing that those who value research will find that this is a good substitute for an equivalent sum on their personal paycheck, while those who do not value research will not and so only those who value research will be attracted to the profession. This proposal may or may not be especially convincing, but the merit of the general point should be obvious: there is more than one way to skin a cat.

6. Conclusion

I have argued that the RAE exercise seems to be encouraging the type and the practices of research typically found in the private sector. Of course, these are early days and the changes that have been detected are not enormous. Nevertheless, they are worrying signs because publicly funded research is usually justified because it complements rather than competes in these respects with the private sector. The source of the problem here, I suspect, is an overreliance on a model of action that ignores or undervalues the part played by norms. I have provided no more than a sketch to support these suspicions. Nevertheless, they connect with the time-honored observation that economists know the price of everything and the value of nothing; and so it is hardly surprising that economics is led astray in areas where people act on the value of things and not their price.

REFERENCES

Arrow, K. 1962. Economic welfare and the allocation of resources for invention. In R. Nelson ed., *The Rate and Direction of Inventive Activity: Economic and Social Factors*. Princeton: Princeton University Press.

Brennan, G. 1996. Selection and the currency of reward. In R. Goodin ed., *The Theory of Institutional Design.* Cambridge: Cambridge University Press.

Dasgupta, P., and P. David. 1994. Towards a new economics of science. *Research Policy* 23:487–521.

Deci, E. 1975. *Intrinsic Motivation.* New York: Plenum Press.

Frank, R., and P. Cook. 1995. *The Winner Take All Society.* New York: Free Press.

Frey, B. 1997. *Not Just for the Money.* Cheltenham: Edward Elgar.

Greif, A. 1994. Cultural beliefs and the organisation of society: a historical and theoretical reflection on collectivist and individualist societies. *Journal of Political Economy* 102:912–50.

Hargreaves Heap, S. 1989. *Rationality in Economics.* Oxford: Basil Blackwell.

———. 1999a. Is self-respect just a another kind of preference? In U. Mäki ed., *The Economic Realm.* Cambridge: Cambridge University Press.

———. 1999b. A note on participatory decision making and social capital. UEA working paper.

HEFCE. 1996. *Funding Method for Research.* Higher Education Funding Council for England, Bristol, July 1996, reference 2/96.

Lee, F., and S. Harley. 1997. Economics divided: the limitations of peer review in a paradigm bound social science. Mimeo. Montfort University.

McNay, I. 1996. *The Impact of the 1992 Research Assessment Exercise on Individual and Institutional Behaviour in English Higher Education.* Summary report. Centre for Higher Education Management, Anglia Polytechnic University.

Rabin, M. 1993. Incorporating fairness into economics and game theory. *American Economic Review* 83:1281–302.

Schelling, T. 1963. *Strategy of Conflict.* Oxford University Press: Oxford.

Segal, Quince and Wicksteed. 1996. *Selective Allocation of Research Funds.* The final report to the Higher Education Funding Council for England.

Slaughter, S., and G. Rhoades. 1996. The emergence of a competitiveness research and development policy coalition and the commercialization of academic science and technology. *Science, Technology, and Human Values* 21:303–39.

Stephan, P. 1996. The economics of science. *Journal of Economic Literature* 34:1199–262.

Sugden, R. 1995. Towards a theory of focal points. *Economic Journal* 105: 533–50.

14

The Importance of Implicit Contracts in Collaborative Scientific Research

Paula E. Stephan and Sharon G. Levin

1. Introduction

Scientific productivity is often examined in the United States at the individual level and often in the context of a career (Bayer and Dutton 1977, Cole 1979, Diamond 1986, Fox 1983, Hargens and Felmlee 1984, Levin and Stephan 1991, Lillard and Weiss 1979, Long 1978, Stephan and Levin 1992, Weiss and Lillard 1982). This contrasts with the focus on the lab as the unit of observation that is common in studies of scientific productivity conducted by many European researchers as well as some U.S. researchers (Knorr-Cetina 1981, Latour and Woolgar 1979, Lynch 1985, Traweek 1988).

Why have researchers in the U.S. focused so extensively on individuals as opposed to groups, and why has this focus persisted despite widespread evidence that science is becoming increasingly a collaborative effort? In this paper we examine these questions. In the process we summarize what is known about the career paths of U.S. scientists and how these career paths are changing. Special emphasis is placed on individuals working in the life sciences, the area of science that has experienced the greatest growth in the United States in recent years, both in resources and in number of individuals receiving graduate degrees.

Section 2 of this paper describes a research tradition that has evolved in the U.S., which emphasizes the individual as the unit of analysis. Section 3 discusses how this approach has been applied to understanding career paths of scientists. We examine what these studies reveal about

Sections 1–4, with minor edits, are taken from the authors' "The Critical Importance of Implicit Contracts in Collaborative Scientific Research," *Revue d'Economie Industrielle* 79, no. 1 (1997): 45–61.

productivity over the life cycle, a research question that has held particular fascination for economists. Factors that contribute to the modest success of these models in explaining productivity over the life cycle are discussed in section 4. We argue that what is particularly lacking in these models is an ability to capture the research process *and* the institutional setting of the process. Knowledge of the research process can arguably best be gained by studying the laboratory as the unit of analysis. The institutional setting of these laboratories, the compensation paid to individuals working in these labs, and the procedures by which funds are raised to support this process are also critical determinants of productivity. These factors are discussed in section 5, where we argue that, despite the increasingly collaborative nature of science, the role the individual scientist plays in assembling resources—especially in the university sector, where the majority of research in the life sciences is produced—means that the individual scientist remains key to understanding productivity. The analogy is drawn between scientists and firms. As the production process in science becomes increasingly complex, the principal investigator increasingly takes on the role of entrepreneur whose job it is to procure resources to sustain the lab and to recruit talent to work in the lab. Section 6 examines the role that implicit contracts play in providing collaborators in the lab.

2. A Research Tradition That Focuses on the Individual

During the past thirty years a research tradition has emerged in the United States that places great emphasis on the analysis of individual-level data from which "general" patterns of behavior are deduced. This micro-based tradition stands in marked contrast to the "case study" method, which provides in-depth analysis of a small number of units and often focuses on the group as a whole, not on individual members of the group. The relative merits of the two methodologies are not discussed here. Neither do we discuss fundamental cultural differences, which place substantially more emphasis in the U.S. than in Europe on the role of the individual in achieving success. Instead we focus on understanding why the micro-based studies have held such fascination for social scientists in the United States and what this has meant for researchers studying scientists.

Three developments or emphases in the United States have fostered this tradition: the development of large longitudinal databases, a head start in the development of information-processing (computer) technology, and an emphasis on statistical models in the social sciences. The development of large longitudinal databases grew out of a widespread interest in the 1960s in understanding the "welfare" and "work experiences" of individuals. Foremost among these are the Panel Study of Income Dynamics, which since its inception in 1967 has followed a sample of five thousand American families, and the National Longitudinal Studies, which collect data on outcomes of specific age groups of the population. Both are university based (the former at the University of Michigan, the latter at Ohio State University). The initial growth and development in the United States of information-processing technology capable of handling more and more data at effectively lower and lower real costs has undoubtedly been an important factor in spurring the research differentiation between the case study approach and the "parameter retrieval" approach. As a result, researchers in the United States have arguably had an absolute advantage as well as a comparative advantage in working with large data sets.[1]

Finally, the heavy focus in the social sciences on the development of statistical models designed to provide efficient estimators of parameters goes hand in hand with the other two factors. Indeed, this focus has developed to the point that many social scientists in the United States could aptly be described as being more interested in the methodologies that they employ than in the answers that the methodologies provide.

3. The Contribution of Micro-Based Studies to Understanding Productivity over the Life Cycle

The micro-based research tradition has been further aided by the development of sophisticated theoretical models that focus on career outcomes. The human capital model, developed in economics, is a case in point and serves here for illustrative purposes.

Ever since the pathbreaking work of Gary Becker (1962) and Theodore Schultz (1963), economists have focused attention on the question of how behavior varies over the life cycle in occupations where investments in human capital (such as education and training) play an important role. The models predict that, due to the finiteness of life, investment behavior declines (eventually) over time.

1. The impact computer technology has had on the social sciences in the United States calls to mind what Derek de Solla Price said in one of his last public lectures: "If you did not know about the technological opportunities that created the new science, you would understandably think that it all happened by people putting on some sort of new thinking cap. . . . [T]he changes of paradigm that accompany great and revolutionary changes may sometimes be caused by inspired thought, but much more commonly they seem due to the application of technology to science" (Price 1986, 247).

Several authors have adapted the human capital framework to develop life-cycle models of scientific productivity. Like their first cousins, these models are driven by the finiteness of life and investigate the implications this has for the allocation of time to research over the scientist's career. The models differ in the assumptions they make concerning the objective function of the scientist but reach somewhat similar conclusions. In its simplest form, the objective is the maximization of income, itself a function of prestige capital (Diamond 1984). In a more complex form, the objective is the maximization of a utility function that includes income as well as research output (Levin and Stephan 1991). Research output is included because of the strong anecdotal evidence that puzzle solving is part of the reward to doing science. The implication of these models is that the stock of prestige capital (reputation) peaks during the career and then declines and that the publishing profile declines over the life cycle. The addition of puzzle solving to the objective function produces the result that research activity is greater, the greater the satisfaction derived from puzzle solving; it also implies that the research profile is flatter, the larger the satisfaction derived from puzzle solving.

The implications of the human capital models for science have been investigated in a number of empirical studies. The dependent variable is generally earnings or publishing activity. In a few instances researchers have adapted the human capital model to study the acceptance of new ideas (Diamond 1980; Hull, Tessner, and Diamond 1978; and Levin, Stephan, and Walker 1995). Here, the premise is that scientists as they age become increasingly vested in their own ideas and hence more and more resistant to alternative theories.

Studies by economists that examine the publishing activity of scientists over the life cycle include those by Diamond (1986), Weiss and Lillard (1982), Levin and Stephan (1991), and Siow (1994).[2] Several classes of problems present themselves in studying research productivity in a life-cycle context. These include measurement, the confounding of aging effects with cohort effects,[3] and the availability of an appropriate

2. For brevity we focus here on studies of publishing activity. For a review of studies of earnings and the acceptance of new ideas see Stephan (1996).

3. Cohort effects can play an important role in affecting productivity. For example, the careers of young scientists educated in the early 1940s were interrupted by World War II; in contrast, scientists educated in the early 1960s in the U.S., when science was experiencing tremendous growth, found themselves on a fast track. But, in a study using cross-sectional data, age is correlated with cohort of the scientist, creating a problem in distinguishing between cohort effects and aging effects. Longitudinal data, which follow a cohort over time, eliminate the contamination of aging effects with cohort effects. But

database. In previous work (Levin and Stephan 1991, Stephan and Levin 1992) we went to great lengths to address many of these issues using data taken from the Survey of Doctorate Recipients, a biennial survey of scientists conducted in the United States by the National Research Council.

Overall, the findings are consistent with the presence of true aging effects. With the exception of particle physicists working in Ph.D.-granting institutions, the age coefficients are statistically significant, often at the 1 percent level. Two general patterns emerge: those in which output declines throughout the career and those in which output initially increases with age and then eventually decreases.

The results of this work, as well as that of others, should not, however, form the basis for concluding that researchers are exceptionally successful in explaining (and thus predicting) productivity over the life cycle. Despite the fact that some indication of an age-publishing relationship is found, the amount of variation explained is usually not large. Diamond, for example, reports R-squares of .09 or less for his research productivity equations; Siow reports R-squares between .05 and .08. The low explanatory power of these models suggests, at a minimum, that other important factors affecting productivity are at play.[4]

4. The Research Process and the Institutional Setting of the Process

The failure of economists to explain scientific productivity satisfactorily in a life-cycle context (and more generally the inability of researchers to predict who will be successful)[5] is undoubtedly related to

longitudinal data confound aging effects with time-period effects and do not permit the researcher to see whether differences exist among cohorts. What is needed is a design that captures the useful aspects of both cross-sectional and longitudinal designs while eliminating the problems they present. One way of doing this is to pool several cross sections. See Levin and Stephan (1991) and Stephan and Levin (1992).

4. The "pseudo" R-squares obtained for the fixed effects model estimated by Levin and Stephan reported above are considerably higher, a fact that is due to the inclusion of "fixed" effects—not because the model is that much richer. Note that these results are for scientists working in the research sector. Stephan and Levin (1992), in a study of Nobel laureates, conclude that although it does not take extraordinary youth to do prize-winning work, the odds decrease markedly by midcareer.

5. Extreme inequality exists in the production of scientific research. In many disciplines, for example, 5 or 6 percent of the scientists who publish at all produce about half of all papers in their discipline (Stephan 1996). Given the magnitude of the resources invested in the education of scientists, from a policy perspective it would be useful to be able to predict who will be successful in order to allocate resources more efficiently.

the fact that the production of scientific knowledge is far more complex than these models assume or data permit and that the complexities have a great deal to do with patterns that evolve over the life cycle.[6] Contrary to what many economists assumed, especially those who placed a heavy emphasis on human capital models, the production of scientific knowledge requires much more than (to use Price's phrase) putting on one's thinking cap. It also requires access to substantial research resources in the form of equipment and colleagues.

One indication of the highly collaborative nature of science is the heavy incidence of coauthorship. Hicks and Katz (1996) document how this has changed over time for UK publications, growing from an average of 2.63 authors per paper in 1981 to 3.34 in 1991. They find that this increase was particularly fueled by a growth in the number of articles with four or more authors. The number of authors does not fully measure the size of the group, however, since it is common practice to have individuals at work on a project who are too far removed from the "science" to receive credit as an author or are insufficiently senior to be considered for authorship. Another reason that authorship is not equivalent to research effort, especially in the life sciences, is the common practice of extending authorship credit in exchange for access to subjects (both of the human and nonhuman variety) or to particular life strains.[7]

Research in the life sciences is especially labor intensive. Some of this labor is required to perform the many repetitive processes that characterize research in the life sciences such as growing microbial cells to make enough biomass for the isolation of a particular molecule for biochemical analysis, purification of DNA from bacteria or yeast, and sequencing of long stretches of DNA. In the United States, Ph.D. students and postdoctorates, who are widely available and command low rates of compensation, perform many of these research tasks.[8] Spiraling spe-

6. Another set of factors affecting productivity that the theoretical models do not capture is characteristics of the individual that lead to higher productivity, such as ability and motivation. While these can be controlled for in a "fixed effects" model by following individuals over time, the actual measurement of these factors is another issue (Stephan and Levin 1992).

7. Groups can also provide synergy, which enhances productivity. The important role such group dynamics play in enhancing productivity is arguably underappreciated by researchers who focus on individuals even when recognizing the trend toward collaboration. Blume (1993) emphasizes the difference between stating that research is done "in groups" versus "by groups."

8. A well-compensated Ph.D. student in the life sciences in the mid-1990s received approximately $16,000 as a graduate research assistant; a postdoctorate received at most $28,000. The labor intensity of many labs no doubt is influenced by the widespread avail-

cialization and an increased emphasis on equipment also require the presence of individuals with unique skills. Interdisciplinary approaches further enlarge the research group (Ziman 1994 and Gibbons et al. 1994).

Research also requires access to equipment. An appreciation of the magnitude of equipment employed in academic research can be obtained by studying the National Science Foundation's data on characteristics of science/engineering equipment in academic settings. The last survey (National Science Foundation 1991) describes the 1988–89 stock of moveable science/engineering equipment in the $10,000 to $999,999 price range at the nation's research-performing colleges, universities, and medical schools.[9] It estimates that the aggregate purchase price of the equipment was about $3.25 billion, expended in the majority of instances during the previous five years. The findings, summarized in Stephan (1996), provide information on incidence of equipment as well as mean price and show, for example, that the mean price of an electron microscope was $119,600; the mean price of an NMR was $146,000. And these averages are undoubtedly biased toward teaching equipment. Sophisticated NMRs and mass spectrometers can easily cost in excess of one million dollars, thereby exceeding the limits of the above survey.

The overwhelming importance of resources to the research process in science means that in many fields access to resources is a necessary condition for doing research. In the United States, many of these resources are located in the university sector, where the lion's share of scientific research, especially in the life sciences, is done. This can be seen from table 14.1, which compares the distribution of articles by sector in the life sciences for the United Kingdom and the United States during the period 1981–94 (private communication with Diana Hicks). Three sectors are given: industry, "other," and university. Percentages add to more than one hundred because "whole counted" articles are reported. Thus, if an article shares an author in industry as well as one in academe, it is attributed to both sectors. In each of the three life science fields we see that the percent attributed to the academic sector in the United States is over 80 and that in clinical medicine it approaches 90. While a large share of the research in the United Kingdom is also done at univer-

ability of graduate students and postdocs. The resistance of the U.S. scientific community to any form of "birth control" in the production of future scientists also undoubtedly relates to the enterprise's interest in keeping a ready supply of cheap labor.

9. Formal collection of the Survey of Academic Research Instrumentation and Instrumentation Needs was suspended by the National Science Foundation in 1997.

TABLE 14.1. Distribution of Whole Counted Articles by Sector, 1981–1994 (percentages)

	U.K.	U.S.	U.K.	U.S.	U.K.	U.S.
	Industry		University		Other	
All fields	37,598 (8)	203,510 (11)	330,085 (72)	1,467,010 (78)	143,739 (31)	559,609 (30)
Clinical medicine	9,301 (6)	42,360 (6)	161,732 (66)	671,782 (88)	7,067 (44)	260,947 (39)
Biomedical research	3,716 (5)	22,129 (6)	50,085 (71)	286,328 (83)	23,752 (34)	92,935 (27)
Biology	1,878 (5)	7,191 (4)	22,618 (59)	145,602 (82)	17,403 (45)	46,330 (26)

SOURCE: J. S. Katz, D. Hicks, F. Narin, and K. Hamilton, SPRU and CHI unpublished data, 1997.

sities, the U.S. academic proportions are considerably higher, especially in clinical medicine and biology.

5. The Institutional Setting of Research in the United States

The university sector in the United States has been characterized by a tenure system that determines, within a period of no more than seven years, whether an individual has the option to remain at the institution or is forced to seek employment elsewhere. The importance of tenure makes it crucial for young scientists to signal to older colleagues that they have the "right stuff" for doing research. A necessary component of this signal is the ability to establish a lab of one's own. And, while startup capital is often provided by the institution, finding the necessary funds to run the lab (not only to buy supplies and equipment but also to hire graduate students, fund postdoctoral positions, and hire technicians) is the responsibility of the individual.[10] A key component of the tenure decision, thus, is whether the individual has been successful in obtaining funding.

The largest source of funding for biomedical research in the United States is the National Institutes of Health (NIH). Its annual budget in 1998 was $13.6 billion, over 80 percent of which went to extramural awards. NIH classifies research grants according to activity and identifies them with activity codes. The traditional research project grant, known

10. Startup funds vary with institution and discipline. In the early 1990s, $250,000 was the approximate upper limit; the average was around $50,000.

as an R01, is the principal vehicle for supporting researchers holding positions in universities, colleges, and other research institutions. Such grants are initiated by a principal investigator (PI), have a term of 3.5 years, have no budget ceiling, and cover research-related expenses such as equipment, supplies, and support-service charges, as well as the salaries of postdoctoral researchers, graduate students, and technicians (National Research Council [NRC] 1994, 27). In recent years, R01 budgets have covered an increasing fraction of the salaries of the principal investigators as well (see discussion below). In fiscal year 1991, the average R01 grant was for a term of 3.8 years with an average total cost of $184,000 (NRC 1994, 27). The percent of R01 applicants who are successful has declined from 31.4 percent in 1984 to 21.3 percent in 1993, reflecting the increased pressure on the system as more and more applications are received and fewer and fewer grants are awarded (NRC 1994, table 2, p. 101).

Extensive preliminary results greatly enhance the probability that the principal investigator will receive R01 funding for the proposed research. This places new investigators at a distinct disadvantage. In response to this, NIH has over the years initiated a number of programs directed at young investigators. The first of these, the R23 program, was begun in 1971. This program was phased out in 1986 and replaced by the R29 program in an effort to support newly independent investigators, as opposed to postdoctoral fellows working in someone else's lab. R29 grants had the distinct feature that they were not renewable. These grants were also extremely competitive—in recent years only about 1 in 4 of the applicants was successful in receiving funding (NRC 1994, table 2, p. 101). In 1998 NIH discontinued the "first" awards because the funding cap that was built into the program was seen as hindering early career development and independence. Instead of changing the cap or initiating a new program, "first" investigators are now encouraged to compete directly for R01 grants.

Not surprisingly, as the probability of getting funding has decreased, the number of applicants who are denied and then resubmit to NIH has increased (what is called an amended application). For example, between 1988 and 1990, the number of R01 applications increased by 27 percent (NRC 1994, 31). Because NIH counts an application only once in a fiscal year, regardless of whether it is amended or "new," this means that the probability that a *reviewed* application is awarded funding is considerably lower than the statistics suggest. In some units of NIH, the success rate of all reviewed applications fell to about 10 percent by the early 1990s (NRC 1994, 33).

One other characteristic of the NIH system crucial to an understanding of the grantsmanship process is that the probability of receiving funding for a continuation (renewal) is significantly higher than the probability of receiving funding for a new application. In 1994, for example, the success rate of new applications was 19.2 percent; for renewals it was 40.7 percent (NIH unpublished data from IMPAC-OER/OD-IRS Program RFMPPHSO). This characteristic is but one indication of the *cumulative* nature of the processes at work in research in the life sciences, where success breeds success.[11] Individuals who receive their degrees at the most prestigious institutions receive the most prestigious postdoctoral appointments. A prestigious postdoctoral appointment greatly facilitates productivity in the early stages of the career and increases the probability that the young life scientist will receive an offer to work at the assistant professor level at a university or medical school.

The probability of getting tenure at such an institution is virtually nonexistent if the researcher does not succeed in obtaining extramural funding to support the lab.[12] This emphasis on the *individual* for funding persists throughout the career of the scientist, at least as long as the scientist wishes to be considered a researcher and direct a lab. Precisely because successful research requires equipment and collaborators working in the lab, the university-based scientist who is committed to being a researcher must continually seek support for the lab that s/he directs. The key role the PI plays in the process is reflected by the fact that labs in academe in the U.S. almost always bear the name of the PI, at least informally.

This does not mean that science is an individual activity in terms of the production of research. At all stages, the level of productivity clearly depends upon the effectiveness of the lab, which is staffed by doctoral students, postdoctorates, and technicians. But in the United States the

11. The failure to satisfactorily model these cumulative processes is another reason life-cycle models of productivity come up short in predicting output.

12. The cumulative process results partly from what Merton (1968) calls the Matthew effect ("the accruing of greater increments of recognition for particular scientific contributions to scientists of considerable repute and the withholding of such recognition from scientists who have not yet made their mark"). It also results from the fact that successful scientists are arguably more productive precisely because they differ in some fundamental way from other scientists. See Stephan and Levin (1992) for a discussion of the distinction between the Matthew effect and the "sacred spark" hypothesis.

funding system that has evolved and the structure of academic appointments means that the *individual* plays a key role. Furthermore, in academe, it is only at the rank of assistant professor or higher that the scientist earns a salary at all commensurate with the training that has been received. And it is only scientists at this rank who can be considered for tenured positions. In short, if a life scientist in the United States wishes to have a career in academe—the sector where the majority of research is performed—the scientist must assume the role of the "rainmaker" for funding, and a prerequisite for rainmaking is a vita that demonstrates past success. It is no surprise that U.S. scientists maintain a strong interest in the receipt of recognition.

Parallels between scientists and firms can be drawn. Stephan (1996) points out that a common finding of research in industrial organization is that entrance by new firms is easy but their survival is not and depends upon their reaching a critical size within a certain time frame. An analogy exists in science, particularly if we think of entry as occurring in graduate school. The majority of entrants survive this phase and a large number continue to the postdoctoral phase. Getting "startup" capital from a dean (or other nonprofit entity) is far harder, and a significant number of scientists never reach the stage where they can direct their own lab. For those who do, the crucial issue then becomes whether this capital can be used to attain (in a specified period of time) the reputation and success required to attract resources in the form of grants and obtain tenure to persist in this effort. This process is made more difficult because funding constraints and priorities, which are set in the majority of cases by the government, are exogenous to the scientist and can change at any time.

The firm analogy does not end here. As the above discussion demonstrates, research is produced by combining labor and capital. As this production process becomes increasingly complex, the PI increasingly takes on the role of the entrepreneur whose job it is to procure resources to sustain the lab—and to recruit talent to work in the lab. The lab retains the identify of the PI and the lab director receives credit for the lab's work, a necessary condition given the role the PI plays in raising the capital (grants) to sustain the lab. The firm analogy continues in the sense that extensive networks exist among labs, just as networks exist among firms, and these networks enhance productivity.

The vast majority of the "employees" in these firms are graduate students and postdocs and the placement of these firms in universities stems from the belief that research and education are joint products.

Moreover, these research entrepreneurs not only want to surround themselves with the young because they are cheap. They want to surround themselves with the young because of the abiding belief among scientists that youth is the fount of ideas.[13] Whether or not this is true, the belief persists.

6. The Role of Implicit Contracts

The human capital model is but one variant of a class of models popular in labor economics that stress that at any one point in time compensation is not necessarily commensurate with productivity. When this occurs, of course, some form of "implicit" contract must be in force between the individual and the firm. Towards the end of the twentieth century a spate of articles were written examining the various forms that these contracts could take (Carmichael 1990, Holmstrom 1981, Lazear 1979, 1981). A common element in this discussion was the notion that some self-correcting mechanism needed to exist for these contracts to work.

The traditional relationship between a graduate student and a professor provides an interesting example of such a contract. The graduate student enters the program, provides some "surplus" for the lab through his or her work as a research assistant, and then leaves the institution to set up an identical type of firm. The reputational incentive of the professor/entrepreneur to not cheat on the arrangement is straightforward—if the student is kept too long, or educated too poorly to be considered "creditworthy" by a future dean, or provided poor information—the professor will cease to be able to attract top graduate students and the source of labor, compensated well below its opportunity cost, will dry up.

This system, which loosely resembles a pyramid scheme, works reasonably well so long as there is a growing supply of venture capital (to continue the analogy described above) for new firms. But for this to occur, funding for science must not only grow, but must grow sufficiently fast to absorb the growing workforce of scientists. Such a tremendous growth in resources is something that the system has been unable to provide, particularly in recent years.

13. There is a widespread belief that science is a young person's game. Most recently this philosophy has been articulated by Edward O. Wilson when he says, in *Consilience,* "We were all young in those days; young scientists have the best ideas and, most important, the most time" (1998, 69). Stephan and Levin (1992) examine the relationship between age and productivity in depth.

But still, the system survives. Three factors have provided breathing room for science. First, in the 1960s and 1970s the baby boomers' entrance into college provided fuel for the system to expand. Second, the concept of "postdoctoral study" was advanced. While not terribly common thirty years ago, the postdoctoral position has now become so common that approximately 50 percent of all Ph.D.s in the life sciences take a postdoctoral position upon receiving their Ph.D. In some subfields it is as high as 70 percent. Recent Ph.D.s take not only one postdoctorate—increasingly they take successive postdoctorates, as can be seen from figure 14.1, which shows the percentage of recent cohorts holding such an appointment.

The postdoctoral position provides relief for the system in several ways. First, by offering employment opportunities for newly minted Ph.D.s, it allows professors to "place" their students more easily. Second, it provides recipients with the opportunity to enhance their research records and thus increase the signal of their creditworthiness. Third, it provides a welcome source of assistance in the lab as science has become increasingly more complex. Fourth, postdoctorates bring new ideas to the lab and help the PI keep the necessary edge to maintain funding from the granting authorities. Finally, and perhaps unwittingly, it diffuses the role that reputation has traditionally played in graduate education. Applicants to graduate school who ask about academic job placements can be told that academe no longer recruits faculty directly from Ph.D. programs, but instead, only considers applicants with postdoctoral experience.

Young immigrant scientists have also played a key role. In contrast to the years before World War II, when immigrant scientists to the U.S. were likely to be mature and come for political reasons, in recent years immigrant scientists have come in order to receive a Ph.D. The increase in foreign nationals, as a percentage of all of those receiving Ph.D.s in the life sciences, is shown in figure 14.2.

The eagerness of foreign nationals to study in the U.S. diffuses even more the role that reputation plays. Rarely do foreign nationals applying to graduate school inquire about job prospects. In an international context their prospects are significantly higher by studying in the U.S. than they would be if they were not to study in the U.S. Thus, the self-correcting mechanism that might result has failed to take place.

The presence of foreign nationals also adds to the supply of postdoctorates. Foreign nationals who receive their Ph.D.s in the U.S. often stay on to take a postdoctoral position after leaving graduate school. A large

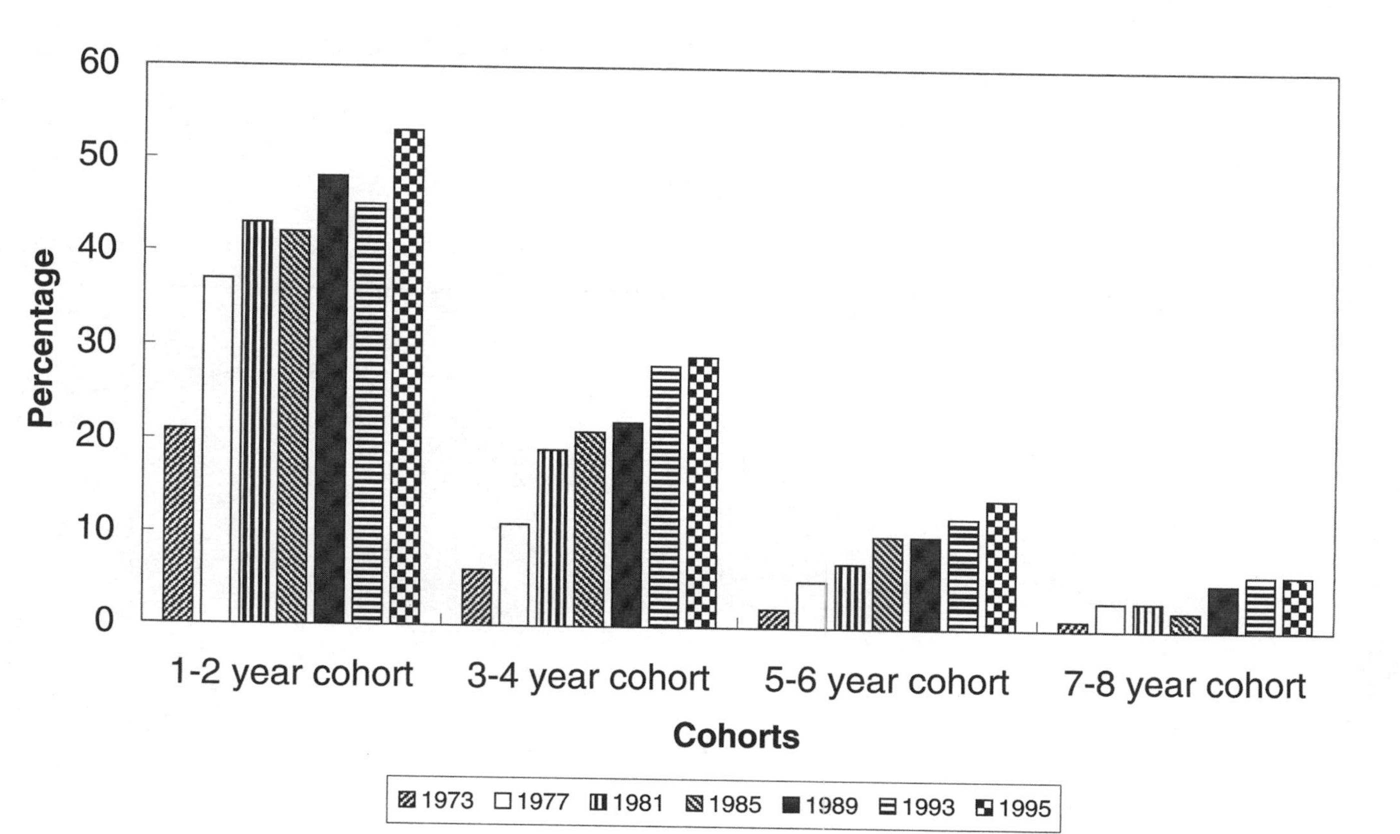

FIGURE 14.1. Percentage of Life Science Doctorates in Postdoctoral Positions by Cohort and Year

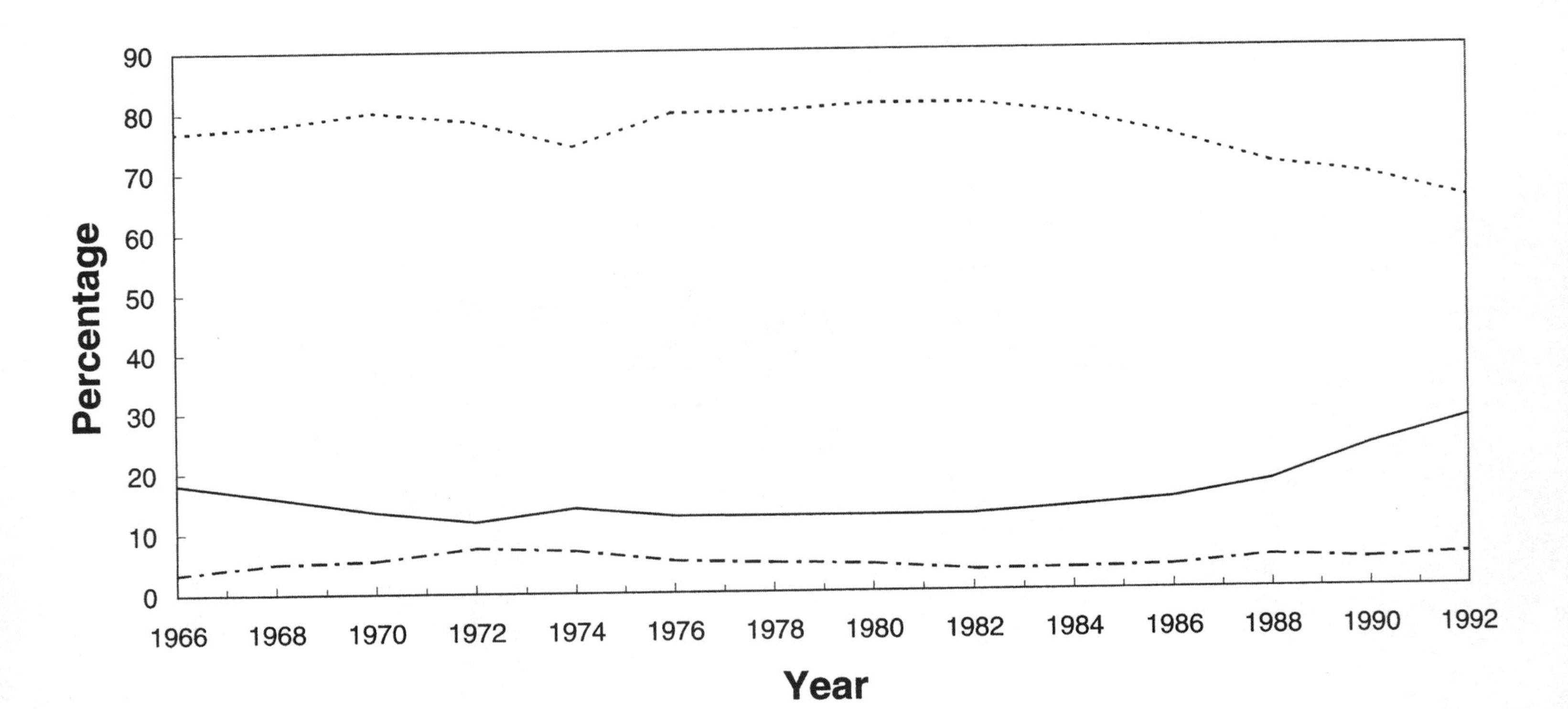

FIGURE 14.2. Nativity/Citizenship of Doctorates in the Life Sciences

number of foreign nationals also come to the U.S. to accept postdoctoral positions after receiving their Ph.D. abroad. This increased supply depresses salaries for postdoctoral positions. It also provides academe the "luxury" of being selective in choosing postdoctoral applicants.

That the system is under increasing stress is seen by another, more recent break in the implicit contract. This break concerns the reported widespread hiring of scientists on soft money in academic institutions. Eleven years of training (seven in graduate school; four in a postdoctoral position) no longer qualify one for a tenure track position and associated risk sharing between the institution and scientist. As institutions—particularly medical schools—in the United States have become increasingly hard-pressed financially, they have reportedly hired more and more scientists into what are referred to as "soft-money" positions—that is, positions that exist only as long as the scientist has funding from external sources that pay both the cost of running the lab and the salary of the scientist directing the lab.[14] At Baylor Medical School, for example, 100 percent of the faculty in the biomedical department receive 80 percent of their salary from grants. This means that if a faculty member fails to maintain funding, not only must the lab be closed, but the faculty member loses his or her position as well at the university. Entrepreneurial skills are not only required to maintain a lab, they are required to maintain employment.

7. Conclusion

Science is increasingly becoming a collaborative process. This fact is indisputable. The backbone of the collaborative nature of U.S. science has been the implicit contract that universities make with graduate students and postdocs. Despite this, the institutional structure of research in the United States, with its heavy emphasis on university-based science, retains a crucial role for the individual scientist in the production of research.

In the United States it is virtually impossible for a scientist to survive and have a career at a university without becoming a principle investigator and directing a lab. The research the PI directs is collaborative, but the majority of the collaborators are graduate students and postdocs—statuses that by their very definition are "temporary."[15] As a result, the

14. This practice is not limited to medical schools. Academic departments in universities now hire scientists into "research assistant professorships," which are soft-money positions.

15. It would be interesting to calculate the proportion of the typical life science research team in the U.S. that is "temporary" and compare this figure to the percent temporary for other countries.

typical lab in the United States experiences substantial turnover among junior scientists, a fact that PIs describe as "healthy" because turnover provides for new points of view. The PI often holds the only permanent position in the lab.

Data in the U.S. are collected for individual scientists, not labs, and privacy laws have historically discouraged the linkage of databases across individuals and indicators. The challenge is to find a meaningful way to study university labs as a unit, instead of associating the research outcome with an individual. The ability to proceed on such a course would enhance our understanding of the research process, an understanding that, in part because of the collaborative nature of research, cannot be adequately illuminated by a life-cycle model that sees the individual, and the finiteness of life, as the driving force. Much could be learned by altering the unit of observation.

ACKNOWLEDGMENTS

The authors wish to thank William D. Amis, Mary Frank Fox, Patrick Mason, and several anonymous referees for helpful comments.

REFERENCES

Audretsch, David B., and Paula E. Stephan. 1997. Knowledge Spillovers in Biotechnology: Sources and Incentives. Unpublished manuscript, Georgia State University, Atlanta.

———. 1996. Company-Scientist Locational Links: The Case in Biotechnology. *American Economic Review* 86 (3): 641–52.

Bayer, Alan E., and Jeffrey C. Dutton. 1977. Career Age and Research Professional Activities of Academic Scientists. *Journal of Higher Education* 48 (3): 259–82.

Becker, Gary. 1962. Investment in Human Capital: A Theoretical Analysis. *Journal of Political Economics* 70 (5, pt. 2): S9–S49.

Blume, Stuart. 1993. Correlates of Creativity. *Science* 259: 107.

Callon, Michel. 1996. Some Elements of an Analysis of Strategic Relations between Firms and University Laboratories. Presented at the European Association for the Study of Science and Technology and Society for Social Studies of Science Meetings, October. Bielefeld, Germany.

Carmichael, H. Lorne. 1990. Efficiency Wage Models of Unemployment: One View. *Economic Inquiry* 28 (2): 269–95.

Cole, Stephen. 1979. Age and Scientific Performance. *American Journal of Sociology* 84 (4): 958–77.

Diamond, Arthur M., Jr. 1980. Age and the Acceptance of Cliometrics. *Journal of Economic History* 40 (4): 838–41.

———. 1984. An Economic Model of the Life Cycle Research Productivity of Scientists. *Scientometrics* 6 (3): 189–96.

———. 1986. The Life Cycle Research Productivity of Mathematicians and Scientists. *Journal of Gerontology* 41 (4): 520–25.

Fox, Mary Frank. 1983. Publication Productivity among Scientists: A Critical Review. *Social Studies of Science* 13 (2): 285–305.

Gibbons, Michael, Camille Limoges, Helga Nowotny, Simon Schwartzman, Peter Scott, and Martin Trow. 1994. *The New Production of Knowledge.* London: Sage.

Hargens, Lowell L., and Diane H. Felmlee. 1984. Structural Determinants of Stratification in Science. *American Sociological Review* 49: 685–97.

Hicks, Diana M., and J. Sylvan Katz. 1996. Where Is Science Going? *Science, Technology, and Human Values* 21 (4): 379–406.

Holmstrom, Bengt. 1981. Contractual Models of the Labor Market. *American Economic Review* 71 (2): 308–13.

Hull, David L., Peter D. Tessner, and Arthur M. Diamond. 1978. Planck's Principle. *Science* 202: 717–23.

Knorr-Cetina, Karin D. 1981. *The Manufacture of Knowledge: An Essay on the Constructivist and Contextual Nature of Science.* Oxford and New York: Pergamon Press.

Latour, Bruno, and Steven Woolgar. 1979. *Laboratory Life: the Social Construction of Scientific Facts.* London and Beverly Hills: Sage.

Lazear, Edward. 1979. Why Is There Mandatory Retirement? *Journal of Political Economy* 87: 1261–84.

———. 1981. Agency Earnings Profiles, Productivity, and Hours Restrictions. *American Economic Review* 71: 606–20.

Levin, Sharon G., and Paula E. Stephan. 1991. Research Productivity over the Life Cycle: Evidence for Academic Scientists. *American Economic Review* 81 (1): 114–32.

Levin, Sharon G., Paula E. Stephan, and Mary Beth Walker. 1995. Planck's Principle Revisited. *Social Studies of Science* 25 (2): 275–83.

Lillard, Lee A., and Yoram Weiss. 1979. Components of Variation in Panel Earnings Data: American Scientists 1960–70. *Econometrica* 47: 437–54.

Long, J. Scott. 1978. Productivity and Academic Position in the Scientific Career. *American Sociological Review* 43: 889–908.

Lynch, Michael. 1985. *Art and Artefact in Laboratory Science: A Study of Work and Shop Talk in a Research Laboratory.* London: Routledge & Kegan Paul.

Mansfield, Robert K. 1995. Academic Research Underlying Industrial Innovations: Sources, Characteristics, and Financing. *Review of Economic Statistics* 77 (1): 55–65.

Merton, Robert K. 1968. The Matthew Effect in Science. *Science* 159: 56–63.

National Research Council. 1994. *The Funding of Young Investigators in the Biological and Biomedical Sciences.* Washington: National Academy Press.

———. 1998. *Trends in the Early Careers of Life Scientists.* Committee on Dimensions, Causes, and Implications of Recent Trends in the Careers of Life Scientists. Washington: National Academy Press.

National Science Foundation. 1991. *Characteristics of Science/Engineering*

Equipment in Academic Settings: 1989–1990. NSF 91-315. Washington: National Science Foundation.

Price, Derek J. de Solla. 1986. *Little Science, Big Science . . . And Beyond.* New York: Columbia University Press.

Rensberger, Boyce. 1995. The Real Science: Raising Enough Money to Keep Going. *Washington Post National Weekly Edition,* January 9, 8–9.

Schultz, T. W. 1963. *The Economic Value of Education.* New York: Columbia University Press.

Siow, Aloysius. 1994. The Organization of the Market for Professors. Working paper. Department of Economics and Institute for Policy Analysis, University of Toronto.

Stephan, Paula E. 1996. The Economics of Science. *Journal of Economic Literature* 34: 1199–235.

Stephan, Paula E., and Sharon G. Levin. 1992. *Striking the Mother Lode in Science: The Importance of Age, Place, and Time.* New York: Oxford University Press.

Traweek, Sharon. 1988. *Beamtimes and Lifetimes: The World of High Energy Physicists.* Cambridge: Harvard University Press.

Weiss, Yoram, and Lee A. Lillard. 1982. Output Variability, Academic Labor Contracts, and Waiting Times for Promotion. *Research in Labor Economics* 5: 157–88.

Wilson, Edward O. 1998. *Consilience: The Unity of Knowledge.* New York: Alfred A. Knopf.

Ziman, John. 1994. *Prometheus Bound: Science in a Dynamic Steady State.* Cambridge: Cambridge University Press.

Zucker, Lynne G., Michael R. Darby, and Jeff Armstrong. 1995. Intellectual Capital and the Firm: The Technology of Geographically Localized Knowledge Spillovers. Working paper 4946. National Bureau of Economic Research, Cambridge, Mass.

15

Digital Diploma Mills

The Automation of Higher Education

David F. Noble

ABSTRACT. In recent years changes in universities, especially in North America, show that we have entered a new era in higher education, one which is rapidly drawing the halls academe into the age of automation. Automation—the distribution of digitized course material online, without the participation of professors who develop such material—is often justified as an inevitable part of the new "knowledge-based" society. It is assumed to improve learning and increase wider access. In practice, however, such automation is often coercive in nature—being forced upon professors as well as students—with commercial interests in mind. This paper argues that the trend towards automation of higher education as implemented in North American universities today is a battle between students and professors on one side and university administrations and companies with "educational products" to sell on the other. It is not a progressive trend towards a new era at all, but a regressive trend, towards the rather old era of mass-production standardization and purely commercial interests.

Recent events at two large North American universities signal dramatically that we have entered a new era in higher education, one which is rapidly drawing the halls of academe into the age of automation. In mid-summer 1998 the UCLA administration launched its historic "Instructional Enhancement Initiative" requiring computer Web sites for all of its arts and sciences courses by the start of the fall term, the first time that a major university has made mandatory the use of computer telecommunications technology in the delivery of higher education. In partnership with several private corporations (including the Times Mirror Company, parent of the *Los Angeles Times*), moreover, UCLA has spawned its own for-profit company, headed by a former UCLA vice chancellor, to peddle online education (the Home Education Network).

Reprinted with permission of David F. Noble and University of Illinois at Chicago. From *First Monday* 3, no. 1 (January 1998), http://www.firstmonday.dk/issues/issue3_1/noble.

In spring 1998 in Toronto, meanwhile, the full-time faculty of York University, Canada's third largest, ended an historic two-month strike, having secured for the first time anywhere formal contractual protection against precisely the kind of administrative action being taken by UCLA. The unprecedented faculty job action, the longest university strike in English Canadian history, was taken partly in response to unilateral administrative initiatives in the implementation of instructional technology, the most egregious example of which was an official solicitation to private corporations inviting them to permanently place their logo on a university online course in return for a $10,000 contribution to courseware development. As at UCLA, the York University administration has spawned its own subsidiary (Cultech), directed by the vice president for research and several deans and dedicated, in collaboration with a consortium of private sector firms, to the commercial development and exploitation of online education.

Significantly, at both UCLA and York, the presumably cyberhappy students have given clear indication that they are not exactly enthusiastic about the prospect of a high-tech academic future, recommending against the initiative at UCLA and at York lending their support to striking faculty and launching their own independent investigation of the commercial, pedagogical, and ethical implications of online educational technology. In fall 1998 the student handbook distributed annually to all students by the York Federation of Students contained a warning about the dangers of online education.

The Classroom vs. the Boardroom

Thus, at the very outset of this new age of higher education, the lines have already been drawn in the struggle which will ultimately determine its shape. On the one side university administrators and their myriad commercial partners, on the other those who constitute the core relation of education: students and teachers. (The chief slogan of the York faculty during the strike was "the classroom vs the boardroom"). It is no accident, then, that the high-tech transformation of higher education is being initiated and implemented from the top down, either without any student and faculty involvement in the decision-making or despite it. At UCLA the administration launched their initiative during the summer when many faculty are away and there was little possibility of faculty oversight or governance; faculty were thus left out of the loop and kept in the dark about the new web requirement until the last moment.

And UCLA administrators also went ahead with this initiative, which is funded by a new compulsory student fee, despite the formal student recommendation against it. Similarly the initiatives of the York administration in the deployment of computer technology in education were taken without faculty oversight and deliberation, much less student involvement. What is driving this headlong rush to implement new technology with so little regard for deliberation of the pedagogical and economic costs and at the risk of student and faculty alienation and opposition? A short answer might be the fear of getting left behind, the incessant pressures of "progress." But there is more to it. For the universities are not simply undergoing a technological transformation. Beneath that change, and camouflaged by it, lies another: the commercialization of higher education. For here as elsewhere technology is but a vehicle and a disarming disguise.

The major change to befall the universities over the last two decades has been the identification of the campus as a significant site of capital accumulation, a change in social perception which has resulted in the systematic conversion of intellectual activity into intellectual capital and, hence, intellectual property. There have been two general phases of this transformation. The first, which began twenty years ago and is still under way, entailed the commoditization of the research function of the university, transforming scientific and engineering knowledge into commercially viable proprietary products that could be owned and bought and sold in the market. The second, which we are now witnessing, entails the commoditization of the educational function of the university, transforming courses into courseware, the activity of instruction itself into commercially viable proprietary products that can be owned and bought and sold in the market. In the first phase the universities became the site of production and sale of patents and exclusive licenses. In the second, they are becoming the site of production of—as well as the chief market for—copyrighted videos, courseware, CD-ROMs, and Web sites.

The first phase began in the mid-1970s when, in the wake of the oil crisis and intensifying international competition, corporate and political leaders of the major industrialized countries of the world recognized that they were losing their monopoly over the world's heavy industries and that, in the future, their supremacy would depend upon their monopoly over the knowledge which had become the lifeblood of the new so-called "knowledge-based" industries (space, electronics, computers, materials, telecommunications, and bioengineering). This focus upon "intellectual capital" turned their attention to the universities as its chief source, implicating the universities as never before in the economic machinery. In the view of capital, the universities had become too important to be left to the universities. Within a decade there was a proliferation of industrial partnerships and new proprietary arrangements, as industrialists and their campus counterparts invented ways to socialize the risks and costs of creating this knowledge while privatizing the benefits. This unprecedented collaboration gave rise to an elaborate web of interlocking directorates between corporate and academic boardrooms and the foundation of joint lobbying efforts epitomized by the work of the Business–Higher Education Forum. The chief accomplishment of the combined effort, in addition to a relaxation of antitrust regulations and greater tax incentives for corporate funding of university research, was the 1980 reform of the patent law which for the first time gave the universities automatic ownership of patents resulting from federal government grants. Laboratory knowledge now became patents, that is intellectual capital and intellectual property. As patent holding companies, the universities set about at once to codify their intellectual property policies, develop the infrastructure for the conduct of commercially viable research, cultivate their corporate ties, and create the mechanisms for marketing their new commodity, exclusive licenses to their patents. The result of this first phase of university commoditization was a wholesale reallocation of university resources toward its research function at the expense of its educational function.

Class sizes swelled, teaching staffs and instructional resources were reduced, salaries were frozen, and curricular offerings were cut to the bone. At the same time, tuition soared to subsidize the creation and maintenance of the commercial infrastructure (and correspondingly bloated administration) that has never really paid off. In the end students were paying more for their education and getting less, and the campuses were in crisis.[1]

The second phase of the commercialization of academia, the commoditization of instruction, is touted as the solution to the crisis engendered by the first. Ignoring the true sources of the financial debacle—

1. Tuition began to outpace inflation in the early 1980s, at precisely the moment when changes in the patent system enabled the universities to become major vendors of patent licenses. According to data compiled by the National Center for Educational Statistics, between 1976 and 1994 expenditures on research increased 21.7% at public research universities while expenditure on instruction decreased 9.5%. Faculty salaries, which had peaked in 1972, fell precipitously during the next decade and have since recovered only half the loss.

an expensive and low-yielding commercial infrastructure and greatly expanded administrative costs—the champions of computer-based instruction focus their attention rather upon increasing the efficiencies of already overextended teachers. And they ignore as well the fact that their high-tech remedies are bound only to compound the problem, increasing further, rather then reducing, the costs of higher education. (Experience to date demonstrates clearly that computer-based teaching, with its limitless demands upon instructor time and vastly expanded overhead requirements—equipment, upgrades, maintenance, and technical and administrative support staff—costs more not less than traditional education, whatever the reductions in direct labor, hence the need for outside funding and student technology fees). Little wonder, then, that teachers and students are reluctant to embrace this new panacea. Their hesitation reflects not fear but wisdom.[2]

The Birth of Educational Maintenance Organizations

But this second transformation of higher education is not the work of teachers or students, the presumed beneficiaries of improved education, because it is not really about education at all. That's just the name of the market. The foremost promoters of this transformation are rather the vendors of the network hardware, software, and "content"—Apple, IBM, Bell, the cable companies, Microsoft, and the edutainment and publishing companies Disney, Simon and Schuster, Prentice-Hall, et al.—who view education as a market for their wares, a market estimated by the Lehman Brothers investment firm potentially to be worth several hundred billion dollars. "Investment opportunity in the education industry has never been better," one of their reports proclaimed, indicating that this will be "the focus industry" for lucrative investment in the future, replacing the health care industry. (The report also fore-

2. Recent surveys of the instructional use of information technology in higher education clearly indicate that there have been no significant gains in either productivity improvement or pedagogical enhancement. Kenneth C. Green, director of the Campus Computing Project, which conducts annual surveys of information technology use in higher education, noted that "the campus experience over the past decade reveals that the dollars can be daunting, the return on investment highly uncertain." "We have yet to hear of an instance where the total costs (including all realistically amortized capital investments and development expenses, plus reasonable estimates for faculty and support staff time) associated with teaching some unit to some group of students actually decline while maintaining the quality of learning," Green wrote. On the matter of pedagogical effectiveness, Green noted that "the research literature offers, at best, a mixed review of often inconclusive results, at least when searching for traditional measures of statistical significance in learning outcomes."

casts that the educational market will eventually become dominated by EMOs—education maintenance organizations—just like HMOs in the health care market.) It is important to emphasize that, for all the democratic rhetoric about extending educational access to those unable to get to the campus, the campus remains the real market for these products, where students outnumber their distance learning counterparts six-to-one.

In addition to the vendors, corporate training advocates view online education as yet another way of bringing their problem-solving, information-processing, "just-in-time" educated employees up to profit-making speed. Beyond their ambitious in-house training programs, which have incorporated computer-based instructional methods pioneered by the military, they envision the transformation of the delivery of higher education as a means of supplying their properly prepared personnel at public expense.

The third major promoters of this transformation are the university administrators, who see it as a way of giving their institutions a fashionably forward-looking image. More importantly, they view computer-based instruction as a means of reducing their direct labor and plant maintenance costs—fewer teachers and classrooms—while simultaneously undermining the autonomy and independence of faculty. At the same time, they are hoping to get a piece of the commercial action for their institutions or themselves, as vendors in their own right of software and content. University administrators are supported in this enterprise by a number of private foundations, trade associations, and academic-corporate consortia which are promoting the use of the new technologies with increasing intensity. Among these are the Sloan, Mellon, Pew, and Culpeper Foundations, the American Council on Education, and, above all, Educom, a consortium representing the management of 600 colleges and universities and a hundred private corporations.

Last but not least, behind this effort are the ubiquitous technozealots who simply view computers as the panacea for everything, because they like to play with them. With the avid encouragement of their private sector and university patrons, they forge ahead, without support for their pedagogical claims about the alleged enhancement of education, without any real evidence of productivity improvement, and without any effective demand from either students or teachers.

In addition to York and UCLA, universities throughout North America are rapidly being overtaken by this second phase of commercialization. There are the stand-alone virtual institutions like University of Phoenix, the wired private institutions like the New School for Social Research, the campuses of state universities like the University of Maryland and the new Gulf-Coast campus of the University of Florida (which boasts of no tenure). On the state level, the states of Arizona and California have initiated their own state-wide virtual university projects, while a consortia of western "Smart States" have launched their own ambitious effort to wire all of their campuses into an online educational network. In Canada, a national effort has been undertaken, spearheaded by the Telelearning Research Network centered at Simon Fraser University in Vancouver, to bring most of the nation's higher education institutions into a "Virtual U" network.

The overriding commercial intent and market orientation behind these initiatives is explicit, as is illustrated by the most ambitious U.S. effort to date, the Western Governors' Virtual University Project, whose stated goals are to "expand the marketplace for instructional materials, courseware, and programs utilizing advanced technology," "expand the marketplace for demonstrated competence," and "identify and remove barriers to the free functioning of these markets, particularly barriers posed by statutes, policies, and administrative rules and regulations."

"In the future," Utah governor Mike Leavitt proclaimed, "an institution of higher education will become a little like a local television station." Start-up funds for the project come from the private sector, specifically from Educational Management Group, the educational arm of the world's largest educational publisher, Simon and Schuster, and the proprietary impulse behind their largesse is made clear by Simon and Schuster CEO Jonathan Newcomb: "The use of interactive technology is causing a fundamental shift away from the physical classroom toward anytime, anywhere learning—the model for postsecondary education in the twenty-first century." This transformation is being made possible by "advances in digital technology, coupled with the protection of copyright in cyberspace."

Similarly, the national effort to develop the "Virtual U" customized educational software platform in Canada is directed by an industrial consortium which includes Kodak, IBM, Microsoft, McGraw-Hill, Prentice-Hall, Rogers Cablesystems, Unitel, Novasys, Nortel, Bell Canada, and MPR Teltech, a research subsidiary of GTE. The commercial thrust behind the project is explicit here too. Predicting a potential fifty billion dollar Canadian market, the project proposal emphasizes the adoption of "an intellectual property policy that will encourage researchers and industry to commercialize their innovations" and anticipates the development of "a number of commercially marketable hardware and software products and services," including "courseware and other learning products." The two directors of the project, Simon Fraser University professors, have formed their own company to peddle these products in collaboration with the university. At the same time, the nearby University of British Columbia has recently spun off the private WEB-CT company to peddle its own educational Web site software, WEB-CT, the software designed by one of its computer science professors and now being used by UCLA. In recent months, WEB-CT has entered into production and distribution relationships with Silicon Graphics and Prentice-Hall and is fast becoming a major player in the American as well as Canadian higher education market. As of the beginning of the 1998 fall term, WEB-CT licensees now include, in addition to UCLA and California State University, the Universities of Georgia, Minnesota, Illinois, North Carolina, and Indiana, as well as such private institutions as Syracuse, Brandeis, and Duquesne.

Education as a Commodity

The implications of the commoditization of university instruction are two-fold in nature, those relating to the university as a site of the production of the commodities and those relating to the university as a market for them. The first raises for the faculty traditional labor issues about the introduction of new technologies of production. The second raises for students major questions about costs, coercion, privacy, equity, and the quality of education.

With the commoditization of instruction, teachers as labor are drawn into a production process designed for the efficient creation of instructional commodities, and hence become subject to all the pressures that have befallen production workers in other industries undergoing rapid technological transformation from above. In this context faculty have much more in common with the historic plight of other skilled workers than they care to acknowledge. Like these others, their activity is being restructured, via the technology, in order to reduce their autonomy, independence, and control over their work and to place workplace knowledge and control as much as possible into the hands of the administration. As in other industries, the technology is being deployed by management primarily to discipline, de-skill, and displace labor.

Once faculty and courses go online, administrators gain much greater direct control over faculty performance and course content than ever before, and the potential for administrative scrutiny, supervision, regimentation, discipline, and even censorship increase dramatically. At the same time, the use of the technology entails an inevitable extension of working time and an intensification of work as faculty struggle at all hours of the day and night to stay on top of the technology and respond, via chat rooms, virtual office hours, and e-mail, to both students and administrators to whom they have now become instantly and continuously accessible. The technology also allows for much more careful administrative monitoring of faculty availability, activities, and responsiveness.

Once faculty put their course material online, moreover, the knowledge and course design skill embodied in that material is taken out of their possession, transferred to the machinery and placed in the hands of the administration. The administration is now in a position to hire less skilled, and hence cheaper, workers to deliver the technologically prepackaged course. It also allows the administration, which claims ownership of this commodity, to peddle the course elsewhere without the original designer's involvement or even knowledge, much less financial interest. The buyers of this packaged commodity, meanwhile, other academic institutions, are able thereby to contract out, and hence outsource, the work of their own employees and thus reduce their reliance upon their in-house teaching staff.

Redundant Faculty in the Virtual University

Most important, once the faculty converts its courses to courseware, their services are in the long run no longer required. They become redundant, and when they leave, their work remains behind. In Kurt Vonnegut's classic novel *Player Piano* the ace machinist Rudy Hertz is flattered by the automation engineers who tell him his genius will be immortalized. They buy him a beer. They capture his skills on tape. Then they fire him. Today faculty are falling for the same tired line, that their brilliance will be broadcast online to millions. Perhaps, but without their further participation. Some skeptical faculty insist that what they do cannot possibly be automated, and they are right. But it will be automated anyway, whatever the loss in educational quality. Because education, again, is not what all this is about; it's about making money. In short, the new technology of education, like the automation of other industries, robs faculty of their knowledge and skills, their control over their working lives, the product of their labor, and, ultimately, their means of livelihood.

None of this is speculation. In fall 1998 the UCLA faculty, at administration request, have dutifully or grudgingly (it doesn't really matter which) placed their course work—ranging from just syllabi and assignments to the entire body of course lectures and notes—at the disposal of their administration, to be used online, without asking who will own it much less how it will eventually be used and with what consequences. At York University, untenured faculty have been required to put their courses on video, CD-ROM, or the Internet or lose their job. They have then been hired to teach their own now automated course at a fraction of their former compensation. The New School in New York now routinely hires outside contractors from around the country, mostly unemployed Ph.D.s, to design online courses. The designers are not hired as employees but are simply paid a modest flat fee and are required to surrender to the university all rights to their course. The New School then offers the course without having to employ anyone. And this is just the beginning.

Educom, the academic-corporate consortium, has recently established their Learning Infrastructure Initiative, which includes the detailed study of what professors do, breaking the faculty job down in classic Tayloristic fashion into discrete tasks, and determining what parts can be automated or outsourced. Educom believes that course design, lectures, and even evaluation can all be standardized, mechanized, and consigned to outside commercial vendors. "Today you're looking at a highly personal human-mediated environment," Educom president Robert Heterich observed. "The potential to remove the human mediation in some areas and replace it with automation—smart, computer-based, network-based systems—is tremendous. It's gotta happen."

Toward this end, university administrators are coercing or enticing faculty into compliance, placing the greatest pressures on the most vulnerable—untenured and part-time faculty, and entry-level and prospective employees. They are using the academic incentive and promotion structure to reward cooperation and discourage dissent. At the same time they are mounting an intensifying propaganda campaign to portray faculty as incompetent, hide-bound, recalcitrant, inefficient, ineffective, and expensive—in short, in need of improvement or replacement through instructional technologies. Faculty are portrayed above all as obstructionist, as standing in the way of progress and forestalling the panacea of virtual education allegedly demanded by students, their parents, and the public.

The York University faculty had heard it all. Yet still they fought vigorously and ultimately successfully to preserve quality education and protect themselves from administrative assault. During their long strike they countered such administration propaganda with the truth about what was happening to higher education and eventually won the support of students, the media, and the public. Most important, they secured a new contract containing unique and unprecedented provisions which, if effectively enforced, give faculty members direct and unambiguous control over all decisions relating to the automation of instruction, including veto power. According to the contract, all decisions regarding the use of technology as a supplement to classroom instruction or as a means of alternative delivery (including the use of video, CD-ROM's, Internet Web sites, computer-mediated conferencing, etc.) "shall be consistent with the pedagogic and academic judgments and principles of the faculty member employee as to the appropriateness of the use of technology in the circumstances." The contract also guarantees that "a faculty member will not be required to convert a course without his or her agreement." Thus, the York faculty will be able to ensure that the new technology, if and when used, will contribute to a genuine enhancement rather than a degradation of the quality of education, while at the same time preserving their positions, their autonomy, and their academic freedom. The battle is far from won, but it is a start.

Student Reactions

The second set of implications stemming from the commoditization of instruction involve the transformation of the university into a market for the commodities being produced. Administrative propaganda routinely alludes to an alleged student demand for the new instructional products. At UCLA officials are betting that their high-tech agenda will be "student driven," as students insist that faculty make fuller use of the Web site technology in their courses. To date, however, there has been no such demand on the part of students, no serious study of it, and no evidence for it. Indeed, the few times students have been given a voice, they have rejected the initiatives hands down, especially when they were required to pay for it (the definition of effective demand, i.e., a market).

At UCLA, students recommended against the Instructional Enhancement Initiative. At the University of British Columbia, home of the WEB-CT software being used at UCLA, students voted in a referendum four-to-one against a similar initiative, despite a lengthy administration campaign promising them a more secure place in the high-tech future. Administrators at both institutions have tended to dismiss, ignore, or explain away these negative student decisions, but there is a message here: students want the genuine face-to-face education they paid for, not a cyber-counterfeit. Nevertheless, administrators at both UCLA and UBC decided to proceed with the their agenda anyway, desperate to create a market and secure some return on their investment in the information technology infrastructure. Thus, they are creating a market by fiat, compelling students (and faculty) to become users and hence consumers of the hardware, software, and content products as a condition of getting an education, whatever their interest or ability to pay. Can all students equally afford this capital-intensive education?

Another key ethical issue relates to the use of student online activities. Few students realize that their computer-based courses are often thinly veiled field trials for product and market development, that while they are studying their courses, their courses are studying them. In Canada, for example, universities have been given royalty-free licenses to Virtual U software in return for providing data on its use to the vendors. Thus, all online activity including communications between students and professors and among students are monitored, automatically logged and archived by the system for use by the vendor. Students enrolled in courses using Virtual U software are in fact formally designated "experimental subjects." Because federal monies were used to develop the software and underwrite the field trials, vendors were compelled to comply with ethical guidelines on the experimental use of human subjects. Thus, all students once enrolled are required to sign forms releasing ownership and control of their online activities to the vendors. The form states "as a student using Virtual U in a course, I give my permission to have the computer-generated usage data, conference transcript data, and virtual artifacts data collected by the Virtual U software . . . used for research, development, and demonstration purposes."

According to UCLA's Home Education Network president John Korbara, all of their distance learning courses are likewise monitored and archived for use by company officials. On the UCLA campus, according to Harlan Lebo of the provost's office, student use of the course Web sites will be routinely audited and evaluated by the administration. Marvin Goldberg, designer of the UCLA WEB-CT software acknowledges that the system allows for "lurking" and automatic storage and retrieval of all online activities. How this capability will be used and by whom is not altogether clear, especially since Web sites are typically being constructed by people other than the instructors. What third parties (besides students and faculty in the course) will have access to the student's communications? Who will own student online contributions? What rights, if any, do students have to privacy and proprietary control of their work? Are they given prior notification as to the ultimate status of their online activities, so that they might be in a position to give, or withhold, their informed consent? If students are taking courses which are just experiments, and hence of unproven pedagogical value, should students be paying full tuition for them? And if students are being used as guinea pigs in product trials masquerading as courses, should they be paying for these courses or be paid to take them? More to the point, should students be content with a degraded, shadow cyber-education? In Canada student organizations have begun to confront these issues head on, and there are some signs of similar student concern emerging also in the U.S.

Conclusion

In his classic 1959 study of diploma mills for the American Council on Education, Robert Reid described the typical diploma mill as having the following characteristics: "no classrooms," "faculties are often untrained or nonexistent," and "the officers are unethical self-seekers whose qualifications are no better than their offerings." It is an apt description of the digital diploma mills now in the making. Quality higher education will not disappear entirely, but it will soon become the exclusive preserve of the privileged, available only to children of the rich and the powerful. For the rest of us a dismal new era of higher education has dawned. In ten years, we will look upon the wired remains of our once great democratic higher education system and wonder how we let it happen. That is, unless we decide now not to let it happen.

16

The Road Not Taken

Revisiting the Original New Deal

Steve Fuller

The Original Rise and Fall of New Deal Science Policy

Nowadays one risks uttering a platitude when saying that research priorities, funding levels, and accountability mechanisms have been distorted by national security concerns during the Cold War. But what would it mean to remove those distortions? First, it would mean recognizing their full extent. The last time this was done was immediately after World War II, during the debate surrounding the establishment of the U.S. National Science Foundation (NSF). But few realize just how many wartime science policy practices were actually normalized by the version of the NSF that came into being (Kleinman 1995). Therefore, we shall have to explore the road not taken in the founding debates, the road that would have led back to the New Deal policies that were abruptly diverted once America entered the war.

Before the two world wars caused the state to take such an active interest in scientific research, academic scientists were primarily teachers who were more concerned with opening students' minds than with pushing back the frontiers of knowledge. Innovative work was, in turn, to be found in the industrial labs and corporate foundations, both of which fostered interdisciplinary work that had no clear place in the university curriculum (Kohler 1991). In those days scientific innovation implied that the novelty had to have some relatively immediate practical uptake. Indeed, until Hiroshima the average American's conception of a "scientist at work" was either the self-taught Thomas Edison or a white-coated industrial chemist—not an academic at all (La Follette 1990: 45–65).

Reprinted with permission of The Open University Press. From *The Governance of Science* (Milton Keynes, UK: Open University Press, 2000), 117–30.

The success of the atomic bomb project changed all that by vividly demonstrating what leading academic scientists could accomplish when given enormous resources with minimal political oversight. The urgent needs of war easily explained this combination of heavy expenditures and high levels of secrecy. Nevertheless, after the war, Vannevar Bush (1945) and the other architects of America's post-war science policy refashioned these conditions as characterizing something called "basic research," the supposed foundations of the scientific enterprise. What had previously been regarded as "unprecedented" research conditions were now normalized as providing the basis of "autonomy," that most sacred of scientific virtues. Ironically, one of the most attractive features of this sort of research, interdisciplinary character, was dropped from the design of the NSF in favour of discipline-based, peer-reviewed research.

However, it would be a mistake to conclude that Bush simply took advantage of an opportunity. In fact, the concept of basic research was integral to the emerging image of the "security state" whose primary activity was preparation for war. Indeed, Bush and his allies were more successful in diminishing science's public accountability by having much defence research reclassified as "basic" than by boosting actual research funding (Reingold 1994: 367–8). Because the "applied research" championed by industry was increasingly consumer oriented and market driven, Bush and his colleagues concluded that a business would invest in research only to the extent that it would be likely to increase its market share. Products too ahead of their times are just as unprofitable as products that fall behind consumer demand. Thus, industry had little incentive to make major revolutionary breakthroughs in research. In contrast, the national security state was dedicated to mastering and overcoming the military capabilities of the Communist Bloc through the design of ever "smarter" weapons, whose potential for "mutually assured destruction" would supposedly be a sufficient threat to end war altogether. This search for what might be called "the military sublime" was the very antithesis of the run-with-the-pack market mentality of business, but was well suited to the sense of ultimate ends that has always animated the basic research mindset. By contrast with society's short-term vision, the military was concerned with technologies that coincided with science's own long-term vision.

Moreover, based on the performance of academic scientists during World War II, the defence establishments of most western powers seemed content to continue to allow scientists to work with minimum oversight. In this way, the military became the main external supporter of basic research in the Cold War era, with 20 percent of all scientists worldwide working in defence-related research by the 1970s (30 percent in the USA and UK) (Proctor 1991: 254). Given this background, the market-oriented science policy analyst should recommend that, in the post–Cold War era, the supply of researchers be reduced to match demand. For the current plethora of researchers is not a "natural" outgrowth of the search for knowledge but a vestige from the upscaling of universities that accompanied the Cold War. This point is often overlooked because Big Science was being "naturalized"—that is, grounded in a rewritten historical sociology and political economy of science—as it was coming to fruition in the 1960s (see Shils 1968; for a systematic critique see Fuller 2000b).

However, Bush's greatest legacy was probably the idea that research funds should be allocated according to the quality of the researcher, a judgement which, in turn, could only be made by other distinguished researchers. Peer review would probably have never enjoyed its sacred status had it not been used to construct the team entrusted with building the bomb. Threatened in common by the Axis powers, the Allied physicists put aside their theoretical differences and focused on collecting colleagues who were most likely to help complete the task in the shortest time. This was a novel use of peer review that Bush and others thought was relevant to resource allocation issues in peacetime.

To be sure, peer review had always been used to evaluate research for professional publication. But to entrust scientists with dividing up the available resources for research, especially once they were no longer focused on a common goal, was to reopen the door to a kind of partisanship that would compromise the integrity of science. The sorts of self-serving judgements that scientists routinely try to suppress when weighing the merits of completed work would arguably re-emerge when evaluating the prospects of work yet to be done and hence compromise the resulting judgements. In historical terms, most of the scientific outsiders whose work was readily recognized by their peers upon completion—certainly Einstein and maybe even Darwin—would have been initially passed over for peer-reviewed funding because of their prior lack of accomplishment and potential threat to established research programmes. Michael Faraday is an especially interesting case because, fully aware of his disadvantaged background, he early volunteered his manual services to Humphry Davy at the Royal Institution's chemistry laboratory, and through apprenticeship managed to rise up within the establishment, so that he became an "institutionalized radical" (Morus 1992).

Bush's moves to extend wartime science policy practices into peacetime did not go unchallenged in the few years that separated the end of World War II and the start of the Cold War. However, the failure of those challenges continues to be felt every time criticism of peer-reviewed funding procedures is answered by moralizing about the evils of "earmarking" or, worse still, "pork barrelling." A project is "earmarked" if its funding comes from outside the official competition, while an earmarked project is "pork barrelled" if it comes specifically from a politically sensitive constituency. These currently reviled terms harken back to an honourable alternative world of science policy that is worth revisiting as a prod to the contemporary imagination, one that specifically targets scientific developments to constituencies who might otherwise fail to achieve them.

The Continuing Legacy of the New Deal

Even before the end of the war, many American state university heads had begun to worry that peer-reviewed funding procedures in peacetime would exaggerate existing resource differences between élite and non-élite institutions that had arisen from selective corporate research and development investment before the war (Mills 1948). As it turns out, the concerns about élitism have been well placed. In recent years, 50 percent of federal science funding in the USA has gone to 33 out of over 2,000 institutions of higher education, virtually the same 33 that receive the lion's share of corporate sponsorship (Office of Technology Assessment 1991: 263–5). One way of forestalling this situation would have been to follow the advice of Chauncey Leake, then provost of the University of Texas, and institute a system of checks and balances (modelled on the U.S. Constitution itself) to prohibit the same scientists from serving on the governing boards of the major journals, professional societies, and private and public funding bodies (U.S. Congress 1945). In effect, a wider cross section of scientists would have been brought into the governance of their fields, creating a broader sense of scientific "peerage." This jury duty model would democratize peer review by counteracting Merton's principle of cumulative advantage, whereby academia's rich get richer and poor get poorer.

More radical proposals would have banished peer review from the funding arena altogether. Peacetime science policy would have resumed the comprehensive welfare programme initiated by the New Deal. One scheme in this vein, favoured by Edwin Land, the inventor of the Polaroid camera's self-developing film, called for the federal government to grant scholarships to science students as seed money for potentially lucrative inventions. Others suggested an "affirmative action" programme for the traditionally disadvantaged southern and western regions of the country, whereby they would receive a disproportionally larger fraction of the funds than regions that had traditionally benefitted from private endowments. There was even talk of a national service programme that would set aside scholarships for clever students from impoverished backgrounds to matriculate at the major universities, on the condition that they would then staff technical training and research facilities in the nation's economically underdeveloped regions (Kevles 1977).

All of the above proposals merit a second look today. Especially noteworthy at the time was the claim that science was a suitable site for redistributionist policies precisely because its modes of enquiry, in principle, could be known and done by anyone. Any differences in scientific performance among universities or regions could be attributed to differences in the quality of the training and research facilities at their disposal, not the quality of the people staffing them. Even opponents of the democratization of science needed to explain how it was possible for many of the greatest scientific achievements to be made by people who were considered "slow" in their day, such as Darwin and Einstein. Rather than allowing only the intellectual élite to pursue scientific careers, it might be better to promote a spread of intelligence so that radical but underdeveloped ideas were not rejected out of hand by the quicker wits equipped with all the "standard counter-arguments." However, this analysis only proved persuasive as long as the fast-paced, high-tech research that characterized World War II was considered an aberration that would be redressed in peacetime by a science policy sensitive to the rate at which scientific innovation could be assimilated by society at large. Unfortunately, the Cold War was declared soon after the Allied victory and this required the rapid deployment of technically trained personnel, a policy that clearly favoured scientists already at the major research centres.

Ironically, despite their claims for being far-sighted thinkers, those who today defend the maintenance, if not increase, of basic research spending neglect the fact that, over the past century, science funding has been subject to recurrent cycles of mobilization (for war) and post-war conversion (for the U.S.A. see Reingold 1994: 367). Indeed, it is often forgotten that in the original NSF (National Science Foundation) hearings, the head of Bell Telephone Laboratories and president of the National Academy of Sciences, Frank Jewett, argued that the state should roll back all of its research investment to pre–World War II levels. The public should then be encouraged to treat scientific research programmes as charities, contributions to which could be the source of tax relief (U.S. Congress 1947). Whatever else one thinks of Jewett's scheme, it would certainly have provided an incentive for members of the public sufficiently wealthy to have incurred a significant tax burden to acquaint themselves with competing research programmes.

The idea of deliberately slowing the pace of scientific progress may seem unthinkable today, but it was treated as a live option when the New Deal sought ways out of the Depression. In that context, science was perceived as a regulatory monster. On the one hand, scientists contributed to a volatile financial climate because their inventions often failed to match their investors' hopes. On the other, when science did deliver on its promises, there was the even greater fear that its results would be monopolized by firms with the power to hold the rest of society in their grip. This problem has reemerged as transnational corporations have found it in their interest to promote more comprehensive intellectual property rights (Drahos 1995). Moreover, once the Depression was in full swing, scientists were chastised for developing new forms of automation that cut the production costs of firms by making workers redundant. Consequently, a key provision of New Deal industrial policy was to curb chronic unemployment by restricting the amount and type of such innovations that could be introduced into the workplace. Some especially zealous regulators went so far as to suggest, on the model of farmers being paid not to raise crops, that scientists be paid not to do certain kinds of socially disruptive research (Proctor 1991: 238). A couple of related ideas would be relevant to our own time. One is that instead of regarding a patent on an innovation as an exclusive (albeit temporary) intellectual property right, the innovation may be taxed in order to discourage the tendency to innovate faster than society can assimilate. The other, of course, is to divert scientists' efforts from research to teaching, given the mass of unread publications that could benefit from more direct exposure.

I have so far neglected one important proposal that was favoured by much of the scientific community to help the state normalize science's relation to the rest of society: unionization. It received considerable attention up to World War II but little afterward. American scientists first acquired a collective identity fighting on the same side in World War I. However, after the German defeat, the career patterns of scientists drifted according to the centrifugal demands of the market, which left the peacetime economy increasingly destabilized. Some scientists took a page from Marx and cast themselves as exploited high-grade workers to whom a union would give the voice to ensure that the fruits of their labour did not do more harm than good. Given that most scientists at the time were themselves from working- to lower middle-class backgrounds, the call for unionization seemed ideologically credible.

Britain provided the exemplar that American scientists followed in the 1920s and 1930s. An especially influential figure in the latter period was the X-ray crystallographer and Marxist historian of science, John Desmond Bernal, whose view of science policy as a form of labour-management relations was influenced by Alfred Mond, head of Imperial Chemical Industries (Jones 1988: 6–14). However, suspicions of latent Communist sympathies slowed the unionization process, until World War II intervened, causing scientists once again to rally around the flag—only this time with some of them co-opted into the management side of military operations, most notably the atomic bomb project (Kuznick 1987). After the war, these scientist-managers became the architects of post-war science policy, the official conduits between science and the public. Although few really believed that the likes of Vannevar Bush spoke for all scientists' interests, florid state-subsidized employment opportunities and a growing economy were no cause for complaint.

Time for a Renewed New Deal? The Deskilling and Casualization of Academic Labour

The conditions that encouraged scientists to see their research interests tied to national military and industrial policies have rapidly changed since the Cold War came to an end. In one important sense the climate has become potentially more conducive to unionization. A good case in point is physics, the standard-bearer of Cold War science policy, which in recent years has been the discipline most strongly pressured to treat its degree holders as "stakeholders." In January 1991, the Council of the American Physical Society (APS) adopted a public position on funding priorities for the first time in its 93-year history, one that favoured broadly based physics research over the more glamorous but élite Superconducting Supercollider (Office of Technology Assessment 1991: 159 n. 40). Three years later, two write-in candidates who worked outside physics were elected to the council: Kevin Aylesworth and Zachary Levine. Their platforms called for restructuring physics education to match the new labour market and for the APS to adopt policies that explicitly represented the interests of all its members. Given that Ph.D. production in physics was still growing at Cold War rates of 10 percent per year, Levine (1994: 17) went so far as to call for "limit[ing] the growth of the physics community in the USA to sustainable levels," and "help[ing] young scientists who wish to make a transition out of physics." This, in turn, sparked a more general debate about whether science degree programmes were broadly enough gauged to enable young scientists to function in work environments that were increasingly interdisciplinary and non-academic (Holden 1995).

It is worth underlining the fact that the depth of the problem of interdisciplinary understanding means that it needs to be addressed during initial training, not simply by "on-the-job experience." A currently popular argument for state-supported research is the "collective store of tacit knowledge" that supposedly results from scientists working with people in other disciplines, government, and industry (Science Policy Research Unit 1996). However, the fact that such knowledge emerges from the research environment after the scientist has already completed his or her education may simply represent a deficiency in training that could be addressed much earlier and hence made more intelligible than the expression "collective store of tacit knowledge" suggests. An experimental paradigm for the promotion of interdisciplinary understanding involved medical students and social work students in a "shared learning" programme during their final year of university study, in which they were required to work together in groups to solve problems of common concern. Being still students, their professional identities and status sensibilities had not yet solidified, which enabled them to succeed at the task at hand and to be receptive to cooperative ventures with members of the corresponding group in the future (Carpenter and Hewstone 1996).

As the state increasingly devolves its research funding functions to the private sector, scientists are returning to the role of economic "wild card" that they played in the 1920s. The vicissitudes of today's biotechnology and cyberspace industries in the world's stock exchanges prove that we are in the midst of encountering a Scylla and Charybdis rather like the one that the New Deal's industrial policy was designed to circumnavigate. But perhaps most significant of these recent developments is the extension of automation beyond the factory shop-floor into the post-industrial heartland of scientific workspaces. Computerized expert systems, the products of the "knowledge engineering" business, are slowly but surely absorbing the work of medical diagnosis, chemical analysis, and engineering design—to name just a few tasks that have traditionally required scientfically trained personnel. It is still an open question whether these machines will take a permanent sizeable chunk out of the scientific workforce or, for that matter, whether they will deskill the workers who remain. A renewed union sensibility in this context might align scientists with technical and support personnel who, despite their different credentials, nevertheless come to have similar experiences with computerized expert systems in the workplace (Fuller 1994b). That we are now in a position to address these matters openly suggests that science is on the verge of becoming a normal part of public policy. To develop this point more fully, consider three senses in which scientists are becoming "deskilled":

1. Scientists lose part of their sphere of discretionary judgement, which, practically speaking, means that they no longer have the final say on the disposition of their own research.

2. Scientists are held accountable to more publicly oriented standards that are neither of their own creation nor necessarily those most likely to promote the interests of their professional community.

3. The kind of knowledge that is lost in (1) is not comparable to the kind that is gained in (2). Administratively designed formal procedures are perceived as a shallow replacement for the scientists' own deeply ingrained and historically sedimented expertise.

All of these concerns have already been extensively articulated in another context, namely the replacement of factory and office workers, as well as professionals, by a kind of computer—the expert system—that can simulate the reasoning processes by which the relevant humans issue judgements in their domain of expertise (Dreyfus and Dreyfus 1987). The emotionally charged debates surrounding this issue typically take it for granted that the promotion of meaningful human labour entails the cultivation of personalized expertise. Yet, such a requirement prevents discussion about the kind of society in which meaningful human labour is to occur. For example, the political theorist Benjamin Barber (1992) has argued that the goal of a democratic society may be better served if each person were encouraged to perform a number of different jobs in his or her lifetime rather than to cultivate obsessively—and to identify exclusively with—one form of labour. In that case, "merely competent" performances subserved to a higher end would be favoured over a jealously guarded expertise pursued as an end in itself. Marx, for example, seemed to have something like this in mind as the utopia of post-revolutionary communism.

Politics aside, another possibility is that the competent/expert distinction has been given more psychological weight than it deserves. Is it so clear that there is a distinctly "cognitive" component to expertise that is, in some sense, sacrificed in the civil society that Barber and other radical democrats envisage? I think not. Scientific judgement appears "expert" only against a certain social background, a major part of which includes a public that allows the scientists virtually complete discretion over how and when they account for their activities, such as whether they explain deviations from normal practice as "errors" or "genius" (Fuller 1994a).

This background also includes supportive, or at least non-obtrusive, peers. Indeed, expertise depends to such an extent on the appearance of consensus among peers that the gloss of expertise can easily invite public scrutiny once two similarly credentialed scientists contest each other outside strictly professional forums. For this reason, scientists who circumvent the peer review process and seek vindication in the media are rarely accepted fully back into the scientific fold (Nelkin 1987). Just as physicians often refuse to testify against their colleagues, one could imagine similar resistance from scientists invited to debate openly the relative merits of their research proposals. In both cases, the motivation would be clear. No matter whose claim to knowledge turned out to be vindicated, the entire discipline's credibility would suffer from public exposure.

Scientists realize that just as credibility is built on circumspection, it is also spent through overexposure and popularization. Likewise, the proliferation of machines that are persuasively portrayed as performing the tasks of experts contributes to an erosion of expertise. This devaluation is often subtle but it can be seen to have operated over a long period. A good indicator is the long-term shift in the image of what it means to "do science." Certain practices, especially those associated with sensory observation, have been removed from the core image of science as they have become mechanized and automated. For example, the need to have a "good eye" for field and experimental work is now an anachronism in most disciplines (Fuller 1993: 137–42). Such mechanized operations are treated as ancillary to the actual conduct of inquiry, the locus of which has shifted to whatever the human continues to do with minimal instrumental mediation. And with the advent of expert systems capable of modelling multiple causal interactions, automation may soon reach the inner sanctum of scientific reasoning—the process of hypothesis testing (Langley et al. 1987).

The history of science and technology is full of alarmists who argue that the removal of restrictions on some jealously guarded form of knowledge will lead to the decline of civilization. However, to put these claims in perspective, it is worth remembering that, for almost every oppressed segment of society, it has been argued that even the spread of literacy would lead to the collapse of authority (Graff 1987). What has been true about these dire predictions is that the relative value of certain forms of knowledge has changed. Once a previously esteemed form of knowledge is made readily available, either because people know it for themselves or it is embodied in technology, the knowledge becomes commonplace and hence irrelevant in determining social worth. Indeed, scientific knowledge is an instance of what the economist Fred Hirsch (1976) originally called a "positional good"—that is, a good whose value is primarily tied to others not having it. In recent years, Hirsch's idea of positional goods has itself been popularized by the French sociologist Pierre Bourdieu (1984) as the differences in training between groups that become the bases for other forms of discrimination. A group is empowered to the extent that the knowledge of its members enables them to do things that members of other groups would like to do but cannot. In Fuller (2000a, chap. 1), I have spoken in terms of possessing social capital, which produces "multiplier effects" as it circulates through the knowledge economy.

Following Hirsch and Bourdieu, then, the net effect of democratization and mechanization is to level old distinctions, invariably spurring new attempts to restrict inquiry by creating new domains of knowledge that, in turn, will provide new opportunities for discrimination. Every deskilling is thus a potential reskilling. For example, value hierarchies are typically so restructured after a scientific revolution that a new paradigm may offer few job prospects for scientists clinging to the old traditions, while providing new opportunities for their more adaptable colleagues. But such a transformation may also come about by more mundane means. With the computerization of the medical workplace, it may soon no longer matter that a diagnostic expert system fails to reproduce the complexity of a physician's reasoning, provided that the system offers good enough diagnoses. Why? Because the shift in the sources of authoritative knowledge—from human to machine—will have been accompanied by a shift in the standards for "worthwhile" knowledge. To put it bluntly, we may decide that we have better things to worry about than the perfectibility of computer performance in that particular domain. If such a response were to become typical then the last half-century of debates over the ability of computers to simulate certain forms of human thought (e.g., Dreyfus 1992) would become beside the point.

How Technology May End Up Redressing the Balance in Academia

My discussion up to this point suggests that expertise is nothing more than a kind of procedural knowledge whose sphere of application has been mystified by those enjoying exclusive access to that knowledge. If so, it would seem to follow that the areas of expertise most vulnerable to the positionality effects noted above are those having the most articulate procedures, perhaps because they are routinely given extensive treatments in textbooks. These areas of expertise form the core of scientific and professional establishments (Hales 1986).

By contrast, among the areas of expertise least vulnerable to problems of easy access are two groups that currently enjoy little respect in science policy forums—namely, people whose knowledge is of too recent vintage to have yet been standardized into procedures and people whose knowledge is too embedded in a specific locale to be of much use as a generalized set of procedures. Speaking abstractly, *novelty* and *locality* are the qualities most resistant to the knowledge engineer trying to design an expert system. Perhaps not surprisingly, these two features of knowledge are valorized by the "postmodern" turn in cultural studies. Indeed, Jean-Francois Lyotard's (1983) seminal essay, which popularized the term "postmodern," originated as a policy study on the likely implications of information technologies on cultural production. Although Lyotard acknowledged Daniel Bell's (1973) "post-industrial" vision as a benchmark for such considerations, it is clear that Lyotard's own image of "knowledge work" in the postmodern condition is one of increasingly politicized knowledge brokerage, and not one of politically sanitized technocratic facilitation. Here I concur with Lyotard, and would suggest that Bell's technocratic ideal is vulnerable to the proliferation of expert systems. A somewhat better image of the emerging knowledge worker is Robert Reich's (1991) "symbolic analyst."

Speaking concretely, novelty and locality are epitomized by, on the one hand, the growing number of recent Ph.D.s whose status as itinerant postdoctoral fellows has earned them the title of "unfaculty" and, on the other hand, the technical and administrative support personnel whose experience with the day-to-day nitty-gritty of the research site typically exceeds that of their more highly credentialled supervisors. These unsung heroes of the research process are underpaid and undervalued, yet they often make all the difference between success and failure: what if the project secretary is unfamiliar with the work rhythms of the researchers, or the lab technician has not been consulted on the overall scheme of the project, or the postdoc's judgement is trusted only in matters of technique but not theory? Just as consumers should be incorporated into decisions taken at the level of production, so too should these "implementers" at the level of project conception. The case of the "unfaculty" is especially acute, as they threaten to become the *Lumpenproletariat* of academia. In 1991 they comprised one out of five academic posts in America's research universities (Office of Technology Assessment 1991: 214–5); by 1998 the proportion had doubled in comparable British institutions (Thomson 1998). Now such staff, who are technically defined as being on "fixed-term contracts" are routinely described as "casualized," as if to draw specific attention to their dispensability.

Economic Competitiveness as the Continuation of Cold War Science Policy by Other Means

We are on the verge of continuing Cold War science policy by other means, as "economic competitiveness" has become the new battleground in the post–Cold War era. For example, Britain's place on the front lines has been ensured by the Office of Science and Technology's inclusion in the Department of Trade and Industry. Not surprisingly, this has been followed by a spate of calls for the state to concentrate its research funding on the universities best equipped to make the biggest international splash, including the National Academies Policy Advisory Group (NAPAG), a consortium of the country's most venerable scientific societies. NAPAG has reckoned the number of British universities that are internationally competitive over a broad range of fields to be 15 out of around 100. They have recommended that research funding be restricted to the most competitive departments, thereby openly endorsing Merton's principle of cumulative advantage. . . . But will this prove to be a sound strategy?

Ironically, the best argument against making competitiveness the premier goal of national science policy comes from Robert Reich, former secretary of labour in the Clinton administration. Reich notwithstanding, the U.S.A. has been otherwise the world's biggest promoter of the competitiveness model. (The main ideological defence of national competitiveness is probably Porter 1990). Reich criticizes the model for presupposing that a nation's inhabitants necessarily benefit from the economic success of firms nominally headquartered within its borders (Reich 1991: 122–5). Applied to science, it follows that the number of papers published or patents registered by a country's leading firms and universities does not automatically translate to benefits for either the scientific community or the larger public of that country. That will depend on the full range of its people's skills, attitudes, and employment settings. Arguably, in the last quarter-century, Japan and Germany have made better use of the knowledge produced in the U.S.A. and UK than the Americans and British have. To be sure, scientists already on the "cutting edge" will always benefit, but that just reinforces the élitism that the competitiveness model breeds. A nation often gains ground on its international competitors at the cost of polarizing its own residents.

This problem can be tackled from either the demand or the supply side of science. The demand-driven strategy is to adjust the training, incentives, and rewards of scientists (and science students) to enable them to appear more attractive to foreign and domestic firms. Despite the lip service often paid to this strategy, its implementation would be prohibitively expensive. This point contributes to the "realism" of NAPAG's approach to research funding. But even with adequate economic resources and political will, the sheer provision of highly skilled, appropriately motivated scientific professionals does not guarantee their gainful employment in what has increasingly become a buyer's market for brains. As the scientific infrastructure of former Second and Third World countries develop at a rate that outpaces the rise in labour costs, firms and universities have seen fit to "outsource" substantial research projects to scientists from these regions. Even First World countries have benefitted from this "globalization of research and development," as when high-grade skills are combined with marginally lower labour costs. Here the U.S.A. and UK may enjoy an advantage over Japan and Germany (NSF 1991: 110).

However, the demand-driven strategy to peacetime science policy suffers from enormous presumptuousness, which in turn reflects the lack of voice that rank-and-file scientists and technical support staff have in setting national science policy. Specifically, the scientific community has never been consulted on its commitment to "competitiveness" as a goal of its activities, be it defined in military, industrial, or even scholarly terms.

The first step toward a supply-driven response to the competitiveness model of science policy would be just such a periodic inventory of the competences and attitudes of the scientific community, including tertiary-sector science students. Without such systematic information, even leaders of the scientific community have been prone to stereotype their constituency. A striking example is the "cry of alarm" issued by the president of the American Association for the Advancement of Science in 1991, which called for the federal government to double its nondefence spending on science. To make his case, he restricted himself to surveying the opinions of tenured and untenured faculty at the top 50 research universities in the U.S.A. (Lederman 1991). Before it was disbanded by the Republican-controlled Congress, the Office of Technology Assessment had begun to take stock of science's missing voices by highlighting statistical data on groups that did not fit the stereotypical scientist: graduate students, women, ethnic minorities and the 20+ percent of the scientific workforce that was migratory and contract-based, otherwise known as the "unfaculty" (Office of Technology Assessment 1991: 205–30). Although these data only scratched the surface, they still suggested that the scientific community was not uniformly sold on the idea of competitiveness as a peacetime goal.

More indirect support for a New Deal for science policy comes from a Carnegie Foundation survey on the effects of the publication-oriented, resource-intensive research university model on academic work conditions across the entire tertiary sector, including private and public universities and liberal arts and community colleges (Boyer 1990). The imperative to become more "research competitive" was shown to be a pervasive source of stress in academics' lives, especially in the harder sciences, as they were drawn away from their preferred image of scholarship as pedagogically relevant, open to interdisciplinary influences and less focused on sheer productivity. The survey found that while academic identity was tied more to disciplinary than specific institutional affiliation, the type of institution in which academics worked shaped their sense of what constituted appropriate scholarship in their discipline. On that basis, accrediting agencies and professional associations should cultivate a plurality of mechanisms for evaluating and promoting academic work that avoids polarizing faculty into a two-tier funding structure that inevitably ranks research over teaching and relegates public service to the sub-intellectual. This would mean, in the British context, that the Office of Science and Technology should be relocated from the Department of Trade and Industry to the Department of Education and Employment.

Contemporary problems often rest on maintaining distinctions that at an earlier point in history did not make a difference. The New Deal approach to national science policy did not presume as strong a distinction between scientists and non-scientists as operates today in, say, the contrast between "peer reviewed" and "earmarked" research. Rather, the idea was to extend scientific peerage, not simply by recruiting more scientists, but by ensuring that science's full constituency was represented on governing boards and policy forums. Were this principle in operation today, it is unlikely that we would witness the repeated complaints that federal science panels are "biased" and "narrow" (Cordes 1996). Public involvement would not seem so alien to scientific autonomy if a wider range of scientists sat on such panels in the first place. In the twentieth century, élitism in science has been driven primarily by military preparedness and, more recently, economic competitiveness. The twenty-first century offers an opportunity for science to regain the lost legacy of the twentieth, whereby science would be driven by the *entire* community of enquirers.

REFERENCES

Barber, B. 1992. *An Aristocracy of Everyone.* New York: Ballantine Books.

Bell, D. 1973. *The Coming of the Post-industrial Society.* New York: Harper & Row.

Bourdieu, P. 1984. *Distinction.* Cambridge, MA: Harvard University Press.

Boyer, E. 1990. *Scholarship Reconsidered: The Priorities of the Professoriate.* Princeton, NJ: Carnegie Foundation for the Advancement of Teaching.

Bush, V. 1945. *Science: The Endless Frontier.* Washington, DC: Office of Scientific Research and Development.

Carpenter, J., and M. Hewstone. 1996. Shared learning for doctors and socal workers: evaluation of a programme. *British Journal of Social Work* 26: 239–57.

Cordes, C. 1996. Critics say membership of federal science panels is too narrow. *Chronicle of Higher Education* 8 March: A26.

Drahos, P. 1995. Information feudalism in the information society. *The Information Society* 11: 209–22.

Dreyfus, H. 1992. *What Computers Still Can't Do.* 2d ed. Cambridge, MA: MIT Press.

Dreyfus, H., and S. Dreyfus. 1987. *Mind over Machine: The Power of Human Intuition and Expertise in the Era of the Computer.* New York: Free Press.

Fuller, S. 1993. *Philosophy of Science and Its Discontents.* 2d ed. New York: Guilford Press.

———. 1994a. The constitutively social character of expertise. *International Journal of Expert Systems* 7: 51–64.

———. 1994b. Why post-industrial society never came: what a false prophecy can teach us about the impact of technology on academia. *Academe* 80 (6): 22–8.

———. 2000a. *The Governance of Science.* Milton Keynes, UK: Open University Press.

———. 2000b. *Thomas Kuhn: A Philosophical History for Our Times.* Chicago: University of Chicago Press.

Graff, H. 1987. *The Legacies of Literacy.* Bloomington, IN: Indiana University Press.

Hales, M. 1986. *Science or Society? The Politics of the Work of Scientists,* 2d ed. London: Free Association Books.

Hirsch, F. 1976. *The Social Limits to Growth.* Cambridge, MA: Harvard University Press.

Holden, C. 1995. Careers '95: the future of the Ph.D. *Science* 270 (6 October): 121–45.

Jones, G. 1988. *Science, Politics, and the Cold War.* London: Routledge and Kegan Paul.

Kevles, D. 1977. The National Science Foundation and the debate over postwar research policy, 1942–1945. *Isis* 68: 5–26.

Kleinman, D. 1995. *Politics on the Endless Frontier: Postwar Research Policy in the United States.* Durham, NC: Duke University Press.

Kohler, R. 1991. *Partners in Science: Foundations and Natural Scientists.* Chicago: University of Chicago Press.

Kuznick, P. 1987. *Beyond the Laboratory: Scientists as Political Activists in 1930s America.* Chicago: University of Chicago Press.

La Follette, M. 1990. *Making Science Our Own: Public Images of Science 1910–1955.* Chicago: University of Chicago Press.

Langley, P., H. Simon, G. Bradshaw, and J. Zytkow. 1987. *Scientific Discovery.* Cambridge, MA: MIT Press.

Lederman, L. 1991. *Science: The End of the Frontier?* Washington, DC: American Association for the Advancement of Science.

Levine, Z. 1994. *Biographical Information and Candidate Statements for the 1994 Election of Councillors to the American Physical Society.* Washington, DC: American Physical Society.

Lyotard, J.-F. 1983. *The Postmodern Condition.* 2d ed. Minneapolis, MN: University of Minnesota Press.

Mills, C. 1948. Distribution of American research funds. *Science* 107: 127–30.

Morus, I. 1992. Different experimental lives: Michael Faraday and William Sturgeion. *History of Science* 20: 1–28.

Nelkin, D. 1987. *Selling Science.* New York: W. H. Freeman.

NSF (National Science Foundation) 1991. *Science and Engineering Indicators.* 10th ed. Washington, DC: U.S. National Science Board.

Office of Technology Assessment. 1991. *Federally Funded Research: Decisions for a Decade.* Washington, DC: U.S. Government Printing Office.

Porter, M. 1990. *The Competitive Advantage of Nations.* London: Macmillan.

Proctor, R. 1991. *Value-Free Science? Purity and Power in the Modern Knowledge.* Cambridge, MA: Harvard University Press.

Reich, R. 1991. *The Work of Nations.* New York: Random House.

Reingold, N. 1994. Science and government in the United States since 1945. *History of Science* 32: 361–86.

Science Policy Research Unit. 1996. *The Relationship between Publicly*

Funded Basic Research and Economic Performance. London: H M Treasury.

Shils, E. (ed.) 1968. *Criteria for Scientific Development: Public Policy and National Goals.* Cambridge, MA: MIT Press.

Thomson, A. 1998. Contract lecturer campaign hot up. *Times Higher Education Supplement.* 20 November.

U.S. Congress. 1945. Senate Subcommittee of the Committee on Military Affairs. *Hearings on Science Legislation.* 79th Congress, 1st sess. (November). Washington, DC: U.S. Government Printing Office.

U.S. Congress. 1947. House Committee on Interstate and Foreign Commerce. *Hearings on the National Science Foundation.* 80th Congress, 1st sess. (March). Washington, DC: U.S. Government Printing Office.

PART VI

The Future of Scientific “Credit”

17

The Republic of Science

Its Political and Economic Theory

Michael Polanyi

My title is intended to suggest that the community of scientists is organized in a way which resembles certain features of a body politic and works according to economic principles similar to those by which the production of material goods is regulated. Much of what I will have to say will be common knowledge among scientists, but I believe that it will recast the subject from a novel point of view which can both profit from and have a lesson for political and economic theory. For in the free cooperation of independent scientists we shall find a highly simplified model of a free society, which presents in isolation certain basic features of it that are more difficult to identify within the comprehensive functions of a national body.

The first thing to make clear is that scientists, freely making their own choice of problems and pursuing them in the light of their own personal judgment, are in fact co-operating as members of a closely knit organization. The point can be settled by considering the opposite case where individuals are engaged in a joint task without being in any way co-ordinated. A group of women shelling peas work at the same task, but their individual efforts are not co-ordinated. The same is true of a team of chess players. This is shown by the fact that the total amount of peas shelled and the total number of games won will not be affected if the members of the group are isolated from each other. Consider by contrast the effect which a complete isolation of scientists would have on the progress of science. Each scientist would go on for a while developing problems derived from the information initially available to all. But

Reprinted with permission of John Polanyi and The University of Chicago Press. From *Knowing and Being* (Chicago: University of Chicago Press, 1969), 49–72.

these problems would soon be exhausted, and in the absence of further information about the results achieved by others, new problems of any value would cease to arise, and scientific progress would come to a standstill.

This shows that the activities of scientists are in fact co-ordinated, and it also reveals the principle of their co-ordination. This consists in the adjustment of the efforts of each to the hitherto achieved results of the others. We may call this a co-ordination by mutual adjustment of independent initiatives—of initiatives which are co-ordinated because each takes into account all the other initiatives operating within the same system.

When put in these abstract terms the principle of spontaneous co-ordination of independent initiatives may sound obscure. So let me illustrate it by a simple example. Imagine that we are given the pieces of a very large jigsaw puzzle, and suppose that for some reason it is important that our giant puzzle be put together in the shortest possible time. We would naturally try to speed this up by engaging a number of helpers; the question is in what manner these could be best employed. Suppose we share out the pieces of the jigsaw puzzle equally among the helpers and let each of them work on his lot separately. It is easy to see that this method, which would be quite appropriate to a number of women shelling peas, would be totally ineffectual in this case, since few of the pieces allocated to one particular assistant would be found to fit together. We could do a little better by providing duplicates of all the pieces to each helper separately, and eventually somehow bring together their several results. But even by this method the team would not much surpass the performance of a single individual at his best. The only way the assistants can effectively co-operate, and surpass by far what any single one of them could do, is to let them work on putting the puzzle together in sight of the others, so that every time a piece of it is fitted in by one helper, all the others will immediately watch out for the next step that becomes possible in consequence. Under this system, each helper will act on his own initiative, by responding to the latest achievements of the others, and the completion of their joint task will be greatly accelerated. We have here in a nutshell the way in which a series of independent initiatives are organized to a joint achievement by mutually adjusting themselves at every successive stage to the situation created by all the others who are acting likewise.

Such self-co-ordination of independent initiatives leads to a joint result which is unpremeditated by any of those who bring it about. Their co-ordination is guided as by "an invisible hand" towards the joint discovery of a hidden system of things. Since its end-result is unknown, this kind of co-operation can only advance stepwise, and the total performance will be the best possible if each consecutive step is decided upon by the person most competent to do so. We may imagine this condition to be fulfilled for the fitting together of a jigsaw puzzle if each helper watches out for any new opportunities arising along a particular section of the hitherto completed patch of the puzzle, and also keeps an eye on a particular lot of pieces, so as to fit them in wherever a chance presents itself. The effectiveness of a group of helpers will then exceed that of any isolated member, to the extent to which some member of the group will always discover a new chance for adding a piece to the puzzle more quickly than any one isolated person could have done by himself.

Any attempt to organize the group of helpers under a single authority would eliminate their independent initiatives and thus reduce their joint effectiveness to that of the single person directing them from the centre. It would, in effect, paralyse their cooperation.

Essentially the same is true for the advancement of science by independent initiatives adjusting themselves consecutively to the results achieved by all the others. So long as each scientist keeps making the best contribution of which he is capable, and on which no one could improve (except by abandoning the problem of his own choice and thus causing an overall loss to the advancement of science), we may affirm that the pursuit of science by independent self-co-ordinated initiatives assures the most efficient possible organization of scientific progress. And we may add, again, that any authority which would undertake to direct the work of the scientist centrally would bring the progress of science virtually to a standstill.

WHAT I have said here about the highest possible co-ordination of individual scientific efforts by a process of self-co-ordination may recall the self-co-ordination achieved by producers and consumers operating in a market. It was, indeed, with this in mind that I spoke of "the invisible hand" guiding the co-ordination of independent initiatives to a maximum advancement of science, just as Adam Smith invoked "the invisible hand" to describe the achievement of greatest joint material satisfaction when independent producers and consumers are guided by the prices of goods in a market. I am suggesting, in fact, that the co-ordinating functions of the market are but a special case of co-ordination by mutual adjustment. In the case of science, adjustment takes place by taking note of the published results of other scientists; while in the case of the market, mutual adjustment is mediated by a system of prices broadcasting current exchange relations, which make supply meet demand.

But the system of prices ruling the market not only transmits information in the light of which economic agents can mutually adjust their actions, it also provides them with an incentive to exercise economy in terms of money. We shall see that, by contrast, the scientist responding directly to the intellectual situation created by the published results of other scientists is motivated by current professional standards.

Yet in a wider sense of the term, the decisions of a scientist choosing a problem and pursuing it to the exclusion of other possible avenues of inquiry may be said to have an economic character. For his decisions are designed to produce the highest possible result by the use of a limited stock of intellectual and material resources. The scientist fulfils this purpose by choosing a problem that is neither too hard nor too easy for him. For to apply himself to a problem that does not tax his faculties to the full is to waste some of his faculties; while to attack a problem that is too hard for him would waste his faculties altogether. The psychologist K. Lewin has observed that one's person never becomes fully involved either in a problem that is much too hard, nor in one that is much too easy. The line the scientist must choose turns out, therefore, to be that of greatest ego-involvement; it is the line of greatest excitement, sustaining the most intense attention and effort of thought. The choice will be conditioned to some extent by the resources available to the scientist in terms of materials and assistants, but he will be ill-advised to choose his problem with a view to guaranteeing that none of these resources be wasted. He should not hesitate to incur such a loss, if it leads him to deeper and more important problems.

THIS IS where professional standards enter into the scientist's motivation. He assesses the depth of a problem and the importance of its prospective solution primarily by the standards of scientific merit accepted by the scientific community—though his own work may demand these standards to be modified. Scientific merit depends on a number of criteria which I shall enumerate here under three headings. These criteria are not altogether independent of each other, but I cannot analyse here their mutual relationship.

(1) The first criterion that a contribution to science must fulfill in order to be accepted is a sufficient degree of plausibility. Scientific publications are continuously beset by cranks, frauds, and bunglers whose contributions must be rejected if journals are not to be swamped by them. This censorship will not only eliminate obvious absurdities but must often refuse publication merely because the conclusions of a paper appear to be unsound in the light of current scientific knowledge. It is indeed difficult even to start an experimental inquiry if its problem is considered scientifically unsound. Few laboratories would accept today a student of extrasensory perception, and even a project for testing once more the hereditary transmission of acquired characters would be severely discouraged from the start. Besides, even when all these obstacles have been overcome, and a paper has come out signed by an author of high distinction in science, it may be totally disregarded, simply for the reason that its results conflict sharply with the current scientific opinion about the nature of things.

I shall illustrate this by an example which I have used elsewhere.[1] A series of simple experiments were published in June 1947 in the *Proceedings of the Royal Society* by Lord Rayleigh—a distinguished Fellow of the Society—purporting to show that hydrogen atoms striking a metal wire transmit to it energies up to a hundred electron volts. This, if true, would have been far more revolutionary than the discovery of atomic fission by Otto Hahn. Yet, when I asked physicists what they thought about it, they only shrugged their shoulders. They could not find fault with the experiment yet not one believed in its results, nor thought it worth while to repeat it. They just ignored it. A possible explanation of Lord Rayleigh's experiments is given in my *Personal Knowledge.*[2] It appears that the physicists missed nothing by disregarding these findings.

(2) The second criterion by which the merit of a contribution is assessed may be described as its scientific value, a value that is composed of the following three coefficients: (a) its accuracy, (b) its systematic importance, (c) the intrinsic interest of its subject-matter. You can see these three gradings entering jointly into the value of a paper in physics compared with one in biology. The inanimate things studied by physics are much less interesting than the living beings which are the subject of biol-

1. M. Polanyi, *The Logic of Liberty* (London: Routledge & Kegan Paul; Chicago: University of Chicago Press, 1951), 12.

2. M. Polanyi, *Personal Knowledge* (London: Routledge & Kegan Paul; Chicago: University of Chicago Press, 1958), 276.

ogy. But physics makes up by its great accuracy and wide theoretical scope for the dullness of its subject, while biology compensates for its lack of accuracy and theoretical beauty by its exciting matter.

(3) A contribution of sufficient plausibility and of a given scientific value may yet vary in respect of its originality; this is the third criterion of scientific merit. The originality of technical inventions is assessed, for the purpose of claiming a patent, in terms of the degree of surprise which the invention would cause among those familiar with the art. Similarly, the originality of a discovery is assessed by the degree of surprise which its communication should arouse among scientists. The unexpectedness of a discovery will overlap with its systematic importance, yet the surprise caused by a discovery, which causes us to admire its daring and ingenuity, is something different from this. It pertains to the act of producing the discovery. There are discoveries of the highest daring and ingenuity, as for example the discovery of Neptune, which have no great systematic importance.

Both the criteria of plausibility and of scientific value tend to enforce conformity, while the value attached to originality encourages dissent. This internal tension is essential in guiding and motivating scientific work. The professional standards of science must impose a framework of discipline and at the same time encourage rebellion against it. They must demand that, in order to be taken seriously, an investigation should largely conform to the currently predominant beliefs about the nature of things, while allowing that in order to be original it may to some extent go against these. Thus, the authority of scientific opinion enforces the teachings of science in general, for the very purpose of fostering their subversion in particular points.

This dual function of professional standards in science is but the logical outcome of the belief that scientific truth is an aspect of reality and that the orthodoxy of science is taught as a guide that should enable the novice eventually to make his own contacts with this reality. The authority of scientific standards is thus exercised for the very purpose of providing those guided by it with independent grounds for opposing it. The capacity to renew itself by evoking and assimilating opposition to itself appears to be logically inherent in the sources of the authority wielded by scientific orthodoxy.

But who is it, exactly, who exercises the authority of this orthodoxy? I have mentioned scientific opinion as its agent. But this raises a serious problem. No single scientist has a sound understanding of more than a tiny fraction of the total domain of science. How can an aggregate of such specialists possibly form a joint opinion? How can they possibly exercise jointly the delicate function of imposing a current scientific view about the nature of things, and the current scientific valuation of proposed contributions, even while encouraging an originality which would modify this orthodoxy? In seeking the answer to this question we shall discover yet another organizational principle that is essential for the control of a multitude of independent scientific initiatives. This principle is based on the fact that, while scientists can admittedly exercise competent judgment only over a small part of science, they can usually judge an area adjoining their own special studies that is broad enough to include some fields on which other scientists have specialized. We thus have a considerable degree of overlapping between the areas over which a scientist can exercise a sound critical judgment. And, of course, each scientist who is a member of a group of overlapping competences will also be a member of other groups of the same kind, so that the whole of science will be covered by chains and networks of overlapping neighbourhoods. Each link in these chains and networks will establish agreement between the valuations made by scientists overlooking the same overlapping fields, and so, from one overlapping neighbourhood to the other, agreement will be established on the valuation of scientific merit throughout all the domains of science. Indeed, through these overlapping neighbourhoods uniform standards of scientific merit will prevail over the entire range of science, all the way from astronomy to medicine. This network is the seat of scientific opinion. Scientific opinion is an opinion not held by any single human mind, but one which, split into thousands of fragments, is held by a multitude of individuals, each of whom endorses the others' opinion at second hand, by relying on the consensual chains which link him to all the others through a sequence of overlapping neighbourhoods.

Admittedly, scientific authority is not distributed evenly throughout the body of scientists; some distinguished members of the profession predominate over others of a more junior standing. But the authority of scientific opinion remains essentially mutual; it is established *between* scientists, not above them. Scientists exercise their authority over each other. Admittedly, the body of scientists, as a whole, does uphold the authority of science over the lay public. It controls thereby also the process by which young men are trained to become members of the scientific profession. But once the novice has reached the grade of an independent scientist, there is no longer any superior above him. His submission to scientific opinion is entailed now in his joining a chain of mutual appreciations, within which he is called upon to bear his equal share of responsibility for the authority to which he submits.

Let me make it clear, even without going into detail, how great and varied are the powers exercised by this authority. Appointments to positions in universities and elsewhere, which offer opportunity for independent research, are filled in accordance with the appreciation of candidates by scientific opinion. Referees reporting on papers submitted to journals are charged with keeping out contributions which current scientific opinion condemns as unsound, and scientific opinion is in control, once more, over the issue of textbooks, as it can make or mar their influence through reviews in scientific journals. Representatives of scientific opinion will pounce upon newspaper articles or other popular literature which would venture to spread views contrary to scientific opinion. The teaching of science in schools is controlled likewise. And, indeed, the whole outlook of man on the universe is conditioned by an implicit recognition of the authority of scientific opinion.

I have mentioned earlier that the uniformity of scientific standards throughout science makes possible the comparison between the value of discoveries in fields as different as astronomy and medicine. This possibility is of great value for the rational distribution of efforts and material resources throughout the various branches of science. If the minimum merit by which a contribution would be qualified for acceptance by journals were much lower in one branch of science than in another, this would clearly cause too much effort to be spent on the former branch as compared with the latter. Such is in fact the principle which underlies the rational distribution of grants for the pursuit of research. Subsidies should be curtailed in areas where their yields in terms of scientific merit tend to be low, and should be channelled instead to the growing points of science, where increased financial means may be expected to produce a work of higher scientific value. It does not matter for this purpose whether the money comes from a public authority or from private sources, nor whether it is disbursed by a few sources or a large number of benefactors. So long as each allocation follows the guidance of scientific opinion, by giving preference to the most promising scientists and subjects, the distribution of grants will automatically yield the maximum advantage for the advancement of science as a whole. It will do so, at any rate, to the extent to which scientific opinion offers the best possible appreciation of scientific merit and of the prospects for the further development of scientific talent.

For scientific opinion may, of course, sometimes be mistaken, and as a result unorthodox work of high originality and merit may be discouraged or altogether suppressed for a time. But these risks have to be taken. Only the discipline imposed by an effective scientific opinion can prevent the adulteration of science by cranks and dabblers. In parts of the world where no sound and authoritative scientific opinion is established, research stagnates for lack of stimulus, while unsound reputations grow up based on commonplace achievements or mere empty boasts. Politics and business play havoc with appointments and the granting of subsidies for research; journals are made unreadable by including much trash.

Moreover, only a strong and united scientific opinion imposing the intrinsic value of scientific progress on society at large can elicit the support of scientific inquiry by the general public. Only by securing popular respect for its own authority can scientific opinion safeguard the complete independence of mature scientists and the unhindered publicity of their results, which jointly assure the spontaneous co-ordination of scientific efforts throughout the world. These are the principles of organization under which the unprecedented advancement of science has been achieved in the twentieth century. Though it is easy to find flaws in their operation, they yet remain the only principles by which this vast domain of collective creativity can be effectively promoted and co-ordinated.

During the last twenty to thirty years, there have been many suggestions and pressures towards guiding the progress of scientific inquiry in the direction of public welfare. I shall speak mainly of those I have witnessed in England. In August 1938, the British Association for the Advancement of Science founded a new division for the social and international relations of science, which was largely motivated by the desire to offer deliberate social guidance to the progress of science. This programme was given more extreme expression by the Association of Scientific Workers in Britain. In January 1943, the Association filled a large hall in London with a meeting attended by many of the most distinguished scientists of the country, and it decided—in the words officially summing up the conference—that research would no longer be conducted for itself as an end in itself. Reports from Soviet Russia describing the successful conduct of scientific research according to plans laid down by the Academy of Science with a view to supporting the economic Five-Year Plans encouraged this resolution.

I appreciate the generous sentiments which actuate the aspiration of guiding the progress of science into socially beneficent channels, but I hold its aim to be impossible and indeed nonsensical.

An example will show what I mean by this impossibility. In January

1945, Lord Russell and I were together on the BBC Brains Trust. We were asked about the possible technical uses of Einstein's theory of relativity, and neither of us could think of any. This was forty years after the publication of the theory and fifty years after the inception by Einstein of the work which led to its discovery. It was fifty-eight years after the Michelson-Morley experiment. But, actually, the technical application of relativity, which neither Russell nor I could think of, was to be revealed within a few months by the explosion of the first atomic bomb. For the energy of the explosion was released at the expense of mass in accordance with the relativistic equation $e = mc^2$, an equation which was soon to be found splashed over the cover of *Time* magazine, as a token of its supreme practical importance.

Perhaps Russell and I should have done better in foreseeing these applications of relativity in January 1945, but it is obvious that Einstein could not possibly take these future consequences into account when he started on the problem which led to the discovery of relativity at the turn of the century. For one thing, another dozen or more major discoveries had yet to be made before relativity could be combined with them to yield the technical process which opened the atomic age.

Any attempt at guiding scientific research towards a purpose other than its own is an attempt to deflect it from the advancement of science. Emergencies may arise in which all scientists willingly apply their gifts to tasks of public interest. It is conceivable that we may come to abhor the progress of science and stop all scientific research, or at least whole branches of it, as the Soviets stopped research in genetics for twenty-five years. You can kill or mutilate the advance of science, you cannot shape it. For it can advance only by essentially unpredictable steps, pursuing problems of its own, and the practical benefits of these advances will be incidental and hence doubly unpredictable.

In saying this, I have *not* forgotten, but merely set aside, the vast amount of scientific work currently conducted in industrial and governmental laboratories.[3] In describing here the autonomous growth of science, I have taken the relation of science to technology fully into account.

BUT EVEN those who accept the autonomy of scientific progress may feel irked by allowing such an important process to go on without

3. I have analysed the relation between academic and industrial science elsewhere in some detail; see *Journal of the Institute of Metallurgy* 89 (1961), 401 ff. Cf. *Personal Knowledge,* 174–84.

trying to control the co-ordination of its fragmentary initiatives. The period of high aspirations following the last war produced an event to illustrate the impracticability of this more limited task.

The incident originated in the University Grants Committee, which sent a memorandum to the Royal Society in the summer of 1945. The document, signed by Sir Charles Darwin, requested the aid of the Royal Society to secure "The Balanced Development of Science in the United Kingdom"; this was its title.

The proposal excluded undergraduate studies and aimed at the higher subjects that are taught through the pursuit of research. Its main concern was with the lack of co-ordination between universities in taking up "rare" subjects, "which call for expert study at only a few places, or in some cases perhaps only one." This was linked with the apprehension that appointments are filled according to the dictates of fashion, as a result of which some subjects of greater importance are being pursued with less vigour than others of lesser importance. It proposed that a co-ordinating machinery should be set up for levelling out these gaps and redundancies. The Royal Society was asked to compile, through its Sectional Committees covering the main divisions of science, lists of subjects deserving preference in order to fill gaps. Such surveys were to be renewed in the future to guide the University Grants Committee in maintaining balanced proportions of scientific effort throughout all fields of inquiry.

Sir Charles Darwin's proposal was circulated by the Secretaries of the Royal Society and the members of the Sectional Committees along with a report of previous discussions of proposals by the Council and other groups of Fellows. The report acknowledged that the co-ordination of the pursuit of higher studies in the universities was defective ("haphazard") and endorsed the project for periodic, most likely annual, surveys of gaps and redundancies by the Royal Society. The members of the Sectional Committees were asked to prepare, for consideration by a forthcoming meeting of the Council, lists of subjects suffering from neglect.

Faced with this request, which I considered at the best pointless, I wrote to the Physical Secretary (the late Sir Alfred Egerton) to express my doubts. I argued that the present practice of filling vacant chairs by the most eminent candidate that the university can attract was the best safeguard for rational distribution of efforts over rival lines of scientific research. As an example (which should appeal to Sir Charles Darwin as a physicist) I recalled the successive appointments to the chair of physics in Manchester during the past thirty years. Manchester had elected to this chair Schuster, Rutherford, W. L. Bragg, and Blackett, in this sequence, each of whom represented at the time a "rare" section of physics: spectroscopy, radioactivity, X-ray crystallography, and cosmic rays, respectively. I affirmed that Manchester had acted rightly and that they would have been ill-advised to pay attention to the claims of subjects which had not produced at the time men of comparable ability. For the principal criterion for offering increased opportunities to a new subject was the rise of a growing number of distinguished scientists in that subject and the falling off of creative initiative in other subjects, indicating that resources should be withdrawn from them. While admitting that on certain occasions it may be necessary to depart from this policy, I urged that it should be recognized as the essential agency for maintaining a balanced development of scientific research.

Sir Alfred Egerton's response was sympathetic, and, through him, my views were brought to the notice of the members of Sectional Committees. Yet the Committees met, and I duly took part in compiling a list of "neglected subjects" in chemistry. The result, however, appeared so vague and trivial (as I will illustrate by an example in a moment) that I wrote to the Chairman of the Chemistry Committee that I would not support the Committee's recommendations if they should be submitted to the Senate of my university.

However, my worries were to prove unnecessary. Already the view was spreading among the Chairmen of the Sectional Committees "that a satisfactory condition in each science would come about naturally, provided that each university always chose the most distinguished leaders for its post, irrespective of his specialization." While others still expressed the fear that this would make for an excessive pursuit of fashionable subjects, the upshot was, at the best, inconclusive. Darwin himself had, in fact, already declared the reports of the Sectional Committees "rather disappointing."

The whole action was brought to a close, one year after it had started, with a circular letter to the Vice-Chancellors of the British universities signed by Sir Alfred Egerton, as secretary, on behalf of the Council of the Royal Society, a copy being sent to the university Grants Committee. The circular included copies of the reports received from the Sectional Committees and endorsed these in general. But in the body of the letter only a small number of these recommendations were specified as being of special importance. This list contained seven recommendations for the establishment of new schools of research, but said nothing about the way these new schools should be co-ordinated with existing activities all over the United Kingdom. The impact of this document on the universities seems to have been negligible. The Chemistry Committee's recommendation for the establishment of "a strong school of analytic chemistry," which should have concerned me as Professor of Physical Chemistry, was never even brought to my notice in Manchester.

I HAVE not recorded this incident in order to expose its error. It is an important historical event. Most major principles of physics are founded on the recognition of an impossibility, and no body of scientists was better qualified than the Royal Society to demonstrate that a central authority cannot effectively improve on the spontaneous emergence of growing points in science. It has proved that little more can, or need, be done towards the advancement of science than to assist spontaneous movements towards new fields of distinguished discovery, at the expense of fields that have become exhausted. Though special considerations may deviate from it, this procedure must be acknowledged as the major principle for maintaining a balanced development of scientific research.[4]

Let me recall yet another striking incident of the post-war period which bears on these principles. I have said that the distribution of subsidies to pure science should not depend on the sources of money, whether they are public or private. This will hold to a considerable extent also for subsidies given to universities as a whole. But after the war, when in England the cost of expanding universities was largely taken over by the state, it was felt that this must be repaid by a more direct support for the national interest. This thought was expressed in July 1946 by the Committee of Vice-Chancellors in a memorandum sent out to all universities, which Sir Ernest Simon (as he then was), as Chairman of the Council of Manchester University, declared to be of "almost revolutionary" importance. I shall quote a few extracts:

> The universities entirely accept the view that the Government has not only the right, but the duty, to satisfy itself that every field of study

4. Here is the point at which this analysis of the principles by which funds are to be distributed between different branches of science may have a lesson for economic theory. It suggests a way in which resources can be rationally distributed between *any* rival purposes that cannot be valued in terms of money. All cases of public expenditure serving purely collective interests are of this kind. A comparison of such values by a network of overlapping competences may offer a possibility for a true collective assessment of the relative claims of thousands of government departments of which no single person can know well more than a tiny fraction.

> which in the national interest ought to be cultivated in Great Britain, is in fact being adequately cultivated in the universities. . . .
>
> In the view of the Vice-Chancellors, therefore, the universities may properly be expected not only individually to make proper use of the resources entrusted to them, but collectively to devise and execute policies calculated to serve the national interest. And in that task, both individually and collectively, they will be glad to have a greater measure of guidance from the Government than, until quite recent days, they have been accustomed to receive. . . .
>
> Hence the Vice-Chancellors would be glad if the University Grants Committee were formally authorised and equipped to undertake surveys of all main fields of university activity designed to secure that as a whole universities are meeting the whole range of national need for higher teaching and research. . . .

We meet here again with a passionate desire for accepting collective organization for cultural activities, though these actually depend for their vigorous development on the initiative of individuals adjusting themselves to the advances of their rivals and guided by a cultural opinion in seeking support, be it public or private. It is true that competition between universities was getting increasingly concentrated on gaining the approval of the Treasury, and that its outcome came to determine to a considerable extent the framework within which the several universities could operate. But the most important administrative decisions, which determine the work of universities, as for example the selection of candidates for new vacancies, remained free and not arranged collectively by universities, but by competition between them. For they cannot be made otherwise. The Vice-Chancellors' memorandum has, in consequence, made no impression on the life of the universities and is, by this time, pretty well forgotten by the few who had ever seen it.[5]

WE MAY sum up by saying that the movements for guiding science towards a more direct service of the public interest, as well as for co-ordinating the pursuit of science more effectively from a centre, have all petered out. Science continues to be conducted in British universities as was done before the movement for the social guidance of science ever started. And I believe that all scientific progress achieved in the Soviet

5. I have never heard the memorandum mentioned in the University of Manchester. I knew about it only from Sir Ernest Simon's article entitled "A Historical University Document," *University Quarterly,* 1–2 (1946–48), 189–92. My quotations referring to the memorandum are taken from this article.

Union was also due—as everywhere else—to the initiative of original minds, choosing their own problems and carrying out their investigations, according to their own lights. This does not mean that society is asked to subsidize the private intellectual pleasure of scientists. It is true that the beauty of a particular discovery can be fully enjoyed only by the expert. But wide responses can be evoked by the purely scientific interest of discovery. Popular response, overflowing into the daily press, was aroused in recent years in England and elsewhere by the astronomical observations and theories of Hoyle and Lovell, and more recently by Ryle, and the popular interest was not essentially different from that which these advances had for scientists themselves.

And this is hardly surprising, since for the last three hundred years the progress of science has increasingly controlled the outlook of man on the universe, and has profoundly modified (for better and for worse) the accepted meaning of human existence. Its theoretic and philosophic influence was pervasive.

Those who think that the public is interested in science only as a source of wealth and power are gravely misjudging the situation. There is no reason to suppose that an electorate would be less inclined to support science for the purpose of exploring the nature of things than were the private benefactors who previously supported the universities. Universities should have the courage to appeal to the electorate, and to the public in general, on their own genuine grounds. Honesty should demand this at least. For the only justification for the pursuit of scientific research in universities lies in the fact that the universities provide an intimate communion for the formation of scientific opinion, free from corrupting intrusions and distractions. For though scientific discoveries eventually diffuse into all people's thinking, the general public cannot participate in the intellectual milieu in which discoveries are made. Discovery comes only to a mind immersed in its pursuit. For such work the scientist needs a secluded place among like-minded colleagues who keenly share his aims and sharply control his performances. The soil of academic science must be exterritorial in order to secure its rule by scientific opinion.

THE EXISTENCE of this paramount authority, fostering, controlling, and protecting the pursuit of a free scientific inquiry, contradicts the generally accepted opinion that modern science is founded on a total rejection of authority. This view is rooted in a sequence of important historical antecedents which we must acknowledge here. It is a fact that the Copernicans had to struggle with the authority of Aristotle upheld by the Roman Church, and by the Lutherans invoking the Bible; that Vesalius founded the modern study of human anatomy by breaking the authority of Galen. Throughout the formative centuries of modern science, the rejection of authority was its battle-cry; it was sounded by Bacon, by Descartes, and collectively by the founders of the Royal Society of London. These great men were clearly saying something that was profoundly true and important, but we should take into account today the sense in which they have meant their rejection of authority. They aimed at adversaries who have since been defeated. And although other adversaries may have arisen in their places, it is misleading to assert that science is still based on the rejection of any kind of authority. The more widely the republic of science extends over the globe, the more numerous become its members in each country, and the greater the material resources at its command, the more there clearly emerges the need for a strong and effective scientific authority to reign over this republic. When we reject today the interference of political or religious authorities with the pursuit of science, we must do this in the name of the established scientific authority which safeguards the pursuit of science.

Let it also be quite clear that what we have described as the functions of scientific authority go far beyond a mere confirmation of facts asserted by science. For one thing, there are no mere facts in science. A scientific fact is one that has been accepted as such by scientific opinion, both on the grounds of the evidence in favour of it and because it appears sufficiently plausible in view of the current scientific conception of the nature of things. Besides, science is not a mere collection of facts, but a system of facts based on their scientific interpretation. It is this system that is endorsed by a scientific authority. And within this system this authority endorses a particular distribution of scientific interest intrinsic to the system; a distribution of interest established by the delicate value-judgments exercised by scientific opinion in sifting and rewarding current contributions to science. Science *is what it is,* in virtue of the way in which scientific authority constantly eliminates, or else recognizes at various levels of merit, contributions offered to science. In accepting the authority of science, we accept the totality of all these value-judgments.

Consider, also, the fact that these scientific evaluations are exercised by a multitude of scientists, each of whom is competent to assess only a tiny fragment of current scientific work, so that no single person is responsible at first hand for the announcements made by science at any time. And remember that each scientist originally established himself as such by joining at some point a network of mutual appreciation extending far beyond his own horizon. Each such acceptance appears then as a submission to a vast range of value-judgments exercised over all the domains of science, which the newly accepted citizen of science henceforth endorses, although he knows hardly anything about their subject-matter. Thus, the standards of scientific merit are seen to be transmitted from generation to generation by the affiliation of individuals at a great variety of widely disparate points, in the same way as artistic, moral, or legal traditions are transmitted. We may conclude, therefore, that the appreciation of scientific merit too is based on a tradition which succeeding generations accept and develop as their own scientific opinion. This conclusion gains important support from the fact that the methods of scientific inquiry cannot be explicitly formulated and hence can be transmitted only in the same way as an art, by the affiliation of apprentices to a master. The authority of science is essentially traditional.

But this tradition upholds an authority which cultivates originality. Scientific opinion imposes an immense range of authoritative pronouncements on the student of science, but at the same time it grants the highest encouragement to dissent from them in some particular. While the whole machinery of scientific institutions is engaged in suppressing apparent evidence as unsound, on the ground that it contradicts the currently accepted view about the nature of things, the same scientific authorities pay their highest homage to discoveries which deeply modify the accepted view about the nature of things. It took eleven years for the quantum theory, discovered by Planck in 1900, to gain final acceptance. Yet by the time another thirty years had passed, Planck's position in science was approaching that hitherto accorded only to Newton. Scientific tradition enforces its teachings in general, for the very purpose of cultivating their subversion in the particular.

I have said this here at the cost of some repetition, for it opens a vista of analogies in other intellectual pursuits. The relation of originality to tradition in science has its counterpart in modern literary culture. "Seldom does the word [tradition] appear except in a phrase of censure," writes T. S. Eliot.[6] And he then tells how our exclusive appreciation of originality conflicts with the true sources of literary merit actually recognized by us:

We dwell with satisfaction upon the poet's difference from his predecessors, especially his immediate predecessors; we endeavour to find

6. T. S. Eliot, *Selected Essays* (London: Faber, 1941), 13.

something that can be isolated in order to be enjoyed. Whereas if we approach a poet without this prejudice, we shall often find that not only the best, but the most individual parts of his work may be those in which the dead poets, his ancestors, assert their immortality most vigorously.[7]

Eliot has also said, in *Little Gidding,* that ancestral ideas reveal their full scope only much later, to their successors:

> And what the dead had no speech for, when living,
> They can tell you, being dead: the communication
> Of the dead is tongued with fire beyond the language of the living.

And this is so in science: Copernicus and Kepler told Newton where to find discoveries unthinkable to themselves.

At this point we meet a major problem of political theory: the question whether a modern society can be bound by tradition. Faced with the outbreak of the French Revolution, Edmund Burke denounced its attempt to refashion at one stroke all the institutions of a great nation and predicted that this total break with tradition must lead to a descent into despotism. In reply to this, Tom Paine passionately proclaimed the right of absolute self-determination for every generation. The controversy has continued ever since. It has been revived in America in recent years by a new defence of Burke against Tom Paine, whose teachings had hitherto been predominant. I do not wish to intervene in the American discussion, but I think I can sum up briefly the situation in England during the past 170 years. To the most influential political writers of England, from Bentham to John Stuart Mill, and recently to Isaiah Berlin, liberty consists in doing what one likes, provided one leaves other people free to do likewise. In this view there is nothing to restrict the English nation *as a whole* in doing with itself at any moment whatever it likes. On Burke's vision of "a partnership of those who are living, those who are dead and those who are to be born," these leading British theorists turn a blind eye. But practice is different. In actual practice it is Burke's vision that controls the British nation; the voice is Esau's, but the hand is Jacob's.

The situation is strange. But there must be some deep reason for it, since it is much the same as that which we have described in the organization of science. This analogy seems indeed to reveal the reason for this curious situation. Modern man claims that he will believe nothing un-

7. Ibid., 14.

less it is unassailable by doubt; Descartes, Kant, John Stuart Mill, and Bertrand Russell have unanimously taught him this. They leave us no grounds for accepting any tradition. But we see now that science itself can be pursued and transmitted to succeeding generations only within an elaborate system of traditional beliefs and values, just as traditional beliefs have proved indispensable throughout the life of society. What can one do then? The dilemma is disposed of by continuing to profess the right of absolute self-determination in *political theory* and relying on the guidance of tradition in *political practice.*

But this dubious solution is unstable. A modern dynamic society, born of the French Revolution, will not remain satisfied indefinitely with accepting, be it only *de facto,* a traditional framework as its guide and master. The French Revolution, which, for the first time in history, had set up a government resolved on the indefinite improvement of human society, is still present in us. Its most far-reaching aspirations were embodied in the ideas of socialism, which rebelled against the whole structure of society and demanded its total renewal. In the twentieth century this demand went into action in Russia in an upheaval exceeding by far the range of the French Revolution. The boundless claims of the Russian Revolution have evoked passionate responses throughout the world. Whether accepted as a fervent conviction or repudiated as a menace, the ideas of the Russian Revolution have challenged everywhere the traditional framework which modern society had kept observing in practice, even though claiming absolute self-determination in theory.

I HAVE described how this movement evoked among many British scientists a desire to give deliberate social purpose to the pursuit of science. It offended their social conscience that the advancement of science, which affects the interests of society as a whole, should be carried on by individual scientists pursuing their own personal interests. They argued that all public welfare must be safeguarded by public authorities and that scientific activities should therefore be directed by the government in the interest of the public. This reform should replace by deliberate action towards a declared aim the present growth of scientific knowledge intended as a whole by no one, and in fact not even known in its totality, except quite dimly, to any single person. To demand the right of scientists to choose their own problems appeared to them petty and unsocial, as against the right of society deliberately to determine its own fate.

But have I not said that this movement has virtually petered out by this time? Have not even the socialist parties throughout Europe endorsed by now the usefulness of the market? Do we not hear the freedom and the independence of scientific inquiry openly demanded today even in important centres within the Soviet domain? Why renew this discussion when it seems about to lose its point?

My answer is that you cannot base social wisdom on political disillusion. The more sober mood of public life today can be consolidated only if it is used as an opportunity for establishing the principles of a free society on firmer grounds. What does our political and economic analysis of the Republic of Science tell us for this purpose?

It appears, at first sight, that I have assimilated the pursuit of science to the market. But the emphasis should be in the opposite direction. The self-co-ordination of independent scientists embodies a higher principle, a principle which is *reduced* to the mechanism of the market when applied to the production and distribution of material goods.

LET ME sketch out briefly this higher principle in more general terms. The Republic of Science shows us an association of independent initiatives, combined towards an indeterminate achievement. It is disciplined and motivated by serving a traditional authority, but this authority is dynamic; its continued existence depends on its constant self-renewal through the originality of its followers.

The Republic of Science is a Society of Explorers. Such a society strives towards an unknown future, which it believes to be accessible and worth achieving. In the case of scientists, the explorers strive towards a hidden reality, for the sake of intellectual satisfaction. And as they satisfy themselves, they enlighten all men and are thus helping society to fulfil its obligation towards intellectual self-improvement.

A free society may be seen to be bent in its entirety on exploring self-improvement—every kind of self-improvement. This suggests a generalization of the principles governing the Republic of Science. It appears that a society bent on discovery must advance by supporting independent initiatives, co-ordinating themselves mutually to each other. Such adjustment may include rivalries and opposing responses which, in society as a whole, will be far more frequent than they are within science. Even so, all these independent initiatives must accept for their guidance a traditional authority, enforcing its own self-renewal by cultivating originality among its followers.

Since a dynamic orthodoxy claims to be a guide in search of truth, it implicitly grants the right to opposition in the name of truth—truth being taken to comprise here, for brevity, all manner of excellence that we recognize as the ideal of self-improvement. The freedom of the individual safeguarded by such a society is therefore—to use the term of Hegel—of a positive kind. It has no bearing on the right of men to do as they please; but assures them the right to speak the truth as they know it. Such a society does not offer particularly wide private freedoms. It is the cultivation of public liberties that distinguishes a free society, as defined here.

IN THIS view of a free society, both its liberties and its servitudes are determined by its striving for self-improvement, which in its turn is determined by the intimations of truths yet to be revealed, calling on men to reveal them.

This view transcends the conflict between Edmund Burke and Tom Paine. It rejects Paine's demand for the absolute self-determination of each generation, but does so for the sake of its own ideal of unlimited human and social improvement. It accepts Burke's thesis that freedom must be rooted in tradition, but transposes it into a system cultivating radical progress. It rejects the dream of a society in which all will labour for a common purpose, determined by the will of the people. For in the pursuit of excellence it offers no part to the popular will and accepts instead a condition of society in which the public interest is known only fragmentarily and is left to be achieved as the outcome of individual initiatives aiming at fragmentary problems. Viewed through the eyes of socialism, this ideal of a free society is conservative and fragmented, and hence adrift, irresponsible, selfish, apparently chaotic. A free society conceived as a society of explorers is open to these charges, in the sense that they do refer to characteristic features of it. But if we recognize that these features are indispensable to the pursuit of social self-improvement, we may be prepared to accept them as perhaps less attractive aspects of a noble enterprise.

These features are certainly characteristic of the proper cultivation of science and are present throughout society as it pursues other kinds of truth. They are, indeed, likely to become ever more marked, as the intellectual and moral endeavours to which society is dedicated enlarge in range and branch out into ever new specialized directions. For this must lead to further fragmentation of initiatives and thus increase resistance to any deliberate total renewal of society.

18

The Instability of Authorship

Credit and Responsibility in Contemporary Biomedicine

Mario Biagioli

In the past decade, the definition of authorship has been the topic of many articles and letters to the editor in scientific and especially biomedical journals. The official position of the ICMJE (International Committee of Medical Journal Editors) has been, and continues to be, that authorship must be strictly individual and coupled with full responsibility for the claims published. But the applicability of ICMJE guidelines has come under increasing debate. Indeed, about a year ago, some journal editors called for a paradigm shift in the definition of authorship, while others argued that "it is time to abandon authorship" altogether.[1] But if the problems with traditional definitions of authorship are at this point dearly laid out and a few proposals put forward, a new comprehensive paradigm has yet to emerge.[2]

Reprinted with permission of The Federation of American Societies for Experimental Biology. From *The FASEB Journal,* Vol. 12 (1998), pp. 3–16.

1. Richard Smith, "Authorship: Time for a Paradigm Shift? The Authorship System Is Broken and May Need a Radical Solution," *Br. Med. J.,* Vol. 314 (5 April 1997), p. 992; Richard Horton, "The Signature of Responsibility," *Lancet,* Vol. 350 (5 July 1997), pp. 5–6. See also Richard Horton and Richard Smith, "Time to Redefine Authorship," *Br. Med. J.,* Vol. 312, 1996. p. 723; Fiona Godlee, "Definition of 'Authorship' May Be Changed," *Br. Med. J.*, Vol. 312 (15 June 1996), pp. 1501–1502; and Evangeline Leash, "Is It Time for a New Approach to Authorship?," *J. Dent. Res.,* Vol. 76, No. 3 (1997), pp. 724–727.

2. The most innovative proposal to date has been put forward by Drummond Rennie, deputy editor (West) of the *Journal of the American Medical Association.* At a conference on scientific authorship held in June 1996 at Nottingham and sponsored by *Lancet,* the *British Medical Journal,* Locknet (an international peer-review research network), and the University of Nottingham, he proposed to replace "author" with "contributor." Contributors should be listed in the byline and the nature of their contribution described in a footnote. In addition, some contributors who are the most familiar with all aspects of the project should be termed "guarantors," and should be in charge of answering any question

Coming to this debate as a historian with a background in the early modern period, I am struck by the similarity between the current emphasis on the coupling of scientific authorship and responsibility and older, premarket definitions of the author. Before the emergence of the figure of the intellectual property holder in the late seventeenth and early eighteenth centuries, the author was construed by the state, the prince, or the church as the individual responsible for the content and publication of a given text.[3] The author was not seen as a creative producer whose work deserved protection from piracy, but as the person upon whose door the police would knock if those texts were deemed subversive or heretical.

Although the Office for Scientific Integrity's (OSI) increasing commitment to democratic due process sets it apart from its inquisitorial ancestors, the current definition of scientific authorship still shares its early modern cousin's relationship to responsibility.[4] Today, if scientists publish dubious claims, they are not accused of *lese majesté* against their absolute ruler or of subverting the church's absolute control on theological doctrines, but they *are* represented as responsible for something that is deemed to be equally absolute: truth.

Although I do not see clear continuities between the sixteenth century and the 1990s, there is something to learn from the genealogy of the figure of the author in science and other fields in reference to how credit and responsibility have been differently defined, joined, or separated in different disciplines after the demise of early absolutist regimes. In particular, the reward system of science and the liberal economy have developed in parallel as two distinct and yet complementary systems

---

that may be elicited by the publication (Evangeline Leash, "Is It Time for a New Approach to Authorship?," p. 726).

3. Michel Foucault, "What Is an Author?," Donald F. Bouchard (ed), *Language, Counter-Memory, Practice* (Ithaca: Cornell University Press, 1977), p. 124. See also Carla Hesse, *Publishing and Cultural Politics in Revolutionary Paris* (Berkeley: University of California Press, 1991), pp. 1789–1810.

4. However, some critics have argued that scientists investigated for misconduct may get less than a fair trial from the Office for Research Integrity (ORI). Among other problems, the ORI mixes investigatorial and prosecutorial tasks and holds hearings (if the defendant so requires) only after it has found misconduct. (Louis M. Guenin, "The Logical Geography of Concepts and Shared Responsibilities Concerning Research Misconduct," *Academic Medicine,* Vol. 71, No. 6 [June 1996], pp. 598–599). Similarly, in his discussion of the "Baltimore case," Daniel Kevles argued that "Imanishi-Kari was, to all intents and purposes, prevented from mounting a genuine defense. The OSI [ORI's institutional ancestor] combined the duties of investigator, prosecutor, judge, and jury, and pursued them all in the manner of the Star Chamber" ("The Assault on David Baltimore," *New Yorker* [27 May 1996], p. 107).

since the seventeenth century, and so the definition of scientific authorship has not been framed by the logic of the reward system of science alone, but by the intersection of these two economies and the way they have carved out different categories of credit and responsibility in relation to each other.

The troubles of scientific authorship have been emphasized by the development of corporate-style contexts of research and professional ethos in the last two decades. The problems, however, were already there. The stress produced by the increased proximity of the complementary economies of science and the market has only highlighted tensions *within* the logic of each system. Authorship is caught between these two tectonic plates, and the pressure is mounting. Although this article cannot prevent earthquakes, I hope it will help locate some of the fault lines and stress points underlying current discussions about the coupling of authorship and responsibility in biomedicine.

Two Complementary Economies of Authorship

In a liberal economy, the objects of intellectual property are artifacts, not nature. One becomes an author by creating something new, something that is not to be found in the public domain. A common view is that copyright is about "original expression," not content or truth.[5] If you paint a landscape, you can claim intellectual property (a form of private property) on the painting (the expression), but not on the landscape itself (the content). Also, copyright does not cover facts or ideas per se. Therefore, though researchers (or journals) can copyright scientific publications and gain some protection from having articles appropriated or reproduced without consent, their rights do not and cannot translate into scientific credit. Saying that they are scientific authors because their papers reflect personal creativity and original expression (the kind of claim one has to make to obtain copyright) would disqualify them as scientists because it would place their work in the domain of artifacts and fiction, not truth. Nor can scientists copyright the content of their claims, because nature is a "fact" and facts (like the landscape represented in a painting) cannot be copyrighted, since they belong to the public domain. In sum, copyright can make scientists authors, but not scientific authors.

Like copyright, patents also reward novelty, since they cover "novel

5. James Boyle, *Shamans, Software, and Spleens: Law and the Construction of Information Society* (Cambridge, Mass.: Harvard University Press, 1996), pp. 51–59.

and nonobvious" claims. But, unlike copyrights, such claims need to be useful to be patentable. Scientists, then, can become "authors" as patent holders but cannot patent theories or discoveries per se (either because they are "useless" or because they are about something that belongs to the public domain).[6] It is becoming increasingly common for scientists (mostly geneticists) to patent natural objects, but they do so by making them potentially useful by carving them out of their state of nature.[7] Nature becomes patentable by being turned into something that is less natural and more useful.

As with copyright, the patent system may provide scientists with an authorship venue, but not with scientific authorship. Scientists can patent useful processes stemming from their research, but scientific authorship is defined in terms of the truth of scientific claims, not of their possible usefulness in the market. In sum, according to definitions of intellectual property, a scientist qua scientist is, literally, a nonauthor. While novel claims are the objects rewarded by both intellectual property law and the reward system of science, the "unit of credit" is dramatically different in these two economies. A new, dramatic discovery that may warrant a Nobel prize cannot be translated, in and of itself, into a patent or a copyright. Likewise, a scientist's copyrights and patents will not earn him or her such an award. It seems, then, that scientific authorship is not "independent" from the logic of the market, but that its definition is complementary to that of market-based authorship as articulated through the copyright or patent systems.

From this complementarity it follows that the primary currency of scientific credit is not money per se, but rewards assigned through peer review (reputation, prizes, tenure, membership in societies, etc.) rather than transacted according to the logic of the market. Intellectual property rights can be exchanged for money because they are a form of private property, and money is the unit of measurement of the value of that form of property. For the same reason, the kind of credit held by a scientific author cannot be exchanged for money because nature (or claims about it) cannot be a form of private property, but belongs in

6. Jeremy Phillips and Alison Firth, *Introduction to Intellectual Property Law,* 3d ed. (London: Butterworths, 1995), pp. 39–42.

7. Eliot Marshall, "Companies Rush to Patent DNA," *Science,* Vol. 275 (7 February 1997), pp. 780–781, provides a review of recent trends. See also "Gene Fragments Patentable, Official Says," *Science,* Vol. 275 (21 February 1997), p. 1055. For an earlier overview on these issues, see Dorothy Nelkin, *Science as Intellectual Property* (New York: MacMillan [for AAAS], 1984).

the public domain. Of course, scientists can operate simultaneously in academic and market economies, but, with the help of university lawyers, they need to keep the boundaries between these two systems as distinct as possible. They can also work in industry or government, in which case authorship may be contractually relinquished in accordance with terms of employment. Again, the logic behind the reward of scientific work with "honorific" credit is not independent from, but complementary to, that of monetary economy.

A number of consequences follow. The first is that authorship credit distributed by the reward system of science has to be attached to a scientist's name and cannot be transferred. It is not transferable because scientific authorship cannot be a form of private property, and only private property (like copyrights and patents) can be transferred from one individual to another. The reasons for attaching authorship to a scientist's name, instead, follow from how the notion of truth is construed by the reward system of science.

Truth, unlike private beliefs, is generally defined as something that ought to be public. The accessibility of truth (or simply true information) in the public domain is that which legitimizes liberal democracy as an egalitarian state form.[8] This assumption also justifies the presence of inequality in the private sphere. If you are not as rich as your neighbor, the story goes, you can't blame it on the fact that you lacked access to the same information that was available to the person next door. Empirically, this reasoning may be questionable, but that does not stop it from being widely used as one of the fundamental justifications for the distinction between the public domain and the private sphere (and private property).

The definition of scientific truth as public is usually presented as an epistemological rather than an economic or legal axiom. Truth is defined as public or, more emphatically, as universal, because it is assumed to be transparent and recognizable by anyone who is competent. Truth should be as public and accessible as its object, nature. Putting on hold, for the moment, the question of whether such a definition reflects actual practice, there are other logical reasons for defining truth as public—reasons that stem from the complementarity between the categories of the reward system of science and of liberal economy.

The relationship between truth and private beliefs is parallel to that between the public domain and private property. Both relationships

8. James Boyle, *Shamans, Software, and Spleens,* pp. 25–34.

hinge on the distinction between private and public—a distinction that is integral to both science and liberal economy. Private property is "private" because it is complementary to the public domain, a vaguely defined category that nevertheless provides the conditions of possibility for private property.[9] Likewise, private beliefs are private because they are complementary to public truth, and are made possible by that very category. For instance, when expressed in material forms such as a literary text, a music score, a painting, or a patent, private beliefs (here broadly construed as any personal thought or conception that deviates from the common stock of knowledge and cultural expressions found in the public domain) become the object of intellectual property. Considered as fictions or artifacts, private beliefs may be "bad" from an epistemological point of view, but are simultaneously very "good" in the eyes of liberal economy because they make intellectual property possible.

Whether or not scientific truth is universal, it still has to be defined as such to maintain the logical coherence of both liberal economy and the reward system of science. A notion of universal scientific truth legitimizes private property defined as the result of specific "deviations" from public, "universal" knowledge, and it confirms the epistemological status of science by virtue of being an activity that is outside of monetary economy and private interests. Universal truth is value-free because it is literally defined as valueless, and yet it is the mother of all property values.

How, then, can scientific credit be defined? Intellectual property is often represented as the result of taking as little as possible from the public domain (the shared "pool" of cultural and natural resources) and transforming it into some kind of "original expression."[10] But a scientist is not represented as someone who transforms reality or produces original expression out of thin air, but as a researcher who, with much work, "detects" something specific within nature—the domain of public and "brute" facts. For that finding to be recognized as true, he or she has to put it back in the public domain (here construed as the "public sphere,"

9. Jessica Litman, "The Public Domain," *Emory Law Journal,* Vol. 39 (1990), p. 999; David Lange, "Recognizing the Public Domain," *Law and Contemporary Problems,* Vol. 44 (1981), p. 147 (1981); James Boyle, "A Politics of Intellectual Property" (unpublished manuscript).

10. Views of copyright framed by beliefs in the romantic figure of the author tend to be more extreme as they present the author's work as coming out of thin air, not from the reelaboration of materials found in the public domain. The origin of this view is discussed in Martha Woodmansee's "The Genius and the Copyright: Economic and Legal Conditions of the Emergence of the 'Author,' " *Eighteenth-Century Studies,* Vol. 17 (1984), pp. 425, 443–444.

which includes, but is not limited to, the community of scientific colleagues). Although this is a loop that begins and ends in some version of the public domain, fundamental changes take place along the way. The starting point is *generic* nature, but the result is a *specific* item of true knowledge about nature. Whereas the production of value in a liberal economy involves a movement between two complementary categories (from generic public domain to specific private property), in science the movement is within the same category (the public domain) and goes from unspecified to specified truth. Both cases involve a transformation from something unspecific to something specific. But if in the case of intellectual property such transition can be legally tracked (as it moves across two different categories), scientific credit is much trickier, because the movement from nature and the public domain to a specific true claim about nature does not cross any recognizable legal threshold. As a result, it cannot be legally tracked or monetarily quantified.

Another way to put it is that, in the case of intellectual property, one can rely on the distinction between the form and content of a work (between "original expression" and the "public domain") to determine authorship and property rights. In science, however, a claim cannot be attributed a "form" (in the legal sense of the term), since that would categorize it as an artifact. But, at the same time, a scientific claim cannot be like nature itself; it cannot just be "content." There is a bit of a paradox here. The transition from unspecified to specific truth cannot be attributed to nature, as nature does not investigate itself. Yet that work should not result in a commercially transactable intellectual property because that would destroy its status as truth. Nevertheless, such a transition has to be marked somehow, not only because scientists deserve fair credit for it, but because it has to be marked in order to exist, to be recognized as a specific truth, not just a chunk of undifferentiated, undescribed nature.

Historically, the solution to this paradox has been to attach scientific credit to the scientist's *name* while construing such credit as nonmonetary.[11] That scientific credit is honorific and attached to a scientist's name

11. Harriet A. Zuckerman, "Introduction: Intellectual Property and Diverse Rights of Ownership in Science," *Science, Technology, and Human Values,* Vol. 13, Nos. 1 and 2 (winter and spring 1988), pp. 7–16; Robert K. Merton, "The Normative Structure of Science," *The Sociology of Science: Theoretical and Empirical Investigations* (Chicago: University of Chicago Press, 1973), pp. 273–275; Robert K. Merton, "Priorities in Scientific Discovery," *The Sociology of Science: Theoretical and Empirical Investigations,* pp. 194–295, 323; Robert Merton and Harriet Zuckermann, "Institutionalized Patterns of Evalua-

is a default solution to a problem posed by the inapplicability of the taxonomies of liberal economy to the case of science and, at the same time, by the need to find a solution that does not delegitimize those taxonomies. Such a definition of scientific credit is the result of a metrological necessity. The scientists' "disinterestedness," therefore, is not the cause for scientific credit being honorific, but a professional value practitioners accept or develop by working in an economy that logically requires their credit to be nonmonetary.

But what also needs to be attached to a scientist's name is responsibility, not just credit. If a true claim about nature were like an artifact, a novel expression, or a piece of literary fiction, responsibility could be negotiated legally. In market environments, an author's responsibility is construed as financial liability, that is, as a matter of property and damages. Also, the legally responsible author may not be the actual producer of those claims, but rather the individual or corporation that paid the producer for his or her labor or rights in those claims. But this cannot apply to true claims about nature because they are in the public domain—a category complementary to that of property and monetary liability. Therefore, in the reward system of science, responsibility for scientific claims falls on the scientist who produced them simply because that individual is the only "hook" on which the movement from unspecified to specified truth can be pinned.

Furthermore, responsibility in science is absolute. It is as absolute as truth because, like scientific credit, truth and responsibility *cannot be quantified.* Responsibility becomes absolute by default. That is why, within this logic, scientific authorship has to be defined in strictly individual terms. If truth and responsibility are absolute, they cannot be attached to a corporate author since that would parcel out something that has to remain absolute. But a new paradox emerges from such a solution: Truth (defined as universal, permanent, absolute, etc.) ends up being hinged on something that is extremely local and ultimately transient—the scientist's name. And such a name needs to be a proper name

---

tion in Science," *The Sociology of Science: Theoretical and Empirical Investigations,* p. 465. Important views on the relationship between scientific and economic credit have been presented in Pierre Bourdieu, "The Specificity of the Scientific Field and the Social Conditions of the Progress of Reason," *Social Science Information,* Vol. 14 (1975), pp. 19–47; and in Warren O. Hagstrom, "Gift Giving as an Organizing Principle in Science," Barry Barnes and David Edge (eds.), *Science in Context* (Cambridge, Mass.: MIT Press, 1982), pp. 21–34.

that is unequivocally connected to a person's body, not to a corporation—a *persona ficta.*

This may cast some light on why scientific fraud is seen as a fundamental aberration, not just a serious problem. Commercial fraud—fraud about property—is certainly not a trivial issue, but it can be handled legally; it can be *quantified* (more or less adequately) in terms of financial damages. Scientific fraud, instead, tends to assume a more ominous status, and does so in part because it *cannot* be properly measured in terms of damages. Fraud shares in the absoluteness of truth and the responsibility it is seen to subvert. Of course, legal and administrative actions can be taken against fraudulent scientists. Universities can fire them, and funding agencies can sue them for misuse of research funds. However, this is a bit like getting Al Capone for tax fraud when he could not be charged with murder. Adapting the False Claim Act of 1865 (developed to curb the delivery of substandard equipment to the army) to sentence scientists with punitive damages up to three times the amount they received from funding agencies shows that the reward system of science cannot prosecute scientific fraud per se, but is forced to step outside itself and adopt the logic of commercial fraud.[12] The emotions stirred by scientific fraud and its moral condemnation as a "crime against the truth" may reflect the fact that while fraud rattles the logic of the reward system of science, its punishment cannot be logically commensurate with the "crime."

In the next section I argue that while the complex relationship between authorship, responsibility, and credit has been highlighted by contingent concerns with scientific misconduct, its roots lie in the tensions between (and within) the economies of science and of the market. After being historically and logically constituted in opposition to each other, these economies are now brought into a closer and uneasy proximity by the development of increasingly large-scale, collaborative, and capital-intensive contexts of research. The grassroots emergence of corporate views of scientific authorship that erode individual responsibility is perhaps the most conspicuous hybrid that has resulted from this process.

12. Paulette V. Walker, "1865 Law Used to Resolve Scientific Misconduct Cases," *The Chronicle of Higher Education,* 26 January 1996, p. A29. Subsequently, some uses of the False Claim Act have been challenged in court (Paulette V. Walker, "Appeals Court Overturns a False-Claim Ruling Against U. of Alabama at Birmingham," *The Chronicle of Higher Education,* 7 February 1997, p. A37).

Big Science and the Reaction against Corporate Authorship

Historically, the debate on authorship and responsibility in biomedicine has developed in response to two distinct trends: the sharp increase of multiauthorship related to the transformation of biomedicine into a "big science," and the emergence of well-publicized cases of scientific fraud (or alleged fraud). That biomedicine is as much about truth as about healing taxpayers' bodies has added further urgency to the problem of responsibility and has made it an unavoidable focus of the debate.

Quantitatively speaking, the scale of multiauthorship in biomedicine still lags behind that of physics.[13] But if articles with hundreds of authors resulting from large multicenter clinical trials are relatively rare, bylines including six or more authors are not. Journal editors and other commentators began to notice this tendency in the 1970s and usually interpreted it as resulting from the need to pool together different skills and specialized knowledge within increasingly large and collaborative research projects.[14] The multiauthorship trend could have opened the door to the acceptance of a corporate notion of authorship as a way to distribute credit in large cooperative research programs, but it clashed with the requirement of individual responsibility.

What did trigger concerns about responsibility was the growing awareness that a given scientific paper may have required the work of a biostatistician, although that person may have had little or nothing to do with the collection of the data he or she eventually analyzed; or that several contributors who may be considered authors (in the sense that they made important contributions to the project) may not be able to defend the work (or perhaps even understand the tasks) accomplished by some of their other colleagues.[15] This state of affairs would pose no

13. The protocols of authorship in large particle-physics experiments are discussed in Peter Galison, "The Collective Author," paper presented at the conference on "What Is a Scientific Author?," Harvard University, 7–9 March 1997. It is interesting that concerns with responsibility and fraud are not as pressing among physicists as among biomedical scientists.

14. Robert S. Alexander, "Editorial: Trends in Authorship," *Circ. Res.,* Vol. 1, No. 4 (July 1953), pp. 281–282; Herbert Dardik, "Multiple Authorship," *Surg. Gynecol. Obstet.,* 1977, Vol. 145, p. 418; R. D. Stroh and F. W. Black, "Multiple Authorship," *Lancet,* 1976, Vol. 2, 1090–1091.

15. Arnold Reiman wrote in 1979 that "[t]he essential criterion [of authorship] is the quality of the intellectual input. A scientific paper is a creative achievement, a record of original productivity, and coauthorship ought to be unequivocal evidence of meaningful

problems in market-based fields where responsibility, credit, and intellectual property rights can be negotiated contractually, but such techniques are not acceptable in science.

While some literary theorists have argued that the emergence of large-scale book markets led to a "death of the author," there has been no parallel movement occasioned by the big science trend in biomedicine.[16] On the contrary, the more collective, corporate, and industrial-style the research contexts become, the more one finds a resistance to accepting the implications this trend is having on the notion of authorship and responsibility. Beyond resistance, there may even be a reaction to the erosion of a notion of individual authorship. For instance, over the years the ICMJE has issued increasingly stricter guidelines about authorship; these guidelines struggle to attach authorship to the "crucial" contributions to a project and to develop taxonomies of credit that would distinguish between authorship (defined as responsibility) and other forms of recognition to be listed not in the authors' byline, but in a separate "acknowledgments" section that, according to some other proposals, could resemble a "film credits" list.[17]

The section on authorship of the ICMJE 1997 *Uniform Requirements for Manuscripts Submitted to Biomedical Journals* reads:

> All persons designated as authors should qualify for authorship. The order of authorship should be a joint decision of the coauthors. Each author should have participated sufficiently in the work to take public responsibility for the content.
>
> Authorship credit should be based only on substantial contributions to (1) conception and design, or analysis and interpretation of data; (2) drafting the article or revising it critically for important intellectual content; and (3) on final approval of the version to be published. Conditions 1, 2, and 3 must all be met. Participation solely in the acquisition of funding or the collection of data does not justify authorship. General

---

participation in the creative effort that produced the paper. To my way of thinking, therefore, the use of coauthorship as a kind of payment for faithful technical assistance or data collection violates this principle." A. S. Reiman, "Publications and Promotions for the Clinical Investigator," *Clin. Pharmacol. Ther.,* Vol. 25 (1979), p. 674.

16. Roland Barthes, "The Death of the Author," *Image-Music-Text,* Stephen Heath (ed.) (New York: Hill and Wang, 1977), pp. 142–148; Walter Benjamin, "The Work of Art in the Age of Mechanical Reproduction," *Illuminations* (New York: Schocken Books, 1969), pp. 217–251.

17. Barbara J. Culliton, "Authorship, Data Ownership Examined," *Science,* Vol. 242 (Nov. 4 1988), p. 658; Eugene Garfield, "The Ethics of Scientific Publication: Authorship Attribution and Citation Amnesia," *Essays of an Information Scientist,* Vol. 5 (Philadelphia: ISI Press, 1983), p. 622; "Editorial: Author!," *Lancet,* Vol. 2 (1982), p. 1199.

> supervision of the research group is also not sufficient for authorship. Any part of an article critical to its main conclusions must be the responsibility of at least one author. Editors may ask authors to describe what each contributed; this information may be published.
>
> Increasingly, multicenter trials are attributed to a corporate author. All members of the group who are named as authors . . . should fully meet the criteria for authorship as defined in the *Uniform Requirements.* Group members who do not meet these criteria should be listed, with their permission, under acknowledgments, or in an appendix.[18]

Overlapping with analyses of the increasingly corporate structure of research, one also finds frequent expressions of concern about the changing ethos of biomedicine. Big science is often equated with big business, an analogy that points as much to the large role of private sector funding as it does to the increased scale of biomedical research.[19] Accordingly, commentators note that the multiauthorship trend reflects not only the increased complexity of modern research, but a growing entrepreneurial ethos. They see it as a problematic response to an increasingly competitive publication-based regime of credit and professional advancement.[20] Under pressure from this complex, entrepreneurial, and competitive environment, practitioners have been alleged to associate authorship more directly with credit than with responsibility, that is, to treat authorship as "a trading chip in an economic game."[21] Those who are concerned with the trading chip attitude about authorship acknowledge that scientists have serious concerns with professional credit and career advance-

18. ICMJE, "Uniform Requirements for Manuscripts Submitted to Biomedical Journals," *JAMA,* Vol. 277, No. 2 (19 March 1997), p. 928.

19. Daniel M. Laskin, "The Rights of Authorship," *J. Oral Maxillofac. Surg.,* Vol. 45 (1987), p. 1; Addeane S. Caelleigh, "Editorial: Credit and Responsibility in Authorship," *Academic Medicine,* Vol. 66, No. 2 (November 1991), pp. 676–677.

20. Marcia Angell, "Publish or Perish: A Proposal," *Annals of Internal Medicine,* Vol. 104 (1986), pp. 261–262; R. L. Engler, J. W. Covell, P. J. Friedman, P. S. Kitcher, and R. M. Peters, "Misrepresentation and Responsibility in Medical Research," *New Engl. J. Med.,* Vol. 317, No. 22 (1987), pp. 1383–1389; Drummond Rennie, Annette Flanagin, "Authorship! Authorship!," *JAMA,* Vol. 271 (1994), pp. 469–471; Jane Smith, "Gift Authorship: A Poisoned Chalice?," *Br. Med J.,* Vol. 309 (1994), pp. 1456–1457; D. W. Shapiro, N. S. Wenger, and M. F. Shapiro, "The Contributions of Authors to Multiauthored Biomedical Research Papers," *JAMA,* Vol. 271 (1994), pp. 438–442.

21. C. C. Conrad, "Authorship, Acknowledgment, and Other Credits," *Ethics and Policy in Scientific Publication,* CBE Editorial Policy Committee (Bethesda: Council of Biology Editors, 1990), pp. 184–187. According to Dardik, "Authorship is akin to success and achievement, and cannot and should not deteriorate into a bargaining tool or commodity" (Herbert Dardik, "Multiple Authorship," p. 418). See also Eugene Garfield, "The Ethics of Scientific Publication: Authorship Attribution and Citation Amnesia," pp. 622–626.

ment. These commentators refer quite explicitly to authorship as the primary "currency" in science but then deplore the "inflation" that excessive multiauthorship might bring to such a currency (not to mention fraud and other unsavory practices that could be elicited by the same pressures toward the accumulation of scientific credit).[22] They criticize the "capitalistic" ethos that seems to be taking over biomedicine but end up casting the threat to science in terms of "inflation." In the end, they use a category that reflects an acceptance of the very market logic they want to resist.

Similarly, the commentators regret, but do not deny, that in practice the quantity rather than quality of publications is often the leading factor in promotion cases, distribution of research funds, etc.[23] While deploring the situation and suggesting improvements (such as quotas on the number of publications submitted for promotion cases), they admit that the big business mentality that has pervaded biomedicine is here to stay. This means that in a large-scale and extremely active professional environment there are material constraints on the time and energy that can be allocated to a review of candidates' or applicants' work on the basis of quality rather than quantity.[24] Practices that have been condemned for their dubious ethics thrive: reliance on "salami science," LPUs (least publishable units), and the rotating distribution of first authorship in a series of related papers published in different disciplinary journals.

Fraud and the Geography of Authorship

Against the background of these changes in the structure of biomedical practice, a series of cases of scientific fraud, or alleged fraud,

22. Daniel M. Laskin, "The Rights of Authorship," *J. Oral Maxillofac. Surg.,* Vol. 45 (1987), p. 1; Addeane S. Caelleigh, "Editorial: Credit and Responsibility in Authorship," *Academic Medicine,* Vol. 66, No. 2 (November 1991), pp. 676–677; Drummond Rennie and Annette Flanagin, "Authorship! Authorship!"; William J. Broad, "The Publishing Game: Getting More for Less," *Science,* Vol. 211 (13 March 1981), pp. 1137–1139; Robert N. Berk, "Irresponsible Coauthorship," *AJR,* Vol. 152 (April 1989), pp. 719–720.

23. David P. Hamilton, "Publishing by—and for?—the Numbers," *Science,* Vol. 250 (7 December 1990), p. 1332; Drummond Rennie and Annette Flanagin, "Authorship! Authorship!" p. 469; "Are Academic institutions Corrupt?" (editorial), *Lancet,* Vol. 342, No. 8867 (7 August 1993), p. 315.

24. Marcia Angell, "Publish or Perish: A Proposal," *Ann. Int. Med.,* Vol. 104 (1986), pp. 261–262; Barbara J. Culliton, "Harvard Tackles the Rush to Publication," *Science,* Vol. 241 (29 July 1988), p. 325; John Maddox, "Why the Pressure to Publish?," *Nature,* Vol. 333 (8 June 1988), p. 493; W. Bruce Fye, "Medical Authorship: Traditions, Trends, and Tribulations," *Ann. Int. Med.,* Vol. 113, No. 4 (15 August 1990), pp. 320, 324–325.

added urgency to debates about scientific authorship.[25] The Darsee and Slutsky cases, and more recently the so-called (or misnamed) "Baltimore case," have received much attention in the popular press, a publicity that has put further pressure both on the scientific community and policy makers. In some instances, senior scientists whose names were on an article's authors' byline argued that they were not responsible for the mistakes or misconduct of their junior colleagues. Accordingly, the blame was to be placed on Darsee, Slutsky, and people like them, not on the honest (if busy) directors of their labs or the department chairs who had agreed to have their name on the articles according to a practice that has since been labeled "honorific" or "gift" authorship.[26] Similar cases and similar justifications continue to this day.[27]

Even though not all instances of scientific fraud can be reduced to situations in which a senior researcher did not take responsibility for the work of a junior associate, it is interesting that this aspect of the problem has received the most attention and that honorific authorship has become a sort of fighting word in debates about scientific misconduct. The high visibility of some of the senior scientists and their institutions (UCSD, Harvard, MIT) does not fully account for the emphasis that has been placed on this aspect of fraud. Although it is not my business to absolve or condemn those involved in these cases, I believe that the primary focus on honorific authorship signals a difficulty in coming to

25. The transcripts from the 31 May 1988 colloquium at NIH on "scientific authorship" open with a statement by Alan Schechter linking the origin of the conference to "a particularly tragic outcome of the investigation of a case of alleged scientific fraud here at NIH." A. N. Schechter, J. B. Wyngaarden, J. T. Edsall, J. Maddox, A. S. Relman, M. Angell, and W. W. Stewart, "Colloquium on Scientific Authorship: Rights and Responsibilities," *The FASEB Journal,* Vol. 3 (February 1989), pp. 209–217. For a summary of the vast literature on the issue of fraud and misconduct, see Marcel C. LaFollette, *Stealing into Print: Fraud, Plagiarism, and Misconduct in Scientific Publishing* (Berkeley: University of California Press, 1992). A recent analysis of the "Baltimore case" appears in Daniel Kevles, "The Assault on David Baltimore," *New Yorker,* 27 May 1996, pp. 94–109, and in his *The Baltimore Case: A Trial of Politics, Science, and Character* (New York: W. W. Norton, 1998).

26. Arnold S. Relman, "Lessons from the Darsee Affair," *N. Engl. J. Med.,* Vol. 308, No. 23 (9 June 1983), p. 1417; Edward J. Huth, "Abuses and Uses of Authorship," *Ann. Int. Med.,* Vol. 104, No. 2 (1986), 266–267; R. L. Engler, J. W. Covell, P. J. Friedman, P. S. Kitcher, and R. M. Peters, "Misrepresentation and Responsibility in Medical Research," *N. Engl. J. Med.,* Vol. 317, No. 22 (1987), pp. 1383–1389; Eugene Braunwad, "On Analyzing Scientific Fraud," *Nature,* Vol. 325 (15 January 1987), pp. 215–216.

27. C. Court and L. Dillner, "Obstetrician Suspended after Research Inquiry," *Br. Med. J.,* Vol. 309 (1994), p. 1459; Jane Smith, "Gift Authorship: A Poisoned Chalice?"

terms with the fact that the practitioners' perceptions of due credit and responsibility may be informed by their location and role within a collaborative project, and that blame, as appropriate as it may be, is not going to eradicate the sociological roots of the problem.

Scientific authorship has a geography as well as a logic. However, the geographical variability of attitudes about authorship is downplayed by the fact that the term science tends to cast an aura of homogeneity on a vast range of diverse disciplines and differently situated individuals and institutions.[28] In science as in society, workers, managers, and lawmakers are not the same people (though they may be citizens of the same state), and such differences are constitutive, not erasable. But scientific culture, because of its emphasis on values such as trust, collegiality, and disinterestedness, has few ways to acknowledge and negotiate these tensions and power differentials. In contrast, liberal economy has abundant categories to explain its litigiousness, and plenty of legal infrastructures to manage it. Therefore, if liberal economy has no problems admitting the sharp economic conflicts at play behind current disputes about intellectual property law, science is inherently ill-equipped to acknowledge that the debate about responsibility, credit, and authorship may reflect struggles among different constituencies.[29]

However, one legacy of the fraud scandals of the early 1980s is precisely the mapping of the different interests and positions of at least three different constituencies: *(1)* Congress and funding agencies; *(2)* universities, research institutions, and academic journals; and *(3)* the practitioners themselves (this group can be further divided into junior and senior researchers).

Some members of Congress, funding agencies, and scientists whose job is to monitor other scientists became concerned about the misuse of research funding and the disrepute that fraud cases bring to biomedical research and to the politicians and institutions that support it.[30] Given

28. D. W. Shapiro, N. W. Wenger, and M. F. Shapiro, "The Contributions of Authors to Multiauthored Biomedical Research Papers," *JAMA,* Vol. 271 (1994), pp. 438–442; Neville W. Goodman, "Survey of Fulfillment of Criteria for Authorship in Published Medical Research," *Br. Med. J.*, Vol. 309 (3 December 1994), p. 1482; S. Eastwood, P. Derish, E. Leash, and S. Ordway, "Ethical Issues in Biomedical Research: Perceptions and Practices of Postdoctoral Research Fellows Responding to a Survey," *Sci. Eng. Ethics,* Vol. 2 (1996), pp. 89–114; Kay L. Fields and Alan R. Price, "Problems in Research Integrity Arising from Misconceptions about the Ownership of Research," *Academic Medicine,* Vol. 68, No. 9, Suppl. (September 1993), pp. S60–S64.

29. James Boyle, *Shamans, Software, and Spleens,* pp. 35–60.

30. Walter W. Stewart and Ned Feder's painstaking analysis of the publications of John Darsee is emblematic of this trend. Its conclusion is that "[s]cientists have to an

their role and interests, it is not surprising that these constituencies identify authorship with responsibility and see honorific authorship as emblematic of how well-funded scientists have become overconfident in their belief that they do not need to earn the freedom from external regulation that has been granted to them but denied to other professions.[31] According to this constituency, science must cleanse itself from misconduct and establish its own policing infrastructures or face the possibility of government regulation. To some extent, this possibility materialized in 1989 with the development of the Office of Scientific Integrity (OSI) and the Office of Scientific Integrity Review (OSIR) within the Public Health Services (PHS), with OSI and OSIR being reorganized into the Office for Research Integrity (ORI) in June 1992.[32] The same year, a United States attorney went so far as to suggest that the legal system could take over the adjudication of claims of scientific misconduct—an option that, to the displeasure of universities, has been increasingly exercised.[33]

Universities, research institutions, and journal editors were quick to respond to these moves, but universities and journals have different stakes in these matters. Universities would like to rely on journals and their editorial practices (refereeing system, etc.) to certify good science, detect misconduct, and possibly alert them about potential problems.[34] Journals, on the other hand, claim that although they do their best to ensure the publication of quality articles written by the authors named

---

unusual degree been entrusted with the regulation of their own activities. Self-regulation is a privilege that must be exercised vigorously and wisely, or it may be lost" ("The Integrity of the Scientific Literature," *Nature,* Vol. 325 [15 January 1987], pp. 207–214).

31. According to Congressman John Dingell, "Scientists need to understand that the best way, perhaps the only way, to avoid the threat of 'science police' is for scientists themselves to show that they have the ability and the will to police themselves. It is a matter of morality, but also of self-interest." J. D. Dingell, "Shattuck Lecture—Misconduct in Medical Research," *New Engl. J. Med.,* Vol. 328, No. 22 (3 June 1993), p. 1614.

32. Donald F. Klein, "Should the Government Assure Scientific Integrity?," *Academic Medicine,* Vol. 68, No. 9, Suppl. (September 1993), pp. S56–S59.

33. Breckinridge L. Willcox, "Fraud in Scientific Research: The Prosecutor's Approach," *Accounting in Research,* Vol. 2 (1992), pp. 139–151; Rex Dalton, "Heat Rises over UCSD 'Misconduct' Charge," *Science,* Vol. 385 (13 February 1997), p. 566; Paulette V. Walker, "2 Lawsuits May Change Handling of Research-Misconduct Charges," *The Chronicle of Higher Education,* 6 June 1997, pp. A27–A28.

34. According to the editors of *JAMA:* "The parent research institutions rely on publications as the coins academics must use to get through the tollgates on their way to academic promotion. And if the promotion committees function well, they weigh as well as count the coins." Drummond Rennie and Annette Flanagin, "Authorship! Authorship!," p. 470. See also Marcel C. LaFollette, *Stealing into Print: Fraud, Plagiarism, and Misconduct in Scientific Publishing,* pp. 156–194.

in the byline, it is not their duty to play judge.[35] After all, their editors are not in the lab and do not have the material resources to push the evaluation process beyond refereeing. In any case, journal editors make decisions based on what scientists themselves (as referees) report to them. When it comes to authorship, most journal editors (in particular those who endorse the ICMJE guidelines) now require all authors to sign a statement such as this:

> AUTHORSHIP RESPONSIBILITY: "I certify that I have participated sufficiently in the conception and design of this work and the analysis of the data (when applicable), as well as the writing of the manuscript, to take public responsibility for it. I believe the manuscript represents valid work, I have reviewed the final version of the manuscript and approve it for publication. Neither this manuscript nor one with substantially similar content under my authorship has been published or is being considered for publication elsewhere, except as described in an attachment. Furthermore, I attest that I shall produce the data upon which the manuscript is based for examination by the editors or their assignees if requested."[36]

Such statements cast journals in a curious role. They end up representing themselves as credit-givers while simultaneously minimizing their responsibility for the credit they give. In the end, it is not clear whether journals are casting themselves as publishers or printers. What I find surprising is not that editors are understandably cautious about their practical ability to certify true knowledge, but that their policies put the onus of assessing scientific authorship and the truth value of claims completely on the scientist's shoulders as if the reward system (of which journals are a crucial element) had little to do with certification.

Insistence on the individuality of authorship and its coupling with complete responsibility is so categorical that it amounts to a demand that authors do what the reward system and peer review should but cannot quite do. Ready or not, practitioners are being volunteered for a

35. A reasonable articulation of this position is in Drummond Rennie, "The Editor: Mark, Dupe, Patsy, Accessory, Weasel, and Flatfoot," *Editorial Policy Committee, CBE, Ethics and Policy in Scientific Publication* (Bethesda, Md.: CBE, 1990), pp. 155–163. See also Don Riesenberg and George D. Lundberg, "The Order of Authorship: Who's on First?," *JAMA,* Vol. 264, No. 14 (10 October 1990), pp. 1857; Helmuth Goepfert, "Responsible Authorship" (editorial), *Head and Neck,* July/August 1989, pp. 293–294; A. N. Schechter, J. B. Wyngaarden, J. T. Edsall, J. Maddox, A. S. Relman, M. Angell, and W. W. Stewart, "Colloquium on Scientific Authorship: Rights and Responsibilities," *The FASEB Journal,* Vol. 3 (February 1989), pp. 209–217 (especially Maddox's statement on p. 214).

36. *JAMA,* Vol. 262, No. 14 (13 October 1989), p. 2005.

sort of "mission impossible." I am not suggesting that referees should be formally coresponsible for the articles they review, but that the current definition of scientific authorship casts peer review not as a system of certification, but as little more than a free (and responsibility-free) consulting service for editors.[37] Although journal editors quite laudably do their best to eradicate the problem of honorific authorship, they do not seem to realize that the reputation of their own journals is constituted through a process that is structurally similar to honorific authorship. If the articles they publish are praised, the journal's credit grows accordingly. But if something goes wrong, the editors can say that they (and the referees) are not responsible for the problem and only the authors are to blame.

This would not be a problem outside of science, where the limits of certification are accepted as a fact of life. For example, the United States Patent Office may grant a patent without checking whether the device or process actually works. Preliminary checks are conducted to detect conspicuous overlaps between a given application and other existing patents, but it is then up to the inventor to find people who would appreciate the value of his or her idea, as it is up to the inventor to defend the patent in court against competing claims. The same can be said about copyrights. In market environments, then, authorship is not absolute, but a resource to be developed (and perhaps defended) through further work, time, and expense.

In contrast, according to the logic of the reward system of science, authorship is as absolute as the truth of the claims on which it rests—a truth that is not to be negotiated in court or through contracts. And authorship credit is construed as something almost instantaneous. You produce a true claim, you take responsibility for it, you publish it, you get credit. Unlike other products, truth does not need to be developed to be recognized. In science, the work for which an author gets credit does not extend past the "filing." The logical function of the peer review system is the certification of truth. It is as though you deposit a check, the bank "reviews" it, and you get the money.

But, in practice, the bank (the peer review system) cannot function so thoroughly and swiftly. A scientist receives full credit for the amount of the "check" she or he deposits, but the funds can be taken back at

37. P. V. Scott and T. C. Smith, "Definition of Authorship May Be Changed: Peer Reviewers Should Be Identified at the End of Each Published Paper," *Br. Med. J.*, Vol. 313 (28 September 1996), p. 821.

any later time if it is contested. The check clears immediately and, at the same time, it never really clears. In a liberal economy, the granting of a copyright or a patent is a way of saying that your check looks potentially good, but that it is up to you to develop its actual value in a market. The limits of certification are acknowledged, but there are a range of tools to manage them.

In practice, the reward system of science is faced with the limits of the peer review system, but cannot fully admit them without jeopardizing its own logic—a logic hinged on the absoluteness of truth. Such a contradiction is not solved, but displaced in time in the hope that it will never express itself, that is, that scientific claims will never be assailed as fraudulent.[38]

Specificity versus Conditions of Possibility

If Congress, universities, and journals couple authorship with total responsibility, the practitioners themselves tend to stress the links between credit, labor, and authorship, while attaching them to a notion of limited responsibility. When practitioners write responses to editorials about authorship policies, they point to the power differentials that frame the debates about authorship.[39] Journal editors assemble themselves in committees, issue guidelines that may reflect their needs and wishes (more than the daily realities of the researchers), and can easily air their opinions in the pages of their own journals.[40] Most indi-

38. The remarkable paucity of analyses of the peer review system before the wave of fraud scandals and the early tendencies to underestimate the frequency of misconduct may reflect a built-in tendency to denial that, far from being arbitrary, is connected to the structural blind spots of the reward system of science. Until recently, the peer review system had been the subject of few sustained analyses (Daryl E. Chubin and Edward J. Hackett, *Peerless Science* [Albany, N.Y.: SUNY Press, 1990], is a notable exception). But several conferences, studies, and publications have emerged since, such as "Guarding the Guardians: Research on Editorial Peer Review," special issue of *JAMA,* Vol. 263, No. 10 (9 March 1990).

39. Michael E. Dewey, "Authors Have Rights Too," *Br. Med. J.*, Vol. 306 (30 January 1993), pp. 118–120. See also comments on Dewey's piece in the *Br. Med. J.,* Vol. 306 (13 March 1993), pp. 716–717, which include remarks like "Authors need to organise themselves to redress the current imbalance of power," "The International Committee of Medical Journals Editors should consider the sort of issue discussed by Dewey," and "[consider] how a mechanism might be set up to allow authors' grievances to be aired." A radical revision of the relationship between journal and contributors has been recently proposed by *Lancet*'s editor (Richard Horton, "The Signature or Responsibility," *Lancet,* Vol. 350 [5 July 1997], p. 6).

40. Occasionally, some editors seem uneasy about this state of affairs. Commenting on a meeting of the ICMJE, the editor of *Lancet* claimed that "medical editors should be banned from assembling in more than twos or threes, lest they conspire to present a

vidual scientists do not have that kind of power. It is almost as though scientists express a wish to "unionize," as they seem to feel they are at the receiving end of authorship policies whose development they do not control.

Expressions of discontent from scientists have become increasingly frequent. In 1988, one could find a letter to the editor stating:

> I wish to comment on the preposterous suggestion, being seriously advanced in some quarters, that all of the authors of a given paper are responsible for all of the material that appears in that paper. If that rule were adopted, it would bring multidisciplinary research to a virtual halt.[41]

Recently, letters to *Science* were peppered by remarks like "it is ridiculous to think that each author can or should be able to vouch for each of the others," or:

> If marriage partners are not held liable for the actions of their spouses, why should we assume that scientific collaborators are liable? In both cases, liability would be tantamount to an assertion of omniscience, and an omniscient scientist would probably be in no need of collaborators.[42]

A third writer voiced frank skepticism about the idea that responsibility for a multiauthored paper be shared by all its authors by saying that:

> [t]his amounts either to banning all papers with more than one author or enshrining a kind of chivalry where scientists agree to destroy their own careers if they happen to work in the same lab as a scientist who commits fraud.[43]

Early in 1997, a report based on a questionnaire circulated through a broad cross section of biomedical practitioners at the University of Newcastle showed that a substantial portion of the scientists could not recall the basic authorship requirements issued by the ICMJE, and when told what they were, found them inapplicable. Many of the respondents

---

homogeneous front to contributors and readers" ("Editorial Consensus on Authorship and Other Matters," *Lancet,* Vol. 2 [1985], p. 595).

41. Avram Goldstein, "Collaboration and Responsibility," *Science,* Vol. 242 (23 December 1988), p. 1623. See also Arnold Friedhoff's letter on the same page.

42. Letters by Jay M. Pasachoff and Craig Loehle in "Responsibility of Co-Authors," *Science,* Vol. 275 (3 January 1997), p. 14.

43. Letter by Tobias I. Baskin, ibid.

(49%) had also experienced situations in which, according to their perception, authorship had been deserved but not awarded.[44]

In sum, researchers seem to favor a notion of authorship that entails limited rather than global responsibility and view authorship as something that should be extended not just to those who allegedly would be able to defend all results, but to anyone who worked at making a trial possible, such as laboratory workers or the many general practitioners who provided and followed patients but may have contributed little or nothing to data analysis.[45]

A perception of authorship as primarily linked to credit and labor, rather than absolute responsibility, is not limited to the "workers" but also to some senior scientists, and it informs the credit arrangements they may adopt (implicitly or explicitly) with their junior colleagues and assistants. For instance, the occurrence of honorific authorship is important evidence of a perspective according to which a director of a lab who provided space, equipment, or prestige and facilitated access to funding and publication is seen (by him- or herself and perhaps by the associates) as an investor. The role of a "remote" lab director is not unlike that of a general practitioner who provided patients but did not necessarily analyze the data or write up the final papers. Perhaps neither the director nor the general practitioner had much to do with the specific results, but they nevertheless made those results possible. In some ways, honorific authorship resembles a common phenomenon of the early modern period: the dedication of a book to the patron who supported the author or, through his high social status, could help him gain legitimation, visi-

44. Raj Bhopal, Judith Rankin, Elaine McColl, Lois Thomas, Eileen Kaner, Rosie Stacy, Pauline Pearson, Bryan Vernon, and Helen Rodgers, "The Vexed Question of Authorship: Views of Researchers in a British Medical Faculty," *Br. Med. J.*, Vol. 314 (5 April 1997), pp. 1009–1012.

45. Domhnall Macauley, "Cite the Workers," *Br. Med. J.*, Vol. 305, (11 July 1992), p. 6845; Ian W. B. Grant, "Multiple Authorship," *Br. Med. J.*, Vol. 298 (11 February 1989), pp. 386–387. See also letters to the editor (*N. Engl. J. Med.*, Vol. 326, No. 16 [16 April 1992], pp. 1084–1985) published in response to J. P. Kassirer and M. Angell, "On Authorship and Acknowledgments," *N. Engl. J. Med.*, Vol. 325 (1991), pp. 1510–1512. A few editors have taken these complaints seriously. An editorial in *Lancet* argued that "[m]any researchers think this definition [ICMJE's] is out of touch with their own research practice. It leans toward being a senior authors' charter, falling short of providing explicit credit for those who actually do research. . . . And, in an era of highly technical, multidisciplinary research, how can all authors be expected 'to take public responsibility for the content'? On balance, the definition seems to fail important tests of relevance and reliability" (Richard Horton, "The Signature or Responsibility," pp. 5–6).

bility, and even some protection from plagiarism for the claims published in that book.[46] If one finds such behavior proper then but not now, this does not mean that early scientists were unethical, but simply that professional ethics is not a matter of ahistorical first principles, but has evolved alongside the reward system of science.

The practitioners who resist a definition of authorship as something that is inherently individual (rather than collective) and tied to absolute (rather than limited) responsibility seem to think in terms of corporate credit and investments—investments they "pay back" by giving authorship credit. Even though it is easy to see how these behaviors clash with the logic of the reward system of science, they accurately reflect the outlooks practitioners develop when operating in large resource-intensive projects that are extended both in space and time. They are "grassroots" and quasi-capitalistic. I say quasi-capitalistic because authorship in academic biomedicine remains a matter of name, not money. It is capitalistic only to the extent that authorship is treated like having stock in a particular project, and responsibility is also treated in a corporate manner. Responsibility in this "take" is not an absolute notion (as the reward system of science would require), but is something limited to one's stock in the project.

A way to summarize the differences between the positions of the ICMJE and those of some of the participants in large projects is to say that the ICMJE focuses on the *responsibility for the specific claims* that emerged from a study, whereas the participants attach authorship to those who have provided the *conditions of possibility* of that study. This kind of demarcation is not new. It reproduces (in logic but not in content) the type of distinctions discussed earlier: those between the public domain and private property, and in the case of science, between unspecified nature/truth and specific truth claims.

The ICMJE states that "participation solely in the acquisition of funding or the collection of data does not justify authorship," that is, they attach authorship only to those tasks that made a *difference,* not just provided a *possibility.*[47] In this logic, the general practitioner who provided patients (or in other settings, an instrument maker, a laboratory assistant, a maintenance technician, or a "remote" lab director) are

46. Mario Biagioli, *Galileo, Courtier* (Chicago: University of Chicago Press, 1993), pp. 103–157; Mario Biagioli, "Etiquette, Interdependence, and Sociability in Seventeenth-Century Science," *Critical Inquiry,* Vol. 22 (1996), pp. 193–238.

47. ICMJE, "Uniform Requirements."

seen as people whose work was not *specific* to that project. They did contribute to its *happening* but not to the fact that the result was X rather than Y. The author, then, is cast as the individual who was *irreplaceable,* someone whose involvement in the study was both necessary and sufficient to its result.

Conceived in those terms, the author would be a sort of bodily counterpart for what has been called the crucial experiment. Accordingly, truth is the outcome of an experiment conceived in a way that only one of its various possible outcomes can be the result of that which is hypothesized as its true cause prior to the conducting of the experiment. The ICMJE's test for authorship seems to translate such a view of natural causality into the domain of human agency. An author is the person, and the only person, who "caused" the outcome of a research project. Of course, more authors can be attached to an article, but the ICMJE guidelines break multiauthorship down to an assembly of separate authors, each fully individual and fully responsible. Coauthorship cannot mean corporate authorship.

But if the ICMJE's position is coherent, it begs at least two further questions. One is whether the view of natural causality entailed by the crucial experiment can be applied to environments where human agency is temporally intermittent and spatially distributed. Unlike gravity acting on all apples all the time, many different people work at different aspects of a research project, often at different times and at different sites. The second is the practical feasibility (rather than conceptual robustness) of a taxonomy of reward that distinguishes true authors from other practitioners eligible only for "acknowledgment" credit.

To begin with the second question, it would appear that, in principle, the widening of categories of scientific credit through the introduction of currencies other than that of authorship could rechannel the pressures that have led to the corporate uses of individual authorship toward the use of other forms of credit giving. But, at present, having one's name in the acknowledgment section or in some other appendix (as requested by the ICMJE) does not do much good to many practitioners, since these credits are not usually retrievable through computer searches. And in biomedicine today, such searches play a crucial role in the production of the author function. Furthermore, such a reform of authorship would work only if accompanied by a serious reeducation not only of the researchers, but also of those who evaluate them for jobs, promotions, and funding.

Compressing Time, Space, and Labor

Going back to the first, more difficult question, I believe that a two-tier system distinguishing between full authors responsible for the truth of the published claims and all the others who provided "only" the conditions of possibility for those claims would introduce not a graduated credit scale, but an incommensurability between two classes of contributors.

The ICMJE guidelines that try to reduce the entire range of collaborative projects distributed in time and space to the model of individual effort and total responsibility reflect a literal extension of the image of the individual author. In doing so, these guidelines fit in a long tradition that has emphasized the agency of the individual author at the expense of other contributions to the knowledge-making process. In literature, the legal concept of the author was developed in the eighteenth century largely as a way to include immaterial objects such as expression and creativity under the category of private property, a category that until then was about material entities.[48] Both writers and publishers were faced with the financial costs of piracy, the result of an early perception of books as objects one bought, claimed as property, and could use in any way one wished (including reproducing them). The figure of the author as the holder of intellectual property rights was developed as a way to limit the property rights of the book buyer by saying that there was more to a book than its materiality, that there was something that could not be relinquished in the act of selling a book. The author, then, was a market construct, one that made both booksellers and writers very happy.

But one can argue that the focus on the individual author as the holder of such newfangled property rights misrepresented the long chain of human agency that produced a literary work. It involved compression and selection. The historical figure of the individual author as romantic genius is the epitome of such misrepresentation through the compression of human agency. Accordingly, the "work" is seen as emerging from an instantaneous act of creativity, not from the time-extended labor of paper makers, font cutters, editors, typesetters, printers, binders, and

48. Mark Rose, *Authors and Owners* (Cambridge, Mass.: Harvard University Press, 1993); Peter Jaszi, "Toward a Theory of Copyright: The Metamorphoses of 'Authorship,' " *Duke Law Journal* (1991), pp. 455–502; Martha Woodmansee, *The Author, Art, and the Market* (New York: Columbia University Press, 1994); Roger Chartier, "Figures of the Author," *The Order of Books* (Stanford: Stanford University Press, 1994), pp. 25–59.

booksellers (not to mention the body of previous literary works from which the author drew his or her "inspiration").[49]

A similar, if less drastic, compression and selection of the chain of human agency is found in the depiction of the figure of the scientific author. Since the emergence of experimental philosophy in the seventeenth century, the notion of the individual author was often constituted through the erasure of the contribution of instrument makers and laboratory technicians who, because of their low social status and credibility, were not perceived as true knowledge makers and whose names were omitted from the published reports.[50]

Historically, then, the author has always been more of an efficient accounting device for intellectual property or scientific credit than an accurate descriptive tool of knowledge-making practices. The tensions produced by author-based forms of accounting have been there since the beginning, and have been made only more conspicuous by the increasing complexity and changing scale of knowledge production (in both the scientific and market economies). The logic behind the two-tier taxonomy of credit one finds in the ICMJE guidelines is homologous to that behind these historical cases. In both cases a line is drawn between the author and those who provided the conditions that made the author's specific results possible. In contemporary biomedicine, the definition of the author does not turn on his or her creativity or original expression, but on his or her responsibility. The parameters that constitute the author are different, but its logic cuts across disciplines and, to a lesser extent, across historical periods.

In fact, the author described in the ICMJE guidelines is not a hyper-individualized, romantic genius. Such a figure worked well to legitimize the author's (or his or her bookseller's) claims of intellectual property by representing literary production as an act that borrowed little or nothing from the surrounding culture. However, the journal editors' primary concern is not the maximizing of intellectual property, but the management of responsibility. Therefore, the editors use the figure of the individual author not as a creative genius, but as the person responsible for

49. Martha Woodmansee, "The Genius and the Copyright: Economic and Legal Conditions of the Emergence of the 'Author,' " *Eighteenth-Century Studies,* Vol. 17 (1984), pp. 425–448. The problems posed by the romantic view of authorship on contemporary intellectual property law are one of the foci of James Boyle's *Shamans, Software, and Spleens.*

50. Steven Shapin, "The House of Experiment in Seventeenth-Century England," *Isis,* Vol. 79 (1988), pp. 373–374; and Steven Shapin, "The Invisible Technician," *American Scientist,* Vol. 77 (November–December 1989), pp. 554–563.

those aspects of the research process that can be represented as constant and stable throughout the process of knowledge production: the conception of the work, the analysis of the collected data, and the writing of the article. In intellectual property law, the individual genius is the one who creates out of nothing, whereas in science the individual author is the one who gives continuity and consistency to a heterogeneous process. Although one emphasizes instantaneity and the other constancy, both figures work as ways of demarcating the final product from its conditions of possibility.

By drawing a wedge between conception and execution, the ICMJE guidelines carve out research practices in two categories: one that is unified, stable, and allegedly laid out since the beginning of the project, and one made up of diverse activities, ideas, and insights that may have developed along the way at different times and places—items that are much more difficult to subject to a neat accounting. The focus on data analysis rather than data collection as a fundamental aspect of authorship is an attempt to compress temporally and spatially diverse labors to an activity that took place in a specific place at a specific time. Similarly, though no one would question that the writing up of the results is a crucial contribution to any scientific project, a text, being an object that is physically well circumscribed and easily accessible, is also very handy for accounting purposes. It is a stable inscription whose content is frozen in time, available in many locations, and yet always the same as opposed to the complex temporally and spatially dispersed activities it is seen as summarizing.

But as reasonable and convenient as this approach may be, it does not guarantee that the conception of the work can always be located in one or a few distinct individuals or that, together with the division of tasks among the various practitioners, it could have been laid out once and for all at the beginning of the project. Similarly, this does not imply that, in principle, the writing of the final paper should be a task rewarded with a kind of credit (authorship) that is incommensurable with (rather than simply more important than) the credit to be given to those responsible for other tasks.

It is not that the "conception of the work" is a convenient fiction, that the focus on the written outcome of research is simply fetishistic in nature, or that the ICMJE guidelines are wrong. The policies proposed are predetermined by the logic of the reward system of science. The point is that the choice of the features deemed to be constitutive of authorship reflects an accounting rationale shaped by a symbiosis with liberal economy. It deploys categories that facilitate the accounting process better than the global description of research practices.

No one, I think, would object to the necessity for accountability or responsibility in science. What is happening, however, is that responsibility is treated as something that preexists and is independent from its accounting protocols. But authorship is not just a *result* of the accounting of human agency and responsibility in the knowledge-making process: it is a category that provides the condition of possibility for such an accounting. *Authorship is both accounted and accounting.* What we take to be authorship in science or intellectual property in a liberal economy are coexistent with the accounting systems that rest on those categories as constitutive assumptions, not as empirical categories that exist independent of the system in which they operate.

Conclusions

Historically, the law has been continuously modified and articulated to manage the emergence of new forms of production and of new interest groups. Unlike various national legal systems, the reward system of science has remarkably fewer tools to adjust to contingent historical changes. In my opinion this is because its logic has been historically tied to an absolute concept of truth and responsibility. However, I believe that an overhaul of the reward system of science in a market direction would not give authorship its desired flexibility.

For one thing, liberal economy and the reward system of science are not independent, but complementary, and are joined at the hip, so to speak, by the hazy category of the public domain. It is the public domain that, in one case, legitimizes liberal democracy and its notion of private property and, in the other case, grounds the notion of truth as universal, transparent, disinterested, etc.

Science and liberal economy both construe value (be it a true scientific claim or intellectual property) as a process of specification that, depending on the economy, is either from the public domain to private property (via the individual's creative expression) or within the public domain from unspecified nature to specified truth claims (via the individual scientist's responsibility). In short, the fundamental dichotomy in liberal economy between public domain and private property is found, mutatis mutandis, in the scientific realm in the distinction between conditions of possibility and specific claims at the roots of credit, authorship, responsibility, and truth.

In both cases the fundamental distinction between specific, individually produced claims and products and the "stuff" that made them possible is both necessary and inherently unstable. Consequently, definitions of authorship in science and in the market—definitions rooted in this distinction—reify such instability. Thus, the conceptual tensions that underlie scientific authorship would not be solved by moving toward more corporate, market-based notions. Although I have a taste for hybrids, in this case a crossbreeding of scientific and market authorship would not join two different and mutually strengthening entities, but two categories that have evolved together (though complementarily) and are both cracking under similar kinds of stress. The crossbreeding would be sterile.

At the same time, authorship policies like those of the ICMJE that reinforce the separation of the economies of science and the market are likely to produce more discontent than sustainable solutions. Erecting stronger boundaries between the two systems is not going to solve the problems, because the problems are *within* each of the two systems. The increased proximity of these different and complementary economies have only enhanced the visibility of previously existing problems.

In sum, I do not think that the conditions for revolutions and new paradigms for scientific authorship are readily available. So much has been hung on it from different sides that, despite its inherent instability, scientific authorship has become virtually unmovable. While there is an implicit awareness that the category of authorship needs to be reconstituted, I think that the proposed solutions find themselves chasing their own tails, often reproducing some of the very tensions they try to solve. And this is not for lack of effort or acumen. Therefore, rather than pursue the chimera of the one conceptually "right" definition, one may take a more pragmatic position by acknowledging that authorship (scientific or not) has always been a matter of compromises and negotiations, and that no new conditions have emerged to change that.

But the logic of compromise begs the question of what are the constituencies that should negotiate it. The current debate and policies, however logically coherent and well intentioned they may be, have a predominantly "top-down" quality to them. This points to the need of appropriate infrastructures to enable a representative number of practitioners with different roles and seniority to participate democratically in the legislation of future authorship protocols. Having been discussed mostly by editors and administrators, authorship has been framed as an administrative problem. It is, instead, an issue whose roots spread so far and wide that its solution may require something of a "constitutional amendment" to the logic of the reward system of science. In the end, the real challenge may be precisely the development of infrastructures to make these broader discussions possible and to provide the conditions of possibility for a workable definition of scientific authorship.

ACKNOWLEDGMENTS

Allan Brandt introduced me to this topic and provided crucial suggestions and comments. I hope he will accept my special thanks and relinquish further claims to rights in this essay. I also wish to thank Rebecca Gelfond, my research assistant, for all the competent help she provided throughout the project, and Jean Titilah for her much-needed editorial assistance. Debbore Battaglia, Sande Cohen, Arnold Davidson, Peter Galison, Michael Han, Barbara Herrnstein-Smith, Michael Gordin, Dan Kevles, and Don MacKenzie have offered important comments and criticism (not all of which, I admit, have found their way into this essay). Finally, I want to thank James Boyle for trying to guide a neophyte through the mazes of the public domain, and Sherry Turkle for having worked through several of the ambiguities of my argument when she had better things to do with her time. This work was supported by a John Simon Guggenheim Fellowship.

19

The Sociology of Scientific Knowledge

Some Thoughts on the Possibilities

D. Wade Hands

Introduction

During the last twenty years the sociology of scientific knowledge (SSK) has emerged as an influential new approach to the study of science. Unlike traditional philosophy of science, which often emphasizes issues such as demarcation, appraisal, and the logic of scientific theory choice, the sociology of scientific knowledge focuses on the inherently social nature of scientific inquiry. According to the SSK, science is practised in a social context, the products of scientific activity are the results of a social process, and scientific knowledge is socially constructed. Although there are a variety of individual points of view within the general framework of the SSK (and the SSK-inspired work in the history of science) these different perspectives are "united by a shared refusal of philosophical apriorism coupled with a sensitivity to the social dimensions of science" (Pickering 1992b: 2). In other words: most of what philosophers of science have said about science is irrelevant, and science is fundamentally social.

Although the SSK raises a number of provocative challenges to traditional epistemology and the philosophy of science, such global philosophical issues are not the primary focus of this paper; some of these issues will surface briefly toward the end of the discussion, but they are not the main theme. The main theme of this paper is economics and what this recent literature on the SSK might mean to economics and economic methodology. In particular I want to address such questions as whether an "economics of science" might not be as important to

Reprinted with permission of Routledge/Taylor & Francis Ltd. From Roger Backhouse, ed., *New Directions in Economic Methodology* (London: Routledge, 1994), 75–106.

the study of science as the sociology of science, and whether the SSK in any sense "leads to" such an economic analysis of science. Such an economics-based investigation into the nature of science would certainly raise a number of questions for economic methodology. While the paper will examine a wide range of issues regarding economics and the SSK, my purpose is only to provide a general discussion of these topics and not to advocate any one particular perspective on the relationship between these two fields.

The paper is arranged in the following way. The first section documents the rise of the SSK and briefly discusses some of its intellectual origins. A few of the differences among the various schools of thought within the SSK will be examined in this section. The second section will consider the issue of whether the SSK might "lead to" an economics of science. In particular the question of "social interests" in science will be discussed and how such "interests" might be economically interpreted. The third section considers some of the existing literature that might be classified as "the economics of science" or "the economics of scientific knowledge." Although this literature is hardly voluminous, it has existed for a long time (one of the early contributions dates from the late nineteenth century), and it is currently expanding. The fourth section considers some of the more general philosophical issues raised by the SSK (and inherited by any economics-based alternative to it): particularly the questions of "circularity" and "reflexivity." In the fifth and final section, I will argue that while the economics of science seems to be a fertile area for additional research, interested economists should recognize that the social study of science (by sociologists or economists) represents a virtual Pandora's box of challenges to the beliefs that most economists hold regarding knowledge, science, and even nature. These wider philosophical implications will leave many economists feeling rather uncomfortable about the whole project of an economic version of the SSK even though it would otherwise seem to be a rather obvious next step for the application of the economic method.

The Rise of the Sociology of Scientific Knowledge

While the recent literature on the SSK draws its intellectual inspiration from a broad range of sources, two important influences can be rather easily identified: the earlier "sociology of science" of the Merton tradition, and the historicist (some would add "relativist") turn within the philosophy of science initiated by authors such as Thomas Kuhn (1962) and Paul Feyerabend (1975).

The origin of the earlier sociology of science is frequently traced to Robert K. Merton's doctoral dissertation in 1935 (Merton 1970). Merton's thesis, in opposition to both "internal" (usually inductivist) histories of science as well as Marxist "external" histories,[1] argued that the development of natural science was promoted by the Puritan ethic of seventeenth-century England. Merton's argument that it was a particular social milieu that brought about the rise of science was quite similar to Max Weber's thesis that it was the Protestant ethic that had spurred on the development of capitalism. It is important to note that Merton's sociology of science was a sociology *of science,* and not necessarily a sociology *of scientific knowledge.* The distinction is very important. Neither Merton nor the other members of the Merton school really questioned the objective validity of our scientific knowledge. Science, for the sociology of *science,* employs a particular scientific method that provides reliable and universal knowledge about the objective world; sociology only enters in an attempt to explain the unique characteristics of the social and institutional context that allows such objective knowledge to be obtained. For this early, Mertonian, sociology of science, "scientific knowledge"—the content of scientific theories—is not inexorably social. There are, for the Merton school, external social factors that promote or impede the development of scientific knowledge, factors that can be studied by the sociology of science, but the objective content of scientific theories exists independently of these social factors. This objective independence of scientific knowledge is not endorsed by many contemporary contributors to the SSK; this is one of the reasons why the recent literature, unlike that of the Merton school, is termed the sociology of *scientific knowledge,* rather than merely the sociology of *science.* Much of the impetus for this (much stronger) claim regarding the constitutively social nature of knowledge itself came from the work of historical philosophers of science like Kuhn and Feyerabend in the 1960s and 1970s.

Thomas Kuhn's (1962) central thesis regarding paradigms, scientific revolutions, and normal science is much too familiar to summarize here, but it is useful to review a few individual parts of Kuhn's argument in order to trace the relationship between his influential ideas on the history of science and the development of the SSK. Kuhn's basic claim is that in mature science the members of a given scientific community are

1. The work of Merton is often characterized as a reaction to the Marxist external histories of science written during the 1920s and early 1930s (Hessen 1931, for example). The Marxist influence on the early development of the sociology of science is emphasized by Collins and Restivo (1983) and Bunge (1991, 1992).

always in the grip of a collectively shared paradigm. In "holding" a certain paradigm what the scientists "see," or do not "see," is determined by the paradigm. Observations are not independent and "theory free," but rather are a product of the paradigm and are "theory laden." During a scientific revolution the scientist's way of seeing, the gestalt, changes; what was once seen "as" one thing, is now seen "as" something else. On this view there are no theory-neutral empirical observations by which scientific theories can be independently judged. Rather it is the scientific theory itself, or more properly the scientific paradigm itself, that actually determines the observations within its domain. Two different paradigms are thus fundamentally "incommensurable"; they constitute two incomparable ways of viewing the world.

Notice how this Kuhnian view of science introduces an irrevocably social element into science. It is not simply that different scientists have different subjective perspectives that taint their observations in various ways; rather each individual scientist "participates in" or "shares" a collective world view—the scientific paradigm—and this collectively held world view determines what they do and do not "see." What is "observed," what is and is not seen as "evidence," becomes a social product; the "world" the scientist participates in, the "world" of science, is socially constructed.[2] This particular aspect of the Kuhnian story—the social construction of the scientist's world—clearly opens the door for a sociological analysis of these scientific worlds (even though such a sociological analysis was not Thomas Kuhn's main interest). Since each individual paradigm constitutes "the facts" in its domain, it cannot be the case that theory choices are made on the basis of the "objective facts." But if it is not the objective evidence that determines the choice between scientific theories then what does? Enter the social studies of science. The scien-

2. It should be noted that Kuhn did not stress an ontological interpretation of this collectively constructed world. For Kuhn, unlike for some of the more radical among the recent sociologists of scientific knowledge, there always seems to be something "out there" that is fixed and unchanged as the scientific paradigm, and thus what the scientists "see," is transformed. When the "duck" in the optical illusion is transformed into a "rabbit" before our very eyes, the paper and the marks on it remain unchanged. As Kuhn stated in "Second Thoughts on Paradigms":

> In *The Structure of Scientific Revolutions,* particularly chap. 10, I repeatedly insist that members of different scientific communities live in different worlds and that scientific revolutions change the world in which a scientist works. I would not want to say that members of different communities are presented with different data by *the same stimuli.* Notice, however, that that change does not make phrases like "a different world" inappropriate. The given world, whether everyday or scientific, is not a world of stimuli. (Kuhn 1977: 309 n. 18, italics added)

tific community, like any other human community, forms a culture; it is a society. This society, this culture, can be examined like any other society, and since the traditional mode of inquiry for the study of society is sociology, the result is the SSK.[3]

While there are very many different individual points of view within the SSK—some inspired more by anthropology than sociology, and some much more radical than others—there is one group that seems to be cohesive enough to be labeled a particular "school" within the SSK. This school is the so-called "strong programme" associated with Barry Barnes, David Bloor, and Steven Shapin.[4]

One of the central theses of the strong programme is that "social interests" determine which scientific theories are successful and which are failures. As Paul Roth has characterized this view:

> The successes of science, both in the laboratory and in the prevailing textbook account, are to be explained by citing those social factors that cause, in a given historical context, a particular scientific theory to triumph (be judged correct) in place of its competitors. More specifically, the considerations determining which scientific theory will prevail, including the standards by which any such theory is deemed better than its alternatives, are tied to perceptions of which theory best rationalizes the interests of the dominant social group. This view differentiates the strong programmers from those . . . (most prominently Karl Mannheim and Robert Merton) who hold that the process of scientific justification is not a form of ideological rationalization and so not to be explained by sociological inquiry. (Roth 1987: 155–6)

Although the strong programme generally emphasizes the macro-social interests or ideologies that "bear on national or dynastic politics" (Bloor 1984: 79), other, more micro-oriented social factors, such as the particular interests of the individual members of a given scientific community, may also be considered.

3. Other philosophical influences on the SSK include L. Wittgenstein's theory of the social/conventional nature of language use and the under-determination thesis associated with the work of Pierre Duhem and W. V. O. Quine. See Bloor (1983) for a discussion of Wittgenstein in this context, and Roth (1987, esp. chap. 7) for a discussion of Duhem and Quine.

4. The literature on the strong programme (and the literature critical of it) has become rather extensive over the last two decades. Major contributions to the strong programme include: Barnes (1977, 1982), Bloor (1976, 1983), and Shapin (1982). The relationship between the strong programme and the discipline of economics is considered in Coats (1984) and Mäki (1992, 1993).

> I mean that the social factors concerned may be ones which derive from the narrowly conceived interests or traditions or routines of the professional community. . . . Much that goes on in science can be plausibly seen as a result of the desire to maintain or increase the importance, status and scope of the methods and techniques which are the special property of a group. (Bloor 1984: 80)

Notice how much the strong programme's view of scientific knowledge differs from the traditional philosophical characterization of science. The traditional view emphasizes the world "out there"—either the real objective "world" of nature (realism) or the "world" of empirical phenomena (instrumentalism)—as the determining factor in our scientific knowledge. For the strong programme it is not the world "out there," but rather it is the particular social context—the social interests present in that context—that determines what beliefs scientists hold, and these beliefs in turn determine what comes to be scientific knowledge. The beliefs that scientists hold are shaped by their social context, the social milieu in which they live and work; since these beliefs determine what comes to be scientific knowledge, the result is a scientific knowledge that is fundamentally social, a product of (and in a certain sense "about") its social context. To explain scientific knowledge within this framework one focuses on the beliefs that scientists hold as the cause of scientific knowledge, and to explain the beliefs that scientists hold one focuses on social context as the cause of those beliefs. This view of science not only elevates the role of "the social" far beyond that which it has traditionally played in the philosophy of science, but also well beyond the role it played in the earlier Mertonian sociology of science.

Although the strong programme is, in certain respects, quite radical, it remains philosophically traditional in at least two ways. First, the strong programme employs a relatively traditional notion of "cause." According to the strong programme the beliefs of scientists have "social causes" in the rather straightforward and commonsense way that any event A might "cause" event B.[5] Second, the strong programme practises a type of sociology that is both empiricist and inductive, thus making it quite traditional in its scientific methodology. According to the strong programme one simply determines the social causes of scientists' beliefs by empirically examining actual science. As Bloor admits, "I am an in-

5. This of course is not to suggest that the notion of "cause" is a philosophically simple notion. It is only to say that the strong programme does not make any original contribution to, nor does it try to make any contribution to, the philosophical discourse on the nature of causality.

ductivist. . . . My suggestion is simply that we transfer the instincts we have acquired in the laboratory to the study of knowledge itself" (Bloor 1984: 83). Such a stance is not only inductivist it is also methodologically monist—it presupposes that social science, in this case the sociology of science, should employ exactly the same methodology as that which has traditionally been characterized as "the method" of natural science.[6] Critics have used both of these traditional aspects of the strong programme as points of attack.

While I will not say much more about the issue of causality,[7] the questions of inductivism and methodological monism will surface again when "reflexivity" is discussed in the penultimate section.

No other group of authors within the SSK forms such a clearly defined "school of thought" as the strong programme; outside the strong programme the SSK is composed of a number of disparate points of view which disagree on a number of issues. Despite this disagreement, if one is willing to live with rather rough-hewn categories, it is possible to characterize one major alternative to the strong programme. This alternative school—the "constructivist" or "ethnographic" approach to the SSK—differs from the strong programme in at least two significant ways.[8] Both of these differences could be considered methodological in a broad sense. First, the constructivist approach differs from the strong programme with respect to the sociological categories and the theoreti-

6. As Mäki states in his recent discussion of the sociology of scientific knowledge and economics:

> It is believed by Bloor that all this amounts to applying the principles of science to science itself. The program is radically pro-science. More particularly, it is based on a naturalistic methodological monism. Unlike some other currents in the sociology of science, Bloor's programme is strongly anti-hermeneutic. (Mäki 1992: 68)

7. One causality question that has been raised regarding the strong programme is the programme's failure to specify any explicit causal mechanism connecting social conditions with the formation of beliefs (see Roth 1987: chap. 8 on "voodoo epistemology"). Another criticism has been that if there actually is an implicit causal mechanism in the strong programme, it is a type of "social behaviorism" that is replete with its own philosophical problems (see Slezak 1991).

8. This alternative school—if "school" is even the appropriate term—is fairly amorphous. It would include such disparate views as the micro-sociological approach of Collins (1985), the ethnographic approaches of Latour and Woolgar (1986), Latour (1987), and Knorr-Cetina (1981), and the "science as practice" approach of Galison (1987) and Pickering (1984). The McMullin (1992) and Pickering (1992a) volumes contain an excellent collection of papers discussing the similarities and differences among these various views. The papers in the McMullin (1988) volume relate the literature on the SSK to other recent developments in the philosophy of science.

cal entities employed in the investigation of science. The strong programme generally (though not exclusively) focuses more on the broad macro-sociological variables at work in the wider society, while constructivist authors tend to focus more on the micro-sociological factors at work in the individual laboratories and other sites of scientific activity. Second, the strong programme and the constructivist approach differ considerably regarding their basic methodology of social inquiry. The strong programme is (as discussed above) narrowly "scientific" in its approach to understanding the social causes of scientific belief; this rather narrow meta-method does not generally characterize the work of those writing from a constructivist point of view. The constructivist authors draw their methodological inspiration from a much wider range of inquiring traditions: the participant-observer approach in anthropology, ethno-methodology, and the hermeneutic tradition to name a few. The constructivist authors are broadly empirical, but it is not the simple inductivism of the strong programme. Those in the constructivist programme seek to understand the social nature of scientific activity and employ a broad range of inquiring frameworks in order to obtain that understanding; it is not simply a matter of applying the natural science method to the study of science as it often is for the strong programme. Constructivist studies in science are generally local, richly detailed, and deeply textured investigations into scientific practice as a life activity; they may focus either on a particular historical episode in science or on contemporary scientific activity, but in either case the result generally involves much more contextual solicitousness than the investigations of the strong programme.[9]

Sociology, Interests, and Economic Opportunity

The SSK seems to leave the door open for an economic analysis of science. If science is done in social communities by individual scientists and we desire to employ social science to help us understand the behaviour of those scientists in that community, then economics seems to be as likely a candidate for the relevant social science as sociology or anthropology. In the SSK, both versions, there is a lot of talk about the "interests" of those in the scientific community, and while economists do not normally use the term "interests," they do in fact explain economic

9. At this point the only self-consciously constructivist study in economics is Weintraub (1991). In addition to this major work on the (constructivist) history of stability theory in general equilibrium, Weintraub has also discussed the constructivist viewpoint in a number of shorter papers (1992, for example).

behaviour on the basis of the "interests" of the agents involved. The economics of science, or the economics of scientific knowledge, seems to be a rather obvious next step in the study of science as the product of a social community of individual agents.[10]

Before taking this obvious next step and considering the economics of science explicitly, I would like to examine a few of the quasi-economic arguments that have been offered from within the SSK. Surprisingly, there are a number of cases within the SSK where, even though the author was not consciously attempting to apply economic analysis to science, the arguments offered do in fact sound very much like economic arguments.[11] This quasi-economic argumentation is emphasized by Uskali Mäki in his recent examination of the SSK and economics:

10. For the remainder of this paper "economic" will mean "microeconomic"; given our desire to explain the behaviour of individual scientist agents (or the intended and/or unintended consequences of their individual actions), macroeconomics does not seem to be particularly useful. The "economics" that will be considered as a possible candidate for studying the economics of science is the standard neoclassical theory of rational choice. As the discussion proceeds it will become obvious that certain strains of neoclassical economics, particularly public choice theory and the microeconomics of the older Chicago school, are better suited to this particular task than other strains such as mathematical general equilibrium theory, but only neoclassical microeconomics (in some form) will be considered.

It is also important to note that "economics" will not include Marxian economics. While this exclusion will probably not surprise (nor bother) most economists, it is important with respect to the SSK. As stated above (see note 1) there is a Marxist tradition in the SSK (or at least the sociology of science) and for Marxist social theory there is little difference between a "sociological" explanation and an "economic" explanation; the economic mode of production determines the social relations. Thus, in this general sense, all of the literature on the Marxist sociology of science is actually on the economics of science broadly defined. Economic explanations are so much associated with Marxian explanations in the SSK that authors often reject "economic" explanations altogether on the basis of the failure of Marxian explanations. For example, the sociologist of science Karin Knorr-Cetina uses the following argument to reject "economic theory" as a source of inspiration for the study of science.

> In economic theory, the notion of capital is linked to the idea of exploitation defined in terms of the appropriation of surplus value, and to the corresponding concepts of class structure and alienation. Without adequate conception of exploitation and class structure, the capitalist model loses its most distinctive characteristics. But how are we to conceive of exploitation and class structure in scientific fields said to be ruled by capitalist market mechanisms? (Knorr-Cetina 1982: 108)

Needless to say, this critique of the economic approach to science has nothing to say (one way or another) about the application of neoclassical microeconomics to the study of science.

11. This distinction, the distinction between sociologists that employ economic-type arguments in the study of science and those who explicitly apply economics to science, is

> It is interesting from our point of view that much of recent sociology of science is built upon analogies drawn from economics. In these suggestions science is viewed as analogous to a capitalist market economy in which agents are maximizing producers who competitively and greedily pursue their self-interest. The point of emphasis in these suggestions is on scientists' action and on the ends involved in that action. (Mäki 1992: 79)

It seems useful to examine a few of these quasi-economic discussions from the existing SSK literature before moving on to the explicit consideration of the economic approach in the next section.

The first case I would like to consider is one that is also discussed by Mäki (1992): Bruno Latour and Steve Woolgar's *Laboratory Life* (1986). This work, a work that is generally (and fittingly) considered one of the more radical positions within the SSK, has a surprisingly large amount of economic argumentation. In chapter 5, where Latour and Woolgar discuss the motivation of scientists, there are many references to the "quasieconomic terms" (p. 190) that scientists, particularly younger scientists, use to describe their own work and professional involvements. The scientists interviewed by Latour and Woolgar repeatedly used the term "credit" to describe that which was being sought through scientific activity as well as that which participation in science would distribute to those who were successful. This scientific "credit" clearly has a component that is direct reward, but the scientists in Latour and Woolgar's study seemed to be motivated by, and interested in, more than simply the direct rewards from credit. Latour and Woolgar expand the notion of credit beyond the simple notion of a reward to a broader issue of professional "credibility." They argue that when scientists are viewed as "engaged in a quest for credibility, we are better able to make sense both of their different interests and of the process by which one kind of credit is transformed into another" (Latour and Woolgar 1986: 200). After elaborating on this expanded notion of credibility (pp. 198–208), Latour and Woolgar embed the concept in a general economic characterization of science.

> Let us suppose that scientists are investors of credibility. The result is the creation of a *market.* Information now has value because, . . . it allows other investigators to produce information which facilitates the

---

of course arbitrary. The distinction requires the assignment of an unobservable methodological intent to the various authors involved. Nonetheless, while it is not without problems, the distinction can be, and will be, usefully maintained in what follows.

> return of invested capital. There is a *demand* from investors for information which may increase the power of their own inscription devices, and there is a *supply* of information from other investors. The forces of supply and demand create the *value* of the commodity, which fluctuates constantly depending on supply, demand, the number of investigators, and the equipment of the producers. Taking into account the fluctuation of this market, scientists invest their credibility where it is likely to be most rewarding. Their assessment of these fluctuations both explains scientists' reference to "interesting problems," "rewarding subjects," "good methods," and "reliable colleagues" and explains why scientists constantly move between problem areas, entering into new collaborative projects, grasping and dropping hypotheses as the circumstances demand, shifting between one method and another and submitting everything to the goal of extending the credibility cycle. (Latour and Woolgar 1986: 206, italics in original)

For Latour and Woolgar it is the market for credibility that determines what scientists work on, what they find interesting, what is considered good work, and ultimately what becomes scientific knowledge.[12]

The second example of an economic-like characterization from within the SSK is Karin Knorr-Cetina's "exchange strategy" representation of the scientific activity in experimental particle physics. In her studies of particle physicists she found two basic strategies; one of these was the very economic-sounding "exchange strategy," which she characterizes in the following way:

12. It should be noted that while Latour and Woolgar's market for credibility sounds very much like the standard neoclassical characterization of a competitive market, their story is, in many respects, quite anti-neoclassical. In the standard neoclassical story of the market the actions of individuals are primary and explanatory—the behaviour of the market prices and quantities is "explained by" and "determined by" the actions of (ontologically primary) individual agents—and not the other way around. Latour and Woolgar are not this individualist, nor are they as linear in their explanatory thinking. For Latour and Woolgar there is a market, and there are credibility-seeking actions by individual scientists, but the latter does not necessarily cause the former. This process involves the market, the scientific culture, the individual scientists, the instruments and a myriad of other factors: all wrapped up in an amalgamated and mutually codetermining ensemble of interdependent influences. Latour and Woolgar certainly have a market in their story, but it is not the standard neoclassical market. As they say:

> This consideration is important because we certainly do not wish to propose a model of behaviour in which individuals make calculations in order to maximize their profits. This would be Benthamian economics. The question of the calculation of resources, of maximization, and of the presence of the individual are so constantly moving that we cannot take them as our points of departure. (1986: 232 n. 10)

> I have defined contingency in terms of a negative relationship of dependence between two desired goals, or research utilities, such that one utility can only be obtained or optimized at the cost of the other. In this situation particle physicists resort to a strategy of commerce and exchange: they balance research benefits against each other, and then "sell off" those which they think that, on balance, they may not be able to afford. Particle physicists refer to this commerce with research benefits as "trade-offs." In the experiment we observed, they traded off tracking particles against electron identification; time needed for calibration against granularity of the detector; performance of the calorimeter against cost; dead time against background reduction; and so on. (Knorr-Cetina 1991: 112–13)

Here again, as in Latour and Woolgar's *Laboratory Life,* scientific activity is described in terms that are quite familiar to economists. It should be noted that this "exchange strategy" was not the only strategy that Knorr-Cetina found among particle physicists—and she also found a totally different strategy among molecular biologists—but nonetheless it clearly is an economic story about the practice of science.

One very important point about these two examples (and perhaps other examples that one might find in the SSK)[13] is they *are not attempts to "apply" economics to science.* Economists often find themselves in the position of trying to model something "as an X"; we might, for example, try to model sticky wages "as a rational response to asymmetric information," or we might try to model the demand for children "as the outcome of a noncooperative game." This is not what is going on in these economic stories about science. The economic argumentation that appears in the literature on the SSK is not a result of the various authors trying to model science "as a competitive market process." The intention of these three authors in particular was certainly not to demonstrate the robustness of the economic method; it was simply to examine the social nature of science through careful ethnographic investigation into the actual practice of science. As we will see in the next section, there are in fact some studies that are motivated by an attempt to "apply" economics to science, but neither Latour and Woolgar nor Knorr-Cetina are such cases. In fact, although it is purely speculation on my part, a reasonable conjecture would be that all three of these authors consider neoclassical

13. Two other sociological studies of science that draw on economic analogies are Bourdieu (1975) and Hagstrom (1965). Both of these studies are discussed in Latour and Woolgar (1986) and Knorr-Cetina (1982). Bourdieu is also discussed in Mäki (1992).

economics to be naively reductionist, narrowly individualist, and in general a quite uninteresting approach to studying (any) social process. The point is, despite the fact that none of these authors intended to apply economics, and perhaps do not even particularly like the discipline, the stories that emerge from their ethnographic investigations of science look very much like the product of economic analysis.[14]

For the third and final example of economic analogies in the SSK I will take a slightly different approach. Rather than simply showing that what was produced in a particular sociological or ethnographic study looks very much like what an economist might say about science, I will discuss a case where a particular study in the SSK has been criticized precisely because the study characterizes the behaviour of scientists in the way that a neoclassical economist would characterize individual behaviour.

Andrew Pickering's *Constructing Quarks* (1984) is an influential sociological history of high-energy physics from 1960 to 1980. The study is self-consciously constructivist and it focuses on the intricate details of the "dynamics of practice" rather than the more general "social interests" that motivate many studies in the SSK. Pickering's basic claim is that scientific activity in particle physics is best understood as "opportunism in context"; he characterizes this opportunism in the following way:[15]

> Perhaps the single most conspicuous departure of *CQ* from the philosophical tradition is that, in *CQ,* I paid great attention to the dynamic aspect of scientific practice. I advanced a general schema for thinking about this dynamics under the slogan of "opportunism in context." The idea was simple enough. Doing science is real work; real work requires resources; different scientists have different degrees of access to such resources; and resources to hand are opportunistically assembled as contexts for constructive work are perceived. My claim, exemplified many times over in *CQ,* was that if one understands scientists as working this way then one can understand, in some detail, why individuals and groups acted as they did in the history of particle physics. (1990: 692)

Many commentators have been critical of Pickering's "opportunism in context" precisely because it sounds so much like economic haggling in the marketplace. Peter Galison, for instance, comments that experimen-

14. The warning in note 11 is still applicable of course.
15. The *CQ* in this quotation refers to *Constructing Quarks* (1984).

tation should not be "parodied as if it were no more grounded in reason than negotiations over the price of a street fair antique" (1987: 277). Similarly, in Paul Roth and Robert Barrett's lengthy critical examination of Pickering's book they make the following remarks about his economic approach:

> Pickering's model of scientific decision-making is thus fundamentally an economic one—scientists invest their expertise in areas promising them the most useful return for this investment. Justification of decision-making is dictated, in this model, by factors completely outside of the purview of traditional philosophy of science. (1990: 594)

While commentators like Roth and Barrett, and Galison, are critical of the economic aspect of Pickering's story, this is not their only, or even their major, concern. In both cases, the fact that Pickering has characterized science as an economic process is a relatively minor infraction compared to the fact that he has almost totally excluded objective reality as a constraint on the experimental behaviour of scientists.[16] However, in the same series of papers that contains Roth and Barrett's criticism, Steve Fuller (1990) attacks Pickering specifically on grounds that are relevant to microeconomics: the way he characterizes the agency of the individual scientists. In Pickering's story the scientists behave essentially in the way that neoclassical agents behave—they make intentional choices on the basis of their beliefs and desires. For Fuller, this argument (common to all microeconomic explanations) presupposes a teleological framework from folk psychology that is just as philosophically suspect as any of the standard philosophical characterizations of the epistemologically moral character of scientists. Fuller says:

> The problem is that, contrary to his own intentions, by attributing agency to the scientists, Pickering has already supposed that they have the sort of *post facto* knowledge that he finds so objectionable in the philosophical accounts. The difference is that his scientists do not foresee hidden entities but hidden opportunities; they are master prognosticators of their own interests, if not the state of the external world. (1990: 671)

While neither Pickering in his original presentation, nor Fuller in his criticism, even mentions neoclassical economics, this criticism of Pickering by Fuller is very relevant to the general question of the economic

16. Pickering clarifies the role of "reality" in his response to Roth and Barrett: Pickering (1990).

analysis of science. The point of Fuller's criticism, the implicit folk psychology of Pickering's story about the behaviour of scientists, is precisely the same criticism that certain philosophers of science, particularly Alexander Rosenberg (1988, 1992), have levelled at microeconomics.[17] Thus, not only is it the case that certain ethnographic and historical studies in the SSK have (without any explicit consideration of economic theory) come to characterize the behaviour of scientists and scientific activity in a very (neoclassical) economic way, the similarities to economics are so great that critics of these "sociological" studies (again without explicit consideration of economics) attack them on exactly the same grounds that economists have recently been attacked by philosophers of science.

In summary, it seems that the SSK exists in some intellectually parallel universe to the universe inhabited by most economists. It is a world in which economic explanations of individual behaviour as well as the social phenomena that emerge from that behaviour (along with some of the criticism of these explanations) clearly exist—not only *do* such explanations exist, but they are generally considered to be both credible and persuasive—and yet it is a world that seems to be totally without economics.[18]

The Economics of Science and/or the Economics of Scientific Knowledge

The previous section makes clear that many of the studies in the SSK describe science in a way that is much like the way that it might be described if it were approached from an explicitly economic perspective. In this section I would like to discuss a few of the attempts to do just that: to approach science from an explicitly economic perspective.

The first work to consider is perhaps the first work ever written on the topic of the economic approach to science: a paper by the American pragmatist philosopher Charles Sanders Peirce (1967; chapter 5 this volume). In this rather amazing paper, Peirce discusses the "economy of research" in a way that not only employs marginal economic analysis

17. This paper on SSK is not the place to attempt a general discussion of the relationship between economics and folk psychology. See Rosenberg (1988: chap. 2; 1992) or Hands (1993: chap. 11) for a more general discussion.

18. One of Pickering's comments is particularly telling in this regard. In his response to Roth and Barrett (R&B) he emphasizes the multifaceted and interdisciplinary nature of his work by saying: "My model of practice—as expressed in *CQ* and as elaborated here—seems to me to touch upon the legitimate interests of a variety of disciplines: sociology, certainly, but also psychology, cognitive science, anthropology and even, to return to R&B, philosophy" (1990: 709).

but does so in a very contemporary manner; this paper, originally published in 1879, would not seem too far out of place in a modern economics journal.[19] Peirce's approach is basically to maximize the utility obtained from various research projects subject to the cost constraint imposed by each project. The first-order conditions require that the marginal utility per dollar of research cost be equated for each of the research projects undertaken; such a result, while certainly not surprising from the viewpoint of modern economics, seems rather astounding for 1879. As Wible (1992b) shows, Peirce's paper really amounts to a modern cost-benefit analysis of research project selection.[20]

Since nothing short of an actual (and extended) quotation from Peirce's paper could possible convey its contemporary style, the following contains most of the first three paragraphs of the paper.

> The doctrine of Economy, in general, treats of the relations between utility and cost. That branch of it which relates to research considers the relations between the utility and the cost of diminishing the probable error of our knowledge. Its main problem is how with a given expenditure of money, time, and energy, to obtain the most valuable addition to our knowledge.
>
> Let r denote the probable error of any result; and write $s = 1/r$. Let $\mathrm{U}r \cdot \mathrm{d}r$ denote the infinitesimal utility of any infinitesimal diminution, $\mathrm{d}r$, of r. Let $\mathrm{V}s \cdot \mathrm{d}s$ denote the infinitesimal cost of any infinitesimal increase, $\mathrm{d}s$, of s. . . . Then the total cost of any series of researches will be
>
> $$\Sigma_i \int \mathrm{V}_i s_i \cdot \mathrm{d}s_i;$$
>
> and their total utility will be
>
> $$\Sigma_i \int \mathrm{U}_i r_i \cdot \mathrm{d}r_i.$$
>
> The problem will be to make the second expression a maximum by varying the inferior limits of its integrations, on the condition that the first expression remains of constant value. (Peirce 1967: 643)

19. I am indebted to James Wible for drawing my attention to this important early contribution of C. S. Peirce. Wible (1992b) discusses Peirce's paper in detail along with the work of Nicholas Rescher, a contemporary philosopher of science influenced by Peirce (see Rescher 1976). Peirce's work has recently appeared in the philosophical discussion surrounding the SSK (see Delaney 1992) and it is also examined in Kevin Hoover (1994).

20. Wible compares Peirce's paper to a 1965 paper by Frederic Scherer on the utility approach to research project evaluation, and finds the resemblance to be "striking and uncanny" (1992b: 14).

As I said, this was a rather amazing paper for the time that it was published.

While Peirce's paper was clearly "way ahead of its time" in many respects, it was, if examined in isolation from the rest of his pragmatist philosophy, only an early contribution to the *economics of science,* rather than an early contribution to the *economics of scientific knowledge.* If we mirror the distinction between the sociology of science and the sociology of scientific knowledge, then the economics *of science* would be the application of economic theory, or ideas found in economic theory, to explaining the behaviour of scientists and/or the intellectual output of the scientific community. That is, given the goals of the individual scientists or those of the scientific community (for example, the "pursuit of truth") the economics of science might be used to explain the behaviour of those in the scientific community or to make recommendations about how those goals might be achieved in a more efficient manner. In this way the economics of science would relate to science in precisely the way that microeconomics has typically related to the firms in the market economy. Peirce's 1879 paper is a very early example of such an economics of science.[21] On the other hand, the economics *of scientific knowledge* (ESK) would involve economics in a philosophically more fundamental way. The ESK would involve economics, or at least metaphors derived from economics, in the actual characterization of scientific knowledge—that is, economics would be involved fundamentally in the epistemological discourse regarding the nature of scientific knowledge. Like the SSK argues that scientific knowledge comes to be constructed out of a social process, the ESK would argue that scientific knowledge comes to be constructed out of an economic process.

Although in isolation Peirce's 1879 paper is more an application of the economics of science than the ESK, Peirce's more general pragmatist philosophy of science does in fact contain elements of an economic theory of knowledge. For Peirce truth is simply that which the community of inquirers converges to over an infinite period of time. His notion of truth is inherently social—a community of inquirers is involved—but it is not merely a product of existing social conditions; it has an indepen-

21. The last paragraph of Pierce's paper is very interesting in this respect.

> It is to be remarked that the theory here given rests on the supposition that the object of the investigation is the ascertainment of truth. When an investigation is made for the purpose of attaining personal distinction, the economics of the problem are entirely different. But that seems to be well enough understood by those engaged in that sort of investigation. (Peirce 1967: 648)

dent existence in that it is the limit (ultimate limit) of a process of inquiry by the inquiring community. As Thomas Haskell has characterized Peirce's position:

> The ultimate consensus to be reached by his community of inquiry is of a very special kind, and his theory of reality, though indubitably social, is not at all relativistic, as twentieth-century analogues have tended to be. Like Thomas Kuhn, he regarded science as the practical accomplishment of a community of researchers. Unlike Kuhn, however, he supposed that the universe was so made that an ultimate convergence of opinion was virtually predestined and that the reality toward which opinion converged was utterly independent, not of thought in general but of what any finite number of human beings thought about it. (Haskell 1984: 205–6)

The economic element in Peirce's view of science becomes clear when he discusses the nature of this "convergence" towards communal truth; his story is basically a competitive story. For Peirce it is the "economy of inquiry" that drives the inquiring community towards truth. The same economic principles that governed the choice of research projects in his 1879 paper, the maximization of return on our collective cognitive investment, propel the scientific community towards its goal. Again quoting Haskell:

> Indeed, the entire process that causes what Peirce called "the most antagonistic views" to converge in the ultimate consensus is strangely reminiscent of the price mechanism in economic markets. There, in accordance with the natural laws of supply and demand, the jockeying of rival consumers and producers looking out for their own interests generates for each commodity a convergence towards its "natural price." In the community of inquiry the clash of erring individuals produces eventually a convergence of opinion about reality. No one in Peirce's community need feel love toward the other members, nor even love of truth, strictly speaking (since no individual's present ideas can be said to correspond with that opinion which the community will ultimately settle on). (1984: 211)[22]

Charles Sanders Peirce thus seems to be an earlier contributor to both the economics of science and the ESK. He discussed the "economy of research" as a way of optimally selecting scientific research projects,

22. Also see Delaney (1992) and Wible (1992b).

but he also integrated this economic argument into his basic philosophy of science and theory of scientific truth.

The second author I would like to discuss, Gerard Radnitzky, is also a philosopher, although a contemporary one, and his view of science is also one that involves economics in both ways: as a tool for explaining what scientists actually do, and as an integral part of his characterization of scientific knowledge. Radnitzky has presented his "economic theory of science" in a series of recent papers: Radnitzky (1986, 1987a, 1987b, 1989). His basic purpose is *"to investigate what may be gained from applying the economic approach, in particular cost-benefit thinking, to the methodology of research"* (Radnitzky 1986: 125, italics in original).[23] Radnitzky argues that by taking cost-benefit analysis (CBA) as our general point of departure it is possible to clarify a number of lingering controversies within the philosophy of science. In particular, he argues that CBA can help illuminate such philosophical questions as: Why do certain scientists hold on to established paradigms even in the face of negative empirical evidence?[24] How are decisions made regarding what is and what is not to count as part of the empirical basis for a scientific theory? And when is it rational to prefer one scientific theory over another? Radnitzky does not say that the economic approach can conclusively "solve" any of these philosophical problems, but rather "that the CBA-frame may provide an organizing scheme that helps the researcher to see what sorts of questions he should take into account when dealing with such problems" (1986: 125).

These particular papers that apply the "economic method" directly to questions of science and economic methodology are really only the tip of the iceberg regarding the importance of economics to Radnitzky's philosophy of science. Radnitzky subscribes to a particular brand of Popperian philosophy; he supports the "critical rationalist" interpretation of Popper's philosophy associated with the work of W. W. Bartley III.[25]

23. Radnitzky defines "cost-benefit analysis" (CBA) quite broadly. He mentions neoclassical economics, Gary Becker in particular, but he also considers the approach to be general enough that it includes Austrian economics as well as Popper's "situational analysis" approach to social science (see Hands 1985, for a discussion of the latter). On the question of the generality of CBA, Radnitzky says:

> Man is a chooser. *All* rational choices involve the weighing up of benefits and costs. Hence, CBA is the core of the economic approach/rational-problem-solving approach. (1986: 127)

24. This question is also considered in Ghiselin (1987).
25. See Bartley (1984, 1990) and Radnitzky and Bartley (1987).

Since the critical rationalist interpretation of Popper is discussed elsewhere there is no reason to reproduce the argument here.[26] All I would like to note is that in the Bartley/Radnitzky characterization, "knowledge" emerges from the competitive process of scientific criticism in the same way that economic welfare emerges from the competitive market process. Their view essentially amounts to an invisible-hand argument for the growth of scientific knowledge and it depends fundamentally on their notion of "criticism" and its role in error elimination. The Bartley/Radnitzky view also depends heavily on arguments from evolutionary epistemology to connect up these competitively emergent theoretical structures with the underlying physical world; their theory, unlike Peirce's for example, endeavors to hold on to scientific realism. In any case, Radnitzky's view, and to a lesser extent critical rationalism more generally, represents a direct application of economic reasoning to questions about the nature of scientific knowledge; it thus constitutes a version of ESK.

Finally, I would like to mention a few of the studies, some quite recent, and some not so recent, where economists, rather than philosophers, have undertaken various exercises in the economics of knowledge. These are primarily studies in the economics of science—applying the tools of economic analysis to the behaviour of a particular type of economic agent: a scientist—although some of these economists also consider issues that might be part of the ESK.

One of these works is Gordon Tullock's *The Organization of Inquiry,* published in 1966. This book, inspired by "six months spent working with Karl Popper" (1966: v) applies an early public choice-type analysis to the study of science. For Tullock one of the identifying characteristics of science is that it is conducted in a scientific community and the book is primarily "devoted to a discussion of this community, the organization which controls inquiry," the main feature of this community is that it is "a system of voluntary co-operation" (1966: 63). Tullock discusses the incentives of and constraints on this community, and also reflects on a few traditional philosophical (particularly Popperian) questions.[27]

26. In addition to the Boland (1994) paper and the works cited therein, the argument is presented in detail in Hands (1993: chap. 11) and also in Wible (1992b).

27. Tullock's comments on the sociology of science are worth quoting:

> I have, for example, read *The Structure of Scientific Revolutions* with profit and pleasure, but it will not be further mentioned in this book. This is not because I regard it as unimportant but because it deals with different problems. In this it is typical. Most of the recent work has been done by people whose basic orien-

Boland (1971)[28] takes a different approach. He uses economics (particularly welfare economics) to attack what he calls "conventionalist" methodologies—methodological approaches that attempt to choose the "best" theory from a set of competing theories. On Boland's view any methodology is considered conventionalist if it recommends choosing scientific theories on the basis of the fact that they are "more simple," or "more general," or "more verifiable," or any other criterion other than truth. In the paper Boland makes an analogy between these conventionalist theory choices and the choices involved in welfare economics. He then uses well-known problems in welfare economics, such as the Arrow impossibility theorem, to show that the conventionalist economic problem is "unsolvable on its own terms" (1971: 105).

Finally, I would like to mention three somewhat recent papers that have directly applied "the economic method" to the practice of science: Diamond (1988) and Wible (1991, 1992a). All three of these papers employ a Gary Becker–based model of rational choice to the problem of science (or a particular problem in science). Diamond (1988) applies Becker's general "economic approach" to the problem of a rational scientist. The rational scientist is one who maximizes a utility function with "scope" and "elegance" as arguments, and faces constraints imposed by time and production functions. Diamond argues that such a "maximization-under-constraints model holds promise of being able to account for scientific progress in a way consistent with the history of science" (1988: 150).

The papers by Wible are of a similar genre. Wible (1991) discusses the question of replication in science by means of Becker's model of the allocation of time, and Wible (1992a) discusses the question of fraud in science by means of the Becker/Ehrlich model of the economics of crime. All three of these papers are replete with formal economic analysis: first-order conditions, qualitative comparative statistics, Kuhn-Tucker conditions, and even optimization under uncertainty. The following quotation from Wible captures the general flavour of the analysis in, and the results available from, these three papers.

> Reconsidering the first-order condition describing the individual scientist's optimization under uncertainty, equation 5 has additional behav-

---

> tation is sociological, while mine is economic. There is no necessary conflict between sociologists and economists, but they do ask rather different questions. (1966: v–vi)

28. Reprinted as chapter 5 of Boland (1989).

> ioral implications. The more negative or the greater the slope becomes in absolute value, the less fraud will be committed by the individual. This will occur if the probability of being discovered, P_f, increases, or if the penalty F_f associated with time spent in illegitimate activities increases. Either of these reduces the incentive to engage in fraudulent activities in science. If the marginal return to fraudulent activities increases relative to legitimate activities, then the individual would be expected to increase the proportion of time spent on illegitimate activities. (1992a: 19)

These papers are clearly *the economics of science* and they are presented in the language and discursive format of contemporary economics. These papers *may not actually be* any more of a direct application of microeconomics to science than the earlier work by Tullock, Boland, or Radnitzky (or even Peirce for that matter), but they certainly *seem* to be, perhaps because of the mathematics, the most direct application of economic analysis to science available in the literature.[29]

The discussion of these quite recent Becker-based models ends the survey portion of the paper. After discussing many of the quasi-economic arguments available in the sociological literature, and also discussing a number of the direct applications of economic theory to science, it is time to return to more philosophical issues.

Caveat Economist: Circularity, Reflexivity, and Epistemic Anomie

Criticisms of the SSK (particularly the strong programme) are legion.[30] From outside the SSK most of this criticism has come from philosophers, but a few philosophically oriented historians of the various scientific disciplines have also been involved. Within the SSK, disagreement has come primarily from one school criticizing the other, although the constructivist approach is sufficiently diverse that serious disagreements often occur between various individuals who share this general point of view. Out of all of the many criticisms that have been, or could

29. The examples I have discussed do not exhaust those in the literature. Mäki (1992) discusses papers by Earl (1983) and Loasby (1986), and Diamond (1992) provides a survey of much of this literature as well as the somewhat related literature on natural science's contribution to economic growth.

30. One of the reasons the strong programme has borne the brunt of the criticism is simply that it has been around longer than most of the constructivist approaches. A second reason for the strong programme being targeted is its rather narrow and carefully delineated methodological stance; critics generally prefer targets that are clearly defined and stable over those that are ill-defined and constantly moving.

be, raised, I would like to focus on only one: the problem of *reflexivity.* The reason for focusing on this one problem is two-fold. First, this seems to be the *major problem;* it gets the most attention in the literature, and many of the other controversies seem to be derived from it. Second, it is also a problem that will trouble any attempt to apply economics to the study of science.

The problem of "reflexivity" arises in the following way. The SSK argues that science is the product of its social context, either the social interests of the scientists involved or other social factors that constitute the social context of science. What scientists observe and the theories they propose are not simply given by the external world "out there," but rather it is constituted by the social context "in here." Now if one accepts this basic claim of the SSK, then should it not also be true for the sociologists who are doing the SSK as well? The sociologists doing the SSK are a community of scientists; if what a community of scientists produces is constituted by its social context, then the output of the SSK will also be a product of, and constituted by, its social context. As Alexander Rosenberg expresses it: "This sort of sociology pulls itself down by its own boot straps" (1985: 380). Focusing particularly on the strong programme, Rosenberg characterizes the problem in the following way:

> Proponents of the "strong program" in the sociology of science [Bloor 1976; Barnes 1977] suggest that science is nothing more than such a social institution, and must be understood as such. But if this argument is correct, it must be self-refuting. If scientific conclusions are always and everywhere determined by social forces, and not by rational considerations, then this conclusion applies to the findings of the sociologist of science as well. (Rosenberg 1985: 379)

Reflexivity is an extremely important issue for the SSK. Many of the advocates of the SSK claim to undermine the hegemony of the natural sciences by showing that what is purported to be objective and "natural" is neither one of these things, but rather simply a product of the social context in which it is produced. If this is true for all human inquiry, then it must be true for the SSK as well; this makes everything socially/context dependent and thus *relative.*[31] This leads many of those writing the SSK

31. I am intentionally giving a rather relativist reading of the reflexivity problem here. It is certainly the case that many authors in the SSK, particularly those in the strong programme, do not see reflexivity as having such relativist implications. For these authors, science proceeds by induction; the fact that most natural scientists act consistently with their social interests has nothing to say about the legitimacy of the sociology of science that correctly applies the inductive method to the study of the behaviour of those in the

to a version of "sociological skepticism" (Kitcher 1992) where individuals are always trapped in the categories of a particular social context (or in Wittgensteinian terms, trapped in a particular language game); individuals are either unable to escape these social categories, or if they can escape, there is no way of rationally deciding among the various frameworks that are available.

Collins and Yearley (1992a, 1992b)—both advocates of a version of constructivism, but critical of sociologists who revel in such relativism—refer to this tendency within the SSK as "epistemological chicken"; they characterize this relativist tendency in the following way:

> In sum, following the lead of the relativists, each new fashion in SSK has been more epistemologically daring, the reflexivists coming closest to self-destruction. Each group has made the same mistake at first; they have become so enamoured of the power of their negative levers on the existing structures as to believe they rest on bedrock. But this is not the case. Though each level can prick misplaced epistemological pretensions, they stand in the same relationship to each other as parallel cultures; no level has priority and each is a flimsy building on the plain. Accepting this we can freely use whatever epistemological "natural attitude" is appropriate for the purpose at hand; we can alternate between them as we will. That is what methodological relativism is all about—the rejection of any kind of foundationalism and its replacement, not by permanent revolution but by permanent insecurity. To reverse the vertical scale of the metaphor, while SSK showed that science did not occupy the high ground of culture, the newer developments must be taken to demonstrate not the failure of SSK, but that there simply is no high ground. (Collins and Yearley 1992a: 308)

The particular solution offered by Collins and Yearley is "social realism"—let social scientists "stand on social things" to explain natural things (1992b: 382)—but other authors have other solutions.[32] Bruno La-

scientific community. The strong programme would admit that the sociologist's tools could be turned on the sociology of science itself; that is just not what they are interested in doing, they are interested in studying natural science.

32. The exchange between Collins and Yearley (1992a, 1992b), Callon and Latour (1992), and Woolgar (1992) offers various interpretations of the reflexivity problem. Collins and Yearley (1992a) discuss three approaches to the problem: "reflexivity" (associated with Woolgar and others), the "French School" (associated with Latour and others), and "discourse analysis" (associated with Mulkay and others). Of these three approaches, the only one that is (thus far) directly relevant to economics is discourse analysis; McCloskey's much discussed work on the "rhetoric of economics" (1985) is basically discourse analysis applied to economics. The Woolgar edited volume (1988) contains a number of different papers that wrestle with the reflexivity problem in a variety of ways.

tour, for example, one of the SSK authors who seems to delight in the problems (opportunities?) of reflexivity, sees the situation quite differently than do Collins and Yearley:

> A few, who call themselves reflexivists, are delighted at being in a blind alley; for fifteen years they had said that social studies of science could not go anywhere if it did not apply its own tool to itself; now that it goes nowhere and is threatened by sterility, they feel vindicated. (Latour 1992: 272)[33]

There are obviously a number of different responses to the reflexivity issue within the SSK. My purpose is not to attempt to defend one of these positions over the others; the purpose is only to point out that *all* of the authors involved in the recent SSK feel impelled to give *some response* to the question of reflexivity and the relativism (that many suggest) it implies. While these questions are clearly relevant to the SSK, it could be argued that if one were only concerned with the *sociology of science,* rather than the SSK, then the problems of reflexivity and relativism would not be relevant (since the cognitive content of scientific theories remains unexamined in the sociology of science). But even here, in the sociology of science, there is a problem if the tools are turned on the carpenter. Even when the focus is only on the culture of science and not on the content of the theories offered, there is clearly a potential problem of regress and circularity when the sociology of science looks at the science of sociology.

Let us now return to the economics of science literature from the previous section; does the reflexivity question also impact the economics of science? The immediate answer is clearly yes. Suppose that one is engaged in a public choice–type analysis of science; for example, sup-

33. Perhaps the most radical reading of reflexivity is given by Steve Woolgar. For Woolgar the primary reason for doing the SSK is its potential for radical reflexivity: the "potential for reevaluating fundamental assumptions of modern thought" (Woolgar 1991: 25). According to Woolgar, the SSK has failed to fully exploit the radical implications of its reflexivity—or perhaps failed because it has not fully exploited the radical implications of reflexivity. For him:

> the radical potential of SSK has been compromised because it has failed to interrogate its concept of hardness and thereby failed to exploit the analytic ambivalence at the heart of its practice. With a few encouraging recent exceptions, SSK has not addressed its own dependence upon conventions of realist discourse. Consequently, SSK fails to address the issue of representation at a fundamental level; it seems set to become another exemplification of the relativist-constructivist formula rather than an occasion for questioning the idea of applying formulae altogether. (Woolgar 1991: 43)

pose that we view individual scientists as acting in their own rational self-interest given the market for professional credibility. This public choice (scientific choice) analysis could just as easily be applied to practising economists—in fact, it could even be applied to the specific public choice economist that was examining the behaviour of scientists. The circularity or reflexivity problem thus occurs in the economics of science just as it does in the sociology of science.

Much of the previously considered work on the economics of science encounters such reflexivity problems. For example, in Gordon Tullock's discussion of science, he characterizes (and implicitly criticizes) fields where teaching responsibilities dictate a larger number of participants than research opportunities can accommodate in the following way:

> One symptom of the existence of this condition is the development of very complex methods of treating subjects which can be readily handled by simple methods. Calculus will be used where simple arithmetic would do and topology will be introduced in place of plane geometry. In many fields of social science these symptoms have appeared. (Tullock 1966: 57)

In this case Tullock is using his economic approach to science to explain the rise of mathematical economics. Now in this particular case Tullock would not recognize a reflexivity problem since he would probably consider himself to be applying the right kind of economics to the question of how the discipline came to pay so much attention to the wrong kind of economics. On the other hand, suppose that Tullock had actually produced a formal model of the behaviour of scientists in an overpopulated field and mathematically deduced his prediction regarding their tendency to formalize. In this case the reflexivity problem would be much more obvious, though no more present, than it was in Tullock's original.

The reflexivity problem also surfaces in the contributions by Wible (particularly 1991 and 1992b). In Wible (1991) the economic approach to science is applied to the question of empirical replication in science, but the particular case he considers is replication in econometrics. Wible is quite explicit about his desire to apply the economics of science to the science of economics.

> While the rational expectations and public choice revolutions have been extremely successful at the theoretical level, there is another domain of economic activity to which the postulate of economic rationality could be applied. This category of human behaviour virtually has been ignored by economic theorists. The domain of human behaviour

> I have in mind is none other than positive economics itself. Economic methodologists and most others who study the professional behaviour of economists usually do not presume that economists are economically rational. (Wible 1991: 165)

There is something very curious about explaining the activity of economists on the basis of an economic theory of behaviour. Should we not then be able to explain the behaviour of the economist who is trying to explain the behaviour of economists on the basis of economic theory, on the basis of economic theory too? Where do such explanations end?[34]

Wible's most recent contribution (1992b) contains a related, and perhaps ever more unsettling, argument. In Wible (1992b) the (neoclassical) economic argument of Radnitzky and Bartley (1987)—the argument that competition in the marketplace of ideas is good for the growth of knowledge—is actually used against neoclassical economics. Knowledge requires competition, but, according to Wible's analysis, there has not been any real competition in the marketplace of economic ideas; rather than competition, economics has been dominated by the monopoly of the neoclassical approach. Thus a neoclassical-based philosophy of science is used to argue for the elimination of the neoclassical hegemony in economics. This seems a bit like throwing oneself out with one's own bath water.

Wible (unlike some of the other contributors to this literature) actually recognizes the self-referentiality and the possible methodological circularity raised by his economic approach. He ends a paper with the following paragraph.

> After presenting the model and its applications, some possible intellectual problems arising from this point of view were raised. Specifically, how far can an economic explanation of economics and other sciences be taken? The logical extreme would be that the rationality of science is completely explainable in terms of economic analysis that economic science itself is primarily an economic phenomenon. The problem with such a position is that it would reduce all philosophies and methodologies of science to economics. Logically this would deprive economics of an independent philosophical or methodological standard of scientific objectivity. Objective knowledge would be impossible. The alternative

34. A similar reflexivity occurs in Grubel and Boland (1986), where mathematical economists are characterized as an interest group that attempts to capture economic rents for its members, and also in chapter 2 of Mayer (1993), where the economics profession's emphasis on formalism is attributed to market failure within the discipline.

> is to recognize that economics construed as rational maximizing behaviour is incomplete and cannot be universally applied—that it can add greatly to an understanding of professional scientific conduct but not explain everything. (Wible 1991: 184)

If one stops just before the last sentence, this quotation contains a clear statement of the "relativity" issue from the SSK. In this case it is a relativity induced by the economic argument that all science is simply the maximizing behaviour of individual scientists, rather than the relativity induced by the sociological argument that all science is socially constituted, but it is relativity nonetheless. Because of the individualism of (neoclassical) economics, the economic approach will characterize science as the product of individual rather than social interests, but the result is a scientific practice that is just as devoid of the traditional cognitive virtues as is the case for the SSK.

None of this discussion of reflexivity or relativism is meant to suggest that the economics of science is not a worthy endeavour, or that economists should be scared off by some of these potentially unsettling epistemological implications of this work. The argument is only that we should *recognize what is at stake and enter into such studies with our philosophical eyes wide open.* Economists interested in the economics of science should be knowledgeable and informed regarding the SSK, and in particular, they should be aware of the potentially quite radical implications of some of this work. If one wants to employ economics, and yet remain safely ensconced within a traditional view that allows for only one universally valid "scientific method," then something akin to the Peirce or Bartley/Radnitzky approaches would seem to be required. On these views knowledge has (some) objective status, but the (universally valid) way of obtaining that knowledge depends on a process that is fundamentally economic. On the other hand, if one is willing to abandon the traditional philosophy of science, and to accept a characterization of knowledge that is fundamentally local, tentative and contingent, then economics, along with a variety of other framework and discursive strategies, could be involved in the inquiry.[35] The point is that whichever of these two general approaches one chooses to pursue, there is serious philosophical work to be done; the economics of science is more than looking at first-order conditions for scientists' optimization problems

35. This latter approach seems to be suggested in that last line of the above quotation from Wible, and it is also one of the main themes in Philip Mirowski (1994).

and/or noting that the marketplace of ideas actually looks like a marketplace. If economists are to make any contribution to human understanding of science or scientific knowledge then they need to scrutinize the sociological literature quite carefully. Economists bring a number of tools to the project that are quite different from the tools of the sociologist; economists think in terms of optimization, information, incentives, and equilibrium; economists also think of individuals creating societies rather than societies creating individuals. While the tools are different, and in some cases perhaps mutually exclusive, neither the general task nor the pitfalls entailed are necessarily different. Those involved in the SSK have travelled through much of this wilderness before us, and to neglect their signposts would surely be a folly.

Conclusion

In this paper I have traced the history of the SSK and discussed the main themes of its two principal schools. I have also discussed a number of places where the SSK produces narratives that sound much like what one would expect from an economic analysis of science. In addition, I have surveyed a number of attempts to explicitly apply economics to science, some by philosophers and some by economists. Finally, I examined the problem of reflexivity and discussed some of the philosophical difficulties it raises for the SSK. I argued that such problems will also be encountered by anyone pursuing the economics of science, and that while such problems are not so great as to deter entry, they need to be recognized and that there is much to learn from the literature on the SSK. In conclusion I would like to note that while the ESK (and the SSK for that matter) represents a great opportunity for those interested in economic methodology and the history of economic thought, this is not an inquiry that will come easy for most economists.

In one respect of course, the economics of science is an inquiry that *should* come easy for economists. As stated above, it seems to be an obvious next step for "the economics of": a veritable gold mine for the Chicago and Virginia schools. There are a myriad of opportunities, not only for Becker-type economics and public choice theory, but also game theory, principal-agent analysis, incentive compatibility, and so on. For years economists have undermined and delegitimized the self-righteousness of politicians—"you are not acting in the societal or national interest, but in your own self-interest"—now the same argument can be applied to scientists. Economics is also a discipline of "unintended consequences" and "invisible hands"; it is not necessary that politicians or scientists be motivated by "higher" values in order to have the emergence of results that are consistent with such higher values. For economists, unlike for most others in modern intellectual life, the ubiquitousness of narrow self-interest in science or elsewhere does not necessarily initiate a wringing of hands or lamentations about lost utopias; it only initiates a conversation about proper prices, compatible incentives, and binding constraints.

Despite all of this, I suspect that the economics of science will be quite difficult for most economists to accept, even those normally engaged in economic methodology. The main problem is that most economists are epistemologically quite traditional. For most economists there is simply a world "out there" and good science represents that world accurately. Such good science has something to do with letting the data decide between theories—confirmation for some authors and falsification for others—but "the facts" in either case. This method is, for most economists, the way real science proceeds; it is the way that economics should proceed, and whatever problems the discipline has are generally a result of failing to follow this method of good science. Such views by economists are worlds away from the "negotiation over facticity" and "the social construction of nature" that pervades much of the SSK. There are many places in the SSK where the a priori distinction between the "natural" and the "social" is totally abandoned; the world is socially coproduced by a variety of actants (agents who may or may not be humans) who engage in negotiation over its construction; in one recent case these actants included the scallops of St Brieuc Bay who actively negotiated with researchers regarding their anchorage.[36] Such stories involving negotiation by non-human actors are unusual of course, even in the more radical versions of the SSK, but the point is simply that almost anything is up for grabs. Not all of the SSK has such a post-modernist flair, but much of it does, and such argumentation cannot be ruled out of court from the general perspective of the social construction of knowledge. As I argued above, any attempt at an ESK will require economists to take the arguments and the insights of the SSK quite seriously. This is what should be done and the prospects are exciting; it will not be easy for many economists to do.

36. See Collins and Yearley (1992a) and Callon and Latour (1992).

ACKNOWLEDGMENTS

I would like to thank Roger Backhouse, Larry Boland, Bruce Caldwell, Bob Coats, Uskali Mäki, Philip Mirowski, Warren Samuels, Paul Wendt, James Wible, and Nancy Wulwick for comments on an earlier draft of this paper.

REFERENCES

Barnes, B. 1977. *Interests and the Growth of Knowledge.* London: Routledge & Kegan Paul.

———. 1982. *T. S. Kuhn and Social Science.* New York: Columbia University Press.

Bartley, W. W., III. 1984. *The Retreat to Commitment.* 2d ed. La Salle, IL: Open Court.

———. 1990. *Unfathomed Knowledge, Unmeasured Wealth.* La Salle, IL: Open Court.

Bloor, D. 1976. *Knowledge and Social Imagery.* London: Routledge & Kegan Paul.

———. 1983. *Wittgenstein: A Social Theory of Knowledge.* New York: Columbia University Press.

———. 1984. The strengths of the strong programme. In J. R. Brown ed., *Scientific Rationality: The Sociological Turn.* Boston: D. Reidel.

Boland, L. A. 1971. Methodology as an exercise in economic analysis. *Philosophy of Science* 38: 105–17.

———. 1989. *The Methodology of Economic Model Building.* London: Routledge.

———. 1994. Scientific thinking without scientific method: Two views of Popper. In R. E. Backhouse ed., *New Directions in Economic Methodology.* London: Routledge.

Bourdieu, P. 1975. The specificity of the scientific field and the social conditions of the progress of reason. *Social Science Information* 14: 19–47.

Bunge, P. 1991. A critical examination of the new sociology of science: Part I. *Philosophy of the Social Sciences* 21: 524–60.

———. 1992. A critical examination of the new sociology of science: Part II. *Philosophy of the Social Sciences* 22: 46–76.

Callon, M., and B. Latour. 1992. Don't throw the baby out with the Bath School! A reply to Collins and Yearly. In A. Pickering ed., *Science as Practice and Culture.* Chicago: University of Chicago Press.

Coats, A. W. 1984. The sociology of knowledge and the history of economics. *Research in the History of Economic Thought and Methodology* 2: 211–34.

Collins, H. 1985. *Changing Order: Replication and Induction in Scientific Practice.* Los Angeles: Sage.

Collins H., and S. Yearley. 1992a. Epistemological chicken. In A. Pickering ed., *Science as Practice and Culture.* Chicago: University of Chicago Press.

———. 1992b. Journey into space. In A. Pickering ed., *Science as Practice and Culture.* Chicago: University of Chicago Press.

Collins R., and S. Restivo. 1983. Development, diversity, and conflict in the sociology of science. *The Sociological Quarterly* 24: 185–200.
Delaney, C. F. 1992. Pierce on the social and historical dimensions of science. In E. McMullin ed., *The Social Dimensions of Science.* Notre Dame, IN: University of Notre Dame Press.
Diamond, A. D. 1988. Science as a rational enterprise. *Theory and Decision* 24: 147–67.
———. 1992. Is there an economics of science? Paper presented at the American Economics Association meetings in Anaheim, CA. January 1993.
Earl, P. E. 1983. A behavioral theory of economists' behavior. In A. S. Eichner ed., *Why Economics Is Not Yet a Science.* London: Macmillan.
Feyerabend, P. K. 1975. *Against Method.* London: New Left Books.
Fuller, S. 1990. They shoot dead horses, don't they? Philosophical fear and sociological loathing in St. Louis. *Social Studies of Science* 20: 664–81.
Galison, P. 1987. *How Experiments End.* Chicago: University of Chicago Press.
Ghiselin, M. T. 1987. The economics of scientific discovery. In G. Radnitky and P. Bernhol eds., *Economic Imperialism: The Economic Approach Applied outside the Field of Economics.* New York: Paragon House.
Grubel, H. G., and L. A. Boland. 1986. On the efficient use of mathematics in economics: some theory, facts and results of an opinion survey. *Kyklos* 39: 419–42.
Hagstrom, W. O. 1965. *The Scientific Community.* New York: Basic Books.
Hands, D. W. 1985. Karl Popper and economic methodology: a new look. *Economics and Philosophy* 1: 83–99.
———. 1993. *Testing, Rationality, and Progress: Essays on the Popperian Tradition in Economic Methodology.* Lanham, MD: Rowman & Littlefield.
Haskell, T. L. 1984. Professionalism *versus* capitalism: R. H. Tawney, Emilie Durkheim, and C. S. Peirce on the disinterestedness of professional communities. In T. L. Haskell ed., *The Authority of Experts: Studies in History and Theory.* Bloomington, IN: Indiana University Press.
Hessen, B. 1931. The social and economic roots of Newton's "Principa." In N. Bukharin et al., *Science at the Crossroads.* (2d ed., 1971). London: Frank Cass & Co.
Hoover, K. D. 1994. Pragmatism, pragmaticism, and economic method. In R. E. Backhouse ed., *New Directions in Economic Methodology.* London: Routledge.
Kitcher, P. 1992. Authority, deference, and the role of individual reason. In E. McMullin ed., *The Social Dimensions of Science.* Notre Dame, IN: University of Notre Dame Press.
Knorr-Cetina, K. 1981. *The Manufacture of Knowledge: An Essay on the Constructivist and Contextual Nature of Science.* Oxford: Pergamon Press.
———. 1982. Scientific communities or transepistemic arenas of research? A critique of quasi-economic models of science. *Social Studies of Science* 12: 101–30.
———. 1991. Epistemic cultures: forms of reason in science. *History of Political Economy* 23: 105–22.

Kuhn, T. S. 1962. *The Structure of Scientific Revolutions.* Chicago: University of Chicago Press.

———. 1977. Second thoughts on paradigms. In T. S. Kuhn, *The Essential Tension: Selected Studies in Scientific Tradition and Change.* Chicago: University of Chicago Press.

Latour, B. 1987. *Science in Action: How to Follow Scientists and Engineers through Society.* Cambridge, MA: Harvard University Press.

———. 1992. One more turn after the social turn. In E. McMullin ed., *The Social Dimensions of Science.* Notre Dame, IN: University of Notre Dame Press.

Latour, B., and S. Woolgar. 1986. *Laboratory Life: The Construction of Scientific Facts.* 2d ed. Princeton: Princeton University Press.

Loasby, B. J. 1986. Public science and public knowledge. *Research in the History of Economic Thought and Methodology* 4: 211–28.

McCloskey, D. N. 1985. *The Rhetoric of Economics.* Madison, WI: University of Wisconsin Press.

McMullin, E. (ed.) 1988. *Construction and Constraint: The Shaping of Scientific Rationality.* Notre Dame, IN: University of Notre Dame Press.

———. (ed.) 1992. *The Social Dimensions of Science.* Notre Dame, IN: University of Notre Dame Press.

Mäki, U. 1992. Social conditioning in economics. In N. de Marchi ed., *Post-Popperian Methodology of Economics: Recovering Practice.* Boston: Kluwer.

———. 1993. Social theories of science and the fate of institutionalism in economics. In U. Mäki, B. Gustafsson, and C. Knudsen eds., *Rationality, Institutions, and Economic Methodology.* London: Routledge.

Mayer, T. 1993. *Truth versus Precision in Economics.* Aldershot: Edward Elgar.

Merton, R. K. 1970. *Science, Technology, and Society in Seventeenth-century England.* New York: Harper & Row [originally published in 1938].

Mirowski, P. E. 1994. What are the questions? In R. E. Backhouse ed., *New Directions in Economic Methodology.* London: Routledge.

Mulkay, M. 1982. Sociology of science in the West. *Current Sociology* 28: 1–116.

Peirce, C. S. 1967. Note on the theory of the economy of research. *Operations Research* 15: 642–8 [originally published in 1879].

Pickering, A. 1984. *Constructing Quarks: A Sociological History of Particle Physics.* Chicago: University of Chicago Press.

———. 1990. Knowledge, practice, and mere construction. *Social Studies of Science* 20: 682–729.

———. (ed.) 1992a. *Science as Practice and Culture.* Chicago: University of Chicago Press.

———. 1992b. From science as knowledge to science as practice. In A. Pickering ed., *Science as Practice and Culture.* Chicago: University of Chicago Press.

Radnitzky, G. 1986. Towards an "economic" theory of methodology. *Methodology and Science* 19: 124–47.

———. 1987a. Cost-benefit thinking in the methodology of research: the "economic approach" applied to key problems of the philosophy of science. In G. Radnitzky and P. Bernholz eds., *Economic Imperialism: The Economic Approach Applied outside the Field of Economics.* New York: Paragon House.

———. 1987b. The "economic" approach to the philosophy of science. *British Journal for the Philosophy of Science* 38: 159–79.

———. 1989. Falsificationism looked at from an "economic" point of view. In K. Gavroglu, Y. Goudaroulis, and P. Nicolacopoulos eds., *Imre Lakatos and Theories of Scientific Change.* Boston: Kluwer.

Radnitzky, G., and W. W. Bartley III. (eds.) 1987. *Evolutionary Epistemology, Rationality, and the Sociology of Knowledge.* LaSalle, IL: Open Court.

Rescher, N. 1976. Peirce and the economy of research. *Philosophy of Science* 43: 71–98.

Rosenberg, A. 1985. Methodology, theory, and the philosophy of science. *Pacific Philosophical Quarterly* 66: 377–93.

———. 1988. *Philosophy of Social Science.* Boulder, CO: Westview Press.

———. 1992. *Economics: Mathematical Politics or Science of Diminishing Returns?* Chicago: University of Chicago Press.

Roth, P. A. 1987. *Meaning and Method in the Social Sciences.* Ithaca, NY: Cornell University Press.

Roth, P. A., and R. Barrett. 1990. Deconstructing quarks. *Social Studies of Science* 20: 579–632.

Shapin, S. 1982. History of science and its sociological reconstructions. *History of Science* 20: 157–211.

Slezak, P. 1991. Bloor's bluff: behaviorism and the strong programme. *International Studies in the Philosophy of Science* 5: 241–56.

Tullock, G. 1966. *The Organization of Inquiry.* Durham, NC: Duke University Press.

Weintraub, E. R. 1991. *Stabilizing Dynamics: Constructing Economic Knowledge.* Cambridge: Cambridge University Press.

———. 1992. Commentary by E. Roy Weintraub. In N. De Marchi ed., *Post-Popperian Methodology of Economics: Recovering Practice.* Boston: Kluwer.

Wible, J. R. 1991. Maximization, replication, and the economic rationality of positive economic science. *Review of Political Economy* 3: 164–86.

———. 1992a. Fraud in science: an economic approach. *Philosophy of the Social Sciences* 22: 5–27.

———. 1992b. Cost-benefit analysis, utility theory, and economic aspects of Peirce's and Popper's conceptions of science. Manuscript.

Woolgar, S. 1981. Interests and explanations in the social study of science. *Social Studies of Science* 11: 356–94.

———. (ed.) 1988. *Knowledge and Reflexivity.* London: Sage.

———. 1991. The turn to technology in social studies of science. *Science, Technology, and Human Values* 16: 20–50.

———. 1992. "Some remarks about positionism: a reply to Collins and Yearley. In A. Pickering ed., *Science as Practice and Culture.* Chicago: University of Chicago Press.

Contributors

Kenneth J. Arrow was born in New York City in 1921 and educated in the city schools and The City College, graduating in 1940. He served in World War II and received his Ph.D. in economics from Columbia University in 1951. He has taught at the University of Chicago, Stanford University, and Harvard University. His research has been primarily in economic theory, especially social choice theory, general equilibrium, behavior under uncertainty, medical economics, and the economics of information and organization. He has also written on operations research, mathematical statistics, and meteorology. He has received several honors, including the Nobel Memorial Prize in Economic Science.

Mario Biagioli is professor of the history of science at Harvard University. He is currently working on a book on the author-function in science and is editing, with Peter Galison, a volume on the history of scientific authorship from the early modern period to "Big Science." He is the author of *Galileo, Courtier* (University of Chicago Press, 1993), and editor of *The Science Studies Reader* (Routledge, 1999). Before joining the Department of History of Science at Harvard in 1995, he taught at UCLA and, as a visiting professor, at Stanford and at the Ecole des Hautes Etudes en Sciences Sociales, Paris.

William A. Brock is Villas Research Professor of Economics at the University of Wisconsin at Madison. His field of interest is microeconomic theory, and his current research focuses on dynamic economic theory. His work has appeared in *Review of Economic Studies, Econometrica,* and *Journal of Economic Theory,* among others. A volume of Professor Brock's selected essays is forthcoming under the title *From Optimal Growth to Non Linear Science in Economics.* Professor Brock is fellow of the Econometric Society, fellow of the American Academy of Arts and Sciences, and member of the National Academy of Sciences.

MICHEL CALLON is professor of sociology at the Ecole Nationale Supérieure des Mines de Paris. Together with Bruno Latour and John Law he took part in the development of the Actor Network Theory. He has published articles and books on the sociology of science and technology and the socioeconomics of innovation. He coedited *Mapping the Dynamics of Science and Technology* and *The Strategic Management of Research and Technology,* and edited *The Laws of the Markets.* He is currently working on the role of associations of patients in the co-production of scientific knowledge and medical therapies.

PARTHA DASGUPTA, who was educated in Varanasi, Delhi, and Cambridge, is the Frank Ramsey Professor of Economics and Chairman of the Faculty of Economics & Politics at the University of Cambridge, and Fellow of St. John's College, Cambridge. During 1991–97 he was chairman of the Beijer International Institute of Ecological Economics of the Royal Swedish Academy of Sciences in Stockholm, and during 1989–92 also professor of economics, professor of philosophy, and director of the Program in Ethics in Society at Stanford University. His research interests have covered welfare and development economics, the economics of technological change, population, environmental and resource economics, the theory of games, and the economics of undernutrition. His publications include *Guidelines for Project Evaluation* (with S. A. Marglin and A. K. Sen; United Nations, 1972); *Economic Theory and Exhaustible Resources* (with G. M. Heal; Cambridge University Press, 1979); *The Control of Resources* (Harvard University Press, 1982); and, most recently, *An Inquiry into Well-Being and Destitution* (Clarendon Press, 1993). He is a fellow of the British Academy, foreign honorary member of the American Academy of Arts and Sciences, foreign member of the Royal Swedish Academy of Sciences, honorary fellow of the London School of Economics, honorary member of the American Economic Association, and member of the Pontifical Academy of Social Sciences. Professor Dasgupta was president of the Royal Economic Society (1998–2001) and president of the European Economic Association (1999).

PAUL A. DAVID is professor of economics at Stanford University, and also (since 1994) senior research fellow of All Souls College, Oxford. David is known internationally for his contributions in American economic history since the colonial era, the economics of science and technology, and economic and historical demography. Two special areas of concentration in his most recent research and publications are the development of "the new economics of science," focusing upon scientific institutions and resource allocation for public-sector R&D, and the role of compatibility standards in the growth and evolution of network industries, with special reference to information technology networks. David has authored more than a hundred journal articles and chapters in books, as well as publishing a number of books under his own name, including *Technical Choice, Innovation, and Economic Growth* (1975), *Reckoning with Slavery* (1976), with several other volumes scheduled to appear. Paul David has served as a consultant to many public and private organizations.

STEVEN N. DURLAUF is professor of economics at the University of Wisconsin at Madison. His fields of interest are macroeconomics, monetary economics, and econometrics, and his recent research focuses on economic complexity, econometrics of social interactions, and income inequality. His publications include "The New Empirics of Economic Growth," (with D. Quah) in *Handbook of Macroeconomics* (J. Taylor and M. Woodford, eds., North Holland, 1999), "The Memberships Theory of Inequality: Ideas and Implications," in *Elites, Minorities, and Economic Growth* (E. Brezis and P. Temin, eds., North Holland, 1999), and "Interactions-Based Models," (with W. Brock) in *Handbook of Econometrics* (J. Heckman and E. Leamer, eds., North Holland, 2001).

PAUL FORMAN (Jahrgang 1937) has long been interested in the scientist's role, and corresponding self-image, as it has varied historically, i.e., with the terms of social integration of the knowledgeable. This interest, first aroused as an undergraduate at Reed College (or was it already as a pupil at Philadelphia's Central High School?), was furthered by graduate study with A. Hunter Dupree and Thomas S. Kuhn at Berkeley. Since joining the Smithsonian Institution in 1972 as a curator responsible for (the history of) modern physics, Forman has combined this interest with others relating more closely to the instruments of science than to the users of them.

STEVE FULLER is professor of sociology at the University of Warwick, England. Originally trained in history and philosophy of science, he is the founder of the research program of "social epistemology," which is the name of a journal and his first book. He has since then published five other books, the most recent of which are *The Governance of Science: Ideology and the Future of the Open Society* (Open University Press, 1999) and *Thomas Kuhn: A Philosophical History for Our Times* (University of Chicago Press, 2000). In 1998 and 1999 he organized two global cyberconferences for the UK's Economic and Social Research Council: one on public understanding of science and another on peer review in the social sciences. He is about to embark on a major study of the impact of intellectual property and knowledge management on academic conceptions of knowledge.

D. WADE HANDS is professor of economics at the University of Puget Sound in Tacoma. His research interests include economic methodology, science theory, and twentieth-century history of economic thought. His papers have appeared in *History of Political Economy, Economics and Philosophy,* and *Philosophy of Science,* among others. He is the editor (with John Davis and Uskali Mäki) of *The Handbook of Economic Methodology* (Edward Elgar, 1998). His most recent book, *Reflection without Rules: Economic Methodology and Contemporary Science Theory,* was published by Cambridge University Press in 2001.

SHAUN P. HARGREAVES HEAP is pro-vice-chancellor at the University of East Anglia. His research is in macroeconomics, rational action in a social and historical context and, most recently, on the economics of television. His publications include *Rationality in Economics* (Basil Blackwell, 1989), *The Theory of Choice* (with others, Basil Blackwell, 1992), *The New Keynesian Macroeconomics* (Edward Elgar, 1992), and *Game Theory: a critical introduction* (with Y. Varoufakis, Routledge, 1995).

PHILIP KITCHER is professor of philosophy at Columbia University. He has written books and articles on issues in the philosophy of mathematics, the philosophy of biology, and the general philosophy of science. He is a past president of the American Philosophical Association (Pacific Division) and a former editor-in-chief of *Philosophy of Science.*

SHARON G. LEVIN is professor of economics and director of graduate studies at the University of Missouri–Saint Louis. Her primary research area is applied microeconomics, especially the economics of science. In recent years her work has focused on issues concerning the quality and composition of the scientific labor force in the United States, including the impacts of immigrant scientists and engineers. Her recent publications have appeared in *Science, Small Business Economics,* and *American Economic Review,* among others, and she is the author of *Striking the Mother Lode in Science* (Oxford University Press, 1992). She was the 1993 recipient of the Chancellor's Award for Excellence in Research and Creativity at the University of Missouri–Saint Louis.

PHILIP MIROWSKI is the Carl Koch Professor of Economics and the History and Philosophy of Science at the University of Notre Dame. He is the author of *More Heat Than Light* (1989) and the editor of *Natural Images in Economics* (1994). He is finishing a book, titled *Machine Dreams,* on the impact of the "cyborg sciences" on late-twentieth-century economics, and a multivolume *Collected Economics Works of William Thomas Thornton.*

RICHARD R. NELSON is an economist by training. Over his career he has taught at Oberlin College, Carnegie Mellon University, Yale University, and Columbia University, where he now is the George Blumenthal Professor of International and Public Affairs, Business, and Law. He also has served as research economist and analyst at the Rand Corporation and at the Council of Economic Advisors. His central interests have been in long-run economic change. Much of his research has been directed toward understanding technological change and how economic institutions and public policies influence the evolution of technology, and how technological change in turn induces economic change more broadly. His work has been both empirical and theoretical. Along with Sidney Winter, he has pioneered in developing a formal evolutional theory of economic change. Over the course of his career, he has been particularly attracted to working with and coordinating relatively large research teams. His National Innovation Systems project involved a team of approximately twenty scholars, and his recent study on "The Sources of Industrial Leadership" involved the coordination of a similar-size group. He was director of the Institute for Social and Policy Studies at Yale University, and now heads Columbia's Public Policy Doctoral Consortium.

DAVID F. NOBLE is professor in the Division of Social Science of York University, Canada. His research interests include social history of higher education, higher education and technology, industry and higher education, history of women and higher education, social history of technology and science, history of labor and industry, history of automation, science, technology, and gender, and, finally, science, technology, and religion. Professor Noble is currently completing *Digital Diploma Mills: The Automation of Higher Education* (Knopf, forthcoming). Other recent books include: *The Religion of Technology: The Divinity of Man and the Spirit of Invention* (Knopf, 1997; Penguin paperback, 1999), *Progress without People: New Technology and Unemployment* (Charles H. Kerr, 1993; Between the Lines, Toronto, 1995), and *A World without Women: The Christian Clerical Culture of Western Science* (Knopf, 1992; Oxford paperback, 1993).

GARY RHOADES is professor and director of the Center for the Study of Higher Education at the University of Arizona. His research focuses on the relationship between science and technology policy, the restructuring of higher education, and the restructuring of professional work on campuses. His work has been supported by the National Science Foundation and published in various journals including *Sociology of Education, Science Technology and Human Values, the Journal of Higher Education,* and *Higher Education.* His recent book is *Managed Professionals: Unionized Faculty and Restructuring Academic Labor* (SUNY Press, 1998).

ESTHER-MIRJAM SENT is associate professor in the Department of Economics and faculty fellow in the Reilly Center for Science, Technology, and Values at the University of Notre Dame. She has published articles in the *Cambridge Journal of Economics, Journal of Economic Methodology, History of Political Economy, Journal of Economic Behavior and Organization,* and *Philosophy of Science,* among other journals. Her book *The Evolving Rationality of Rational Expectations: An Assessment of Thomas Sargent's Achievements* (Cambridge University Press, 1998) was awarded the 1999 Gunnar Myrdal Prize of the European Association for Evolutionary Political Economy. She has recently started a research project on Herbert Simon.

SHEILA SLAUGHTER is professor of higher education in the Center for the Study of Higher Education at the University of Arizona. Her research areas are science and technology policy, political economy of higher education, academic freedom, and women in higher education. She is the author (with Larry Leslie) of *Academic Capitalism: Politics, Policies and the Entrepreneur-*

ial University (Johns Hopkins University Press, 1997) and has contributed to several journals, including *Journal of Higher Education* and *The Review of Higher Education.* In 1999, she received a National Science Foundation grant, with Jennifer Croissant and Gary Rhoades, for "Universities in the Information Age: Changing work, organization and values in academic science and engineering." She received the Association for the Study of Higher Education Research Achievement Award in 1998.

PAULA E. STEPHAN is professor of economics and associate dean at the Andrew Young School of Policy Studies at Georgia State University. A labor economist by training, her recent research focuses on issues in science and technology. She is interested both in the careers of scientists and engineers as well as the process by which knowledge moves across institutional boundaries in the economy. She has published in such journals as the *American Economic Review, The Journal of Economic Literature, Economic Inquiry, Science,* and the *Social Studies of Science.* Stephan has received funding for research from the North Atlantic Treaty Organization, the National Science Foundation, the Alfred P. Sloan Foundation, the U.S. Department of Labor, the Exxon Education Foundation, the U.S. Department of Labor, and the Andrew Mellon Foundation. She has served on numerous committees of the National Research Council.

STEPHEN TURNER is graduate research professor of philosophy at the University of South Florida. He has written extensively in the history and philosophy of social science. Most recently, he edited *The Cambridge Companion to Weber* (2000). His work on science has been concerned primarily with the funding, politics, and organization of science, notably the rise of federal funding of geology in the nineteenth century. He has been appointed at Virginia Tech, Boston University, and the University of Notre Dame, and was the Simon Honorary Visiting Professor at the University of Manchester.

JAMES R. WIBLE is the Carter Professor of Economics at the University of New Hampshire. He has authored a book, *The Economics of Science,* and twenty articles on macroeconomics, the history of economic thought, and economic methodology. He received his Ph.D. from Penn State and has taught at Penn State, Maine, Northeastern, and New Hampshire. During the spring semester of 1998, he was a visiting scholar in economics at Duke University. In 1993–94, he was a Faculty Scholar at the University of New Hampshire. While in graduate school for economics, he spent a year as a fellow in Penn State's Interdisciplinary Graduate Program in the Humanities. In that program, he studied philosophy of science, American pragmatism, and cognitive psychology.

JOHN ZIMAN is emeritus professor of physics at the University of Bristol. He studied at Oxford and lectured at Cambridge before becoming professor of theoretical physics at Bristol in 1964. His researches on the theory of the electrical and magnetic properties of solid and liquid metals earned his election to the Royal Society in 1967. Voluntary early retirement from Bristol in 1982 was followed by a period as visiting professor at Imperial College, London, and from 1986 to 1991 as founding director of the Science Policy Support Group. Since 1994 he has been the convenor of the Epistemology Group, which studies the evolution of knowledge and invention. He was chairman of the Council for Science and Society from 1976 to 1990, and has written extensively on various aspects of the social relations of science and technology.

Index